The Reticuloendothelial System

A COMPREHENSIVE TREATISE

Volume 2
Biochemistry and Metabolism

The Reticuloendothelial System
A COMPREHENSIVE TREATISE

General Editors:
Herman Friedman, *University of South Florida, Tampa, Florida*
Mario Escobar, *Medical College of Virginia, Richmond, Virginia*
and
Sherwood M. Reichard, *Medical College of Georgia, Augusta, Georgia*

Volume 1   MORPHOLOGY
Edited by Ian Carr and W. T. Daems

Volume 2   BIOCHEMISTRY AND METABOLISM
Edited by Anthony J. Sbarra and Robert R. Strauss

Volume 3   PHYLOGENY AND ONTOGENY
Edited by Nicholas Cohen and M. Michael Sigel

Volume 4   PHYSIOLOGY
Edited by Sherwood M. Reichard and James P. Filkins

Volume 5   PHARMACOLOGY
Edited by Andor Szentivanyi and Jack R. Battisto

Volume 6   IMMUNOLOGY
Edited by Joseph A. Bellanti and Herbert B. Herscowitz

Volume 7   IMMUNOPATHOLOGY
Edited by Noel R. Rose and Benjamin V. Siegel

Volume 8   CANCER
Edited by Herman Friedman and Ronald B. Herberman

Volume 9   RES AND ALLERGIC DISEASES
Edited by Peter Abramoff and S. Michael Phillips

Volume 10   INFECTION
Edited by John P. Utz and Mario R. Escobar

The Reticuloendothelial System

A COMPREHENSIVE TREATISE

Volume 2
Biochemistry and Metabolism

Edited by

ANTHONY J. SBARRA

*St. Margaret's Hospital for Women
and
Tufts University School of Medicine
Boston, Massachusetts*

and

ROBERT R. STRAUSS

*Albert Einstein Medical Center
Philadelphia, Pennsylvania*

PLENUM PRESS • NEW YORK AND LONDON

Library of Congress Cataloging in Publication Data

Main entry under title:

The Reticuloendothelial system.

Includes index.
CONTENTS: v. 1. Carr, I., Daems, W. T., and Lobo, A. Morphology.—v. 2. Sbarra,
A. J. and Strauss, R. R. Biochemistry and metabolism.
1. Reticuloendothelial system. 2. Macrophages. I. Friedman, Herman, 1931-
II. Escobar, Mario R. III. Reichard, Sherwood M. [DNLM: 1. Reticuloendothelial
system. WH650 R437]
QP115.R47   591.2'95   79-25933
ISBN 0-306-40292-0 (v. 2)

© 1980 Plenum Press, New York
A Division of Plenum Publishing Corporation
227 West 17th Street, New York, N.Y. 10011

Printed in the United States of America

Contributors

ROBERT C. ALLEN • U.S. Army Institute of Surgical Research and Clinical Investigation Service, Brooke Army Medical Center, Fort Sam Houston, Texas

BERNARD M. BABIOR • Department of Medicine and Blood Research Laboratory, Tufts-New England Medical Center, Boston, Massachusetts

A. J. M. BALM • Department of Blood Cell Chemistry, Central Laboratory of the Netherlands Red Cross Blood Transfusion Service, Amsterdam; and Laboratory of Clinical and Experimental Immunology, University of Amsterdam, Amsterdam, The Netherlands

R. E. BASFORD • Department of Biochemistry, University of Pittsburgh School of Medicine, Pittsburgh, Pennsylvania

GERALD I. BYRNE • Division of International Medicine, Cornell University Medical College, New York, New York

PETER DENSEN • Department of Medicine, University of Virginia School of Medicine, Charlottesville, Virginia

PETER ELSBACH • Department of Medicine, New York University School of Medicine, New York, New York

RICHARD B. JOHNSTON, JR. • Department of Pediatrics, National Jewish Hospital and Research Center and University of Colorado School of Medicine, Denver, Colorado

THOMAS C. JONES • Division of International Medicine, Cornell University Medical College, New York, New York

SEYMOUR J. KLEBANOFF • Department of Medicine, University of Washington, Seattle, Washington

GERALD L. MANDELL • Department of Medicine, University of Virginia School of Medicine, Charlottesville, Virginia

VICTOR A. NAJJAR • Division of Protein Chemistry, Tufts University School of Medicine, Boston, Massachusetts

PIERLUIGI PATRIARCA • Istituto di Patologia Generale, Università di Trieste, Italy

MARILYN C. PIKE • Laboratory of Immune Effector Function, Howard Hughes Medical Institute, Division of Rheumatic and Genetic Diseases, Department of Medicine, Duke University Medical Center, Durham, North Carolina

DOMENICO ROMEO • Istituto di Chimica Biologica, Università di Trieste, Italy

D. ROOS • Department of Blood Cell Chemistry, Central Laboratory of the Netherlands Red Cross Blood Transfusion Service, Amsterdam; and Laboratory of Clinical and Experimental Immunology, University of Amsterdam, Amsterdam, The Netherlands

FILIPPO ROSSI • Istituto di Patologia Generale, Università di Padova, Sede di Verona, Italy

JULIUS SCHULTZ • Papanicolaou Cancer Research Institute, Miami, Florida; and Department of Biochemistry, University of Miami Medical School, Coral Gables, Florida

RALPH SNYDERMAN • Laboratory of Immune Effector Function, Howard Hughes Medical Institute, Division of Rheumatic and Genetic Diseases, Department of Medicine, Duke University Medical Center, Durham, North Carolina

JOHN K. SPITZNAGEL • Department of Microbiology, Emory University, Atlanta, Georgia

RUNE L. STJERNHOLM • Department of Biochemistry, Tulane Medical School, New Orleans, Louisiana

JERROLD WEISS • Department of Medicine, New York University School of Medicine, New York, New York

JAN MACIEJ ZGLICZYŃSKI • Institute of Medical Biochemistry, Nicolaus Copernicus Academy of Medicine, 31-034 Krakow, Poland

Foreword

This comprehensive treatise on the reticuloendothelial system is a project jointly shared by individual members of the Reticuloendothelial (RE) Society and biomedical scientists in general who are interested in the intricate system of cells and molecular moieties derived from these cells which constitute the RES. It may now be more fashionable in some quarters to consider these cells as part of what is called the mononuclear phagocytic system or the lymphoreticular system. Nevertheless, because of historical developments and current interest in the subject by investigators from many diverse areas, it seems advantageous to present in one comprehensive treatise current information and knowledge concerning basic aspects of the RES, such as morphology, biochemistry, phylogeny and ontogeny, physiology, and pharmacology as well as clinical areas including immunopathology, cancer, infectious diseases, allergy, and hypersensitivity. It is anticipated that by presenting information concerning these apparently heterogeneous topics under the unifying umbrella of the RES attention will be focused on the similarities as well as interactions among the cell types constituting the RES from the viewpoint of various disciplines. The treatise editors and their editorial board, consisting predominantly of the editors of individual volumes, are extremely grateful for the enthusiastic cooperation and enormous task undertaken by members of the biomedical community in general and especially by members of the American as well as European and Japanese Reticuloendothelial Societies. The assistance, cooperation, and great support from the editorial staff of Plenum Press are also valued greatly. It is hoped that this unique treatise, the first to offer a fully comprehensive treatment of our knowledge concerning the RES, will provide a unified framework for evaluating what is known and what still has to be investigated in this actively growing field. The various volumes of this treatise provide extensive in-depth and integrated information on classical as well as experimental aspects of the RES. It is expected that these volumes will serve as a major reference for day-to-day examination of various subjects dealing with the RES from many different viewpoints.

Herman Friedman
Mario R. Escobar
Sherwood M. Reichard

Introduction

The biochemistry of the reticuloendothelial system (RES) is a truly broad and wide ranging topic. The RES consists of a complex of interacting cells and their supporting structure important in many biologic activities. It is now widely accepted that the major cell types in the RES are the mononuclear phagocytes, a group of cells with similar morphology, function, and origin, ranging from those cells found in the blood to those present in the tissue and lymphoid organs, as well as other cells which belong to the RES, including those which are generally not classified as mononuclear phagocytes. For example, polymorphonuclear leukocytes have been studied quite extensively. It is practically impossible to examine the biochemistry of macrophages per se without making comparisons with these and other cells. There are many activities shared by these cell classes. It is also important to examine cells which are normal (resident) cells, as well as those which have been stimulated, at least in terms of biochemical and physiological activities. Nevertheless, it is widely assumed that phagocytosis per se, i.e., ingestion of particles, as well as microbicidal and tumoricidal activity are directly related to biochemical activation and function.

This volume, the second in a series on the RES, presents a broad range of topics involving biochemical activities of the cell constituents of the RES as well as related cells. Among the subjects discussed are attachment of recognition factors involving these cells in biologic activity and their biochemical reactions. Chemotactic factors are also important, since the cells of the RES not only recognize such factors but also produce such factors. Among the biochemical activities of the RES are those involving carbohydrate metabolism, lipid metabolism, amino acid metabolism, etc. These topics are covered in this volume. Furthermore, the biochemistry of microbicidal and tumoricidal activity is quite important and is discussed in detail. In addition, the myeloperoxidase system as well as oxygen-independent systems are presented. The relationship of free-radical production by RE cells and the role of such free radicals in microbicidal activity is also discussed in this volume. The relationship of such biochemical activities to health and diseases are also discussed, especially in regards to effects on macrophage activities and the RES in general.

It is important to note that recent renewed interest concerning all aspects of cellular immunology has directed increasing attention to the biochemical activities of cells involved in immune responsiveness, including those required for antibody formation as well as cellular immune responses. In this regard, subcellular factors are now known to be involved in immunity including lymphokines and monokines, as well as small pharmacologically active agents such as those exemplified by tuftsin, a small peptide which appears to activate cells of the RES. This topic is also included.

It is apparent that intimate knowledge of biochemical and metabolic activities of the diverse cells which are involved in immune responsiveness in general and specific microbicidal and tumoricidal activity in particular is necessary before fruther advances can be made in regards to developing specific agents which affect in either a positive or negative manner these cell activities. It is anticipated that this volume, which brings together in a single source many important aspects of RE cell biochemistry and metabolism, will be a valuable starting point for many investigators and students who wish to familiarize themselves with this important topic.

A distinguished group of authors has been assembled to present the most up-to-date information available concerning this subject. It seems certain that further fundamental analyses of the biochemistry and metabolism of RE cells will ultimately yield new and powerful means of prediction, manipulation, and control of the RES in terms of applications to bioscience and medicine in general. There seems to be little question that the immense literature reviewed in this volume indicates that the field of biochemistry as applied to the RES has had an extraordinary influence on understanding the functional activities of this cell system and holds much promise for further developments.

Herman Friedman
Mario Escobar
Sherwood Reichard

Preface

We have been assigned the task of discussing the various biochemical activities of cells belonging to the reticuloendothelial system. Having at our disposal a number of experts that are eminently qualified to help us, it would appear that our charge is not difficult. However, there is a problem; what exactly is the reticuloendothelial system (RES)? What cell types are in this system? What do we know about the biochemistry of these cells? Are the mononuclear phagocytes in this system? According to Aschoff, they are; however, within the past few years, van Furth and others have advocated that these cells are truly not able to fulfill the criteria of the RES. On the basis of common origin, morphology, and function, van Furth has classified these mononuclear phagocytes and macrophages as belonging to the mononuclear phagocyte system (MPS). Further, the polymorphonuclear phagocyte, certainly the most studied cell, at least from a biochemical point of view, has not been claimed in either the RES or the MPS.

We will not, in this volume, attempt to resolve this conflict in classification. Our overall objective will be to present information regarding the biochemical activities of different phagocytic cells which are able to perform a physiological function, i.e., to protect the host against infection. The polymorphonuclear leukocyte is the first phagocytic cell that was systematically studied from a biochemical point of view. Without doubt, the bulk of our knowledge is restricted to this cell. It has served, and is serving, as a model for all other studies. The biochemical activity of other cell types has also been studied, but to a lesser degree, and this will also be discussed.

Our purpose, in this volume, will be to call on experts that will discuss the biochemical and physiological functional activities of a number of cells that have been shown to participate in host–parasite interactions. This is the common denominator for our choice of material covered in this volume.

Drs. Pike and Snyderman in the initial chapter discuss chemotaxis. They point out that the direct migration of leukocytes in response to chemotactic factors results from the culmination of a complex series of biochemical events. These are discussed. Drs. Jones and Byrne review attachment and recognition factors associated with the interaction between microbes and mammalian phagocytic cells. Dr. Najjar describes a tetrapeptide, present in leukokinin, which can bind specifically to neutrophils to effect a prompt stimulation of their phagocytic activity. The physiological function of the tetrapeptide called "tuftsin" is clearly noted. Dr. Stjernholm reviews and summarizes for us early, as well as recent work, on the carbohydrate metabolism of the circulating leukocyte. Dr. Elsbach does likewise with the lipid metabolism of the cell. In addition, Dr. Elsbach points out specific areas where more information in this aspect of cell metabolism is needed.

Dr. Basford, in his chapter, reviews the literature on the physiologic role of glutathione in phagocytizing leukocytes. He concludes from his studies that adequate levels of glutathione, glutathion reductase, and, perhaps, glutathione peroxidase are required to protect the leukocyte from damage due to excess H_2O_2 or O^-_2 not utilized in the killing of phagocytized organisms. Drs. Rossi, Patriarca, and Romeo discuss, at length, the "metabolic burst" accompanying phagocytosis. They feel that the unravelling of the events involved are not only important in relation to host-defense but they also may aid in the elucidation of the many complex molecular mechanisms of broad biological interest. Drs. Roos and Balm, in their chapter, discuss the neutrophil, the monocyte, and macrophage. They focus on the oxidative intracellular killing of microorganisms, nonoxidative mechanism of intracellular microbicidal activity, and extracellular cytotoxicity. They conclude that the neutrophil is the simple cell, the monocyte is an immature cell, and the macrophage is a highly differentiated, multifunctional cell. Altogether, these phagocytes comprise a very effective host-defense system. In the next chapter, Dr. Schultz traces for us the historical development of myeloperoxidase; the early and the most recent advances are presented. On the same subject and in the next chapter, Dr. Jan Zgliczynski characterizes myeloperoxidase from neutrophils and peroxidase from different cell types. It is noted that peroxidase activity is present in almost all types of host-defense cells: The chemical and biological activity of peroxidase is well-described in these two chapters.

Dr. Klebanoff, lucidly describes the myeloperoxiase-mediated cytotoxic systems. Dr. Robert Allen describes free radical production by RE cells and notes their physiological functions. Dr. Babior, describes the role of oxygen radicals in microbial killing by phagocytes. He develops the concept that oxygen radicals, known to be lethal to living systems, are turned to good use by phagocytic cells in their activities against pathogenic microorganisms.

Dr. John Spitznagel discusses in his chapter the many different ways that polymorphonuclear leukocytes may exert their antimicrobial activities. He points out that in addition to the oxygen dependent systems, there are also those that operate quite independently of it. Dr. Spitznagel considers the various antimicrobial systems that act intracellularly and independently of molecular oxygen. He further presents selected evidence for their role in defense against infection in animals and humans. Drs. Densen and Mandell develop and discuss the concept that phagocytic cells have evolved a multifaceted mechanism for killing microbes. However, microorganisms have evolved equally effective means of avoiding or neutralizing phagocyte microbicidal functions. In the final chapter, Dr. Johnston reviews the known biochemical defects of human neutrophils, monocytes, and macrophages that predispose to disease, in particular, to recurrent infections.

We are certain that in these chapters the respective authors have presented useful and recent data concerning the biochemical and physiological activities of phagocytic cells. We are equally certain that there is some unavoidable overlap and some controversy.

We are grateful to each of the authors for their contributions. There is no question that they have done a splendid job and in our judgment, deserve congratulations.

Anthony J. Sbarra
Robert Strauss

Contents

3. Biochemistry and Physiology of Tuftsin Thr-Lys-Pro-Arg

Victor A. Najjar

4. Carbohydrate Metabolism

Rune L. Stjernholm

5. Lipid Metabolism by Phagocytic Cells

Peter Elsbach and Jerrold Weiss

6. Glutathione Metabolism in Leukocytes

R. E. Basford

7. Metabolic Changes Accompanying Phagocytosis

Filipo Rossi, Pierluigi Patriarca, and Domenico Romeo

8. The Oxidative Metabolism of Monocytes

D. Roos and A. J. M. BALM

9. Myeloperoxidase

JULIUS SCHULTZ

10. Characteristics of Myeloperoxidase from Neutrophils and Other Peroxidases from Different Cell Types

Jan Maciej Zgliczyński

11. Myeloperoxidase-Mediated Cytotoxic Systems

SEYMOUR J. KLEBANOFF

12. Free-Radical Production by Reticuloendothelial Cells

ROBERT C. ALLEN

13. The Role of Oxygen Radicals in Microbial Killing by Phagocytes

Bernard M. Babior

14. Oxygen-Independent Antimicrobial Systems in Polymorphonuclear Leukocytes

John K. Spitznagel

References   366

15. **Antimicrobial Functions of Phagocytes and Microbial Countermeasures**

    PETER DENSEN and GERALD L. MANDELL

    1. Introduction   369
    2. Cellular Development, Morphology, and Physiology   370
        2.1. Neutrophils   370
        2.2. Monocytes-Macrophages   373
        2.3. Eosinophils   374
    3. Microbial Mechanisms of Resistance to the Microbicidal Activity of Phagocytes   375
        3.1. Avoiding Recognition   375
        3.2. Inhibition of Chemotaxis   376
        3.3. Inhibition of Attachment   378
        3.4. Inhibition of Ingestion   380
        3.5. Depression of the Metabolic Burst   384
        3.6. Alteration of Degranulation   384
        3.7. Entry of Pathogens into Cells   388
        3.8. Resistance to Bactericidal Activity   388
        3.9. Escape from the Phagosome   389
    4. Conclusions   390
    References   390

16. **Biochemical Defects of Polymorphonuclear and Mononuclear Phagocytes Associated with Disease**

    RICHARD B. JOHNSTON, JR.

    1. Introduction   397
    2. Defects of Chemotaxis   398
        2.1. Primary Cellular Abnormalities   398
        2.2. Secondary Cellular Abnormalities   403
    3. Defects of Phagocytosis   406
        3.1. Actin Dysfunction   406
    4. Defects of Microbicidal Activity   407
        4.1. Chronic Granulomatous Disease   407
        4.2. Glucose-6-Phosphate Dehydrogenase (G-6-PD) Deficiency   411

Biochemical and Biological Aspects of Leukocyte Chemotactic Factors

MARILYN C. PIKE and RALPH SNYDERMAN

1. INTRODUCTION

The degradation of antigen by the immune system is the result of a complex series of events beginning with detection of the foreign material by recognition elements of the immune system and culminating with the ingestion of the antigen by phagocytic cells. One mechanism by which phagocytes may accumulate at antigenic sites is chemotaxis, or the directed migration of cells in response to a gradient of a chemoattractant substance. Much data has accumulated over the past twenty years concerning the nature of the chemical substances which produce the unidirectional migration of leukocytes *in vitro*. Microbial or viral materials, upon interaction with humoral and cellular immune recognition components can lead to the production of chemotactic substances. Some antigenic materials may possess intrinsic chemotactic properties themselves.

Although many chemotactic factors have been described, little is known of their biological significance or of the actual biochemical events which are triggered in leukocytes by the chemoattractants. The directed migration of these cells does appear to require the recognition of chemical signals by the cell membrane, the activation of energy-forming processes within the cell, and the translation of this energy by cytostructural elements into directed movement. Several lines of evidence also suggest that the specific binding of chemotactic factors to cellular membrane receptors, the methylation of membrane-associated molecules, divalent cation fluxes, activation of the hexose monophosphate shunt, and the polymerization of actomyosin-like molecules within the cell are all required for unidirectional movement. The following report will review what

MARILYN C. PIKE and RALPH SNYDERMAN • Laboratory of Immune Effector Function, Howard Hughes Medical Institute, Division of Rheumatic and Genetic Diseases, Department of Medicine, Duke University Medical Center, Durham, North Carolina 27710.

is known of many of the described chemotactic substances and the evidence available concerning the biochemical processes which lead to the directed migration of leukocytes.

2. CHEMOTACTIC FACTORS

The development of quantitative *in vitro* techniques for the study of leukocyte chemotaxis has permitted the investigation of the biochemical nature of chemotactic factors (Ward *et al.*, 1965; Keller and Sorkin, 1965; Snyderman *et al.*, 1968, 1970, 1972; Goetzl and Austen, 1972; Gallin *et al.*, 1974). While many agents can be shown to affect cell motility *in vitro,* in most cases the biological role of such factors in inflammation is poorly understood. The following review will focus upon chemotactic factors for which there is either some evidence of a biological role or upon factors which have enabled the study of physiological aspects of leukocyte chemotaxis.

2.1. COMPLEMENT (C)-DERIVED CHEMOTACTIC FACTORS

The first indication that C may be involved in the production of chemotactic activity in serum was provided by Boyden (1962). The role of isolated C components in the chemotaxis of neutrophils was later reported by Ward *et al.* (1965, 1966). These investigators found that chemotactic activity was generated by the sequential activation of partially purified C components. Chemotactic activity was reported to become manifest upon the formation of an activated macromolecular complex of C5, C6, and C7. Cleavage of partially purified C3 by plasmin was also reported to produce a chemotactic factor with a molecular weight of $\sim$ 6000 (Taylor and Ward, 1967). Using whole serum activated with endotoxin, Snyderman *et al.* (1968) showed that the generation of a neutrophil chemotactic factor was dependent upon the presence of C5. The chemotactic factor generated in guinea pig serum upon incubation with endotoxin had a molecular weight of $\sim$ 15,000. Interestingly, such sera contained no substantial chemotactic activity which could be attributed to a $C\overline{567}$ complex. Shin *et al.* (1968), Snyderman *et al.* (1968, 1969), and Jensen *et al.* (1969), went on to show that the low-molecular-weight chemotactic activity produced in guinea pig serum was due to the cleavage product of the fifth component of C, C5a. The cleavage of highly purified C5 with either trypsin or $EAC\overline{1423}$ resulted in generation of a 15,000-molecular weight fragment which possessed not only chemotactic activity but anaphylatoxic activity as well (Shin *et al.*, 1968). The chemotactic activity associated with this fragment was abrogated by incubation of the factor with antibody to C5 but not to C3 (Snyderman *et al.*, 1969). Ward and Newman (1969) found that cleavage of human C5 with trypsin also generated chemotactic activity for neutrophils.

Additional evidence was presented indicating that the $C\overline{567}$ complex was not required for the generation of chemotactic activity in whole serum. Normal levels of chemotactic activity were shown to be generated by C activation in the sera of rabbits congenitally devoid of C6 (Stetcher and Sorkin, 1969; Snyderman *et al.*, 1970). These results demonstrated that chemotactic activity could be generated at the level of C5 or earlier. Since no activity was generated in C5-deficient mouse serum (Snyderman *et al.*, 1968, 1971a), C activation through C3 was not sufficient to produce chemotactic activity. It was, moreover, shown that the chemotactic factor produced in C-sufficient mouse serum was antigenically related to C5 and had a molecular sieve elution profile identical to C5a (Snyderman *et al.*, 1971a). C5 was thus shown to be essential for chemotactic factor generation.

Activation of serum by many agents has been shown to produce chemotactic activity derived from C5. C5a activity can be generated in sera by treatment with immune complexes, cobra venom factor, endotoxin, zymosan, staphylococcal protein A, or by an enzyme released by cultured kidney cell infected with herpes simplex virus (Snyderman *et al.*, 1968, 1970; Shin *et al.*, 1969a; Brier *et al.*, 1970; Pike and Daniels, 1975; Moeller *et al.*, 1978).

C5a is a potent chemotactic factor not only for neutrophils, but for monocytes, macrophages, basophils, and eosinophils as well (Kay, 1970; Snyderman *et al.*, 1971b, 1972; Kay and Austen, 1972). The biologic importance of C5a as a chemotactic factor produced at inflammatory foci *in vivo* has also been established. Peritoneal fluids obtained from guinea pigs or mice which had been injected intraperitoneally (i.p.) with inflammatory stimuli contained chemotactic activity which was shown to have a molecular weight of ~ 15,000, and whose chemotactic activity was inhibitable by antibody to C5 (Snyderman *et al.*, 1971a). Maximal activity was recovered from the exudates approximately 2 hr after injection of the inflammatory stimulus, after which time polymorphonuclear leukocytes (PMNs) began to accumulate in the peritoneal cavity. Additional evidence supporting the role of C5a as a major chemotactic factor produced *in vivo* was that mice congenitally devoid of the fifth component of C were markedly deficient in mobilizing PMNs in response to an i.p. injection of endotoxin as compared to C5-sufficient mice (Snyderman et al., 1971a). C5a has also been found in the synovial fluids of patients with rheumatoid arthritis (Ward and Zvaifler, 1971). When injected into the skin of humans or guinea pigs, C5a induces the local accumulation of PMNs and macrophages (Jensen et al., 1969). Kazmierowski *et al.* (1977) have recently shown that, following bronchoalveolar lavage of a primate's lung, an appreciable number of PMNs accumulate in respiratory fluids. Such fluids were found to contain chemotactic activity of molecular weight 15,000 which was inhibitable by antisera to C5 and was, therefore, concluded to be C5a.

As stated before, cleavage products of C3 have also been implicated as chemotactic agents. Chemotactically active fragments produced from C3 by the action of plasmin or proteolytic enzymes derived from bacteria have been reported (Taylor and Ward, 1967; Chapitis *et al.*, 1971). Bokisch *et al.* (1969) found that C3 subjected to the action of trypsin or $C\overline{42}$ resulted in the generation of a peptide which possessed both neutrophil chemotactic activity and was anaphylatoxic. Other workers have found that the biological activity of C3a is anaphylatoxic but not chemotactic (Shin *et al.*, 1969b), and little has been reported concerning any *in vivo* chemotactic activity associated with C3 fragments. Recent studies by Fernandez *et al.* (1976, 1978) would appear to discount any previous studies associating chemotactic activity with C3a. These workers have determined the entire primary structure of C3a and have studied both its biological and physical properties using very pure preparations. Such preparations of C3a possessed no chemotactic activity for macrophages or neutrophils, but did contain anaphylatoxic activity. Similarly, pure preparations of C5a had chemotactic activity at concentrations as low as 10^{-8} M. Since C5a and C3a have strikingly similar physical properties and, thus, are very difficult to separate by standard chromatographic techniques, previous reports purporting C3a as a chemotactic factor were probably due to contamination of these preparations with C5a.

2.2. CELLULAR-DERIVED CHEMOTACTIC FACTORS

Lymphocytes, when incubated with mitogens or specific antigens, produce and release soluble mediators of inflammation termed lymphokines. Some of these substances have been shown to be chemotactic for neutrophils, monocytes, macrophages, and basophils. Ward *et al.* (1969) reported that guinea pig lymphocytes, when exposed to specific antigens, produced a chemotactic factor of molecular weight 43,000 for homologous macrophages. Stimulated lymphocyte culture supernatants also contained a neutrophil chemotatic factor which could be distinguished from migration inhibition factor and the macrophage chemotactic factor (Ward *et al.*, 1970). A lymphokine which is chemotactic for lymphocytes has also been described but has yet to be characterized (Ward *et al.*, 1970).

Using human lymphocytes, it was shown that a substance with a molecular weight of 12,500 was produced when leukocytes from purified-protein-derivative-sensitive (PPD-sensitive) humans were incubated with this antigen (Snyderman *et al.*, 1972; Altman *et al.*, 1973). The same or similar substance was produced when lymphocytes were incubated with mitogens and chemotactic activity could be detected within the first 6 hr in culture. The chemotactic factor was termed lymphocyte-derived chemotactic factor or LDCF. Production of LDCF, which has been shown to be isoelectrically and antigenically distinct from C5a (Altman *et al.*, 1974), does not require cell division by lymphocytes but does require new protein synthesis (Snyderman *et al.*, 1977a). Supernatants of mixed lymphocyte cultures (MLC) from HLA-nonidentical individuals also contain LDCF (Snyderman *et al.*, 1977a). LDCF has been shown to be produced by both B and T lymphocytes (Altman *et al.*, 1974; Mackler *et al.*, 1974; Wahl *et al.*, 1974). T cells require the presence of macrophages in culture for LDCF synthesis, but B cells apparently do not (Wahl *et al.*, 1975). *In vivo* studies have suggested that a chemotactic factor for macrophages is produced at sites of delayed hypersensitivity. Extracts of guinea pig skin from sites of delayed hypersensitivity reactions contained macrophage chemotactic activity, while extracts of normal skin did not (Cohen *et al.*, 1973). A method for studying the kinetics of delayed hypersensitivity reactions *in vivo* was developed in this laboratory. Specific antigen was introduced through indwelling silicon plastic catheters which had been implanted in the peritoneal cavities of sensitized guinea pigs (Postlethwaite and Snyderman, 1975). After various times, peritoneal fluid samples were withdrawn and analyzed for inflammatory cell content and for chemotactic activity. It was found that, concomitantly with the appearance of macrophages in the peritoneal cavity in response to antigen challenge, a 12,500-molecular-weight factor was present which possessed chemotactic activity for homologous macrophages. The chemotactic factor produced *in vivo* could not be distinguished from LDCF produced by antigen-stimulated lymphocytes *in vitro*. These findings suggest that a chemotactic lymphokine is produced *in vivo* and may be responsible for the accumulation of macrophages at sites of delayed hypersensitivity reaction.

2.3. BACTERIAL CHEMOTACTIC FACTORS AND SYNTHETIC FORMYLATED PEPTIDES

Many strains of bacteria produce substances which are chemotactic for neutrophils and macrophages. Keller and Sorkin (1967) found that culture filtrates of *Staphylococcus albus* and *Escherichia coli* contained chemotactic activity for neutrophils but no biochemical characterization of the active products was done. Ward *et al.* (1968) found chemotactic activity in the culture supernatants from *S. albus*, α-hemolytic streptococci, *Streptococcus pneumoniae*, *E. coli*, and *Proteus mirabilis*, and the factor obtained from *S. pneumoniae* cultures was found to have a molecular weight of $\sim$ 3600. Walker *et al.* (1969) reported that a chemotactic factor was produced by *S. aureus* and had a molecular weight of $>$ 10,000 since it was nondialyzable. Tempel *et al.* (1970) found that the chemotactic activities associated with three different types of bacteria that grow in the oral cavity have a molecular weight of $<$ 10,000. As is indicated by the above studies, the bacterial factors described by different laboratories could not be attributed to a single common entity.

While studying the nature of bacterial chemotactic factors, Schiffman *et al.* (1975) made the important discovery that *N*-terminal blocked methionyl di- and tripeptides are chemotactic for neutrophils and macrophages. Substances such as these peptides may be analagous to those derived from the *N*-terminal regions of newly synthesized bacterial proteins. Formylation of the NH_2-terminus is necessary for the chemotactic activity of these peptides. These investigators, therefore, speculated that the chemotactic response of mammalian leukocytes to formylated amino acids may represent a simple recognition system that detects the presence of microbial agents, since eukaryotic cells largely initiate protein synthesis with nonacylated methionine.

The finding that *N*-formylated methionyl peptides are chemotactic for leukocytes has permitted the synthesis of highly purified, structurally defined substances, which were heretofore unavailable, for studying the physiology and biochemistry of the interaction of chemotactic factors with motile cells. Preliminary studies have been done correlating the structure–function relations of the formylated peptides in inducing chemotaxis and lysosomal enzyme release. Showell *et al.* (1976) demonstrated that for maximum chemotactic activity, the positively charged terminal amino acid group must be neutralized and that N-formylation was preferable to N-acylation, supporting the findings of Schiffman *et al.* (1975). Methionine is not absolutely required in the NH_2-terminal position for chemotactic activity of formylated oligopeptides. An amino acid of similar hydrophobic character such as norleucine can be substituted. In the second position, a neutral amino acid with a nonpolar side chain is required for maximal activity. The most active oligopeptides were those containing a phenylalanine residue in the third position. It has not yet been determined whether phenylalanine enhances activity because of its aromaticity or stereo-specificity. It is also not known whether the enhanced activity associated with phenylalanine is due to its being in the third position from the NH_2 terminal or because it is itself the terminal residue of the peptide. The structure–function relationships of several formylated oligopeptides were the same for chemotaxis and lysosomal enzyme release (Showell *et al.*, 1976). The availability of the N-formylated peptides with varying potencies has allowed the first direct demonstration of a specific chemotactic factor receptor on the surface of neutrophils (see below).

2.4. LIPID-ASSOCIATED CHEMOTACTIC FACTORS

Turner *et al.* (1975a,b) made the first observation that oxidized components of polyenoic fatty acids are chemotactic for neutrophils. These investigators found that exposure of arachidonic acid to air oxidation or platelet lipoxygenase resulted in a chemotactic byproduct, 12-L-hydroxy-5,8,10,14-eicosatetraenoic acid (HETE). These findings were confirmed by Goetzl *et al.* (1977), who found that HETE was not only chemotactic for neutrophils, but for eosinophils as well. HETE is not as effective a chemotactic agent for mononuclear cells as it is for PMNs. Goetzl and Gorman (1978) found that another oxygenation product of arachidonic acid, 12-L-hydroxy-5,8,10-heptadecatrienoic acid (HHT), was also chemotactic for neutrophils and eosinophils. Both HETE and HHT differ from other chemotactic factors in that, when incubated with PMNs, they enhance the migration of these cells to other chemotactic substances (Goetzl and Gorman, 1978), while other factors tend to "deactivate" or "desensitize" the chemotactic response. Lipid chemotactic factors have been isolated from the peritoneal cavities of passively sensitized rats after challenge with antigen (Valone and Goetzl, 1978), suggesting that lipid chemotactic factors may have *in vivo* relevance.

3. BIOCHEMICAL AND PHYSIOLOGIC CONSEQUENCES OF THE INTERACTION OF CHEMOTACTIC FACTORS WITH LEUKOCYTES

Chemotactic factors produce a wide spectrum of physiologic and metabolic changes in leukocytes, some of which are or appear to be associated with the chemotactic response. Other changes induced by these chemotactic agents such as lysosomal enzyme release and cell aggregation are effects which may be distinct from chemotactic events. Chemotaxis appears to require initial recognition of a chemotactic factor by the leukocyte membrane. This is followed by transmission of this event to cytoskeletal elements within the cell which then results in polarized movement. In the following section, we will review the known biochemical consequences of the interaction of leukocytes with chemotactic factors and their relationship to actual cellular migration and other functions of leukocytes.

3.1. SPECIFIC LEUKOCYTE MEMBRANE RECEPTORS FOR CHEMOTACTIC FACTORS

The availability of highly active, N-formylated chemotactic peptides has allowed the detection and partial characterization of specific receptors for these substances on the membranes of human and rabbit PMNs (Williams *et al.*, 1977; Aswanikumar *et al.*, 1977). Direct binding studies to cellular receptors were carried out in our laboratory using tritiated N-formyl-methionyl-leucyl-phenylalanine of high specific radioactivity ([^{3}H]-FMLP) (Williams *et al.*, 1977). This particular peptide has the most potent chemotactic activity of any of the N-formylated peptides yet synthesized. Its EC_{50} is 5×10^{-9} M for human PMNs. The binding characteristics of [^{3}H]-FMLP to PMNs fulfilled all the criteria for the demonstration of specific receptors. The binding of [^{3}H]-FMLP to human PMNs was saturable and of high specificity, with a calculated equilibrium dissociation constant for the interaction of [^{3}H]-FMLP with its binding site of 12–14 nM. [^{3}H]-FMLP binding was rapid ($t_\frac{1}{2}$ = 2 min) and readily reversed by the addition of a large excess of unlabeled FMLP to an equilibrated mixture of [^{3}H]-FMLP and PMNs. The specificity of the [^{3}H]-FMLP binding site was investigated by comparing the relative potencies of a series of N-formylated-methionyl peptides as chemotactic agents with their ability to compete for the [^{3}H]-FMLP binding site (Figure 1). It was found that the order of potency of the various peptides for the aforementioned responses was exactly the same, indicating that the [^{3}H]-FMLP binding sites have the specificity expected of receptor sites which mediate the PMN response to the chemotactic peptides.

The ability of other substances to inhibit or compete for the [^{3}H]-FMLP binding site was tested. Sodium azide (0.01 M) or the protease inhibitor, tosyl-L-pheylalanyl chloromethane, had no effect on [^{3}H]-FMLP binding. Concentrations of C5a, which were tenfold higher than that necessary to give a half-maximal chemotactic response, did not compete for the [^{3}H]-FMLP binding site. Indeed, a C5a specific binding site which is apparently distinct from the N-formylated peptide site, has been identified on the surface of human PMNs (Chenoweth and Hugli, 1977). An inhibitor of the chemotactic response to formylated methionyl peptides, f-phe-met, behaved as a competitive antagonist of [^{3}H]-FMLP binding. Human erythrocytes and column-purified lymphocytes demonstrated little or no specific binding of [^{3}H]-FMLP (Williams *et al.*, 1977). Others have also demonstrated the presence of N-formylated peptide receptors on the surfaces of rabbit exudate PMNs (Aswanikumar *et al.*, 1977).

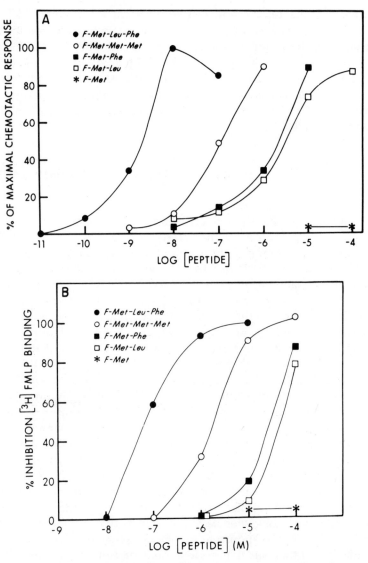

FIGURE 1. Specificity of [³H]-FMLP binding sites towards chemotactically active peptides. (A) Chemotactic activity of N-formylmethionyl peptides. The chemotactic response was measured for each indicated concentration of peptide. The maximal response to 10^{-8} M f-met-leu-phe was 102 cells per high power field. (B) Effects of N-formylmethionyl peptides on [³H]-FMLP binding. [³H]-FMLP was incubated with human PMNs in the presence of the indicated concentrations of the various peptides. Specific binding was determined and the percent inhibition of [³H]-FMLP binding caused by each concentration of unlabeled peptide was computed. Taken from Williams *et al.* (1977).

These findings demonstrate that specific receptor sites for high potency chemotactic factors do exist on the surface of leukocytes. Whether other factors such as oxidized lipids, lymphocyte-derived chemotactic substances, or denatured proteins initiate chemotaxis by binding to specific receptors has not yet been reported. It seems reasonable to hypothesize that high potency chemotactic factors act via distinct specific receptors and low-potency factors may exert their effect in the absence of a specific molecular interaction with membrane components (i.e., via hydrophobic interactions).

The recognition of PMNs of the high potency chemotactic factors seems to be analogous to the initiation of physiologic responses by hormones in hormonally responsive tissues. The further availability of highly specific, radioactively labeled, purified chemotactic factors should continue to advance our knowledge of the leukocyte chemotactic response.

3.2. REQUIREMENT OF S-ADENOSYL-METHIONINE-MEDIATED METHYLATION FOR CHEMOTAXIS

The biochemical events in leukocytes which are triggered immediately and directly as a consequence of the interaction of a chemotactic substance with its receptor are as yet poorly defined. In bacteria, the complexing of chemoattractants with their corresponding chemoreceptors results in the methylation of specific membrane proteins, and this event is required for the chemotactic response (Kort et al., 1975; Springer et al., 1977). Until recently, the role of methylation in eukaryotic cell motility was unknown. Adenosine deaminase (ADA), the enzyme which catalyzes the conversion of adenosine to inosine, has been shown to play an important modulating role in the S-adenosyl-methionine (SAM)-mediated methylation pathway (Kredich and Martin, 1977) (Figure 2). Specific inhibition of ADA plus the addition of adenosine and L-homocysteine to lymphoblasts caused a marked elevation in intracellular S-adenosyl-homocysteine (SAH), a potent inhibitor of SAM-mediated methylation (Kredich and Martin, 1977). The possibility that ADA and methylation also play a role in modulating chemotaxis was suggested by the finding that a continuous macrophage cell line, P388D1, while having other biological functions intact, lacked both random and directed migratory responses and had very low levels of ADA (Synderman et al., 1977b). Considering this association of ADA with macrophage migration and the requirement of methylation for bacterial chemotaxis, we investigated the role of methylation in monocyte chemotaxis and phagocytosis (Pike et al., 1978).

The chemotaxis of human peripheral blood monocytes was quantified using FMLP as the chemoattractant. Monocytes were suspended in medium containing the inhibitor of ADA, erythro-9-(2-hydroxy-3-nonyl) adenine (EHNA) plus various concentrations of adenosine and L-homocysteine thiolactone, and tested for their chemotactic responsiveness. In the presence of 10 μM EHNA plus 0.1 mM or 1.0 mM adenosine, monocyte chemotaxis was inhibited by 18 and 26% respectively (Figure 3). The further addition of 0.1 mM L-homocysteine thiolactone increased the inhibition to 82 and 97%. L-Homocystine in combination with EHNA and adenosine produced somewhat less inhibition of chemotaxis than did L-homocysteine thiolactone, and D-homocystine had no effect at all. The addition to monocytes of EHNA alone, adenosine alone, or homocysteine alone produced no effect on the chemotactic response. The inhibition of chemotaxis

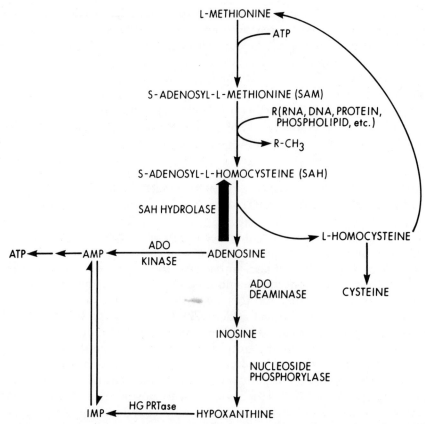

FIGURE 2.   Pathways of adenosine metabolism and *S*-adenosyl-L-methionine-dependent methyla-tion. ado, Adenosine; HGPRTase, hypoxanthine-guanine phosphoribosyl transferase; R, methyl accepting group. Taken from Pike *et al.* (1978).

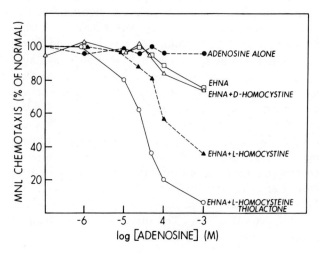

FIGURE 3. Effects of EHNA, L-homocysteine thiolactone, D-ho-mocystine and L-homocystine on human monocyte chemotaxis as a function of adenosine concentra-tion. Percent of normal = (E/C × 100) where E is the response of cells incubated with the various substances and C is the reponse of cells incubated in medium alone. Taken from Pike *et al.* (1978).

produced by EHNA, adenosine, and L-homocysteine thiolactone was not only dependent on the adenosine concentration but on the L-homocysteine thiolactone concentration as well. The effect of these agents could be reversed by washing the cells prior to quantification of chemotaxis, but not by adding 10 μM uridine or 0.1 mM methionine.

Intracellular SAH levels were then measured in isolated monocytes treated with similar doses of EHNA, adenosine, and L-homocysteine. The incubation of monocytes with 10 μM EHNA and 1.0 mM adenosine produced a fivefold increase in intracellular SAH. The most marked effects on monocyte SAH levels were noted, however, when cells were incubated with EHNA, adenosine, and 0.1 mM L-homocysteine thiolactone. Under these conditions, SAH was elevated by as much as 1500-fold.

Since SAH is a potent inhibitor of SAM-mediated methylation reactions, we next determined the effect of EHNA, adenosine, and L-homocysteine thiolactone on monocyte protein carboxy-O-methylation. The addition to cells of 10 μM EHNA plus 50 μM or 0.1 mM adenosine inhibited carboxy-O-methylation in these cells by 25% and 43%, respectively. The further addition of 0.1 mM L-homocysteine thiolactone to such monocyte preparations greatly augmented the inhibition of methylation to 76 and 84%, respectively. EHNA or adenosine alone produced no inhibition of monocyte methylation. The addition to monocytes of 10 μM EHNA and 0.1 mM L-homocysteine thiolactone in the absence of adenosine, conditions which produced no inhibition of chemotaxis, caused a 42% inhibition of carboxy-O-methylation, indicating that this particular type of methylation must be inhibited by at least this amount before effects on monocyte chemotaxis are noted. The nature of the molecules whose methylation is required for chemotaxis was suggested by the finding that protein synthesis is minimally, if at all, required for a chemotactic response. These experiments indicate that the methylation of preformed proteins, phospholipid molecules, or other small molecules such as catechols, but not the methylation of DNA or RNA, may be required for the monocyte chemotactic response.

We also determined whether methylation was required for another function of monocytes, erythrophagocytosis. Monocytes were suspended in EHNA plus various concentrations of adenosine with and without L-homocysteine thiolactone and tested for their ability to phagocytosize $Na^{51}CrO_4$-labeled opsonized sheep erythrocytes. Adenosine alone, at concentrations of 0.1 μM to 1.0 mM or 10 μM EHNA alone produced no effect on phagocytosis. EHNA in combination with adenosine produced an enhancement of the phagocytic response, a maximum of 90% occurring at 10^{-5} M adenosine. The further addition to the monocytes of 0.1 mM L-homocysteine thiolactone prevented the enhancement of phagocytosis noted in the presence of adenosine and EHNA, but, more importantly, did not depress the phagocytic response even under conditions where chemotaxis was inhibited by as much as 97%.

These findings for the first time demonstrate a requirement of SAM-mediated methylation for the chemotaxis of one type of eukaryotic cell, the human monocyte. In addition, it appears that the phagocytic response of these cells is either independent of methylation or much less sensitive to inhibition of such reactions by SAH.

3.3.   IONIC EVENTS IN CHEMOTAXIS

Macrophages and neutrophils have been shown to contain substantial quantities of the muscle contractile proteins, actin, myosin, a troponic-C-like molecule and tropomyosin [Tatsumi *et al.*, 1973; Stossel and Pollard, 1973; D. M. Watterson and R. Snyderman, unpublished (1977)]. These proteins most likely play a role in both the directed and random migration of leukocytes. Since the control of the assembly of these molecules in muscle tissues is dependent upon ion fluxes, many laboratories have examined the ionic events initiated by the interaction of chemotactic factors with leukocytes. The role of calcium fluxes in neutrophil chemotaxis was studied in this laboratory using lanthanum ion (La^{3+}) which inhibits transmembrane calcium movement (Boucek and Snyderman, 1976). The addition of La^{3+} to human neutrophils caused a dose-dependent inhibition of the chemotaxis of these cells in response to endotoxin-activated human serum (AHS), with almost complete inhibition of the chemotactic response occurring at 10^{-3} M La^{3+}. The inhibitory effects of La^{3+} on neutrophil chemotaxis could not be attributed to cell death and were reversed by increasing the external Ca^{2+} concentration.

The role of Ca^{2+} flux in chemotaxis was then determined by incubating PMNs with ^{45}Ca until equilibrium was attained. At this time, AHS, heat-inactivated normal serum, the chemotactic peptide f-met-phe, or medium alone were added to the equilibrated mixture of PMNs and ^{45}Ca, and the incubation continued. At various times thereafter, cell samples were deposited into and washed in La^{3+} containing medium which effectively seals the cell against further Ca^{2+} influx or efflux, and the amount of radiolabel contained in the cells was then determined. The results of these experiments indicated that there was a fourfold increase in ^{45}Ca influx in cells treated with AHS as compared to cells incubated with heated normal serum. f-Met-phe caused a similar, although less dramatic, increase in intracellular Ca^{2+}. The addition of various concentrations of La^{3+} to the incubation mixture resulted in a dose-dependent inhibition of AHS-induced Ca^{2+} influx which correlated well with the inhibition of PMN chemotaxis by La^{3+}. These studies suggest that calcium influx is required for chemotaxis, and the amount of calcium which enters the cell is adequate to activate a contractile process.

Other laboratories have studied changes in cellular ion fluxes induced by chemotactic factors. Gallin *et al.* (1975) and Gallin and Gallin (1977) have used standard electrochemical techniques to determine whether the interaction of chemotactic factors with macrophages triggered membrane potential changes. Of cells treated with C5a, 80% responded with membrane potential changes. These changes were characterized by either a depolarization followed by a hyperpolarization (26%), a hyperpolarization alone (65%), or a depolarization without hyperpolarization (8%). These C5a-evoked potential changes could be blocked by the addition of Mg^{2+}-EGTA, suggesting that Ca^{2+} was necessary for manifestation of the response. Through analysis of reversal potentials, these investigators concluded that the interaction of chemotactic factors with macrophages includes a small transient depolarization related to an influx of Ca^{2+}

and, perhaps, sodium, and a much larger hyperpolarization related to potassium efflux (Gallin *et al.*, 1978a,b). These same investigators found that chemotactic-factor-induced potential changes preceded pseudopod formation in macrophages.

Ion fluxes induced by FMLP in rabbit neutrophils have been studied by Naccache *et al.* (1977). FMLP induced a large, rapid increase in membrane permeability to ^{22}Na in these cells and smaller, delayed enhancements of ^{42}K and ^{22}Na efflux. Depending on the concentration of extracellular calcium, FMLP produced either an increase or decrease in the cell-associated steady-state level of calcium.

From the above findings, it is clear that chemotactic factors alter the permeability of leukocyte membranes to ions. The actual mechanism by which sodium, potassium, and calcium contribute to triggering the chemotactic response, however, remains uncertain.

3.4. CYCLIC NUCLEOTIDE CHANGES BY CHEMOTACTIC FACTORS

Modulation of the leukocyte chemotactic response may be controlled, at least in part, by cAMP and cGMP. The evidence for this hypothesis is that agents which increase cGMP levels such as ascorbic acid, serotonin, phenylephrine, carbachol, and phorbol myristate acetate enhance the chemotactic responses of human monocytes and neutrophils, while agents which enhance intracellular cAMP levels such as isoproterenol, prostaglandin E_1, the ionophore A23187, and polystyrene beads depress the chemotaxis of these cells (Estensen *et al.*, 1973; Rivkin *et al.*, 1975; Sandler *et al.*, 1975). Hatch *et al.* (1977) have demonstrated that chemotactic factors including trypsinized human complement, f-met-ala, and a bacterial chemotactic factor cause cellular accumulation of cGMP in leukocytes but no alteration in cAMP levels. Gallin *et al.* (1978a) have shown that agents which increase cAMP concentrations in cells inhibit the enhanced locomotion of human monocytes produced by serotonin or ascorbate. These changes in monocyte locomotion correlated well with the inhibition by these agents of serotonin or ascorbate-induced elevations of cGMP. These findings would suggest that cyclic nucleotide alterations do indeed affect chemotaxis, but the actual mechanism by which they accomplish this remains unclear.

3.5. OTHER LEUKOCYTE METABOLIC CHANGES INDUCED BY CHEMOTACTIC FACTORS

The metabolism of leukocytes is dramatically altered by their interaction with chemotactic factors, but not all the changes are necessarily associated with the actual directed migratory response. Chemotactic factors are known to produce the release of lysosomal enzymes from cytocholasin-B-treated cells (Goldstein *et al.*, 1973; Becker and Showell, 1974). The concentration of chemotactic factor needed to produce this response is generally higher than that necessary to result in directed migration; however, the structure–activity relationships of some chemotactic factors such as the formylated peptides are the same both for chemotaxis and lysosomal enzyme release (Showell *et al.*, 1976)

Chemotactic substances also produce *in vitro* granulocyte aggregation. Craddock *et al.* (1977) have suggested that the chemotactic factors produced by interaction of complement with dialyzer cellophane membranes in patients undergoing hemodialysis may account for the acute granulocytopenia produced in the first hour of dialysis. O'Flaherty *et al.* (1977) have reported that chemotactic factors injected *in vivo* in rabbits produce neutropenia. These authors suggest that chemotactic-factor-induced hypermargination or aggregation of neutrophils in blood vessels may be the mechanism responsible for this phenomenon.

The *N*-formylated-methionyl peptides have been shown to produce chemiluminescence in leukocytes (Hatch *et al.*, 1978; Lehmeyer *et al.*, 1979). Chemiluminescence is presumably a measure of superoxide and myeloperoxidase-dependent reactions which are necessary for the killing of phagocytized bacteria in PMNs. The f-met peptides exhibited the same structure–activity relationships in producing chemiluminescence and chemotaxis, again demonstrating that the two different activities are elicited through interaction of the peptides with a common cellular receptor.

4. CONCLUSIONS

The directed migration of leukocytes in response to chemotactic factors results from the culmination of a complex series of biochemical events, the details of which are currently being elucidated. It is not possible to hypothesize a sequence of occurrences produced in leukocytes by chemotactic factors which ultimately result in directed migration (Figure 4). The initial recognition of at least some chemotactic gradients by leukocytes appears to occur through complexing of the chemoattractant with specific membrane receptors. The chemotactic factor, by combining with the receptor, may induce a perturbation in the leukocyte membrane perhaps through a conformational change in the receptor molecule and rearrangement of membrane lipids. This perturbation may cause local permeability changes in the membrane for divalent and perhaps univalent ions. The influx of ions, particularly Ca^{2+}, may activate regulatory molecules of the cell's contractile protein system. In muscle cells, it is known that Ca^{2+} binds to the regulatory subunit, TN-C, of the molecule, troponin, which allows the actomyosin bridge to form, resulting in the contraction-producing "power stroke." A troponin-C-like molecule has been found in human leukocytes, along with actin and myosin, so it is not unreasonable to assume that a similar contractile process operates in leukocytes.

It is not yet clear how SAM-mediated methylation fits into the contractile mechanism. Several possibilities, which are currently under investigation, are (1) that methylation allows or facilitates binding of the chemotactic factor to its receptor; (2) that methylation of membrane elements may be required for the formation of "ion channels" which would be necessary for the observed de-

---

I. Recognition of chemotactic gradients through:
   1. Complexing of chemotactic factor with specific membrane receptors
   2. Nonspecific perturbation of membrane by chemotactic factor, e.g., hydrophobic interactions

   ↓

II. Translation of external chemical stimuli to interior of leukocyte through:
   1. Membrane depolarization
   2. ?Methylation of specific membrane elements
   3. Ion fluxes
   4. Activation of metabolic processes

   ↓

III. Polarized contraction of cell in direction of chemotactic gradients induced by:
   Reorganization [and possibly methylation] of muscle-like contractile elements
   a. Microfilaments
   b. Microtubules

FIGURE 4.   Requirements for chemotaxis.

polarization and hyperpolarization of the cells; (3) that methylation of membrane phospholipids may be required for new receptor or new membrane synthesis; or (4) that methylation of contractile proteins may allow their assembly or somehow facilitate formation of a "rachet." Indeed, myosin, actin, and the troponin-C-like protein have an unusual amino acid composition characterized by trimethyl-N-lysine and/or N-methyl-histidine.

The varied effects produced in leukocytes by chemotactic substances would indicate that these factors act on these cells as hormones do on hormonally responsive tissues. Further investigations of the interaction of chemotactic factors with leukocytes should facilitate not only the understanding of directed cell movement but also the biochemical control mechanisms which govern the responses of these cells to extracellular chemical stimuli.

REFERENCES

Altman, L. C., Snyderman, R., Oppenheim, J. J., and Mergenhagen, S. E., 1973, A human mononuclear leukocyte chemotactic factor: Characterization, specificity and kinetics of production by homologous leukocytes, *J. Immunol.* **110**:801.

Altman, L. C., Mackler, B. F., and Chassey, B. M., 1974, Physiochemical characterization of chemotactic lymphokines produced by human thymus derived (T) and bone marrow derived (B) lymphocytes, *J. Reticuloendothel. Soc.* **16** (Suppl.):15a.

Aswanikumar, S., Corcoran, B., Schiffman, E., Day, A. R., Freer, R. J., Showell, H. J., and Pert, C. B., 1977, Demonstration of a receptor on rabbit neutrophils for chemotactic peptides, *Biochem. Biophys. Res. Commun.* **74**:810.

Becker, E. L., and Showell, H. J., 1974, The ability of chemotactic factors to induce lysosomal enzyme release. II. The mechanism of release, *J. Immunol.* **112**:2055.

Bokisch, V. A., Müller-Eberhard, H. J., and Cochrane, C. G., 1969, Isolation of a fragment (C3a) of the third component of human complement containing anaphylatoxin and chemotactic activity and description of an anaphylatoxin inactivator of human serum, *J. Exp. Med.* **129**:1109.

Boucek, M. M., and Snyderman, R., 1976, Calcium influx requirement for human neutrophil chemotaxis: Inhibition by lanthanum chloride, *Science* **193**:905.

Boyden, S., 1962, The chemotactic effect of mixtures of antibody and antigen on polymorphonuclear leukocytes, *J. Exp. Med.* **115**:453.

Brier, A. M., Snyderman, R., Mergenhagen, S. E., and Notkins, A. L., 1970, Inflammation and herpes simplex virus: Release of a chemotaxis-generating factor from infected cells, *Science* **170**:1104.

Chapitis, J., Ward, P. A., and Lepow, I. H., 1971, Generation of chemotactic activity from human serum and purified components of complement by *Serratia* proteinase, *J. Immunol.* **107**:317.

Chenoweth, D. E., and Hugli, T. E., 1978, Demonstration of specific C5a receptor on intact human polymorphonuclear leukocytes, *Proc. Natl. Acad. Sci. USA* **75**:3943.

Cohen, S., Ward, P. A., Toshida, T., and Burek, C. L., 1973, Biologic activity of extracts of delayed hypersensitivity skin reacion sites, *Cell. Immunol.* **9**:363.

Craddock, P. R., Hammerschmidt, D., White, J. G., Dalmasso, A. P., and Jacob, H. S., 1977, Complement (C5a)-induced granulocyte aggregation *in vitro*: A possible mechanism of complement-mediated leukostasis and leukopenia, *J. Clin. Invest.* **60**:260.

Estensen, R., Hill, H. R., Quie, P. G., Hogan, N., and Goldberg, N. A., 1973, Cyclic GMP and cell movement, *Nature* **245**:458.

Fernandez, H. N., and Hugli, T. E., 1976, Partial characterization of human C5a anaphylatoxin. I. Chemical description of the carbohydrate and polypeptide portions of human C5a, *J. Immunol.* **117**:1688.

Fernandez, H. N., Henson, P. M., Otani, A., and Hugli, T. E., 1978, Chemotactic response to human C3a and C5a anaphylatoxins: I. Evaluation of C3a and C5a leukotaxis *in vitro* and under simulated *in vivo* conditions, *J. Immunol.* **120**:109.

Gallin, E. K., and Gallin, J. I., 1977, Interaction of chemotactic factors with human macrophages: Induction of transmembrane potential changes, *J. Cell Biol.* **75**:277.

Gallin, J. I., Clark, R. A., and Kimball, H. R., 1974, Granulocyte chemotaxis. An improved *in vitro* assay employing ^{51}Cr labeled granulocytes, *J. Immunol.* **110**:233.

Gallin, E. K., Wiederhold, M. L., Lipsky, P. E., and Rosenthal, A. S., 1975, Spontaneous and induced membrane hyperpolarizations in macrophages, *J. Cell. Physiol.* **86**:653.

Gallin, J. I., Sandler, J. R., Clyman, R. I., Manganiello, V. C., and Vaughn, M., 1978a, Agents that increase cyclic AMP inhibit accumulation of cGMP and depress human monocyte locomotion, *J. Immunol.* **120**:492.

Gallin, J. I., Gallin, E. K., Malech, H. L., and Cramer, E. B., 1978b, Structural and ionic events during leukocyte chemotaxis, in: *Leukocyte Chemotaxis: Methods, Physiology, and Clinical Implications* (J. I. Gallin and P. G. Quie, eds.), pp. 123–157, Raven Press, New York.

Goetzl, E. J., and Austen, K. F., 1972, A method for assessing the *in vitro* chemotatic response of neutrophils utilizing ^{51}Cr-labeled human leukocytes, *Immunol. Commun.* **1**:421.

Goetzl, E. J., and Gorman, R. R., 1978, Chemotactic and chemokinetic stimulation of human eosinophil and neutrophil polymorphonuclear leukocytes by 12-L-hydroxy-5,8,10-heptadecatrienoic acid (HHT), *J. Immunol.* **120**:526.

Goetzl, E. J., Woods, J. M., and Gorman, R. R., 1977, Stimulation of human eosinophil and neutrophil polymorphonuclear leukocyte chemotaxis and random migration by 12-L-hydroxy-5,8,10,14-eicosatetraenoic acid, *J. Clin. Invest.* **59**:179.

Goldstein, I., Hoffstein, S., Gallin, J., and Weissmann, G., 1973, Mechanisms of lysosomal enzyme release from human leukocytes: Microtubule assembly and membrane fusion induced by a component of complement, *Proc. Natl. Acad. Sci. USA* **70**:2916.

Hatch, G. E., Nichols, W. K., and Hill, H. R., 1977, Cyclic nucleotide changes in human neutrophils induced by chemoattractants and chemotactic modulators, *J. Immunol.* **119**:450.

Hatch, G. E., Gardner, D. E., and Menzel, D. B., 1978, Chemiluminescence of phagocytic cells caused by N-formylmethionyl peptides, *J. Exp. Med.* **147**:182.

Jensen, J., Snyderman, R., and Mergenhagen, S. E., 1969, Chemotactic activity: A property of guinea pig C5 anaphylatoxin, in: *Cellular and Humoral Mechanisms in Anaphylaxis and Allergy* (H. Z. Movat, ed), pp. 265–278, Karger, Basel.

Kay, A. B., 1970, Studies on eosinophil leukocyte migration, II. Factors specifically chemotactic for eosinophils and neutrophils generated from guinea-pig serum by antigen-antibody complexes, *Clin. Exp. Immunol.* **7**:723.

Kay, A. B., and Austen, K. F., 1972, Chemotaxis of human basophil leukocytes, *Clin. Exp. Immunol.* **11**:557.

Kazmierowski, J. A., Gallin, J. I., and Reynolds, H. Y., 1977, Mechanism for the inflammatory response in primate lungs: Demonstration and partial characterization of an alveolar macrophage-derived chemotactic factor with preferential activity for polymorphonuclear leukocytes, *J. Clin. Invest.* **59**:273.

Keller, H. U., and Sorkin, E., 1965, On the chemotactic and complement fixing activity of gamma-globulins, *Immunology* **9**:241.

Keller, H. U., and Sorkin, E., 1967, Studies on chemotaxis. V. On the chemotactic effect of bacteria, *Int. Arch. Allergy Appl. Immunol.* **31**:505.

Kort, E. N., Goy, M. F., Larsen, S. H., and Adler, J., 1975, Methylation of a membrane protein involved in bacterial chemotaxis, *Proc. Natl. Acad. Sci. USA* **72**:3939.

Kredich, N. K., and Martin, D. W., Jr., 1977, Role of S-adenosyl-homocysteine in adenosine mediated toxicity in cultured mouse T lymphoma cells, *Cell* **12**:931.

Lehmeyer, J. E., Snyderman, R., and Johnson, R. B., Jr., 1979, Stimulation of neutrophil oxidative metabolism by chemotactic peptides: Influence of calcium ion concentration and cytochalasin B and comparison with stimulation by phorbol myristate acetate, *Blood* **54**:35.

Mackler, B. F., Altman, L. C., Rosenstreich, D. L., and Oppenheim, J. J., 1974, Induction of lymphokine production by EAC and of blastogenesis by soluble mitogens during human B cell activation, *Nature (London)* **249**:834.

Moeller, G. R., Terry, L., and Snyderman, R., 1978, The inflammatory response and resistance to endotoxin in mice, *J. Immunol.* **120**:116.

Naccache, P. H., Showell, H. J., Becker, E. L., and Sha'afi, R. I., 1977, Transport of sodium, potassium, and calcium across rabbit polymorphonuclear leukocyte membranes: Effect of chemotactic factor, *J. Cell. Biol.* **73**:428.

O'Flaherty, J. T., Kreutzer, D. L., and Ward, P. A., 1977, Neutrophil aggregation and swelling induced by chemotactic agents, *J. Immunol.* **119**:232.

Pike, M. C., and Daniels, C. A., 1975, Production of C5a upon the interaction of staphylococcal protein A with human serum *Fed. Proc.* **34**:853A.

Pike, M. C., Kredich, N. K., and Snyderman, R., 1978, Requirement S-adenosyl-methionine mediated methylation for human monocyte chemotaxis, *Proc. Natl. Acad. Sci. USA* **75**:3928.

Postlethwaite, A. E., and Snyderman, R., 1975, Characterization of chemotactic activity produced *in vivo* by a cell mediated immune reaction in the guinea pig. *J. Immunol.* **114**:274.

Rivkin, I. J., Rosenblatt, J., and Becker, E. L., 1975, The role of cyclic AMP in the chemotactic responsiveness and spontaneous mobility of rabbit peritoneal exudate neutrophils. The inhibition of neutrophil movement and the elevation of cyclic AMP levels by catecholamines, prostaglandins, theophylline and cholera toxin, *J. Immunol.* **115**:1126.

Sandler, J. A., Gallin, J. I., and Vaughn, M., 1975, Effects of serotonin, carbamyl choline and asboric acid on leukocyte cGMP and chemotaxis, *J. Cell Biol.* **67**:480.

Schiffman, E., Corcoran, B. A., and Wahl, S. M., 1975, N-Formyl-methionyl peptides as chemoattractants for leukocytes, *Proc. Natl. Acad. Sci. USA* **72**:1059.

Shin, H. S., Snyderman, R., Friedman, E., Mellors, A., and Mayer, M. M., 1968, Chemotactic and anaphylatoxic fragment, cleaved from the fifth component of guinea pig complement, *Science* **162**:361.

Shin, H. S., Gewurz, H., and Snyderman, R., 1969a. Reaction of cobra venom factor with guinea pig complement and generation of an activity chemotactic for polymorphonuclear leukocytes, *Proc. Soc. Exp. Biol. Med.* **131**:203.

Shin, H. S., Snyderman, R., Friedman, E., and Mergenhagen, S. E., 1969b, Cleavage of guinea pig C3 by serum-treated endotoxic lipopolysaccharide, *Fed. Proc.* **28**:485.

Showell, H. J., Freer, R. J., Zigmond, S. H., Schiffman, E., Aswanikumar, S., Corcoran, B., and Becker, E. L., 1976, The structure–activity relations of synthetic peptides as chemotactic factors and inducers of lysosomal enzyme secretion for neutrophils, *J. Exp. Med.* **143**:1154.

Snyderman, R., Gewurz, H., and Mergenhagen, S. E., 1968, Interactions of the complement system with endotoxic lipopolysaccharide. Generation of a factor chemotactic for polymorphonuclear leukocytes, *J. Exp. Med.* **128**:259.

Snyderman, R., Shin, H. S., Phillips, J. K., Gewurz, H., and Mergenhagen, S. E., 1969, A neutrophil chemotactic factor derived from C'5 upon interaction of guinea pig serum with endotoxin, *J. Immunol.* **103**:413.

Snyderman, R., Phillips, J. K., and Mergenhagen, S. E., 1970, Polymorphonuclear leukocyte chemotactic activity in rabbit serum and guinea pig serum treated with immune complexes. Evidence for C5a as the major chemotactic factor, *Infect. Immun.* **1**:521.

Snyderman, R., Phillips, J. K., and Mergenhagen, S. E., 1971a, Biological activity of complement *in vivo:* Role of C5 in the accumulation of polymorphonuclear leukocytes in inflammatory exudates, *J. Exp. Med.* **134**:1131.

Snyderman, R., Shin, H. S., and Hausman, M. S., 1971b, A chemotactic factor for mononuclear leukocytes, *Proc. Soc. Exp. Biol. Med.* **138**:387.

Snyderman, R., Altman, L. C., Hausman, M. S., and Mergenhagen, S. E., 1972, Human mononuclear leukocyte chemotaxis: A quantitative assay for mediators of humoral and cellular chemotactic factors, *J. Immunol.* **108**:857.

Snyderman, R., Meadows, L., and Amos, D. B., 1977a, Characterization of human chemotactic lymphokine production induced by mitogens and mixed leukocyte reactions using a new microassay, *Cell. Immunol.* **30**:225.

Snyderman, R., Pike, M. C., Fischer, D., and Koren, H., 1977b, Biological and biochemical activities of the continuous macrophage cell lines, P388D1 and J774.1, *J. Immunol.* **119**:2060.

Springer, M. S., Goy, M. F., and Adler, J., 1977, Sensory transduction in *Escherichia coli:* Two complementary pathways of information processing that involve methylated proteins, *Proc. Natl. Acad. Sci. USA* **74**:3312.

Stetcher, V. J., and Sorkin, E., 1969, Studies on chemotaxis. XII. Generation of chemotactic activity for polymorphonuclear leukocytes in sera with complement deficiencies, *Immunology* **16**:231.

Stossel, T. P., and Pollard, T. D., 1973, Myosin in polymorphonuclear leukocytes, *J. Biol. Chem.* **248**:8288.

Tatsumi, N., Shibata, N., and Okamura, Y., 1973, Actin and myosin A from leukocytes, *Biochem. Biophys. Acta* **305**:433.

Taylor, F. B., and Ward, P. A., 1967, Generation of chemotactic activity in rabbit serum by plasminogen-streptokinase mixtures, *J. Exp. Med.* **126**:149.

Tempel, T. R., Snyderman, R., Jordan, H. V., and Mergenhagen, S. E., 1970, Factors from saliva and oral bacteria, chemotactic for polymorphonuclear leukocytes: Their possible role in gingival inflammation, *J. Periodontol.* **41**:3/71.

Turner, S. R., Campbell, J. A., and Lynn, W. S., 1975a, Polymorphonuclear leukocyte chemotaxis towards oxidized lipid components of cell membranes, *J. Exp. Med.* **141**:1437.

Turner, S. R., Tainer, J. A., and Lynn, W. S., 1975b, Biogenesis of chemotactic molecules by the arachidonate lipoxygenase system of platelets, *Nature* **257**:680.

Valone, R. H., and Goetzl, E. J., 1978, Immunologic release in the rat peritoneal cavity of lipid chemotactic and chemokinetic factors for polymorphonuclear leukocytes, *J. Immunol.* **120**:102.

Wahl, S. M., Iverson, G. M., and Oppenheim, J. J., 1974, Induction of guinea pig B-cell lymphokine synthesis by mitogenic and non-mitogenic signals to Fc, Ig and C3 receptors, *J. Exp. Med.* **140**:1631.

Wahl, S. M., Wilton, J. M., Rosenstreich, D. L., and Oppenheim, J. J., 1975, The role of macrophages in the production of lymphokines by T and B lymphocytes, *J. Immunol.* **114**:1296.

Walker, W. S., Bartlett, R. L., and Kurtz, H. M., 1969, Isolation and partial characterization of a staphylococcal leukocyte cytotaxin, *J. Bacteriol.* **97**:1005.

Ward, P. A., and Newman, L. J., 1969, A neutrophil chemotactic factor from human C'5, *J. Immunol.* **102**:93.

Ward, P. A., and Zvaifler, N. J., 1971, Complement-derived leukotactic factors in inflammatory synovial fluids of humans, *J. Clin. Invest.* **50**:606.

Ward, P. A., Cochrane, C. G., and Müller-Eberhard, H. J., 1965, The role of serum complement in chemotaxis of leukocytes *in vitro*, *J. Exp. Med.* **122**:327.

Ward, P. A., Cochrane, C. G., and Müller-Eberhard, H. J., 1966, Further studies on the chemotactic factor of complement and its formation *in vivo*, *Immunology* **11**:141.

Ward, P. A., Lepow, I. H., and Newman, L. J., 1968, Bacterial factors chemotactic for polymorphonuclear leukocytes, *Am. J. Pathol.* **52**:725.

Ward, P. A., Remold, H. G., and David, J. R., 1969, Leukotactic factors produced by sensitized lymphocytes, *Science* **163**:1079.

Ward, P. A., Remold, H. G., and David, J. R., 1970, The production from antigen-stimulated lymphocytes of a luekotactic factor distinct from migration inhibitory factor, *Cell. Immunol.* **1**:162.

Williams, L. T., Snyderman, R., Pike, M. C., and Lefkowitz, R. J., 1977, Specific receptor sites for chemotactic peptides on human polymorphonuclear leukocytes, *Proc. Natl. Acad. Sci. USA* **74**:1204.

2

Attachment and Recognition Factors in the Interaction between Microbes and Mononuclear Phagocytes

THOMAS C. JONES and GERALD I. BYRNE

1. INTRODUCTION AND HISTORICAL ASPECTS

Removal of microbes from the vascular system and from tissue sites is handled by cells of the monocyte–macrophage system, the mononuclear phagocytes. *In vivo*, this process is a complex one, referred to as "clearance." Difficulties in interpreting removal of microbes or carbon particles from the vascular system are magnified by factors such as organ blood flow, changes in plasma constituents, and changes in cell populations (Saba, 1970). In spite of this complexity, studies have clearly shown the importance of mononuclear phagocytes in recognizing and removing particles from the mammalian host (Spector, 1970; North, 1970). These cells are effective because of their specialized surfaces for recognition of foreign antigens, as well as damaged and serum coated particles, and because of their functional ability to then ingest and digest the particles. This review will focus on those factors occurring at the mononuclear phagocyte surface which affect this attachment and recognition process.

The conflict between Metchnikoff (1905) and Ehrlich (1900) concerning whether phagocytic cells or humoral factors are primary in the defense against invading microbes initiated our understanding of these events, but their conflict must be put aside during this discussion. The events at the cell surface clearly include and are entirely dependent upon the cell, the surface of the microbe, and the fluid environment. The interaction between serum factors and cells was well described in the early part of this century by Wright and Douglas (1903); later

THOMAS C. JONES and GERALD I. BYRNE • Division of International Medicine, Cornell University Medical College, New York, New York 10021. Supported in part by N.I.H. Grants AI-70754, AI-10821, AI-05643 and the Rockefeller Foundation.

more detailed comparative studies were made by Mudd and Mudd (1933). Correlation between *in vivo* and *in vitro* observations were made in the thirties, such as by observation of tubercle bacilli (Lurie, 1932).

Numerous *in vitro* studies in the past few decades described the variation in interactions between microbes and phagocytes. For instance, Hirsch and Strauss (1964) classified microbes into those requiring no serum factors to enhance phagocytosis by polymorphonuclear leukocytes, those that required antibody alone, and those that required heat-labile factors as well as antibody. Careful description of the physiology of the macrophage by Karnovsky (1962), Cohn and Benson (1965), and others, and the definition of particle attachment to cells by Rabinovitch (1970) and understanding of cell cytophilic substances (Berkin and Benacerraf, 1966), paved the way for the explosion of information about the dynamic events at the macrophage surface which has occurred in the past decade. These studies required the development of aritificial *in vitro* systems. Questions concerning the relevance of each *in vitro* model to *in vivo* observations will continue to be raised. For instance, does the contact angle of a colony of organisms with a fluid-phase surface document that hydrophobic microbial surfaces determine virulence (van Oss *et al.*, 1975), or is this merely an interesting correlate? Are the dynamics of phagocyte function more like the situation *in vivo* when studied in suspension, or when studied after adherence to a glass surface? Do studies of the affinity of monomeric immunoglobulin interaction with Fc receptors (Unkeless and Eisen, 1975) give us information applicable to antigen–antibody complexes (Unkeless, 1977)? Can equations applicable to enzyme kinetics (Stossel, 1975), physical attractive and repulsive forces between particles (Curtis, 1967), and molecular affinities (Bell, 1978) be applied to complex microbe and phagocyte surfaces? This review is written on the assumption that many of these *in vitro* systems can be applied to *in vivo* situations. However, careful assessment of their application must be made repeatedly and generalizations must be avoided. In this context, these data may indicate new insight into usefulness, as well as potential artifacts of clearance studies.

2. THE PHAGOCYTIC EVENT

Phagocytosis is an endocytic process during which the cell plasma membrane surrounds a particle, then undergoes changes which lead to plasma membrane fusion, and results in a particle-filled phagosome. Features of this process such as submembrane events associated with membrane movement (Stossel, 1975), lysosomal fusion (Silverstein *et al.*, 1977), microbicidal systems (Klebanoff and Hamon, 1975), and particle digestion are discussed elsewhere in this volume or in other reviews. Phagocytosis is a continuous process, with numerous changes occuring simultaneously to accomplish the interiorization of the particle. The separation of this process into phases of attachment and ingestion (Jones, 1975a) or recognition and ingestion (Stossel, 1975) has been done for purposes of analysis.

The early events of phagocytosis can be separated into five stages each occurring at different rates (Figure 1). (1) Changes occur in substances in the fluid environment. For example, complement may be activated or "consumed" by antigen–antibody interactions. This will affect the rate of complement-mediated recognition processes at the cell surface. (2) Changes occur in the microbial or particle surface. Antigen-specific IgG molecules interact with antigen at a specific rate (Kabat, 1976), leading to rapid change in the microbe surface. Changes induced by proteolytic enzymes or microbial autolysis occur at different rates. These differences affect the rate of attachment to the phagocyte surface. (3) The affinity of molecules for attachment sites on the phagocyte surface determine attachment. For example, monomeric IgG2a has a greater affinity for its Fc receptor than does IgG2b (Unkeless and Eisen, 1975). (4) Following attachment, changes in the interaction with surface receptors for recognition may or may not occur. Examples of the separation of these events are reviewed below. Events associated with Fc–Fc receptor interaction during particle recognition appear different from those associated with receptors which recognize altered proteins (Jones and Yang, 1977). Alterations may occur in the phagocyte surface which affect this recognition process. "Activated" macrophages have more Fc receptors than do nonactivated cells (Rhodes, 1975); they possess new antigens and they can ingest IgM-complement-coated red blood cells (Bianco *et al.*, 1975). The subsequent events of phagocytosis will not be detailed here. In Figure 1 they are summarized as (5) the ingestion stage. They include complex steps of spreading attachment (Griffin *et al.*, 1975), submembrane structural alterations (Stossel, 1975), changes in energy metabolism, membrane fusion, lysosomal enzyme synthesis, and secretion (Silverstein *et al.*, 1977) and stimulation of microbicidal systems (Klebanoff and Hamon, 1975). In Figure 1, the first four stages are indicated as stages in equilibrium (a, b, c), whereas the final step of ingestion (d) is not. This is based on various studies of relative binding affinities between molecules, and between molecules and particles. There have been no studies which suggest such an equilibrium between recognition and ingestion. Many studies of phagocytosis have, in reality, analyzed the collision rate between

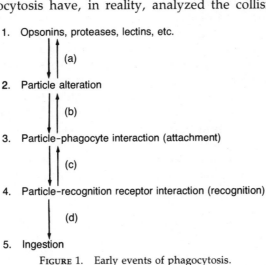

FIGURE 1.   Early events of phagocytosis.

particles and cells, some the rate of attachment or affinity, and a few the recognition event. Measurement of the ingestion process itself has been discussed frequently, but this has usually not been measured. The ingestion event is affected by factors such as temperature. Measurement of digestion has been done by measuring degradation of radioisotope-labeled microbial products. By comparison to events of attachment and recognition, digestion is a slow process. The end result is the phagocytic event, a truly impressive means of defense against microbial invasion.

3. KINETICS OF PARTICLE–MACROPHAGE INTERACTIONS

3.1. FACTORS AFFECTING ATTACHMENT AND RECOGNITION

Particle–phagocyte interaction is influenced by various conditions. The numerous studies by Rabinovitch (1970) have been reviewed previously. He demonstrated that fresh red blood cells do not adhere readily to macrophages whereas aged red cells do. Particles attach to freshly explanted phagocytes at a slower rate than they do to cells in culture 24 hr. Immunoglobulin enhances attachment to macrophages but not to fibroblasts. The various factors which may contribute attachment have also been reviewed (Jones, 1975a). They include the electrochemical forces at the cell surface, hydrophobic and hydrophilic characteristics of the surface, zeta potential, and covalent bonding. All of these have been examined as potentially important repulsive or adhesive factors between two differing particle surfaces.

When two particles collide in a fluid phase they separate again unless some force maintains the two surfaces together (Curtis, 1967). Often the frequency of collision between particles and phagcytes is the determinant of the phagocytic rate. Obviously, in systems in which collisions occur infrequently [such as in a Maaløe tumble assay (Maaløe, 1947) containing too few cells or microbes], the attachment of particles to cells is low. Conversely, in systems with massive numbers of one particle or another, interactions between the two will be rapidly saturated. The potential for an artifactual decrease in collision events must be considered in any assay of phagocyte attachment since this introduces a variable unrelated to the surface of the cell. In studies of mycoplasma–macrophage interaction, we noted that when complement was added to anti-mycoplasma antibody in a tumbled suspension of organisms and macrophages, the rate of attachment actually decreased from 3.5% of [^{3}H]thymidine counts per minute to 2.0% (Jones and Yang, 1977). This was very likely due to complement damage to the organisms which removed them from the assay and was, therefore, unrelated to the effects of complement on the attachment event per se.

Studies of attachment have been made using organisms which attach to the phagocyte surface, but which do not initiate recognition. They include mycoplasma (Jones and Yang, 1977) gonococci (Roberts, 1977) spirochetes (Brause and Roberts, 1978), Concanavalin-A-coated bacteria (Allen *et al.*, 1971) and IgM- and complement-coated red blood cells (Bianco *et al.*, 1975). Measurement of the kinetics of attachment of other particles to phagocytes required that subsequent events of phagocytosis be blocked by reduced temperature or inhibitors of macrophage metabolism. These manipulations may under certain conditions affect the cell surface. Mycoplasmas attach easily to mouse peritoneal macrophages spread for 48 hr on a glass surface, but not to macrophages in suspension. Organisms labeled with [^{3}H]thymidine attached at a rate of 0.1% of the suspension particles/minute to freshly explanted cells, but at a rate more than twice as fast to cells on a glass surface for 48 hr. The mechanism of this attachment was nonspecific and nonimmunologic. Under the same conditions, IgG-coated mycoplasmas attached at the same rate to cells recently attached to a glass surface as to cells spread on glass for 48 hr. This indicated that the macrophage surface changed during glass attachment to allow increased adhesion between particles by one mechanism, but another mechanism (the rate of Fc–Fc receptor interaction) did not change. When the macrophages were maintained in suspension for 48 hr, the rate of attachment of mycoplasma did not change. If these changes in the macrophage surface were due to increased synthesis of plasma membrane, then an increased rate of attachment to Fc receptors should also have been seen. This change in the cell surface was, therefore, due to a change in the affinity between particles and the macrophage in connection with glass adherence of the cell and cell spreading. Studies have not been made comparing these results with the rate of attachment to "activated" macrophages. Since enhanced spreading is a feature of these cells, it is likely that nonspecific attachment to activated cells is increased.

In order to study kinetics of the *recognition* phase of phagocytosis uninfluenced by the rate of particle collision, it is more important to use a particle already attached to the phagocyte surface than to alter the particle in such a way that recognition occurs. This has been done with the mycoplasma model and with the model of antibody-coated red blood cells. Analysis of the ingestion of surface particles suggested that measurement of particle ingestion by metabolic or morphologic changes measured the rate of recognition between components rather than the ingestion process itself (Jones and Yang, 1977). Correlates of ingestion measure the rate of recognition between particles and phagocytes.

3.2. MEASUREMENT OF MICROBE RECOGNITION

Methods for evaluating the rate of interaction between a particle and a surface have been developed from studies of molecular interactions. Such models have only recently been used to evaluate microbe–phagocyte interaction. Calculation of binding affinities, use of Scatchard plots, and calculation of association and dissociation rates have been most commonly used. Comparison of particle–cell interaction to enzyme kinetics has also been done. Comparisons have been made using the equations of adsorption isotherms, the rate at which a product is removed from the environment by adsorption to a surface. This model bears close analogy with the situation of microbe–phagocyte surface interaction.

The Scatchard equation allows derivation of an equilibrium constant for a particular reaction in which the amount of bound and unbound product can be measured. Scatchard plots (Scatchard, 1949) have been frequently used to plot the binding of hormones (Erdmann *et al.*, 1976; Jacobs and Cuatrecasas, 1976) or drugs to cellular receptors. The equation $r/c = K_n - K_r$ is used, where r is the amount of bound material in micrograms and c the moles of unbound material. K is the equilibrium constant and n is the limiting value of r if c is large. The ratio of bound to unbound ligand versus the concentration of receptor-bound ligand is plotted. When linear plots are seen, two pieces of information can be obtained. The point at which the extrapolated curve intercepts the abscissa (concentration of receptor bond ligand) represents the maximum number of ligands that can bind. From this the relative concentration of receptors may be determined. The number of antigen-binding sites on antibodies has been determined by this method (Eisen, 1974). A Scatchard plot will also indicate whether receptor sites are homogeneous.

Studies of ouabain binding (Erdmann *et al.*, 1976) and insulin binding (Jacobs and Cuatrecasas, 1976) demonstrated that equilibrium binding yielded nonlinear Scatchard plots. These results could mean that there were more than a single class of binding sites on the host cell (Weidemann *et al.*, 1970). The binding may be negatively cooperative (De Meyts and Roth, 1975). Nonspecified binding may play a role (Cuatrecasas and Hollenberg, 1975), or ligand–ligand interactions may be occurring (Jacobs *et al.*, 1975). A lack of linearity may also be the result of instability of membrane preparations, i.e., receptors do not remain homogeneous for the length of time of the assay because the membrane preparations become denatured. On the other hand, nonlinearity may be related to function, i.e., receptors may be mobile and binding results in lateral motion and changes in configuration of receptors. True Scatchard plots require freely reversible reactions in equilibrium, but simulated Scatchard plots may be applied to the attachment of particles to phagocytes. Homogeneity or heterogeneity may be indicated, and the relative concentrations of binding sites may be estimated.

A simulated Scatchard plot of the binding of *Chlamydia* to L cells is shown in Figure 2. The curve is nonlinear presumably because of host cell membrane damage. It is known that numbers of *Chlamydia* somewhat in excess of 100 per host cell cause immediate host cell toxicity (Moulder *et al.*, 1976). It has been suggested that this toxicity is a result of membrane damage occurring during attachment. If the curve is arbitrarily divided into two linear portions and extrapolated to the abscissa, then the low-dose portion intercepts at a point representing an inoculum size that is in good agreement with the minimal dose of *Chlamydia* required for inducing immediate toxicity. The high-dose portion intercepts at a point that may represent the maximal number of *Chlamydia* that will bind to L-cell membranes whether they are damaged or not. This number would therefore represent the maximal number of binding sites.

Results are no doubt quite relative and are an oversimplification of the complex events taking place during attachment. Nonetheless, Scatchard analysis may be quite helpful in interpreting the binding of particles to the phagocyte surface.

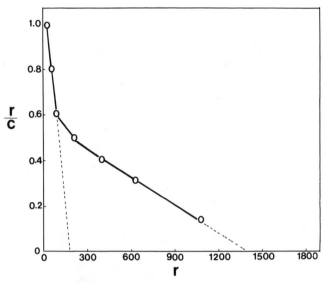

FIGURE 2.   Simulated Scatchard analysis of chlamydial binding to L cells. C, Initial concentration of parasite added per host cell; r, parasites bound after 30 min of incubation. Use of radioisotopically labeled *Chlamydia* allowed careful measurement of concentrations.

Comparisons of the rate of the reaction to enzyme kinetics have recently been recommended by Weisman and Korn (1967) and by Stossel (1973). In this case a resulting reaction which removes the particle and cell surface (such as ingestion) is the reaction product (A + B→AB→A′B). In the equation, A is the particle, B is the phagocyte membrane receptor, AB the attached particle, and A′B the interiorized particle. The reaction must be saturable. The standard equation

$$\frac{1}{V} = \frac{K_m}{V_{max}} \cdot \frac{1}{S} \cdot \frac{1}{V_{max}}$$

is used. V is the reaction product formed, S the concentration of reactant, and V_{max} the rate of the reaction at saturation. Calculation of the K_m of this reaction allows further evaluation of the reaction under some conditions. For instance, Stossel (1973) determined the rate of ingestion of particles in the presence and absence of opsonins by varying the particle number. After documenting first-order kinetics of the reaction, he analyzed the process by use of calculation of enzyme reactions. Because the interaction he observed involved particle collision, recognition, and ingestion, interpretations of the data in functional terms is likely to be oversimplified. However, he clearly demonstrated the utility of kinetic evaluation of the reaction. When mycoplasma attached to mouse peritoneal macrophages were evaluated in a similar manner, a quantitative comparison of the effects of antibody and trypsin on initiating the recognition event could be made (Jones and Yang, 1977). Figure 3 shows the use of such analysis to evaluate the reaction. The measured event is the metabolic change associated with recog-

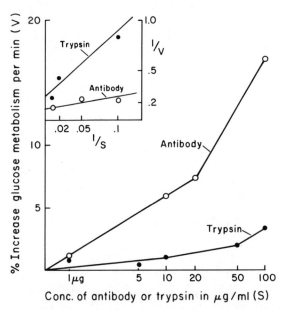

FIGURE 3. Comparison of the rate of change of [^{14}C]-1-glucose to ^{14}CO$_2$ metabolism depending on the concentration of anti-mycoplasma antibody or trypsin. The concentration of antibody is recorded in micrograms, where 1 mg = 1:10,000 dilution of anti-mycoplasma serum. The insert shows the relationship between the reciprocals of the rate of glucose metabolism (V) and concentration of antibody or trypsin (S). From Jones and Yang (1977).

nition and ingestion, i.e., the increased metabolism of glucose to carbon dioxide. The particle number and phagocyte membrane mass are constant and the concentrations of antibody or proteolytic enzyme are varied. This reaction is known to be saturable. In these studies, the rate of recognition (V_{max}) and the affinity between reactants (K_m) appeared to differ. It is premature to generalize from these observations; however, such measurements are helpful in carefully evaluating and comparing mechanisms of the phagocytic process.

Comparison of particle cell interaction to adsorption isotherms has also been useful for evaluating attachment. The equation $x/m = C$ is used, where x is the amount of material absorbed to m, the surface, and C is the concentration of the material. The logarithm of x/m should yield a straight line, allowing relative comparison of affinities between particles and surface.

Recently, attempts have been made to analyze the relative forces between particles under different conditions. These studies will also provide new insights into particle–cell or cell–cell adhesive reactions. Analysis of forces is even more complex than Scatchard plots, for instance, since the equations differ for solutes, sphere-shaped particles, or disk-shaped particles (Bell, 1978).

All of these measurements can make only relative comparisons, because molecular concentrations are not available. Nevertheless, they are useful under certain controlled situations for evaluating particle–cell attachment or recognition.

4.   MODELS OF MICROBE–RECEPTOR INTERACTIONS

4.1.   LECTIN BINDING

Lectins, proteins derived from certain plants, are able to bind with high affinity to carbohydrate-containing substances. Recently, attention has focused on these substances as a mechanism of attachment between plants and nitrogen-fixing bacteria (Dazzo *et al.*, 1976). These studies showed that a clover seed lectin, trifolin, binds a 2-deoxyglucose moiety on the surface of the symbiotic bacterium, *Rhizobium trifolii*. This allows preferential adsorption of these infective organisms to the plant root. Albersheim *et al.* (1977) showed a specific interaction between the bacterial lipopolysaccharide and the lectin of the plant symbiotic with that bacterium.

It is known that plant lectins also bind to carbohydrate groups on the surface of macrophages (Goldman and Raz, 1975; Edelson and Cohn, 1974) and other susceptible cells (Berlin *et al.*, 1975). Vladovsky and Sachs (1975) showed that neuraminidase treatment of transformed hamster fibroblasts resulted in increased binding of soybean agglutinin, decreased binding of wheat germ agglutinin, and no change in the binding of the jack bean agglutinin, Concanavalin A (Con A). Thus, receptors containing sialic acid were required for soybean agglutinin binding, but not wheat germ agglutinin, or Con A. The sugar moiety, mannoside, was required for Con A binding and α-methyl mannoside competitively inhibited the lectin binding. Con A bound to mouse peritoneal macrophages in a specific and dose-dependent way (Goldman and Raz, 1975). Rabbit alveolar macrophages underwent metabolic stimulation similar to that seen during phagocytosis (Berlin *et al.*, 1975). Edelson and Cohn (1974) demonstrated that Con A was pinocytized, and lysosomal fusion did not occur unless α-methyl-mannoside was added. Lectin binding can serve as a model to study antigen binding, and particle binding to macrophages because lectins induce similar endocytic phenomena when bound to the surface of various classes of macrophage (Kabat, 1976).

4.2. IMMUNOGLOBULIN G AND COMPLEMENT BINDING

The most completely studied model of the interaction between a microbe and a phagocytic cell is the interaction between the opsonized organism (IgG antibody) and the Fc receptor site. This is certainly one of the most important mechanisms of microbe–phagocyte interactions. No organism has been identified which has evolved to use the Fc receptor as a specific attachment site for the microbe. However, the host has developed this mechanism for changing the microbe surface to one which allows an increased rate of attachment and recognition, that is, the binding affinities between antigen–antibody complexes and the phagocyte surface are greater than those of monomeric IgG antibodies and of the unopsonized microbe surface.

Detailed reviews of the composition of the Fc portion of the IgG antibody and the portion or portions most likely involved in antibody–receptor interaction have been presented elsewhere (Silverstein *et al.*, 1977; Ciccimarra *et al.*, 1975). The important site of the Fc portion of the IgG molecule is a peptide region in the C_H3 domain of the heavy chain. Intact disulfide bonds are also required. Carbohydrate residues of the molecule are not involved. The receptors for the Fc portion of the IgG molecule are divided into trypsin-resistant and trypsin-sensitive types. The binding of antigen–antibody complexes to trypsin-resistant receptors occurs with an affinity of 7×10^6–2.4×10^7 liters/mole at 37°C. Each macrophage has 5–8×10^5 of these receptors (Unkeless, 1977). Binding is not dependent on host cell metabolism or the functioning of microfilaments. Interaction of immunoglobulin G with the trypsin-resistant receptor initiates the ingestion process. Figure 4 shows an electron micrograph of a mycoplasma minutes after addition of antimycoplasma IgG. The mononuclear cell microvillus protrusions are seen partially around the particle. This event is in part one of "spreading attachment" to Fc receptors (Griffin *et al.*, 1975) and in part the "triggering" of submembrane mechanical events (Stossel, 1975).

Trypsin-sensitive Fc receptors bind in a selective manner subclasses of immunoglobulin (mouse IgG2a, human IgG1 and IgG3, guinea pig IgGa). These immunoglobulins have a high affinity for mononuclear cell Fc receptors in monomeric configurations and they are termed cytophilic antibodies. The relative importance of these two kinds of receptors in terms of function is still under active discussion (Silverstein *et al.*, 1977). The trypsin-resistant receptor–antigen–antibody complex interaction appears far more significant in endocytosis of opsonized particles. This may be because binding of IgG to antigen initiates a clustering of molecules which induces a change in configuration with higher affinity for the trypsin-resistant receptors.

Receptors for the third component of complement have also been identified and these may also be important in initial attachment and recognition of coated particles to mononuclear phagocytes. Proteases cleave the heavy chain of C3 to a particle of 140,000–170,000 molecular weight, C3b or C3d. These molecules have high binding affinity for mononuclear phagocytes. The receptor to which they attach is trypsin-sensitive and is clearly separate from the Fc receptor (Bianco and Nussenzweig, 1977). The complement receptor is of particular interest because it acts as an *attachment* site on the normal or unstimulated cell. On the stimulated or "activated" cell, this attachment process may continue to recognition of the complement–antigen complex and then ingestion. Thus, a change in the macrophage surface and in the complement receptor has been associated with "activation" of the cell. This change initiates mechanisms which convert the attachment site into one that "recognizes" the complex. The model, therefore, allows examination of the kinetics of attachment and recognition of particles, and particularly the changes in mononuclear cell membrane which are associated with the recognition process. Some investigators have suggested that reorganization in the mononuclear cell membrane allows the IgM ligand rather than the C3b to initiate recognition and ingestion (Ehlenberger and Nussenzweig, 1977). The change in the membrane of activated macrophages is now under investigation to clarify these points.

There are two types of membrane receptors for C3, CRI and CRII. The CRI receptor (also termed *b receptor* and *immune-adherence receptor*) binds C3b and C4b, but does not bind C3d. CRII (d receptor) binds C3b and C3d. Binding to complement receptors (in contrast to binding to Fc receptors) is temperature dependent. The chemical nature of complement receptors remains unknown. They contain a protein portion documented by their sensitivity to proteolytic enzymes. The difficulties in identifying the receptors have been recently reviewed (Bianco, 1977).

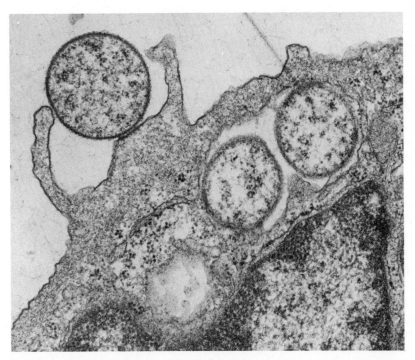

FIGURE 4. Electron photomicrograph of *Mycoplasma pulmonis* on and within a mouse peritoneal macrophage. Cells fixed 5 min after the addition of anti-mycoplasma antibody to macrophages infected with mycoplasmas. Several mycoplasma organisms are seen within phagosomes. One organism on the surface has microprojections of macrophage membrane surrounding it, showing the early steps of phagocytosis. ×33,000. From Jones and Hirsch (1971).

Functional significance of these complement receptors on mononuclear phagocytes may be seen in the ability to control the traffic of immune complexes at the cell surface (attachment) under one condition (e.g., when IgM immunoglobulin and complement are present, but cells are not in an "activated" state) and initiate ingestion of particles under another condition (i.e., when cells are "activated" and IgM immunoglobulin *is* present). An *in vivo* model supporting these *in vitro* observations has been described (Schreiber and Frank, 1972). IgM-coated RBCs when injected into normal guinea pigs were sequestered temporarily, then released into the circulation. When the guinea pigs were first infected with BCG (an agent which can activate macrophages), the IgM-coated cells were removed from the circulation by mononuclear phagocytes.

It is clear that cooperation between Fc and complement receptors can occur. One hundred and fifty IgG molecules are required by a mononuclear phagocyte to ingest a red blood cell when C3b is present, but 6000 are required when complement is absent (Ehlenberger and Nussenzweig, 1977). This is felt to be entirely due to the ability of C3b to promote enhanced contact between the red cell and the phagocyte because other purely physical agents which promote attachment, such as centrifugation and dextran, can have a similar effect. The C3b–C3 receptor appears much more active in enhancing binding affinity for attachment than the Fc–Fc receptor. On the other hand, the Fc–Fc receptor initiates the recognition step leading to ingestion. When C3b enhances the binding affinity between a particle and the cell, the Fc–Fc receptor acts more efficiently allowing recognition and ingestion to proceed at lower IgG concentrations.

5.   SPECIFIC EXAMPLES OF MICROBE–MACROPHAGE INTERACTIONS

Attachment and recognition of microorganisms at the surface of macrophages falls into three patterns: immunologically specified, unspecified (nonspecific), and parasite specified.

5.1.   IMMUNOLOGICALLY SPECIFIED: PNEUMOCOCCUS

Opsonization is crucial for clearance and intracellular destruction of virulent microbes which have surface properties that interfere with attachment and/or recognition by phagocytes. For example, virulent pneumococci are surrounded by a carbohydrate capsule and are resistant to phagocytosis. Nonvirulent strains are unencapsulated, and are readily attached and recognized by macrophages (Wood, 1946). However, opsonized pneumococci are attached, recognized, and ingested more readily than are nonvirulent strains (van Oss and Gillman, 1972). Thus, opsonization (immunological attachment and recognition) appears to be essential for the uptake and subsequent destruction of virulent pneumococci by macrophages.

Until the discovery that Fc and C3 receptors on the surface of macrophages mediated immunological attachment and recognition, it was not clear just why opsonized microbes were taken up more readily than their unopsonized counterparts. It has been suggested that attachment is primarily based upon the "wettability" of surface (van Oss *et al.*, 1975). The role of charge distribution and other physical factors such as van der Waals forces and zeta potential have also been implicated (Jones, 1975a; Stossel, 1975). Any of these parameters may play a role in the initial accessibility of particles in solution to the phagocyte surface. We are now aware of the importance of specific molecular recognition between molecules, i.e., receptors. These appear especially important in maintaining firm attachment, and in attaining attachment that progresses to recognition stages of ingestion of the opsonized particles (Silverstein *et al.*, 1977).

5.2.   NONSPECIFIC: GRAM-NEGATIVE BACTERIA, MYCOPLASMA, SPIROCHETES

Numerous microbes are attached, recognized, ingested and destroyed within phagolysosomes in the absence of opsonizing serum. This type of attachment and recognition is often called *nonspecific* because specific attachment loci on the surface of the macrophage or the microbe have not been directly implicated. *Nonspecific* does not mean that such components are not present. It is clear that rapid ingestion of these organisms can occur whether they are attached by immunologic or nonspecific means.

Structures resembling lectins present on the surface of microbes are one means of initiating so-called nonspecific attachment and recognition (Goldman, 1977). When yeast cells were added to hamster fibroblast cultures, attachment did not occur. However, if the yeast cells were first coated with the jack bean lectin, Con A, then attachment occurred, and this proceeded to ingestion. Bar-Shavit *et al.* (1977) have reported that mannose monosaccharides on the surface of mouse peritoneal macrophages and human polymorphonuclear leukocytes were required for the nonspecific attachment of *Escherichia coli* and *Salmonella typhi*. Inhibition of attachment resulted when the microbes were incubated with D-mannose, methyl α-D-mannopyranoside, or yeast mannan, prior to incubation with macrophages. Inhibition did not result when the bacteria were incubated with any other sugar. Attachment of opsonized bacteria was not inhibited by incubation with mannose. These data indicate that certain gram-negative bacteria contain lectin-like substances on their surfaces which bind with high affinity to glycoproteins containing accessible mannose residues.

Mycoplasma pulmonis is an example of a microbe which attaches to but is not recognized by mouse macrophages (Figure 5) (Jones and Hirsh, 1971). That is, attachment does not proceed to ingestion. These microbes remain bound to the macrophage surface, and actually divide in this microenvironment. When immunoglobulin is added to cultures containing attached mycoplasmas, recognition results in ingestion via immunologically specified means. Recognition also results when mycoplasmas are treated with trypsin. The latter recognition presumably is of the nonspecific variety, although attachment loci were not specifically identified. It has been reported that a trypsin-sensitive membrane protein promotes attachment of *Mycoplasma pneumoniae* to the surface of hamster tracheal organ cultures (Hu *et al.*, 1977).

Brause and Roberts (1978) demonstrated that viable *Treponema pallidum* organisms attached to human mononuclear phagocytes, but were not ingested by these cells. Attachment appeared to occur via one or both poles of the spirochete. Heat-killed treponemes did not attach, and neither normal nor hyperimmune serum promoted ingestion. Thus, in the case of this microbe, nonspecific attachment was not transformed into immunologic recognition in the presence of specific antibody. The explanation for this unusual ability to resist opsonization is unclear. Figure 6 shows the treponeme attached to the surface of a mononuclear phagocyte.

5.3. PARASITE SPECIFIED: *CHLAMYDIA, TOXOPLASMA*

Certain obligate intracellular parasites gain entry into macrophages by mechanisms that resemble phagocytosis. For example, the obligate intracellular prokaryote *Chlamydia* (Moulder, 1969) and the obligate intracellular protozoan *Toxoplasma* (Jones and Hirsch, 1972) grow within vacuoles in the cytoplasm of

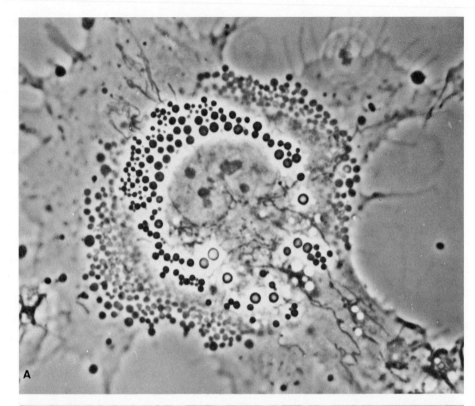

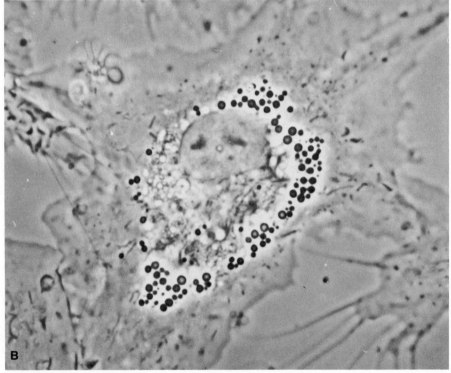

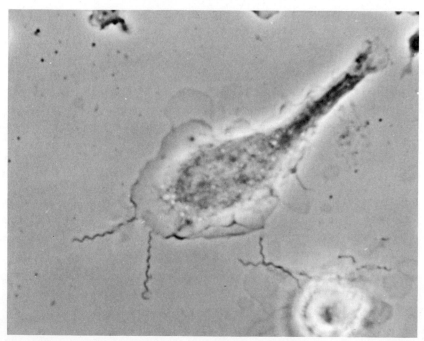

FIGURE 6.   Phase contrast photomicrograph of macrophage infected with *Treponema pallidum*. Two spirochetal organisms are seen attached by one pole to the spread membrane of the macrophage. These organisms remained attached to the macrophage surface in this manner even after addition of anti-treponemal serum. Courtesy of Dr. B. D. Brause.

macrophages. These vacuoles originate during the endocytic process but do not fuse with lysosomes, and are thus referred to as parasitophorous vacuoles. Although the intracellular fate of these vacuoles is different than for phagocytic vesicles, the process of attachment and recognition appear to be quite similar; the process is viewed as "parasite specified."

Attachment of *Chlamydia psittaci* appears to be mediated by the interaction of heat-labile components on the surface of the parasite with trypsin-sensitive host cell components. The ease with which the chlamydial attachment and recognition ligand is destroyed by heat suggests that it is all or part protein. It is resistant to detergents and proteases indicating that it is an integral part of the chlamydial cell envelope. Each chlamydial cell specifies only its own attachment and recognition, that is, the chlamydial surface ligand is not a generalized promotor of phagocytosis, since the presence of infectious chlamydia does not stimulate the attachment and uptake of heat-inactivated chlamydia (Bryne and

FIGURE 5.   Phase-contrast photomicrographs of mouse peritoneal macrophages 24 hr after infection with *M. pulmonis*. (A) A cell covered by the 1-μ phase-dense mycoplasmas. The mycoplasmas should be distinguished from the black refractile lipid bodies around the nucleus. (B) An infected cell 1 hr after addition of 20 μg/ml of trypsin. Mycoplasmas have all been interiorized and appear as phase-lucent grapelike clusters. The macrophage surface is clear of mycoplasma. From Jones *et al.* (1972).

Moulder, 1978). Parasite viability is not required for parasite-specified attachment and recognition, ultraviolet-light-inactivated chlamydia are taken up with the same avidity as infectious microbes. The process of attachment and ingestion may be separated temporally, and it is the recognition–ingestion stage of the process that appears to be the rate-limiting step (Byrne, 1978). A *C. psittaci* cell attached to an L cell is shown in Figure 7.

C. psittaci when coated with antibody is no longer capable of attachment to L cells, but is internalized at an enhanced rate by mouse macrophages (Gardner, 1977; Wyrick and Brownridge, 1978). A similar situation was seen in the case of vaccinia virus (Silverstein, 1975) and toxoplasmas (Jones, 1975b). In each case the antibody-coated parasites were degraded within phagolysosomes, whereas uncoated parasites underwent normal intracellular development.

Toxoplasma gondii also facilitates attachment to macrophages by mechanisms that are not yet clear. Specialized parasite structures may be involved, perhaps associated with a positively charged surface coat which is continuously being

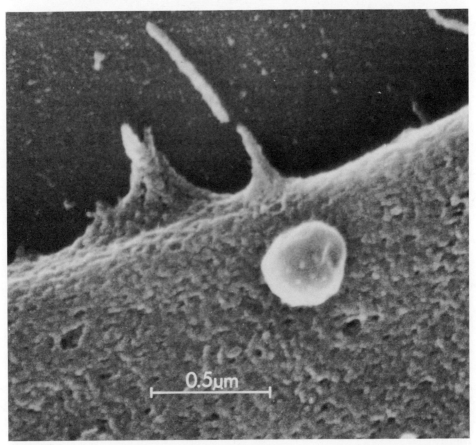

FIGURE 7.   Scanning electron micrograph of a chlamydial elementary body attached to the surface of an L cell. ×50,000.

renewed by the parasite (Jones, 1979). A similar mechanism has been described for entry of *Plasmodium lophurae* into erythrocytes; histone-like proteins have been identified which promote endocytosis (Kilejean, 1976).

Based on observations of the interaction between microbes and phagocytes it is clear that the attachment and recognition stages of phagocytosis differ according to the microbe being ingested and that both microbe and macrophage surface factors are critical to the events. Receptors that are involved in opsonization have been identified and the process of characterizing the interactions between these receptors and immunoglobulins and complement is underway. Macrophage and microbe surface factors also play a role in nonspecific attachment. Lectinlike substances on the surface of the microbe or the cell are good examples of this kind of attachment. Parasite-specified attachment may lead to a different recognition event whereby parasites end up in vacuoles with which lysosomal fusion does not occur and the microbes avoid degradation.

6. INTERACTIONS OF LYMPHOKINES AT THE MACROPHAGE MEMBRANE

Lymphokines are soluble substances produced by sensitized lymphocytes that mediate various biological effects on macrophages and polymorphonuclear cells. A variety of lymphokine activities (migration inhibition factor, chemotactic factor, activating factor, cytotoxic factor, proliferation inhibitory factor, interferon, and transfer factor) (Remold and David, 1974) have been identified, but the mechanisms of their action have not been determined (David, 1975). It is clear that soluble products secreted by sensitized lymphocytes alter the responsiveness of macrophages in numerous ways. The studies of Mackaness (1969) concerning lymphocyte–macrophage interaction during *Listeria* infection, and the identification of a lymphocyte product that prevents macrophage migration from a capillary tube (George and Vaughan, 1962) serve as examples. Mouse peritoneal macrophages normally support intracellular *Toxoplasma* growth. When a soluble lymphocyte product, produced by previously sensitized spleen cells or nonadherent peritoneal cells is added to macrophages, *Toxoplasma* growth is inhibited.

Though lymphokines are not known to mediate recognition events between microbes and phagocytes, their role in control of microbial replication of intracellular organisms is now clear. Since they appear to initiate their action by molecular interactions at the macrophage plasma membrane, attachement and recognition of these molecules is included in this chapter.

6.1. EXAMPLES OF LYMPHOKINE EFFECTS

The process of macrophage activation by lymphoyte products has been reviewed by David (1975). However, it is not known yet how lymphokines chemically modify macrophages. Remold (1973) reported that carbohydrates on the plasma membrane were essential for macrophage inhibition factor (MIF) activity. These investigators demonstrated that α-L-fucose blocked the effects of MIF, and macrophages incubated with fucosidase no longer responded to MIF. These data suggested that α-L-fucose was an essential portion of the macrophage MIF surface receptor. Remold (1974) has also found that macrophage-associated serine esterase activity influences the interaction of MIF and macrophages. Esterase inhibitors were added to macrophage culture, and these cells were then more sensitive to MIF than were the controls. Thus, enzymes on the macrophage surface may play a role in regulating the effects of lymphocyte products (such as MIF) on macrophage function.

Lymphokines produced by cells stimulated with *Toxoplasma* (Jones *et al.*, 1977), or mycobacterial antigen (Cahall and Youmans, 1975) interact with macrophages such that subsequent inhibition of microbial replication within those cells occurs. The characteristics of this interaction at the macrophage surface is just now being defined. A major problem in these studies is that the lymphokine has been only partially characterized. Both toxoplasma and mycobacterial products are proteins or glycoproteins with molecular weights in the range of 20,000–80,000. The receptor for toxoplasma lymphokine on the macrophage surface is sensitive to neuraminidase.

It is known that binding of lymphokines to the surface of macrophages results in intracellular changes associated with inhibition of microbial multiplication. For example, increased levels of cyclic AMP and decreased levels of cyclic GMP have been observed in lymphokine-treated murine macrophages. Under other conditions of lymphokine–macrophage interaction cyclic GMP has been shown to increase (Kurland *et al.*, 1977). Much more needs to be learned about the interaction between lymphokines and mammalian cell membranes. It is possible that the activity will be similar to models of peptide–membrane interaction described for cholera or diphtheria toxins.

6.2. POTENTIAL MODELS FOR LYMPHOKINE ACTION

An examination of other molecules that mediate intracellular changes following binding to the surface of sensitive cells may yield insight concerning how lymphokines bind to and mediate changes with macrophages. Two systems which have been well-studied include the attachment of cholera and diphtheria toxins to susceptible cells.

Diphtheria toxin (Pappenheimer, 1977) and cholera toxin (Finkelstein, 1975) are remarkably similar in that each of these toxins is composed of two polypeptide subunits. Subunit A is the active or toxic component, and subunit B recognizes receptors on the surface of sensitive cells. Toxoids which lack component B are not effective because they do not attach to sensitive cells, and, similarly, toxoids which lack component A are not effective because they lack the active subunit.

Since this discussion deals with attachment, we will not review the actions of the subunit A except to indicate that the A components of both cholera and diphtheria toxins are believed to penetrate into the cell. The mechanism whereby this happens is not clear. Diphtheria toxin results in a discontinuance of protein synthesis due to the irreversible binding of polypeptide A to elongation factor 2. Cholera toxin activates adenyl cyclase which results in increased levels of cyclic AMP.

It is apparent that for diphtheria toxin, the B fragment is required for attachment, and initiates steps resulting in the interiorization of the A fragment. It is not clear how this process occurs, but rate analyses have suggested that endocytosis is not the primary mechanism (Pappenheimer, 1977). The entire toxin molecule is a single polypeptide chain of molecular weight 62,000, and the amino acid sequence has been determined (DeLange *et al.*, 1976). The B fragment (binding fragment) is generated by addition of reducing agent (breaking disulfide linkages) to toxin preparations that had been nicked at an exposed loop of 14 amino acids. Its molecular weight is 40,000, and it remains very unstable after being separated from A fragment.

Mutants have been obtained that lack either functional A or functional B fragments. One isolate lacks that portion of the B fragment containing the disulfide bridge nearest the carboxyl terminus, and although the molecule retains full enzymatic activity, it is nontoxic because it does not bind to susceptible cells. (Susceptible cells are defined as those which have "receptors" that interact with the B fragment.) Specific antibody directed against toxoid does prevent binding.

Cell receptors for diphtheria toxin are not affected by the action of trypsin, chymotrypsin, pronase, lysozyme, neuraminidase, or hyaluronidase, but phospholipase C treatment causes susceptible cells to be less sensitive to the effects of diphtheria toxin. The number of receptor- or toxin-binding sites on HeLa cells have been calculated to be 4000 (Boquet and Pappenheimer, 1974). The chemical nature of the receptor is unclear; it is not similar to the glycoprotein that binds abrin and ricin or the ganglioside which binds cholera toxin.

Attachment of diphtheria toxin may serve as a model for other surface-bound "signal" proteins. Initial attachment at the surface promotes internalization of a second component. This model may provide clues to the mechanism of action of such moieties as lymphokines and interferon at the macrophage surface.

Cholera toxin is a protein of molecular weight 84,000 (Finkelstein, 1975) which may be separated into A and B regions by boiling in SDS or by treatment with urea at a low pH. It is clear that the B region of the molecule is required for binding to the cell surface. This portion of the molecule has molecular weight 56,000, and is composed of 5 to 6 subunits of molecular weight 8–14,000 each. Treatment of susceptible cells with B fragments prevents subsequent toxicity of the entire molecule indicating that membrane receptor sites are bound to the B fragment. The receptor with which the B fragment interacts is a neuraminidase-insensitive glycolipid (Cuatrecasas, 1973). A monosialosyl ganglioside called GM, has been shown to specifically interact with B fragments, and the susceptibility of cells to toxin depends upon the presence of GM in the membrane (Hollenberg *et al.*, 1974). Subsequent to binding, the A fragment is thought to be transmitted into the cell, leaving the B fragment bound to the surface.

Thus, for both toxins alteration of cellular function results following initial interaction of a component of the toxin specifically designed for attachment functions. The moieties on the susceptible cells that are recognized are not known at this time, but presumably this interaction allows transmission of the A subunits to intracellular loci where they carry out their specific functions. The presence of two components, one mediating attachment and one responsible for activity, is a model that may prove worthwhile in the study of lymphokine activities on macrophages.

7. SUMMARY

Attachment and recognition factors associated with the interaction between microbes and mammalian phagocytic cells have been reviewed. Changes in the fluid environment, the particle and the phagocyte surface occur simultaneously and at different rates. In order to evaluate the early stages of phagocytosis careful separation of these changing processes is necessary. Under certain conditions, evaluation of the early events of phagocytosis may be aided by application of measurement techniques such as Scatchard plots, enzyme kinetics, adsorption isotherms, and equations of adhesive forces.

The role of receptor and ligand in microbe–macrophage and lymphokine–cell interactions has recently been emphasized. Several examples of such receptor–ligand interactions include lectin, immunoglobulin, and complement binding.

Microbes attach to macrophages by different mechanisms which can be divided into three general categories: (1) immunologically specified, (2) nonspecific, or (3) parasite specified. Immunologically specified attachment and recognition are promoted by immunoglobulins and complement. Nonspecific attachment describes those receptors which have not been identified; one example is attachment promoted by lectin–carbohydrate interactions. Parasite-specified attachment includes examples of obligate intracellular microbes which require attachment and entry into mammalian cells for survival.

Lymphokines appear to act at the cell membrane, by as yet undefined molecule–receptor interactions which then lead to appropriate intracellular changes. Examples of moieties which promote attachment of similar molecules include the B fragments of diphtheria and cholera toxins. Further study of the biochemistry of these molecules may aid in determining dynamics of lymphokine–macrophage interaction.

REFERENCES

Albersheim, P., Ayers, A. R., Valent, B. S., Ebel, J., Hahn, M., Wolpert, J., and Carlson, R., 1977, Plants interact with microbial polysaccharides, *J. Supramol. Struct.* **6**:599.

Allen, J. M., Cook, G. M. W., Poole, A. R., 1971, Action of Concanavalin A on the attachment stage of phagocytosis by macrophages, *Exp. Cell Res.* **68**:466.

Bar-Shavit, Z., Ofek, J., Goldman, R., Mirelman, D., and Sharon, N., 1977, Mannose residues on phagocytes as receptors for the attachment of *Escherichia coli* and *Salmonella typhi*, *Biochem. Biophys. Res. Comm.* **78**:455.

Bell, G. I., 1978, Models for the specific adhesion of cells to cells, *Science* **200**:618.

Berken, A., and Benacerraf, B., 1966, Properties of antibodies cytophilic for macrophages, *J. Exp. Med.* **123**:119.

Berlin, R. D., Oliver, J. M., Ukena, T. E., and Yin, H. H., 1975, The cell surface, *N. Engl. J. Med.* **292**:515.

Bianco, C., 1977, Plasma membrane receptors for complement, in: *Biological Amplification Systems in Immunology* (N. K. Day and R. A. Good, eds.), pp. 9–84. Plenum Press, New York.

Bianco, C., and Nussenzweig, V., 1977, Complement receptors, in: *Contemporary Topics in Molecular Immunology* (R. R. Porter and G. L. Ada, eds.), Vol. 6, pp. 145–176, Plenum Press, New York.

Bianco, C., Griffin, F. M., Jr., and Silverstein, S. C., 1975, Studies of the macrphage complement receptor: Alteration receptor function upon macrophage activation, *J. Exp. Med.* **141**:1278.

Boquet, P., and Pappenheimer, A. M., Jr., 1974, Interaction of diphtheria toxin with mammalian cell membranes, *J. Biol. Chem.* **251**:5770.

Brause, B. D., and Roberts, R. B., 1978, Attachment of virulent *Treponema pallidum* to human mononuclear phogocytes, *Br. J. Ven. Dis.* **54**:218.

Byrne, G. I., 1978, Kinetics of phagocytosis of *Chlamydia psittaci* by mouse fibroblasts (L cells): Separation of the attachment and ingestion stages, *Infect. Immun.* **19**:607.

Byrne, G. I., and Moulder, J. W., 1978, Parasite-specified phagocytosis of *Chlamydia psittaci* and *Chlamydia trachomatis* by L and HeLa Cells, *Infect. Immun.* **19**:598.

Cahall, D. L., and Youmans, G. P., 1975, Molecular weight and other characteristics of mycobacterial growth inhibiting factor produced by spleen cells obtained from mice immunized with viable attenuated mycobacterial cells, *Infect. Immun.* **12**:841.

Ciccimarra, F., Rosen, F. S., and Meiler, E., 1975, Localization of the IgG effector site for monocyte receptors, *Proc. Natl. Acad. Sci. USA* **72**:2081.

Cohn, Z. A. and B. Benson, 1965, The differentiation of mononuclear phagocytes. Morphology, cytochemistry and biochemistry. *J. Exp. Med.* **121**:153.

Cuatrecasas, P., 1973, Gangliosides and membrane receptors for cholera toxin, *Biochemistry* **12**:3558.

Cuatrecasas, P., and Hollenberg, M. D., 1975, Binding of insulin and other hormones to non-receptor materials: Saturability, specificity and apparent "negative cooperativity," *Biochem. Biophys. Res. Comm.* **62**:31.

Curtis, A. S. G., 1967, *The Cell Surface: Its Molecular Role in Morphogenesis*, p. 80, Logos Academic Press, London.

David, J. R., 1975, Macrophage activation by lymphocyte mediators, *Fed. Proc.* **34**:1730.

Dazzo, F. B., Napoli, C. A., Hubbell, D. H., 1976, Adsorption of bacteria to roots as related to host specificity in the rhizobium–clover symbiosis, *Appl. Environ. Microbiol.* **32**:166.

DeLange, R. J., Drazia, R. E., and Collier, R. R., 1976, Amino acid sequence of fragment A, an enzymically active fragment from diphtheria toxin, *Proc. Natl. Acad. Sci. USA* **73**:69.

DeMeyts, P. and Roth, J., 1975, Cooperativity in ligand binding: A new graphic analysis *Biochem. Biophys. Res. Comm.* **66**:1118.

Edelson, P. J., and Cohn, Z. A., 1974, Effects of Concanavalin A on mouse peritoneal macrophages. I. Stimulation of endocytic activity and inhibition of phagolysosome formation, *J. Exp. Med.* **140**:1364.

Ehlenberger, A. G., and Nussenzweig, V., 1977, The role of membrane receptors for C3b and C3d in phagocytosis, *J. Exp. Med.* **145**:357.

Ehrlich, P., 1900, On immunity with special reference to cell life, *Proc. R. Soc. London* **66**:424.

Eisen, H. N., 1974, Antibody–antigen reactions, in: *Microbiology* (D. D. Davis, R. Dulbecco, H. N. Eisen, H. S. Ginsberg, and W. B. Wood, eds.), 2nd ed., pp. 359–404, Harper and Row, New York.

Erdmann, E., Phillip, G., and Tanner, G., 1976, Ouabain-receptor interactions in $(Na^+ + K^+)$–ATPase preparations. A contribution to the problem of nonlinear Scatchard plots, *Biochim. Biophys. Acta* **455**:287.

Finkelstein, R. A., 1975, Immunology of cholera, *Curr. Top. Microbiol. Immunol.* **69**:137.

Gardner, M., 1977, Abstracts, 77th Annual Meeting of the American Society for Microbiology. D-6, American Society for Microbiology, Washington, D.C.

George, M., and Vaughan, J. H., 1962, *In vitro* cell migration as a model for delayed hypersensitivity, *Proc. Soc. Exp. Biol. Med.* **111**:514.

Goldman, R., 1977, Lectin-mediated attachment and ingestion of yeast cells and erythrocytes by hamster fibroblasts, *Exp. Cell Res.* **104**:325.

Goldman, R., and Raz, A., 1975, Concanavalin A and the *in vitro* induction in macrophages of vacuolation and lysosomal enzyme synthesis, *Exp. Cell Res.* **96**:393.

Griffin, F. M., Jr., Griffin, J. A., Leider, J. E., and Silverstein, S. C., 1975, Studies on the mechanism of phagocytosis. I. Requirements for circumferential attachment of particle-bound ligands to specific receptors on the macrophage plasma membrane, *J. Exp. Med.* **142**:1263.

Hirsh, J. G., and Strauss, B., 1964, Studies on heat-labile opsonin in rabbit serum, *J. Immunol.* **92**:145.

Hollenberg, M. D., Fishman, P. H., Bennett, V., and Cuatrecasas, P., 1974, Cholera toxin and cell growth: Role of membrane gangliosides, *Proc. Natl. Acad. Sci. USA* **71**:4224.

Hu, P. C., Collier, A. M., and Baseman, J. B., 1977, Surface parasitisms by *Mycoplasma pneumoniae* of respiratory epithelium, *J. Exp. Med.* **145**:1328.

Jacobs, S., and Cuatrecasas, P., 1976, The mobile receptor hypothesis and "cooperativity" of hormone binding. Application to insulin, *Biochim. Biophys. Acta* **433**:482.

Jacobs, S., Chang, K., and Cuatrecasas, P., 1975, Estimation of hormone receptor affinity by competitive displacement of labeled ligand: Effect of concentration of receptor and ligand, *Biochem. Biophys. Res. Comm.* **66**:687.

Jones, T. C., 1975a, Attachment and ingestion phases of phagocytosis, in: *Mononuclear Phagocytes* (R. van Furth, ed.), pp. 269–282, Blackwell, Oxford.

Jones, T. C., 1975b, Phagosome-lysosome interaction with *Toxoplasma*, in: *Mononuclear Phagocytes* (R. van Furth, ed.), pp. 595–608, Blackwell, Oxford.

Jones, T. C., 1979, Entry and development of toxoplasmas in mammalian cells, in: *Microbiology 1979* (D. Schlesinger, ed.), pp. 135–139, American Society for Microbiology, Washington, D.C.

Jones, T. C., and Hirsch, J. G., 1971, The Interaction *in vitro* of *Mycolasma pulmonis* with peritoneal macrophages and L cells, *J. Exp. Med.* **133**:231.

Jones, T. C., and Hirsch, J. C., 1972, The Interaction between *Toxoplasma gondii* and mammalian cells. II. The absence of lysosomal fusion with phagocytic vocuoles containing living parasites, *J. Exp. Med.* **136**:1173.

Jones, T. C., and Yang, L., 1977, Attachment and ingestion of mycoplasmas by mouse macrophages. I. Kinetics of the interactions and effects on phagocyte glucose metabolism, *Am. J. Pathol.* **87**:331.

Jones, T. C., Masur, H., Len, L., and Fu, T. L. T., 1977, Lymphocyte-macrophage interaction during control of intracellular parasitism, *Am. J. Trop. Med. Hyg.* **26**:187.

Jones, T. C., Yeh, S., and Hirsch, J. G., 1972, Studies on attachment and ingestion phases of phagocytosis of *Mycoplasma pulmonis* by mouse peritoneal macrophages, *Proc. Soc. Exp. Biol. Med.* **139**:464.

Kabat, E. A., 1976, *Structural Concepts in Immunology and Immunochemistry*, 2nd ed., p. 95, p. 167, Holt, Rinehart and Winston, New York.

Karnovsky, M. L., 1962, Metabolic basis of phagocytic activity, *Physiol. Rev.* **42**:143.

Kilejean, A., 1976, Does a histidine-rich protein from *Plasmodium lophurae* have a function in merozoite penetration? *J. Protozool* **23**:272.

Klebanoff, N., and Hamon, C. B., 1975, Antimicrobial systems of monuclear phagocytes, in: *Mononuclear Phagocytes* (R. van Furth, ed.), pp. 507–532, Blackwell, Oxford.

Kurland, J. I., Hadden, J. W., and Moore, M. A., 1977, Role of cyclic nucleotides in the proliferation of committed granulocyte-macrophage progenitor cells, *Cancer Res.* **37**:4534.

Lurie, N. B., 1932, The correlation between the histological changes and the fate of living tubercle bacilli in the organs of tuberculous rabbits, *Exp. Med.* **55**:31.

Maaløe, O., 1947, On the dependence of the phagocytosis-stimulating action of immune serum on complement, *Acta Pathol. Microb. Scand.* **23**:34.

Mackaness, G. B., 1969, The influence of immunologically committed lymphoid cells on macrophage activity *in vivo*, *J. Exp. Med.* **129**:973.

Metchnikoff, E., 1905, *Immunity in Infective Diseases*, Cambridge University Press, London.

Moulder, J. W., 1969, A model for studying the biology of parasitism: *Chlamydia psittaci* and mouse fibroblasts (L Cells), Bioscience **19**:875.

Moulder, J. W., Hatch, T. P., Byrne, G. I., and Kellogg, K. R., 1976, Immediate toxicity of high multiplicities of *Chlamydia psittaci* for mouse fibroblasts (L cells), *Infect. Immun.* **14**:277.

Mudd, E. B. H., and Mudd, S., 1933, The process of phagocytosis. The agreement between direct observation and deductions from theory, *Gen. Physiol.* **16**:625.

North, R. J., 1970, The Relative importance of blood monocytes and fixed macrophages to the expression of cell mediated immunity to infection, *J. Exp. Med.* **132**:521.

Pappenheimer, A. M., Jr., 1977, Diphtheria toxin. *Annu. Rev. Biochem.* **46**:69.

Rabinovitch, M., 1970, Phagocytic recognition, in: *Mononuclear Phagocytes*, 1st ed. (R. van Furth, ed.), p. 299–320, Blackwell, Oxford.

Remold, H. G., 1973, Requirements for α-L-fucose in the macrophage membrane receptor for MIF, *J. Exp. Med.* **138**:1065.

Remold, H. G., 1974, The enhancement of MIF activity by inhibition of macrophage associated esterases, *J. Immunol.* **112**:1571.

Remold, H. R., and David, J. R., 1974, Migration inhibition factor and other mediators in cell-mediated immunity, in: *Mechanisms of Cell Mediated Immunity* (R. T. McCluskey and A. Cohen, eds.), pp. 25–42, John Wiley, New York.

Rhodes, J., 1975, Macrophage heterogeneity in receptor activity: The activation of macrophage Fc receptor function *in vivo* and *in vitro*, *J. Immunol.* **114**:976.

Roberts, R. B., 1977, Gonococci-leukocyte interactions, in: *The Gonococcus* (R. B. Roberts, ed.), pp. 333–354, John Wiley and Sons, New York.

Saba, T. M., 1970, Physiology and physiopathology of the reticuloendothelial system, *Arch. Intern. Med.* **126**:1031.

Scatchard, G., 1949, The attractions of proteins for small molecules and ions, *Ann. N.Y. Acad. Sci.* **51**:660.

Schreiber, A. D., and Frank, M. M., 1972, Role of antibody and complement in the immune clearance and destruction of erythrocytes. I. *In vivo* effects of IgG and IgM complement-fixing sites, *J. Clin. Invest.* **51**:575.

Silverstein, S. C., 1975, The role of mononuclear phagocytes in viral immunity, in: *Mononuclear Phagocytes* (R. van Furth, ed.), pp. 557–568, Blackwell, London.

Silverstein, S. C., Steinman, R. M., and Cohn, Z. A., 1977, Endocytosis, *Annu. Rev. Biochem.* **46**:669.

Spector, W. G., 1970, The macrophage in inflammation, *Sem. Hematol.* **3**:132.

Stossel, T. P., 1973, Quantitative studies of phagocytosis: Kinetic effects of actions and of cations and of heat-labile opsonin, *J. Cell Biol.* **58**:346.

Stossel, T. P., 1975, Phagocytosis: Recognition and ingestion, *Sem. Hematol.* **12**:83.

Unkeless, J. C., 1977, The presence of two Fc receptors on mouse macrophages: Evidence from a variant cell line and differential trypsin sensitivity, *J. Exp. Med.* **145**:931.

Unkeless, J. C., and Eisen, H. N., 1975, Binding of monomeric immunoglobulins to Fc receptors of mouse macrophages, *J. Exp. Med.* **142**:1520.

van Oss, C. J., and Gillman, C. F., 1972, Phagocytosis as a surface phenomenon, *J. Reticuloendothel. Soc.* **12**:283.

van Oss, C. J., Gillman, D. F., and Neumann, A. W., 1975, *Phagocytic Engulfment and Cell Adhesivenesses*, Marcel Dekker, New York.

Vlodavsky, I., and Sachs, L., 1975, Lectin receptors on the cell surface membrane and the kinetics of lectin-induced cell agglutination, *Exp. Cell Res.* **93**:111.

Weidemann, M. J., Erdelt, H., and Klingenberg, M., 1970, Adenine nucleotide translocation of mitochondria. Identification of carrier sites, *Eur. J. Biochem.* **16**:313.

Weisman, R. A., and Korn, E. D., 1967, Phagocytosis of latex beads by acanthamoeba. I. Biochemical properties, *Biochemistry* **6**:485.

Wood, B., 1946, Studies of the mechanisms of recovery in pneumococcal pneumonia, *J. Exp. Med.* **84**:387.

Wright, A. E., and Douglas, S. R., 1903, An experimental investigation of the role of blood fluids in connection with phagocytosis, *Proc. Roy. Soc. London* **72**:357.

Wyrick, P. B., and Brownridge, E. A., 1978, Growth of *Chlamydia psittaci* in macrophages, *Infect. Immun.* **19**:1054.

3

Biochemistry and Physiology of Tuftsin Thr-Lys-Pro-Arg

VICTOR A. NAJJAR

1. INTRODUCTION

The discovery of the tetrapeptide tuftsin (Thr-Lys-Pro-Arg) came about in an unexpected manner during our early studies on the mechanism of phagocytosis by polymorphonuclear granulocytes (Fidalgo *et al.*, 1967a,b).

For the past few years we have been engaged in exploring the physiological role of cytophilic γ-globulins. These are separable types of γ-globulins that bind specially to the outer surface of various blood cells. Erythrophilic fractions bind to red cells (Fidalgo *et al.*, 1967a,b; Najjar *et al.*, 1972) and affect the life span of the erythrocyte *in vivo*. Other fractions, equally specific, bind to blood platelets (Constantopoulos and Najjar, 1974) and granulocytic leukocytes (Fidalgo and Najjar, 1967a,b), thrombophilic and leukophilic respectively. More recently, it was shown that blood monocytes and lymphocytes also bind γ-globulin onto the outer membrane (Saravia *et al.*, 1978). Leukophilic γ-globulin stimulates the phagocytic activity of the neutrophilic granulocyte, and is necessary for its survival *in vitro*. It was during the latter study that we came upon the tetrapeptide. It became apparent in due course that the granulocyte possessed the ability to cleave this polypeptide from the membrane-bound leukophilic γ-globulin. The freed peptide is then capable of exerting, on a molar basis, a stimulatory effect on the phagocytic activity of the cell to the full extent observed with the intact molecule. It was, thus, clear that the cytophilic γ-globulin molecule was merely a carrier molecule.

VICTOR A. NAJJAR • Division of Protein Chemistry, Tufts University School of Medicine, Boston, Massachusetts 02111. This was supported by the Department of Health, Education, and Welfare, Public Health Service Grant 5 R01 AI09116; National Science Foundation Grant PCM76-23008; and The National Foundation March of Dimes 1-556.

2.   MECHANISM OF RELEASE OF TUFTSIN FROM THE CARRIER LEUKOKININ

In a series of reports, it was shown that leukokinin, a leukophilic γ-globulin fraction (Thomaidis *et al.*, 1967), binds specifically to blood neutrophils to effect a prompt stimulation of their phagocytic activity. During the process, the tetrapeptide tuftsin is cleaved and utilized by the cell. The leukokinin, relieved of the tetrapeptide, is now rendered inactive. In order to maintain a continuous stimulation, the inactivated leukokinin is shed and fresh tuftsin carrying leukokinin now binds to the free cell receptor to make available further quantities of tuftsin and effect further stimulation of phagocytosis. In this manner the blood phagocyte maintains a state of continuous stimulation. The final cleavage of the tetrapeptide from leukokinin is made by the specific enzyme leukokininase. This enzyme is present on the outer surface of the cell membrane with its active site directed outwards. This was well documented by the demonstration that the live intact and motile cell is highly efficient in stripping off all available tuftsin from donor leukokinin without internalization of the carrier molecule. It was possible to isolate and characterize tuftsin-deficient leukokinin after incubation with live cells. The tuftsinless molecule had the same molecular weight and physical properties as unstripped leukokinin except that it was biologically inactive (Satoh *et al.*, 1972). The enzyme leukokininase responsible for releasing tuftsin has been characterized and its properties delineated (Nishioka *et al.*, 1973a).

The tetrapeptide is present in the heavy chain of leukokinin. It is located in the Fc portion in proximity to the carbohydrate binding region at residues 289–292 of the sequence H—Val-His-Asn-Ala-Lys-*Thr-Lys-Pro-Arg*-Glu-Gln-Gln-Tyr—OH (Edelman *et al.*, 1969). For the complete release of this tetrapeptide from the carrier molecule, it is first cleaved at the arginylglutamyl (Arg-Glu) bond by a splenic enzyme, tuftsin endocarboxy peptidase, to produce leukokinin-S (S for spleen) with a free carboxyl end of tuftsin. The second cleavage at the lysylthreonyl bond takes place after leukokinin-S binds to the blood and tissue granulocyte through the action of its membrane enzyme leukokininase. In the absence of the spleen, therefore, no release of tuftsin can occur and phagocytosis is impaired (see below).

3.   FATE OF THE TETRAPEPTIDE TUFTSIN

Once tuftsin is released from the carrier molecule it presumably enters the cell and is finally hydrolyzed completely to its individual amino acids. It is not known whether the tetrapeptide exerts its action directly through its receptor membrane sites or after it enters the cell, if indeed it does enter.

Receptor sites exist on the membrane that bind tuftsin with high affinity. The addition of appropriate and not excessive quantities of the tetrapeptide to granulocyte preparations results in an immediate disappearance of tuftsin from the medium. The receptor site must include neuraminic acid. Treatment of live cells with highly purified neuraminidase removes 135×10^6 molecules of neuraminic acid per cell and abolishes completely the biological effect of the tetrapeptide without affecting the viability of the cell. It was estimated that per cell approximately 30×10^6 molecules of sialic acid are linked by 2–3' keto-linkage and the remaining 105×10^6 exist in the 2–6' linkage (Constantopoulos and Najjar, 1973a).

Several enzymes in the cytosol cleave tuftsin rapidly into Lys-Pro-Arg, Lys-Pro, Pro-Arg, and the individual amino acids. There are no enzymes in the granules that are capable of hydrolysis of tuftsin. However, cell membranes of macrophages and granulocytes show a mild aminopeptidase activity and also release the tripeptide which is a strong inhibitor of tuftsin activity and in that capacity may play a regulatory role (Rauner *et al.*, 1976; Najjar *et al.*, 1980).

4. CHEMISTRY OF THE TETRAPEPTIDE TUFTSIN

Tuftsin is a highly basic tetrapeptide, a property which made its isolation somewhat cumbersome. It strongly adheres to filter paper and to Sephadex columns which render its recovery rather incomplete. It was finally isolated from leukokininase digests of leukokinin preparations as well as tryptic digests of whole γ-globulin or the leukophilic fraction (Nishioka *et al.*, 1972, 1973a). Once the sequence was determined as Thr-Lys-Pro-Arg, it was readily synthesized (Nishioka *et al.*, 1973b). The synthetic product proved to be as effective and possessed the same specific activity in stimulating phagocytosis as the natural product.

5. CHEMICAL SYNTHESIS OF TUFTSIN

5.1. SOLID PHASE SYNTHESIS

Various methods have been used which produced biologically active preparations. In our laboratory, the tetrapeptide was synthesized by the solid phase method of Merrifield (1964; Najjar and Merrifield, 1966) in which arginine was esterified to a chloromethyl resin, a polystyrene divinyl benzene polymer. The guanido function of arginine was protected by a nitro group and the α-amino by the tertiary butyloxycarbonyl (BOC) group. The latter was acid cleaved and the incoming BOC proline was coupled to the nitroarginine by dicyclohexylcarbodiimide (DCC). The procedure of BOC cleavage and DCC coupling was repeated with lysine and finally with threonine. The ϵ-NH$_2$ group of the former residue was protected by a carbobenzoxy group and the hydroxyl function of threonine by an ether linkage to a benzyl group.

After cleavage of the tetrapeptide from the resin with HBr in acetic acid, the NO$_2$ group of arginine was reduced with H$_2$ and palladium on barium sulfate as catalyst. The product contained minor impurities which were powerful inhibitors of its biological activity. These were removed by chromatography on Aminex columns (Nishioka *et al.*, 1973b).

5.2. SYNTHESIS WITH POLYMERIC REAGENTS

Fridkin and his associates synthesized tuftsin using the classical approach with a yield of 39% and the activated polymeric method (Fridkin *et al.*, 1977) with a yield of 80%. Both methods yielded highly active peptides (Spirer *et al.*, 1975). The more interesting polymeric method involved the coupling of a carboxyl-activated residue esterified to 3-nitro-4-hydroxy benzyl-polystyrene (solid phase), to the recipient amino function of a residue in solution. By contrast, in the Merrifield technique, the coupling of the carboxy-activated amino acid is free in solution whereas the amino acceptor residue is fixed on a solid support.

With this technique, which was successfully used in other peptide syntheses at the Weizmann Institute (Patchornik *et al.*, 1973; Kalir *et al.*, 1974), Fridkin *et al.* (1975, 1976, 1977) were able to synthesize tuftsin of very high purity.

5.3. SYNTHESIS BY CLASSICAL METHODOLOGY WITH VARYING REAGENTS

Yajima *et al.* (1975) synthesized tuftsin with the usual chemical method, in solution; using the soluble activated ester of 5-chloro-8-hydroxyquinoline and the N^α protecting group *p*-methoxybenzyloxycarbonyl group. Other protecting groups were tosyl for N^G of arginine and carbobenzoxy for N^ϵ lysine. Cleavage of all groups was successfully preformed in trifluoromethanesulfonic acid. An active tetrapeptide was obtained.

We have successfully used the organic acid to cleave the tetrapeptide from the supporting resin as well as all protecting functions except the nitro group of arginine which was partially removed (Chaudhuri and Najjar, 1978).

Konopinska *et al.* (1976, 1977) also synthesized tuftsin in solution using the classical approach. The carboxy-terminal *O*-benzyl ester, the *N*-protecting group, and the NO_2 arginine side chain protecting groups were all removed by catalytic hydrogenation.

Martinez and Winternitz (1976) reported the synthesis of tuftsin with the trimeric phosphonitrilic chloride as condensing agent, as well as by the active ester of *o*-nitrophenyl group. They obtained active peptides with either method for *in vitro* as well as *in vivo* experiments (Martinez and Winternitz, 1976; Martinez *et al.*, 1977).

Vičar and colleagues (1976) also synthesized the tetrapeptide by combining synthetic fragments of the tetrapeptide rather than by sequential synthesis as above. They utilized the Zervas reagent (Zervas and Hamalides, 1965), 2-nitrobenzenesulfenyl chloride, for α-amino-blocked residues and characterized tuftsin by circular dichroism studies at pH values of 2.0, 4.9, and 8.5. This Zervas reagent has been used for the synthesis of the octadecapeptide brandykininyl-bradykinin (Najjar and Merrifield, 1966).

6. METHODS OF ASSAY

Tuftsin can be assayed before or after release from the carrier leukokinin.

6.1. TUFTSIN ASSAY IN LEUKOKININ

Assay of tuftsin bound to leukokinin has been accomplished using homologous or autologous buffy coat cells. A leukokinin-rich fraction can be prepared by phosphocellulose fractionation (Thomaidis *et al.*, 1967). Alternatively, whole γ-globulin prepared from inactivated serum or inactivated serum itself can be used as a source of tuftsin. Essentially, cells are incubated with target particles which may be bacteria (Najjar, 1974) or yeast (Fridkin *et al.*, 1977) in appropriate isotonic buffers at appropriate temperature and duration. Cells are then stained and scored for the cell fraction that is bearing particles or the number of particles ingested by a given number of cells.

6.2. ASSAY OF FREE TUFTSIN AFTER RELEASE FROM LEUKOKININ

Tuftsin can be released from leukokinin by the specific granulocyte membrane enzyme leukokininase. This can be prepared from rabbit peritoneal granulocyte or from buffy coat (Fidalgo and Najjar, 1967a,b). Incubation of leukokinin or whole γ-globulin with the enzyme results in quick release of tuftsin which is completely stable as no destructive enzymes contaminate leukokininase. Tuftsin is then extracted with ethanol, dried, and dissolved in buffer.

Tuftsin can be more conveniently released from leukokinin, by limited trypsin digestion of whole γ-globulin (Najjar and Constantopoulos, 1972; Nishioka *et al.*, 1973a; Najjar, 1974) or serum (Spirer *et al.*, 1977a). The tetrapeptide can then be extracted with alcohol and prepared as above. The free tetrapeptide can now be assayed by one of several methods:

It can be done in the conventional manner and a semiquantitative measure of tuftsin can be obtained from a standard assay curve with known amounts of tuftsin (Constantopoulos and Najjar, 1972; Najjar and Constantopoulos, 1972; Erp and Fahrney, 1975).

The phagocytic cells from a buffy coat preparation can be made to settle and adhere onto glass cover slips placed in falcon dishes and then washed. Such immobilized cells can then be treated with samples containing tuftsin followed by target particles (Fridkin *et al.*, 1977).

6.3. REDUCTION OF NITROBLUE TETRAZOLIUM

Nitroblue tetrazolium has been widely used for studies on the activation of the hexosemonophosphate shunt following phagocytosis which results in the reduction of the dye in a manner that parallels the extent of phagocytosis (Baehner and Nathan, 1968; Park *et al.*, 1968). The dye forms complexes with fibrinogen and heparin which are then phagocytized by the cells (Segal and Levy, 1973). The extent of phagocytosis is deduced from the amount of reduced dye estimated colorimetrically.

All these methods of assay, while easy to use and require no special reagents nevertheless yield at best semiquantitative data. They are quite useful in assessing the absence or diminution of tuftsin activity such as one encounters in instances of complete absence of tuftsin as in cases of tuftsin deficiency or after loss of splenic function. These methods may not be sufficiently quantitative for assaying slight or moderate variation in tuftsin levels as might occur in certain diseases such as leukemia, sickle cell anemia, Hodgkin disease, and other debilitating diseases (Najjar and Constantopoulos, 1972; Constantopoulos *et al.*, 1972, 1973a,b).

6.4. RADIOIMMUNOASSAY OF TUFTSIN

Spirer *et al.* (1977a) have successfully devised a quantitative radioimmunoassay method and have been able to measure tuftsin blood levels utilizing only 2 ml of serum.

Tuftsin was synthesized as above, but before cleavage of the side chain groups, carbobenzoxy-p-aminophenylacetic acid was coupled to the free amino groups of threonine. This was followed by complete deprotection of all protected functions with hydrogen fluoride. The p-aminophenylacetyl-tuftsin was then diazotized and linked to bovine serum albumin or ribonuclease and the protein–tuftsin conjugate used for raising antibodies in rabbits. Antibodies so obtained bound [^{125}I]-p-aminophenylacetyl-tuftsin at dilution of 1:1500. With appropriate standard curves obtained by competitive binding, they could secure quantitative values of tuftsin in serum samples. The assay proved to be highly specific for the tripeptide sequence -Lys-Pro-Arg-OH which was the antigenic determinant. With this technique, normal subjects gave values of about 278 ± 13 ng·ml^{-1} (Spirer *et al.*, 1977a).

The method was also found to be highly reproducible with an interassay correlation index of 0.97. The specificity was also tested with a series of related and unrelated oligopeptides. LH-RH, the antiallergic peptide Asp-Ser-Asp-Pro-Arg, TRH, liver cell growth factor Gly-His-Lys and synthetic Leu-Arg-Pro-Gly gave no cross reactions. Tuftsin analogs on the other hand, showed considerable cross reactions. These were [ω-NO$_2$^{4}]-tuftsin, p-aminophenylacetyl-tuftsin, des-Thr- tuftsin, acetyl- tuftsin, tyrosyl- tuftsin, [Ala^1]- tuftsin, [Ser^1]- tuftsin, and [Lys^1]-tuftsin. The strongest binding analog was the tripeptide des-Thr^1-tuftsin. Synthetic tuftsin caused a 50% binding of the radioactive analog at a value of 140 ng·ml^{-1}.

7. BIOLOGICAL ACTIVITY OF TUFTSIN *IN VITRO*

The isolation and purification of the peptide was guided by its biological activity, namely the stimulation of phagocytosis by blood neutrophils with *Staphylococcus aureus* as the target particle. The question arose as to whether the granulocyte was simply rendered more active resulting in an increase in the collision frequency with the target particle. Alternatively, the phagocyte, through the action of tuftsin, could be rendered more efficient at the level of the engulfing process (Najjar, 1974, 1976). It is possible that both processes are involved.

7.1. STIMULATION OF PHAGOCYTOSIS

Blood neutrophils and tissue macrophages are capable of a basal level of phagocytosis of particulate substances. Upon the addition of tuftsin, their activity increases to a level over twice that of the basal value. The concentration that is required for approaching maximal activity for both phagocytic cells is about $2-4 \times 10^{-7}$ g·ml^{-1} (Figure 1). These values were obtained equally well using *Staphylococcus aureus* (Najjar *et al.*, 1968, 1972) or yeast cells (Fridkin *et al.*, 1976, 1977). Both techniques were based on the number of phagocytes that contain particles. However, similar data were obtained if the assay was carried out by monitoring the total number of particles engulfed by a given number of granulocytes. Figure 1 gives the dose-response curves of tuftsin with *Staphylococcus aureus* and yeast cells as the target particles.

7.2. STIMULATION OF MOTILITY AND VIABILITY

The tetrapeptide tuftsin stimulates the migration of granulocytes upwards in erect glass capillary tubes (Figure 2). This rate is dose dependent. After 6 hr, the migration of granulocytes from the buffy coat reached 310 μm. In the presence of tuftsin, migration attained a level of 430 and 700 μm at 5 and 25 nmol · ml^{-1}, respectively (Table 1).

Table 1 shows the rate of migration under various conditions. It is clear that the rate of unidirectional migration at tuftsin concentration of 5 nmol·ml^{-1} showed a 2 hr lag and migration ceased after 5 hr. On the other hand, at 25 nmol·ml^{-1} stimulation lagged for only 1 hr and unidirectional migration continued unabated well beyond 6 hr of incubation. The free amino acid constituents of tuftsin added together did not show any stimulation above control rates. The addition of the pentapeptide analog Thr- Lys- Pro- Pro- Arg, which

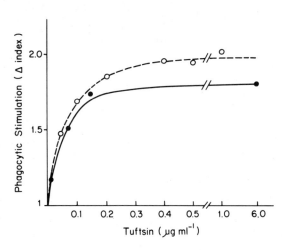

FIGURE 1. Dose-response effect of tuftsin. This is a composite figure of phagocytosis assay using *Staphylococcus aureus* (Constantopoulos and Najjar, 1972) and yeast cells (Fridkin *et al.*, 1977). Both reactions involved blood polymorphonuclear leukocytes. Note the similarity of responses in both cases. (For details refer to original article.)

inhibits tuftsin stimulation of phagocytosis, was also strongly inhibitory to tuftsin-stimulated migration. At about one third the concentration of tuftsin the analog was completely inhibitory and the migration proceeded at the basal rate.

It should be stressed that the basal activities of blood granulocytes, whether expressed as phagocytosis or as migration, are not at all affected by this analog; only the stimulation is abolished. This indicates that, in the absence of tuftsin, such as in cases of splenectomy, a basal level of activity always remains. The patient will more or less manage well except for instances of overwhelming infections with bacteria not readily susceptible to phagocytosis. Similarly, in cases with tuftsin deficiency (Constantopoulos *et al.*, 1972; Inada *et al.*, 1977; Najjar, 1977), only a few cases at an early age show frequent or continuous infections. The great majority of patients so far observed have only mild infections. Nevertheless, the mere occurrence of infections indicates that the basal activity does not suffice to ward off infection in animals and in man (Eraklis *et al.*, 1967; Shinefield *et al.*, 1966).

It should be noted here that tuftsin appears to increase the viability of the cells. After 2–3 hr, control cells without tuftsin cease to migrate on glass, remain

FIGURE 2.   Vertical migration of polymorphonuclear leukocytes. Fresh heparinized dog blood was washed three times with four volumes of Hanks' solution. 0.1 ml of packed cells was mixed with 0.1 ml of Hanks' solution containing tuftsin. The mixture was drawn into microhematocrit tubes, sealed, centrifuged, and incubated open-side up at 37°C. After 2 hr the migration in tuftsin tubes was demonstrably greater than in the control tubes with upward swarming of PMN cells. This picture was taken after 18 hr incubation. The first three tubes on the left are control tubes, the next three and the last three tubes to the right contain 25 and 5 nmol/ml of tuftsin, respectively. Note that the PMN cell front, in tubes containing 25 nmol/ml, swarmed a distance of 3.3 mm. The tubes containing 5 nmol/ml showed no swarming, but a considerable number of cells, not visible in the figure, moved 1.2 mm. PMN cells in control tubes leveled off early at 0.3 mm in 3 hr (Nishioka *et al.*, 1972).

TABLE 1.   RATE OF GRANULOCYTE MIGRATION UNDER VARIOUS CONDITIONS[a]

Additions to Hanks' medium, ml^{-1}	Rate of granulocyte migration in glass microtubes (μm)					
	Hours					
	1	2	3	4	5	6
None[b]	190	250	308	300	306	310
Tuftsin amino acids each 25 nmol[c]	220	255	310	304	310	310
Tuftsin 5 nmol[d]	—	252	350	412	425	430
Tuftsin 25 nmol	180	353	485	585	625	700
Tuftsin 25 nmole + analog 8 nmol[e]	180	248	315	308	305	300

[a] Freshly drawn heparinized dog blood was washed three times with Hanks' medium in 0.1% glucose. Six siliconized microhematocrit tubes were used for each solution to be tested. The tubes were filled by capillary action with the various solutions to be tested at the concentration indicated and followed by the packed cells. The tubes were sealed at the cell end and centrifuged. These were then fixed vertically in an incubator at 37°C equipped with a microscope and an ocular millimeter scale for continuous observation without disturbing the tubes.
[b] Hanks' medium alone.
[c] The constituent amino acids that make up tuftsin, namely, threonine, lysine, proline, and arginine.
[d] The tetrapeptide tuftsin, Thr-Lys-Pro-Arg.
[e] Tuftsin 25 nmol·ml^{-1} plus the pentapeptide analog Thr-Lys-Pro-Pro-Arg 8 nmol·ml^{-1}.

quiescent for a few hours, and finally drop off the glass capillary tube or lyse. However, in the presence of tuftsin they survive much longer and have been observed to maintain their mobility and viability after 16 hr (Nishioka *et al.*, 1973b; Najjar, 1974, 1976).

7.3.   REDUCTION OF NITROBLUE TETRAZOLIUM

It has been postulated that in the act of engulfing particulate matter, there is an activation of the hexosemonophosphate shunt (Humbert *et al.*, 1973). The activation by methylene blue does not require any phagocytic activity while that by endotoxin is secondary to phagocytic stimulation (Karnovsky, 1968). The effect of the latter is of considerable interest.

Spirer *et al.* (1975) showed that tuftsin stimulated the reduction of nitroblue tetrazolium by granulocytes. This reduction was found to be similar in extent to that produced by endotoxin and appears to be initiated by the same mechanism.

It has recently come to light that the tetrazolium dye is in fact phagocytized by the granulocytes. The dye is in the form of a complex with fibrinogen or heparin (Segal and Levy, 1973). Consequently, it remains undecided whether tuftsin simply stimulates phagocytosis of the dye or acts as a direct activator of the shunt (Spirer *et al.*, 1975).

One of the interesting by-products of the study by the Israeli group (Spirer *et al.*, 1975) relates to the mechanism by which endotoxin stumulates the reduction of nitroblue tetrazolium. They found that the activation of the shunt which resulted in the reduction of the tetrazolium dye by tuftsin and endotoxin is not additive. Analogs of tuftsin that inhibit the effect of tuftsin also inhibit the effect of endotoxin. This may indeed relate to common or overlapping receptor sites on the outer membrane of the cell. It would be of interest to investigate the possibility that sialic acid on the granulocyte membrane may be necessary to the biological action of endotoxin as it has been shown to be for tuftsin (Constantopoulos and Najjar, 1973a). Should this prove to be the case, then a study of the competitive binding of tuftsin and endotoxin to the membrane receptors would be of considerable interest.

7.4.   CHEMOTACTIC EFFECT OF TUFTSIN

When buffy coat is prepared in an appropriate buffer in a vertically positioned capillary tube, the granulocytes move with equal probability in all directions, but appear to gain distance only in an upward direction. Because their motion is random rather than directed, they tend to display a diffuse distribution. On the other hand, if tuftsin is present in the upper buffer layer, these cells move upward unidirectionally in a band of concentrated cells. Below this band of cells, the concentration of granulocytes is markedly diminished. The concentration of tuftsin below and at the band is minimal because it is quickly used up by the cells. Above the band, tuftsin concentration is undiminished and cells move upwards towards tuftsin (Nishioka *et al.*, 1973b). Nishioka (personal communication) has shown clearly that tuftsin is chemotactic with the agarose gel technique. On plastic surfaces, tuftsin is not chemotactic. The explanation may reside in the finding that tuftsin actually inhibits sticking of granulocytes to plastic surfaces (Najjar, 1974).

An interesting observation has been reported where it was shown that synthetic tuftsin exhibited a significant stimulation of chemotaxis in the *E. coli* system (Yajima *et al.*, 1975).

7.5.   EFFECT OF TUFTSIN ON MuLV PRODUCTION

Another biological activity of singular importance has recently been reported. It was shown by Luftig *et al.* (1977) and Orozlan *et al.* (1978) that the sequence from the amino terminal end of viral protein p12 of Raucher murine leukemia virus (R-MuLV) is Pro- Thr- Leu- Thr- Ser- Pro- Leu- Asn- *Thr- Lys- Pro- Arg*-Pro-Gln-Val-Leu-Pro-Asp-X-Gly—. One readily observes that the sequence includes the tetrapeptide tuftsin, Thr-Lys-Pro-Arg, starting at the ninth residue from the amino terminal. This finding, in addition to other considerations of tuftsin effects on the cleavage of the *gag* gene product p65-70, led Luftig *et al.* (1977) to a study of the possible biological activity of tuftsin on the MuLV infected cells. Tuftsin added in concentrations up to 100 μM to cultures of these cells resulted in a threefold increase in virion-associated reverse transcriptase. They concluded that R-MuLV contains a p12 protein which can potentially enhance membrane activity and lead to increased virion production.

The occurrence of tuftsin in leukokinin and other fractions of γ-globulin (see above) as well as in p12 of R-MuLV may be of special biological significance. It was shown in several laboratories that the sequence Thr-Lys-Pro-Arg is uniquely competent in stimulating phagocytosis (Nishioka *et al.*, 1973a,b; Constantopoulos and Najjar, 1972; Erp and Fahrney, 1975; Fridkin *et al.*, 1976, 1977; Konopinska *et al.*, 1977). Any variations or deletions thereof, even involving homologous residues, results in either an agonist with decreased activity, an inactive analog, or a powerful tuftsin antagonist. It should be appreciated that the chances of encountering the tetrapeptide in another protein is less than 20^4 (one chance in 160,000). This is obtained on the premise that all 20 amino acids occur at the same frequency. However, the frequency cannot be greater than the least frequent amino acid. In tuftsin, all residues are below average frequency. Consequently, the occurrence of this sequence in p12 of R-MuLV strongly suggests a more general biological role. Since tuftsin has been implicated in membrane function (phagocytosis, chemotaxis, and motility), one could presume, as Luftig *et al* (1977) did, that the tetrapeptide in p12 has a membrane function that somehow leads to increased virion production.

It is of interest that the tripeptide (Thr-Lys-Pro, representing the first three residues, occurs in histone IV (Ogawa *et al.* 1969) and human C reactive protein (Osmand *et al.*, 1977).

7.6. SPECIFICITY OF THE BIOLOGICAL EFFECT OF TUFTSIN

As mentioned earlier, the tetrapeptide is released from leukokinin which binds specifically to polymorphonuclear granulocytes. These cells promptly cleave off the peptide to render to available for phagocytic stimulation. In view of this, it would be expected that the tuftsin effect would be specific towards granulocytes, and indeed it is. It has no detectable effects on the erythrocyte, platelet, or lymphocyte. However, it does stimulate the phagocytic activity of lung and peritoneal macrophages of mice and guinea pigs (Constantopoulos and Najjar, 1972). This stimulatory effect crosses the species barrier and is exerted on all phagocytic granulocytes from human, dog, rabbit, guinea pig (Najjar, 1974, 1977; Fridkin *et al.*, 1976, 1977) and cows (Erp and Fahrney, 1975). It should be stressed at this point that the structure of tuftsin is highly specific. As we shall see below, any alteration or replacement of the residues, even with homologous amino acids, either lowers the activity or renders it inhibitory.

8. BIOLOGICAL ACTIVITY OF TUFTSIN ANALOGS *IN VITRO*

Several analogs have been synthesized and tested for biological activity. In our laboratory, only a few were prepared (Najjar, 1974). In order to determine whether any of the terminal residues were redundant, we synthesized Thr-Lys-Pro, Lys-Pro-Arg and Lys-Pro-Pro-Arg. These were completely inactive indicating that both the threonyl and the arginyl residues are essential for the activity of the tetrapeptide. The addition of another proline residue next to that present in tuftsin yielded a very powerful inhibitor. Thr-Lys-Pro-Pro-Arg was completely inhibitory to the motility of the cell and to its phagocytic activity even in the presence of threefold excess of tuftsin (Nishioka *et al.*, 1973b; Satoh *et al.*, 1972). In fact, impurities arising during synthesis of tuftsin have elution characteristics on Aminex columns very close to those of tuftsin. These impurities are very strongly inhibitory to the biological activity of tuftsin.

Fridkin *et al.* (1976, 1977) prepared several analogs that were fully characterized by electrophoretic, chromatographic mobility, and optical rotation at the [α] D line. These were Lys-Pro-Arg, Thr-Lys-Pro-Arg(NO$_2$), Ala-Lys-Pro-Arg, Lys-Lys-Pro-Arg, Ser-Lys-Pro-Arg, Val-Lys-Pro-Arg, N^α-Acetyl-Thr-Lys-Pro-Arg, *p*-aminoacetyl-Thr-Lys-Pro-Arg, and Tyr-Thr-Lys-Pro-Arg. Where the amino terminal was replaced by Lys or Ser and tested with the yeast cell assay, there was some activity, approximately 20–30% of the value obtained with tuftsin. Most of the rest were inhibitory. [ω-NO$_2$Arg^4]-Tuftsin were inert (Tyr-tuftsin was not tested). With the exception of [Lys^1]-tuftsin and [ω-NO$_2$^{4}]-tuftsin, these analogs inhibited tuftsin activity to various extents. When the biological activity was assayed with the reduction of nitroblue tetrazolium, none were able to activate the hexosemonophosphate shunt and reduce the dye, while tuftsin was effective in the same test under the same conditions. Two analogs, Lys-Pro-Arg and Ala-Lys-Pro-Arg were inhibitory to the dye reduction activity of the tetrapeptide. Thus with nitroblue tetrazolium, tuftsin uniquely stimulated dye reduction (Spirer *et al.*, 1975).

Konopinska *et al.* (1976) synthesized several oligopeptides and compared their activity with that of tuftsin using staphylococcus as target particle. The carboxy terminal Arg was replaced by Ala or Lys. Both showed approximately half the activity of tuftsin. Similarly reduced activity was obtained where Thr was replaced by Val, Pro by Ala, Lys by Ala, and Lys by Orn as well as, in some peptides, where two or more residue replacements were made, such as Thr-Ala-Ala-Arg, Thr-Orn-Pro-Ala, and Thr-Ala-Arg-Lys. Almost inactive were Thr-Pro-Lys-Ala, Thr-Lys-Lys-Ala, Thr-Lys-Ala-Ala, and Thr-Gly-Gly-Lys.

It should be obvious that the net positive charge as such does not relate to the stimulatory activity. Furthermore, oligopeptides of lysine with a varying number of residues 3, 4, 6, and 20 were devoid of any activity (Fridkin *et al.*, 1977). Based on their results, Konopinska *et al.* (1976) proposed a conformation for tuftsin or its active analogs in solution that appears essential for some level of activity. This is derived from the knowledge, obtained from X-ray (Ueki *et al.*, 1971), infrared spectroscopy (Deber, 1974), and nuclear magnetic resonance studies (Kopple *et al.*, 1975), that tetrapeptides tend to form easily a 4 → 1 hydrogen bonded β-turn which tuftsin can perform. This is shown in Figure 3. Note the hydrogen bonding between the nitrogen of the peptide bond of Arg and the carbonyl group of Thr. In addition, the figure indicates that the structure can be stabilized by ionic bonding between the −COO$^-$ of Arg and NH$_3$$^+$ of Thr. However, nmr studies by Blumenstein *et al.* (1979) did not reveal a β turn.

A summary of the various analogs of tuftsin and their activity is found in Table 2.

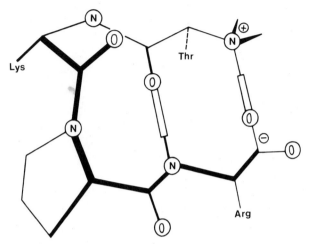

FIGURE 3.   Possible conformation of tuftsin in solution (Konopinska *et al.*, 1976).

9.   BIOLOGICAL ACTIVITY OF TUFTSIN *IN VIVO*

9.1.   ABSENCE OF SYSTEMIC EFFECTS AFTER TUFTSIN ADMINISTRATION

The primary interest in the pursuit of studies of tuftsin has been the expectation that by stimulating phagocytosis in blood and tissues, it could function as a general antiinfectious agent. It would not be restricted to combating any particular bacterial disease, but would increase immunity to infectious bacteria through the general stimulation of phagocytes (Najjar, 1977) and consequent enhancement of their destruction through activation of the hexosemonophosphate shunt (Spirer *et al.*, 1975). Consequently, it is important to assess the systemic and toxic effects that the tetrapeptide might have incident to its therapeutic use. The tetrapeptide is a highly basic compound, and it was feared that it might possess vasoactive properties akin to those of bradykinin, another basic peptide which has very marked vasomotor effects. This did not prove to be the case. Tuftsin was injected intravenously in rats up to the amount of 25 mg per kilogram of body weight. Blood pressure, heart rate, respiratory rate, and electrocardiograms were recorded. There was no change whatsoever in any of these parameters (Najjar, 1974).

9.2.   ANTIINFECTIOUS EFFECTS OF TUFTSIN

9.2.1.   Bactericidal Effects in Normal Mice

Martinez *et al.* (1977) studied the bactericidal activity of macrophages from the peritoneal cavity of mice injected intraperitoneally with *Listeria monocytogenes*. This was done with and without the injection of tuftsin at 10 or 20 mg per kilogram of body weight. It was found that after 15 min of incubation the control yielded 5% killing of bacteria, whereas in the presence of tuftsin, 50% and 69% of the injected bacteria were killed at levels of 10 and 20 mg of tuftsin,

TABLE 2. THE BIOLOGICAL ACTIVITY OF TUFTSIN AND ITS ANALOGS AS MEASURED WITH VARIOUS PHAGOCYTIC CELLS

Tuftsin analogs	Relative activity[a]	Biological assay: phagocytosis		Dye reduction	References
		Granulocytes	Macrophages		
H-Thr-Lys-Pro-Arg-OH	1.0	Human, dog, rabbit, bovine, guinea pig	—	—	Constantopoulos and Najjar (1972); Nishioka et al. (1973); Najjar (1974); Erp and Fahrney (1975); Fridkin et al. (1976); Konopinska et al. (1976, 1977); Inada et al. (1977)
	1.0	—	Mice peritoneal, rabbit lung	—	Constantopoulos and Najjar (1972); Martinez et al. (1977)
	1.0	—	—	Nitroblue tetrazolium	Spirer et al. (1975); Fridkin et al. (1976)
H-Thr-Lys-Pro-Lys-OH	0.3–0.5	Human, dog, guinea pig	—	—	Rauner et al. (1976); Konopinska et al. (1976, 1977)
H-Lys-Pro-Arg-OH	0.0	Human, dog	—	—	Nishioka et al. (1973a)
	Inhibition[b]	—	—	Nitroblue tetrazolium	Fridkin et al. (1977)
H-Thr-Lys-Pro-OH	0.0	Human, dog	—	—	Nishioka et al. (1972, 1973b)
H-Lys-Pro-Pro-Arg	Inhibition	Human, dog	—	—	Najjar (1976)
H-Thr-Lys-Pro-Pro-Arg	Inhibition	Human, dog	—	—	Nishioka et al. (1972, 1973b)
H-Lys-Lys-Pro-Arg-OH	0.0–0.3	Human	—	—	Fridkin et al. (1976, 1977)
H-Ser-Lys-Pro-Arg-OH	0.0–0.2 Inhibition	Human	—	—	Fridkin et al. (1976, 1977)
H-Tyr-Lys-Pro-Arg-OH	0.0	Human	—	—	Fridkin et al. (1976, 1977)
H-Val-Lys-Pro-Arg-OH	Inhibition	Human	—	—	Fridkin et al. (1976, 1977)
	0.0	—	—	Nitroblue tetrazolium	Fridkin et al. (1976)

Analog	Relative activity[a]		Method	Reference
H-Ala-Lys-Pro-Arg-OH	0.5	Guinea pig	—	Konopinska et al. (1976)
	Inhibition	Human	—	Fridkin et al. (1977)
	Inhibition	—	Nitroblue tetrazolium	Fridkin et al. (1977)
H-Thr-Lys-Pro-Ala-OH	0.7	Guinea pig	—	Konopinska et al. (1976, 1977)
H-Thr-Lys-Pro-Arg(NO₂)	0.0	Human	—	Fridkin et al. (1976, 1977)
	0.0	—	Nitroblue tetrazolium	Fridkin et al. (1976)
H-Thr-Lys-Ala-Arg-OH	0.5	Guinea pig	—	Konopinska et al. (1976, 1977)
H-Thr-Orn-Pro-Arg-OH	0.5	Guinea pig	—	Konopinska et al. (1976, 1977)
H-Thr-Orn-Pro-Ala-OH	0.4	Guinea pig	—	Konopinska et al. (1976, 1977)
H-Thr-Ala-Val-Arg-OH	0.7	Guinea pig	—	Konopinska et al. (1976, 1977)
H-Thr-Ala-Arg-Lys-OH	0.6	Guinea pig	—	Konopinska et al. (1976, 1977)
H-Thr-Gly-Gly-Lys-OH	0.1	Guinea pig	—	Konopinska et al. (1976, 1977)
H-Thr-Lys-Lys-Ala-OH	0.7	Guinea pig	—	Konopinska et al. (1976, 1977)
H-Thr-Lys-Ala-Ala	0.1	Guinea pig	—	Konopinska et al. (1976, 1977)
Acetyl-Thr-Lys, Pro-Arg-OH	Inhibition	Human	—	Fridkin et al. (1977)
	0.0	—	Nitroblue tetrazolium	Fridkin et al. (1977)
p-Aminophenylacetyl-Thr-Lys-Pro-Arg-OH	Inhibition	Human	—	Fridkin et al. (1977)
	0.0	—	Nitroblue tetrazolium	Fridkin et al. (1977)
(Lys)₃, ₄, ₆, ₂₀	0.0	—	Nitroblue tetrazolium	Fridkin et al. (1977)

[a] Relative activity of 1.0 is taken as the activity obtained in the particular laboratory for the tetrapeptide tuftsin. That obtained by the analog is the fraction obtained relative to tuftsin in the same laboratory.
[b] Inhibition denotes that the analog inhibits that particular activity of tuftsin obtained with the method indicated.

respectively. After 30 min, 63% and 72% were killed with the two respective doses of tuftsin. The full course of 60 min is shown in Figure 4.

9.2.2. Bactericidal Effects in Leukemic Mice

It was shown that the effect of tuftsin bacterial killing in leukemic mice was very good even though the level did not approach that obtained in normal mice. Furthermore, the livers and spleens of control leukemic mice were incapable of checking the growth of bacteria 1 hr after injection, since in most cases, depending on the bacterium, cell number increased sometimes over twofold of the count at zero time. However, following the injection of tuftsin, 20 mg·kilo^{-1} to leukemic mice the bacterial count fell to 57% of that of the control. This is, again, a high rate of bactericidal activity caused by the administration of tuftsin.

9.2.3. Blood Clearing Effect of Tuftsin *in Vivo*

The rate of blood clearing in normal mice is depicted in Figure 5. The figure shows the clearing of *Staphylococcus aureus*. Blood was withdrawn at intervals and scored for viable cells. Injections of tuftsin were made intraperitoneally (i.p.) immediately after intravenous injections of bacteria. Blood was then withdrawn at various times from 0–60 min. It is quite obvious that the rate of clearing upon injection of 10 mg or 20 mg·kilo^{-1} body weight in normal mice was noticeably

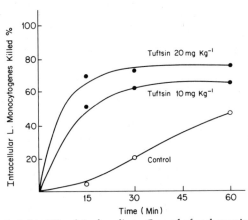

FIGURE 4.   The bactericidal activity of tuftsin. Time course of the effect of tuftsin on the intracellular bactericidal activity of mouse peritoneal macrophages against *Listeria monocytogenes* (Martinez *et al.*, 1977).

One milliliter of a suspension of *L. monocytogenes* $1–2 \times 10^6 \cdot ml^{-1}$ in gelatin Hanks' medium with 10% fetal serum containing an amount of tuftsin corresponding to 0, 10, 20 mg·kilo^{-1} of body weight, was injected intraperitoneally (i.p.) into mice. After 5 min, the animal was sacrificed and 2 ml of saline solution was injected i.p. to harvest the macrophages. Cell suspensions were washed twice with cold gelatin Hanks' solution and centrifuged at 110g for 4 min. The number of macrophages was adjusted to $6–8 \times 10^6 \cdot ml^{-1}$ of medium. One ml of each was incubated at 37°C. At intervals thereafter of 0, 15, 30, and 60 min, the viable bacteria were determined. For this purpose, the cells were centrifuged, 4 min at 110g and 1 ml of distilled water was added to the sedimented cells containing cold bovine serum albumin. This was then frozen (-170°C) and thawed (37°C) 3 times, and counted. Appropriate dilutions were then plated and viable bacteria scored.

The figure represents a plot of the percent of bacteria killed as a function of time, for control, and for tuftsin at 10 mg and 20 mg·kg^{-1} of body weight. This was adapted from a table of values given by Martinez *et al.* (1977). The table included complete bacterial counts. Not shown in the graph, is the count at zero time. The control without tuftsin yielded a count of intracellular bacteria per 10^6 macrophages of 35,550, at the beginning of incubation, 5 min after i.p. injection of *L. monocytogenes*. By contrast, the corresponding values following tuftsin injection were 60, 442 and 61,515 at 10 and 20 mg·kg^{-1}, respectively. This indicates the stimulation of phagocytosis *in vivo* by tuftsin.

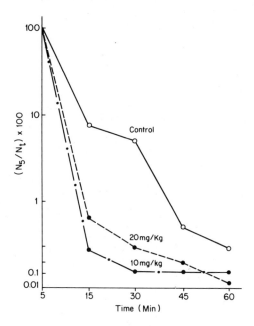

FIGURE 5. Effect of tuftsin on bacterial clearing (Martinez *et al.*, 1977). Mice of strain CD-1 were injected intravenously (i.v.) with 0.25 ml of a culture of *S. aureus*. This was followed by an i.p. injection of tuftsin at 10 mg and 20 mg·kg^{-1} body weight. At the indicated times, blood was obtained by cardiac puncture. The blood samples of the three animals were mixed, treated appropriately (as in Figure 4) and inoculated on nutrient agar, and counted. The number of bacteria was calculated on the basis of counts·ml^{-1} of blood. This type of experiment was repeated three times.

It was possible to assess the number of bacteria in the liver and spleen of these animals. In such cases, four animals were sacrificed by cervical dislocation after 5, 10, and 60 min. The organs were removed, homogenized in isotonic NaCl. The suspension was diluted and grown on nutrient agar and colonies counted. (For details see legend to Figure 4.)

greater than the control without tuftsin. Other bacteria were tested (*Listeria monocytogenes, Escherichia coli,* and *Serratia marcescens*) and yielded the same overall result.

Blood clearing of bacteria in leukemic mice was also studied. Tuftsin administration definitely stimulated bacterial clearing compared to the control; however, the rate of clearing was still lower than that obtained in normal mice.

9.2.4. Tuftsin Effect on Animal Survival

Based on animal survival, it was found (Martinez, 1976) that when mice were injected with near-lethal doses of pneumococcus organisms only 10% of the control animals survived while fully 50% survived when given tuftsin 20 mg·kilo^{-1} i.p. However, when mice were inoculated with myxovirus, the opposite was true. About 40% of the control survived, whereas only 10% of the tuftsin-injected mice survived. The latter results may be explained on the basis that the phagocytes were incapable of coping with the virus as they were with the pneumococcus, and the increased fatality may result from increased virus uptake by the cells. However, the possibility that virus production was augmented is equally reasonable in view of the results obtained by Luftig *et al.* (1977).

10. EFFECT OF TUFTSIN ON THE IMMUNOGENIC FUNCTION OF MACROPHAGES

In a recent cooperative study led by Feldman, Fridkin, and Spirer, it was reported (Tzehoval *et al.*, 1978) that tuftsin augments considerably the antigen-specific macrophage-dependent "education" of T lymphocytes.

Adherant macrophages were incubated with keyhole limpet hemocyanin

(KLH) 50 μg/ml at 37°C for 4 hr with or without tuftsin at concentrations of 1–20 × 10^{-8} M. Excess antigen was then removed by thorough washing. Spleen cells were then added to the antigen-fed macrophage monolayer and incubated overnight. The nonadherent cells were reseeded on fresh dishes to allow the adherence of any contaminating macrophages. The free cells were then exposed to X-ray (1000 rads) and injected into the footpads of syngeneic mice. After 7 days, a cell suspension was prepared from the popliteal lymph nodes and cultured in an appropriate medium for 3 days with and without KLH antigen. At this point, 2 μCi of tritiated thymidine were added. After 4 hr of incubation, the incorporated radioactivity was measured in a scintillation counter.

It was found that tuftsin produced a significant, concentration-dependent stimulation of the immunogenic function of macrophages. A maximal stimulatory activity of 700% of control was attained at 1.6 × 10^{-8} M per 20 × 10^6 macrophages. This activity could be obtained only when tuftsin is presented at the same time of antigen processing. There was also a clear demonstration that this high stimulatory potency of tuftsin is specific and depended on the integrity of the amino acid sequence. Modification of either terminal leads to reduction loss or inhibition of macrophage activation. Replacement of threonine by serine or its deletion resulted in an inhibitory activity. Substitution of the guanido function of the arginine residue with methyl or a nitro group, or its replacement by lysine resulted in an inactive or inhibitory analog. However, amidation of arginine or its replacement by D-arginine did not affect activity. However, augmentation of activity was obtained if glycine was added to the carboxy terminal, or if alanine replaces threonine at the amino terminal. Furthermore, Thr-Lys-Pro, Thr-Lys-Arg, and Thr-Lys-Leu-Arg greatly reduced or abolished activity altogether.

11. EFFECT OF TUFTSIN ON THE ACTIVATION AND THE TUMORICIDAL ACTIVITY OF MACROPHAGES

One of the major subjects of inquiry in the reticuloendothelial system has been the killing effect of nonspecifically activated macrophages on tumor cells. The activating effect of tuftsin on macrophages is exerted in quantities rivaling hormonal concentrations. We are currently engaged in studying the effect of tuftsin and its analogs on the ability of macrophages to destroy cancer cells. My former colleague already has preliminary evidence showing a definite effect on cancer cells. Dr. Nishioka (personal communications) conducted experiments both *in vivo* and *in vitro*. Ten control mice received 1 × 10^5 L1210 leukemia cells by intraperitoneal injection. The experimental group (10 mice) was injected with the same number of cells, followed by 0.2 μg of tuftsin. While all mice which received tumor cells alone died in 11 days (mean survival 9.2 days), those that received leukemia cells plus tuftsin showed 50% survival for over 30 days.

Our results with tuftsinyltuftsin are also encouraging (Najjar *et al.*, 1980). Following the i.p. injection of 5 × 10^3 L1210 cells and 20 μg of tuftsinyltuftsin per mouse, after 21 days, 25 control mice receiving saline showed 48% survival, whereas 25 mice receiving the octapeptide showed 68% survival. The surviving mice, 11 and 17 respectively, were further injected with 1×10^5 L1210 cells. On day 35, 17% of the peptide-treated mice and 50% of the control died, giving an overall survival for the total experiment of 24% for the control and 56% for the peptide-injected. In another experiment where only 1×10^4 L1210 cells were injected only once, there was an overall survival of 20% for the control group and 80% for the peptide-treated group.

In vitro cytotoxicity enhancement was shown by Nishioka using [³H]proline release assay. Peritoneal exudate cells of DBA mice and L1210 leukemia cells were incubated at a ratio of 50:1 respectively. Tuftsin 10 μg/ml was used to activate the macrophages. Activated macrophages showed statistically significant enhanced cytotoxicity (32%) as compared to the control.

Cytotoxicity enhancement of purified human neutrophils by tuftsin was also examined with an established human melanoma cell line, 26-5. With [³H]proline-labeled L1210 cells, in the range of tuftsin concentration of 2–20 μg, 13–23% of cytotoxicity enhancement by tuftsin has been obtained.

12. TUFTSIN RECEPTOR SITES

Fridkin, Spirer, and their colleagues prepared radioactive tuftsin through the incorporation of tritium-labeled arginine. This was done by coupling N-hydroxysuccinimide ester of the protected tripeptide Thr-Lys-Pro (Stabinsky *et al.*, 1978a) to radioactive arginine. By so doing, Stabinsky *et al.* (1978b) were able to study the binding of tuftsin to human blood cells. They showed that binding to polymorphonuclear (PMN) leukocytes and monocytes was specific, rapid, saturable, and reversible. Dissociation constants (K_d) for these phagocytes were 130 and 125 nM, with binding sites per cell approximately 50,000 and 100,000 respectively. The K_d values are remarkably close to the K_m observed for phagocytosis (Constantopoulos *et al.*, 1972; Fridkin *et al.*, 1977) (See Figure 1). Lymphocytes showed only threshold binding and no binding could be detected to erythrocytes.

Fudenberg's groups (Nair *et al.*, 1978) also studied binding of labeled tuftsin with ¹⁴C or ¹²⁵I by appropriate addition of labeled residues at either C or N terminals. They found that neutrophils bound 72 ± 10% of labeled tuftsin in the incubation medium. Incorporation of the label at the N terminal reduced binding to 10%. The binding was specific. Equally specific binding was obtained for monocytes and lymphocytes, although the extent of binding was lower than that obtained for PMN leukocytes. In a study of the effect of tuftsin on leukocyte migration, Horsmanheimo *et al.* (1978) showed that, with the agarose migration test, monocyte migration was significantly enhanced. No effect on PMN migration was observed. However, Nishioka (personal communication) observed a definite stimulation using the same test. This contrasts with the large stimulation of PMN migration in glass tubes so readily demonstrable (Nishioka *et al.*, 1973a). It is quite possible that small impurities in the tuftsin samples used might account for the differences observed by the two laboratories (see below).

13.   CONGENITAL TUFTSIN DEFICIENCY

This is a functional deficiency due to the presence of a mutant tuftsin peptide (Constantopoulos *et al.*, 1972; Constantopoulos and Najjar, 1973b; Najjar, 1975, 1977; Inada *et al.*, 1977). This peptide has not yet been identified chemically. However, it has been characterized by its ability to inhibit the biological effect of the normal tetrapeptide tuftsin *in vitro*, and it appears to possess physical characteristics similar to Thr-Glu-Pro-Arg- (Najjar *et al.*, 1980).

13.1.   DIAGNOSTIC CHARACTERISTICS

13.1.1.   Clinical

The symptoms and signs vary considerably between patients, as well as with the age of the patient. The manifestations arise from a general inability to cope with infections. In its severest form it occurs in early childhood. There is widespread skin infection with an eczematous rash throughout. Draining lymph glands are large and suppurative. There is a persistent upper respiratory infection. Pneumonitis and bronchitis are not uncommon. The microorganisms involved are usually staphylococcus, pneumococcus, and monilia.

In the adult, the disease is in its mildest form. Repeated mild and localized skin infections may occur. There is often a high frequency of upper respiratory infection that is sometimes of long duration. In general, these are mere exaggerations of the usual infections that occur in the population. Although it is expected (see below) that infections such as pneumonitis, meningitis, septicemia, etc., may be very severe and protracted, these have been rarely encountered in the few families reported (Najjar, 1977; Inada *et al.*, 1977). In order to verify this aspect, extensive studies should be carried out on a large number of cases with severe generalized infection with the object of obtaining a correlation index between tuftsin deficiency and severe infections as has been repeatedly observed between the severity of infections and the absence of the spleen (Erikson *et al.*, 1968; Kevy *et al.*, 1968; King and Shumaker, 1952; Eraklis *et al.*, 1967).

There have been to date seven families involved. Six were identified, of which four have been reported by Constantopoulos and his colleagues (Najjar, 1977), primarily in Massachusetts. One family in Morioka, Iwate, Japan was described by Inada *et al.* (1977). In each case, the father or the mother show the deficiency by laboratory analysis. The deficient parents give a mild history of repeated infection or none at all. One case of intermittent thrombocytopenic purpura with repeated severe infections having either primary tuftsin deficiency or due to splenic impairment, has been identified by radioimmunoassay by Spirer *et al.* (1977a,b); assays on different occasions showed absence of tuftsin. The patient gave a history of recurrent infections of the skin and lungs with no previous evidence of imparied cellular or humoral immunity.

Other than the knowledge that the deficiency is familial, little can be said of the genetic aspect. The number of families is small and the methodology used to detect the deficiency has not been quantitative enough to identify heterozygotes. The symptomatology is not sufficiently unique to identify the disease among grandparents and great-grandparents.

13.1.2. Laboratory Findings

There are two important findings that render a diagnosis likely: (1) There should be no tuftsin activity in the patients serum or γ-globulin extracts; (2) the extracted peptide may show strong inhibitory effects on the activity of normal serum extracts or synthetic tuftsin. Thus far all the cases studied, our own (Najjar, 1977) and that of Inada *et al.* (1977) in Japan, have shown a strong inhibition of the biological action of tuftsin. However, it is likely that extracts of some mutant patients may not be inhibitory although the majority might be, as demonstrated by the fact that most analogs show an inhibitory effect (Fridkin *et al.*, 1976, 1977).

14. ACQUIRED TUFTSIN DEFICIENCY

Deficiency in leukokinin was observed in dogs (Najjar *et al.*, 1968) and rats (Likhite, 1975) following experimental splenectomy. Since leukokinin deficiency denotes tuftsin deficiency, a survey of deficiency in splenectomized human patients was undertaken. It has been known for sometime that children with splenectomy had a high rate of severe infections (Erikson *et al.*, 1968). This was equally true in splenectomized animals (Shinefield *et al.*, 1966). Similarly, patients in whom splenic function has been curtailed also exhibit a high rate of severe infections. These are examplified by sickle cell anemia (Seeler *et al.*, 1972) and myelogenous leukemia (Constantopoulos *et al.*, 1973a). These cases show an absence of tuftsin activity (Najjar *et al.*, 1972; Constantopoulos *et al.*, 1973b). However, they do not possess an inhibitory peptide as does congenital tuftsin deficiency. In the absence of a functional spleen, tuftsin cannot be cleaved at its carboxy terminal arginine. Consequently, leukokininase is incapable of releasing a free tetrapeptide even if it could cleave tuftsin at the threonine amino peptide bond. Thus, while a normal tetrapeptide exists in the leukokinin of such cases (Satoh *et al.*, 1972), it cannot be released, a necessary step in order to exercise its full function.

Assay of the level of tuftsin in cases after elective splenectomy revealed virtual absence of any activity (Najjar and Constantopoulos, 1972; Constantopoulos *et al.*, 1973b; Najjar, 1975, 1976). More accurate measurements by radioimmunoassay (Spirer *et al.*, 1977a,b) confirmed this fact. The serum levels of tuftsin in 20 cases of elective splenectomy (118 ± 7.89 ng·ml^{-1}) were significantly lower than in 35 normal subjects (255.71 ± 10.27 ng·ml^{-1}). Similarly, in cases where the spleen, though not removed, shows evidence of severe impairment of function through thrombosis or leukemic infiltration the levels of tuftsin activity are either low or absent (Constantopoulos *et al.*, 1973a).

We have observed that long after splenectomy, there is no obvious deficiency in any of the γ-globulin phosphocellulose fractions (I–IV), (Judge and Najjar, 1972; Najjar, 1975) complement components 3' and 5', as well as various immunological components of γ-g and γ-M (Najjar and Constantopoulos, 1972). The only true difference in the γ-globulin resulting from the absence of the spleen is that there is no leukokinin-S, i.e., tuftsin is not cleaved at the Arg-Gln bond. Consequently, it is not surprising that analysis of γ-globulins from normal and splenectomized subjects for the presence of the tetrapeptide yield values that are quite similar (Table 3). It is of interest that the content of tuftsin in γ-globulin varies only slightly from one individual to another both in normals and splenectomized subjects.

Subjects who had experienced rupture of the spleen which necessitated splenectomy appear to have leukokinin-S with normal tuftsin levels (Constantopoulos *et al.*, 1973b). The reason may be that following traumatic rupture, splenic tissue spills into the peritoneal cavity where it is implanted and grows to various proportions (Cahalane and Kiesselbach, 1970). In a recent study of splenectomized patients, Spirer *et al.* (1977a,b) showed by radioimmunoassay, that 50 cases of splenectomy after rupture of the spleen had serum levels of 234.34 ± 10.22 ng $\cdot$ ml^{-1}. The latter approaches normal values of 255.71 ± 10.27 ng$\cdot$ml^{-1}, as indicated above.

These findings may indeed explain the frequency of severe infections in cases of elective splenectomy, as compared to their absence in parallel cases after rupture of the spleen.

15. CONCLUSIONS

Based on the work performed in several laboratories, it is possible to arrive at certain conclusions, some of which are tentative:

TABLE 3.   QUANTITATIVE ESTIMATION OF TUFTSIN (THR-LYS-PRO-ARG) IN γ-GLOBULIN OF NORMALS AND SPLENECTOMIZED (ELECTIVE) SUBJECTS[a]

Normals		Splenectomized	
Subjects	nmol/mg	Subjects	nmol/mg
V.N.	5.6	J.M.	5.2
B.M.	7.7	Gl.O.	4.7
A.G.	6.3	Ga.O.	5.3
N.M.	4.7	R.H.	6.3
S.S.	5.5	H.M.	7.1
L.V.	5.7	B.K.	5.3

[a] γ-Globulin was prepared from heat-inactivated fresh serum by three precipitations in 0.33 saturation ammonium sulfate. It was then thoroughly dialyzed against 0.1 M Tris-HCl buffer, pH 8.1. 200 mg of protein were subjected to performic acid oxidation and completely digested with 1% trypsin in the same buffer. The pH was kept at 8.1 throughout by appropriate additions of Tris. It was purified on Dowex followed by Aminex (Nishioka *et al.*, 1972, 1973a). No corrections were made for recovery. However, all samples were run in an identical manner utilizing the same chromatography column throughout.

a. The tetrapeptide (L-threonyl-L-lysyl-L-prolyl-L-arginine) tuftsin is the functional unit of the carrier molecule leukokinin.

b. It stimulates all the known biological activities of the two major phagocytic cells; the granulocyte and the macrophage. Half-maximal stimulation is obtained at approximately 100 nM.

c. Tuftsin displays specific binding sites on the membrane surface of macrophages and granulocytes with a dissociation constant for either cell nearly identical to the Km of its phagocytic stimulation, 125–130 nM.

d. In addition to phagocytic stimulation, it also enhances pinocytosis, cell motility, and cell longevity.

e. Tuftsin also stimulates macrophage bactericidal activity to a pronounced degree.

f. Blood clearing of several types of bacteria is greatly enhanced in mice by the injection of 10–20 mg·kg^{-1} of body weight.

g. A notable activity of tuftsin involves the processing and presentation of antigen. The overall effect is to augment, in a highly specific manner, the macrophage-dependent programing or "education" of T lymphocytes.

h. A preliminary study of the stimulation of the tumoricidal activity of the macrophages by tuftsin appears to be very encouraging. Tuftsin at very low dosage, effected a 50% survival of DBA mice injected with 1×10^5 L1210 leukemia cells. This is quite impressive in view of the "AA" lethality of this strain of leukemia cells.

i. These biological effects of tuftsin can, therefore, be translated into a general antiinfection effect with marked enhancement of the potential for recovery from bacterial diseases.

j. Another activity of tuftsin which was not predictable is the stimulation of production in tissue culture of Rauscher murine leukemia virus. In this connection, it is of particular interest that one of the viral proteins p12 displays the complete tuftsin sequence 9–12 residues from the amino terminal. This is unlikely to be a random occurrence, and it could have biological meaning perhaps related to membrane budding and excretion of virus.

k. The high specificity of tuftsin resides in its structure. Any substitution or derivatization so far attempted results mostly in loss of activity or marked reduction, and, in many instances, inhibition of tuftsin activity in any of the parameters so far studied.

l. Finally, the conversion of leukokinin to leukokinin-S by a spleen enzyme that hydrolyzed the tuftsinyl–glutamyl bond (-Arg-Glu-) followed by leukokininase action represents yet another form of biological posttranslational modification of protein.

m. Based on all the considerations enumerated above, it is our contention that tuftsin appears to be a membrane-active peptide. If so, it would predictably stimulate most if not all functional activities of cell membranes.

REFERENCES

Baehner, R. L., and Nathan, D. G., 1968, Quantitative nitroblue tetrazolium test in chronic granulomatous disease, *N. Engl. J. Med.* **278**:971.

Blumenstein, M., Layne, P. P., and Najjar, V. A., 1979, Nuclear magnetic resonance studies on the structure of the tetrapeptide tuftsin, L-threonyl-L-lysyl-L-prolyl-L-arginine, and its pentapeptide analogue L-threonyl-L-lysyl-L-prolyl-L-arginine, *Biochemistry* **18**:5247.

Cahalane, S. F., and Kiesselbach, N., 1970, The significance of the accessory spleen, *J. Pathol.* **100**:139.

Chaudhuri, M. K., and Najjar, V. A., 1978, The solid phase synthesis of tuftsin and its analogs, *Anal. Biochem.* **95**:305.

Constantopoulos, A., and Najjar, V. A., 1972, Tuftsin, a natural and general phagocytosis stimulating peptide affecting macrophages and polymorphonuclear granulocytes, *Cytobios* **6**:97.

Constantopoulos, A., and Najjar, V. A., 1973a, The requirement for membrane sialic acid in the stimulation of phagocytosis by the natural tetrapeptide, tuftsin, *J. Biol. Chem.* **248**:3819.

Constantopoulos, A., and Najjar, V. A., 1973b, Tuftsin deficiency syndrome a report of two new cases, *Acta Paediatr. Scand.* **62**:645.

Constantopoulos, A., and Najjar, V. A., 1974, The physiological role of the lymphoid system. The binding of autologous thrombophilic γ-globulin to human platelets, *Eur. J. Biochem.* **41**:135.

Constantopoulos, A., Najjar, V. A., and Smith, J. W., 1972, Tuftsin deficiency; A new syndrome with defective phagocytosis, *J. Pediatr.* **80**:564.

Constantopoulos, A., Likhite, V., Crosby, W. H., and Najjar, V. A., 1973a, Phagocytic activity of the leukemic cell and its response to the phagocytosis stimulating tetrapeptide, tuftsin, *Cancer Res.* **33**:1230.

Constantopoulos, A., Najjar, V. A., Wish, J. B., Necheles, T. H., and Stolbach, L. L., 1973b, Defective phagocytosis due to tuftsin deficiency in splenectomized subjects, *Am. J. Dis. Child.* **125**:663.

Deber, C. M., 1974, Evidence for β-turn analogs in proline peptides in the solid state. An infrared study, *Macromolecules* **7**:47.

Edelman, G. M., Cunningham, B. A., Gall, W. E., Gottlieb, P. D., Rutishauser, U., and Waxdal, M. J., 1969, The covalent structure of an entire γG immunoglobulin molecule, *Proc. Natl. Acad. Sci. USA* **63**:78.

Eraklis, A. J., Kevy, S. V., Diamond, L. K., and Gross, R. E., 1967, Hazard of overwhelming infection after splenectomy in childhood, *N. Engl. J. Med.* **276**:1225.

Erikson, W. D., Burgert, E. O., and Lynn, H. B., 1968, The hazard of infection following splenectomy in children, *Am. J. Dis. Child.* **116**:1.

Erp, E. E., and Fahrney, D., 1975, Chromatographic characterization and opsonic activity of bovine erythrophilic and leukophilic γ-globulins, *Arch. Biochem. Biophys.* **168**:1.

Fidalgo, B. V., and Najjar, V. A., 1967a, The physiological role of the lymphoid system. III. Leucophilic γ-globulin and the phagocytic activity of the polymorphonuclear leucocyte, *Proc. Natl. Acad. Sci. USA* **57**:957.

Fidalgo, B. A., and Najjar, V. A., 1967b, The physiological role of the lymphoid system. VI. The stimulatory effect of leucophilic γ-globulin (leucokinin) on the phagocytic activity of human polymorphonuclear leucocytes, *Biochemistry* **6**:3386.

Fidalgo, B. V., Katayama, Y., and Najjar, V. A., 1967a, The physiological role of the lymphoid system. V. The binding of autologous (erythrophilic) γ-globulin to human red blood cells, *Biochemistry* **6**:3378.

Fidalgo, B. A., Najjar, V. A., Zukoski, C. F., and Katayama, Y., 1967b, The physiological role of the lymphoid system. II. Erythrophilic γ-globulin and the survival of the erythrocyte, *Proc. Natl. Acad. Sci. USA* **57**:665.

Fridkin, M., Kalir, R., Warshawsky, A., and Patchornik, A., 1975, Chemistry, structure and biology, in: *Peptides* (R. Walter and J. Meienhofer, eds.), pp. 395–401, Ann Arbor Science Publisher, Ann Arbor.

Fridkin, M., Stabinsky, Y., Zakuth, V., and Spirer, Z., 1976, Synthesis, structure-activity relationships and radioimmunoassay of tuftsin, in: *Peptides 1976* (A. Loffet, ed.), pp. 541–549, Editions de l'Université de Bruxelle, Brussels.

Fridkin, M., Stabinsky, Y., Zakuth, V., and Spirer, Z., 1977, Tuftsin and some analgos. Synthesis and interaction with human polymorphonuclear leukocytes, *Biochim. Biophys. Acta* **496**:203.

Horsmanheimo, A., Horsmanheimo, M., and Fudenberg, H. H., 1978, Effect of tuftsin on migration of polymorphonuclear and mononuclear human leukocytes in leukocyte migration agarose test, *Clin. Immunol. Immunopathol.* **11**:251.

Humbert, J. R., Gross, G. P., Vatter, A. E., and Hathaway, W. E., 1973, Nitroblue tetrazolium reduction by neutrophils. Biochemical and ultrastructural effects of methylene blue, *J. Lab. Clin. Med.* **82**:20.

Inada, K., Nemoto, N., Nishijima, A., Wada S., Hirata, M., and Yoshida, M., 1977, A case suspected of tuftsin deficiency, *Proceedings of the International Symposium on Phagocytosis*, Tokyo.

Judge, J. F. X., and Najjar, V. A., 1972, The occurrence of Thr-Lys-Pro-Arg tuftsin in human γ-globulin, unpublished.

Kalir, R., Fridkin, M., and Patchornik, A., 1974, (4-Hydroxy-3-nitro) benzylated polystyrene. An improved polymeric nitrophenol derivative for peptide synthesis, *Eur. J. Biochem.* **42**:151.

Karnovsky, M. L., 1968, The metabolism of leukocytes, *Semin. Hematol.* **5**:156.

Kevy, S. V., Tefft, M., Vawter, G. F., and Rosen, F. S., 1968, Hereditary splenic hypoplasia, *Pediatrics* **42**:752.

King, H., and Shumaker, H. B., Jr., 1952, Splenic studies: I. Susceptibility to infection after splenectomy performed in infancy, *Ann. Surg.* **136**:239.

Konopinska, D., Nawrocka, E., Siemion, I. Z., Szymaniec, S., and Slopeck, S., 1976, Synthetic and conformational studies with tuftsin and its analogues, in: *Peptides 1976* (A. Loffet, ed.), pp. 535–539, Editions de l'Université de Bruxelle, Brussels.

Konopinska, D., Nawrocka, E., Siemion, I. Z., Slopeck, S., Szymaniec, S., and Klonowska, E., 1977 Partial sequences of histones with tuftsin activity, *Int. J. Pept. Protein Res.* **9**:71.

Kopple, K. D., Go, A., and Philipauskas, D. R., 1975, Studies of peptide conformation, evidence for β structures in solutions of linear tetrapeptides containing proline, *J. Am. Chem. Soc.* **97**:6830.

Likhite, V. V., 1975, Opsonin and leukophilic γ-globulin in chronically splenectomized rats with and without heterotopic autotransplanted splenic tissue, *Nature* **253**:742.

Luftig, R. B., Yoshinaka, Y., and Oroszlan, S., 1977, Sequence relationship between Rauscher leukemia virus (RLV) P65–70 (*gag* gene product) and tuftsin as monitored by the P65–70 proteolytic factor, (Abstract v1562), *J. Cell Biol.* **25**:397a.

Martinez, J., 1976, Applications du chlorure de phosphonitrile en synthèse peptidique, Doctoral thesis, Academie de Montpellier, Université des Sciences Techniques de Languedoc, Montpellier.

Martinez, J., and Winternitz, F., 1976, Synthesis of tuftsin with the phosphonitrilic chloride trimer and by the O-nitrophenyl ester procedure (1976), in: *Peptides 1976* (A. Loffet, ed.), pp. 551–553, Editions de l'Université de Bruxelle, Brussels.

Martinez, J., Winternitz, F., and Vindel, J., 1977, Nouvelles synthèses et propriétés de la tuftsine, *Eur. J. Med. Chem. - Chim. Ther.* **12**:511.

Merrifield, R. B., 1964, Solid-phase peptide synthesis. III. An improved synthesis of bradykinin, *Biochemistry* **3**:1385.

Nair, R. M. G., Ponce, B., and Fudenberg, H. H., 1978, Interactions of radiolabeled tuftsin with human neutrophils, *Immunochemistry* **15**:901.

Najjar, V. A., 1974, The physiological role of γ-globulin, *Adv. Enzymol.* **41**:129.

Najjar, V. A., 1975, Defective phagocytosis due to deficiencies involving the tetrapeptide tuftsin, *J. Pediatr.* **87**:1121.

Najjar, V., 1976, The physiological role of membrane γ-globulin interaction, in: *Biological Membranes* (D. Chapman and D. F. H. Wallach, eds.), pp. 191–240, Academic Press, New York.

Najjar, V. A., 1977, Molecular basis of familial and acquired phagocytosis deficiency involving the tetrapeptide, Thr-Lys-Pro-Arg, tuftsin, *Exp. Cell Biol.* **46**:114.

Najjar, V. A., 1980, Biological and biochemical characteristics of the tetrapeptide tuftsin, Thr-Lys-Pro-Arg, in: *Macrophages and Lymphocytes* (M. R. Escobar and H. Friedman, eds.), pp. 131–147, Plenum Press, New York.

Najjar, V. A., and Constantopoulos, A., 1972, A new phagocytosis-stimulating tetrapeptide hormone, tuftsin and its role in disease, *J. Reticuloendothel. Soc.* **12**:197.

Najjar, V. A., and Merrifield, R. B., 1966, Solid phase peptide synthesis. VI. The use of the o-nitrophenylsulfenyl group in the synthesis of the octadecapeptide bradykinylbradykinin, *Biochemistry* **5**:3765.

Najjar, V. A., Fidalgo, B. V., and Stitt, E., 1968, The physiological role of the lymphoid system. VII. The disappearance of leukokinin activity following splenectomy, *Biochemistry* **7**:2376.

Najjar, V. A., Nishioka, K., Constantopoulos, A., and Satoh, P. S., 1972, The function and interaction of erythrophilic γ-globulin with autologous erythrocytes, *Sechste Internationales Symposium über Structur and Funktion der Erythrozyten* (S. M. Rapoport and F. Jung, eds.), pp. 615–627, Akademie Verlag, East Berlin.

Najjar, V. A., Chaudhuri, M. K., Konopinska, D., Beck, B. D., Layne, P. P., and Linehan, L., 1980, Biology of tuftsin: A possible role in cancer suppression and therapy, in: *Augmenting Agents in Cancer Therapy: Current Status and Future Prospects* (M. A. Chirigos and E. Hersh, eds.), Raven Press, New York, (in press).

Nishioka, K., Constantopoulos, A., Satoh, P., and Najjar, V. A., 1972, The characteristics, isolation and synthesis of the phagocytosis stimulating peptide tuftsin, *Biochem. Biophys. Res. Commun.* **47**:172.

Nishioka, K., Constantopoulos, A., Satoh, P. S., Mitchell, W. M., and Najjar, V. A., 1973a, Characteristics and isolation of the phagocytosis-stimulating peptide, tuftsin, *Biochim. Biophys. Acta* **310**:217.

Nishioka, K., Satoh, P. S., Constantopoulos, A., and Najjar, V. A., 1973b, The chemical synthesis of the phagocytosis stimulating tetrapeptide tuftsin (Thr-Lys-Pro-Arg) and its biological properties, *Biochim. Biophys. Acta* **310**:230.

Ogawa, Y., Quagliarotti, G., Jordan, J., Taylor, C. W., Starbuck, W. C., and Busch, H., 1969, Structural analysis of glycine-rich, arginine-rich histone, *J. Biol. Chem.* **244**:4387.

Oroszlan, S., Henderson, L. E., Stephenson, J. R., Copeland, T. D., Long, C. W., Ihle, J. N., and Gilden, R. V., 1978, Amino- and carboxyl-terminal amino acid sequences of murine leukemia virus *gag* gene coded proteins, *Proc. Natl. Acad. Sci. USA* **76**:5010.

Osmand, P., Gewurz, H., and Friedenson, B., 1977, Partial amino-acid sequences of human and rabbit C-reactive proteins: Homology with immunoglobulins and histocompatibility antigens, *Proc. Natl. Acad. Sci. USA* **74**:1214.

Park, B. H., Fikrig, S. M., and Smitwich, E. M., 1968, Infection and nitroblue-tetrazolium reduction by neutrophils, *Lancet* **2**:532.

Patchornik, A., Fridkin, M., and Katchalski, E., 1973, Use of polymeric reagents in the synthesis of linear and cyclic peptides, in: *The Chemistry of Polypeptides* (P. G. Katsoyannis, ed.), pp. 315–333, Plenum Press, New York.

Rauner, R. A., Schmidt, J. J., and Najjar, V. A., 1976, Proline endopeptidase and exopeptidase activity in polymorphonuclear granulocytes, *Mol. Cell. Biochem.* **10**:77.

Saravia, N. G., Derreberry, S., and Robinson, J. P., 1978, The association of serum proteins with peripheral blood cells at low ionic strength. Ultrastructural examination. *Mol. Cell. Biochem.* **20**:167.

Satoh, P. S., Constantopoulos, A., Nishioka, K., and Najjar, V. A., 1972, Tuftsin, threonyl-lysyl-prolyl-arginine, the phagocytosis stimulating messenger of the carrier cytophilic γ-globulin leucokinin, in: *Chemistry and Biology of Peptides* (J. Meinhoffer, ed.), pp. 403–408, Ann Arbor Science Publisher, Ann Arbor.

Seeler, R. A., Metzger, W., and Mufson, M. A., 1972, *Diplococcus pneumoniae* infections in children with sickle cell anemia, *Am. J. Dis. Child.* **123**:8.

Segal, A. W., and Levy, A. J., 1973, The mechanism of the entry of dye into neutrophils in the nitroblue tetrazolium (NBT) test, *Clin. Sci. Mol. Med.* **45**:817.

Shinefield, H. R., Steinberg, C. R., Kaye, D., 1966, The effect of splenectomy on the susceptibility of mice inoculated with *Diplococcus pneumoniae*, *J. Exp. Med.* **123**:777.

Spirer, Z., Zakuth, V., Golander, A., Bogair, N., and Fridkin, M., 1975, The effect of tuftsin on the nitrous blue tetrazolium reduction of normal human polymorphonuclear leukocytes, *J. Clin. Invest.* **55**:198.

Spirer, Z., Zakuth, V., Bogair, N., and Fridkin, M., 1977a, Radioimmunoassay of the phagocytosis stimulating peptide tuftsin in normal and splenectomized subjects, *Eur. J. Immunol.* **7**:69.

Spirer, Z., Zakuth, V., Diamant, S., Mondorf, W., Stefanescu, T., Stabinsky, Y., and Fridkin, M., 1977b, Decreased Tuftsin concentrations in patients who have undergone splenectomy, *Br. Med. J.* **2**:1574.

Stabinsky, Y., Fridkin, M., Zakuth, V., and Spirer, T., 1978a, Snythesis and biological activity of tuftsin and of [O=C-THR¹] tuftsin, *Int. J. Pept. Protein Res.* **12**:130.

Stabinsky, Y., Gotlief, P., Zakuth, V., Spirer, Z., and Fridkin, M., 1978b, Specific binding sites for the phagocytosis stimulating peptide tuftsin on human polymorphonuclear leukocytes and monocytes, *Biochem. Biophys. Res. Commun.* **183**:599.

Thomaidis, T. S., Fidalgo, B. V., Harshman, S., and Najjar, V. A., 1967, The physiological role of the lymphoid system. IV. The separation of γ-globulin into physiologically active components by cellulose phosphate chromatography, *Biochemistry* **6**:3369.

Tzehoval, E., Segal, S., Stabinsky, Y., Fridkin, M., Spirer, Z., and Feldman, M., 1978, Tuftsin (an Ig-associated tetrapeptide) triggers the immunogenic function of macrophages: General implications to activation of programmed cells, *Proc. Natl. Acad. Sci. USA* **75**:3400.

Ueki, T., Bando, S., Ashida, T., and Kakudo, M., 1971, The structure of *o*-bromocarbobenzoxy-glycyl-L-prolyl-L-leucyl-glycyl-L-proline ethyl acetate monohydrate: A substrate of the enzyme collagenase, *Acta Cryst.* **27**:2219.

Vičar, J., Gut, V., Frič, I., and Blaba, K., 1976, Synthesis and properties of tuftsin, L-threonyl-L-lysyl-L-prolyl-L-arginine, *Coll. Czech. Chem. Commun.* **41**:3467.
of tuftsin, *Chem. Pharm. Bull.* **23**:371.

Yajima, H., Ogawa, H., Watanabe, H., Fujii, N., Kurobe, M., and Miyamoto, S., 1975, Studies on Peptides. XLVIII. Application of the trifluoromethanesulphonic acid procedure to the synthesis of tuftsin, *Chem. Pharm. Bull.* **23**:371.

Zervas, L., and Hamalides, C., 1965, New methods in peptide synthesis. II. Further examples of the use of the *O*-nitrophenylsulfenyl group for the protection of amino groups, *J. Am. Chem. Soc.* **87**:99.

4

Carbohydrate Metabolism

RUNE L. STJERNHOLM

1. INTRODUCTION

The circulating leukocyte is a convenient cell for biochemical study of metabolic interactions. With the exception of the red cell and the platelet, leukocytes are the most available of all body cells. In the last two decades numerous new techniques have been developed, which allow isolation of pure populations of cells. Progress has been made in the search for differences between normal white cells and cells under impact of disease. This work has increased our knowledge of the body's defense mechanisms and provided us with new insights into the biochemistry of the white cells, not only in terms of enzymology and chemical composition but also in terms of cell physiology and cellular function.

2. EARLY INVESTIGATIONS

Before 1950 several attempts were made to measure metabolic activity in leukocytes isolated from peritoneal exudates or from whole blood. Respiration, glucose uptake, and lactate production were quantified, but the data were sometimes conflicting if not irreproducible. The main reasons for these difficulties was that the method of isolation many times damaged the cells and the incubation medium used differed greatly between the various investigators. Agglutination and clumping was commonly observed which implicated cell injury and cell death. To overcome these difficulties, artificial buffer systems were introduced which may have kept the leukocyte in single cell suspensions but invariably produced side effects which led to divergent results.

From the early work it was clear that drastic changes in pH, exposure to EDTA, heparin, prolonged centrifugation, or even agitation of cells in nonsiliconized glassware could be harmful to the leukocytes.

RUNE L. STJERNHOLM • Department of Biochemistry, Tulane Medical School, New Orleans, Louisiana 70112.

In spite of these difficulties many interesting observations were made. A very perceptive summary prepared by Esmann (1962) shows that the early investigators reported that leukocytes metabolize glucose at a rate of 2–7 μmol/hr/10^8 cells with concomitant lactate production of 5–16 μmol/hr/10^8 cells. The variations in the results are best explained by the different experimental conditions such as method of isolation of cells, use of suspending medium, leukocyte concentration, time of incubation, and substrate concentration.

3. GLYCOLYSIS

The most thorough study of glycolysis was made by Beck and Valentine (1952a,b) and Beck (1955, 1958a). These investigators used a homogenate of human leukocytes. The K_m and the V_{max} for most of the glycolytic enzymes were determined, and it was concluded that hexokinase and phosphofructokinase were rate limiting and regulated the metabolic flow from glucose to lactate. In several elegant experiments, they demonstrated a stimulation of glycolysis by addition of cofactors and phosphorylated intermediates. Furthermore, the stoichiometric relationship between glucose uptake and lactate production was observed and validated under numerous conditions. Beck was among the first investigators to use ^{14}C-labeled glucose as substrate. Using [U-^{14}C]glucose, Beck demonstrated that 86% of the utilized glucose was recovered as lactate.

Work by Wagner and Yorke (1952) had already demonstrated the phosphorylated intermediates would accumulate during anaerobic incubations of leukocyte homogenates. Under the conditions of these experiments, it was also observed that the inorganic phosphate (P_i) concentration would decrease. Some of the intermediates were identified as glucose-1-P, glucose-6-P, fructose-6-P, fructose diphosphate, and phosphoglyceric acid.

Since leukocytes respire and contain mitochondria it was reasonable to assume that the oxygen uptake was required by the respiratory chain and thus needed for oxidative phosphorylation. It soon became evident that respiration was a far more complex phenomenon than could be explained by a mitochondrial electron carrier system. It was fashionable in the 1950s decade to study both Pasteur and Crabtree effects in all living systems, and many biochemists gave their fair attention to these vexing problems of the white blood cells. The underlying thought was, of course, to learn about control and regulation of substrate metabolism. In addition, it was believed that aberrations of metabolic control mechanisms could give insight or lead to an understanding of leukemic conditions (Luganova and Seits, 1959). The literature is vast, but the results were not easy to interpret. The Pasteur effect is still an unsolved problem and a challenge to any biochemist. Lack of interest in the Pasteur effect, as observed or not observed in leukocytes in health and disease, stems from the fact that simple measurements of respiration, glucose utilization, NAD$^+$/NADH ratios, and P_i levels no longer provide the sophisticated informations needed to solve the problem. The general principle implicated in the Pasteur effect is a sequence of allosteric feedback, involving ATP, ADP, AMP, citrate, glucose-6-P, etc. The targets for allosteric regulation are not only phosphofructokinase, NAD-dependent isocitrate dehydrogenase, and pyruvate kinase, but also hexokinase and/or control of catalyzed transport of glucose across the cell membrane. It should be pointed out that pyruvate kinase regulation is important only in tissues capable of gluconeogenesis, and white blood cells lack the enzymes necessary for the reversal of glycolysis.

From all the experiments measuring O_2 uptake, it is clear that respiration is low in comparison to lactate production. A rough calculation shows the ratio of glycolysis to respiration to be about 25. Thus, only 4% of the glucose utilized is metabolized to CO_2 and water (Esmann, 1962). A similar calculation based on [6-^{14}C]glucose and [1-^{14}C]acetate showed that human neutrophils converted about 2% of glucose utilized to CO_2 and H_2O (Stjernholm et al., 1969).

White blood cells, as we will see later, contain mitochondria and a functional citric acid cycle, but the cells derive the bulk of their energy from glycolysis via substrate phosphorylation. One of the reasons for the poor utilization of the electron carrier system, is the sparse appearance of mitochondria in white blood cells. As a consequence, pyruvate dehydrogenase which is a mitochondrial enzyme becomes rate limiting. Abels et al. (1944) were actually unable to demonstrate pyruvate dehydrogenase activity in either intact leukocytes or homogenates. Perona and Zatti (1959), however, observed inhibition of this activity in the presence of arsenite using neutrophils of guinea pig exudates. The arsenite is an inhibitor of the lipoic acid component of the enzyme complex, as first described by Hager and Gunsalus (1953).

Glyconeogenesis is not expected to occur in leukocytes. The activities of pyruvate carboxylase, PEP-carboxykinase and fructosediphosphatase are minute or nonexistent (Noble et al., 1961).

Glycerol can be metabolized by the white blood cell (Noble et al., 1960; Tibbling, 1970). Although the most logical entry into the glycolytic scheme would involve glycerol kinase and glycerolphosphate dehydrogenase to dihydroxyacetone phosphate, there is evidence that in leukocytes of exudates from the rat, the metabolism involves "free" dihydroxyacetone, which then is phosphorylated to the corresponding phosphate intermediate by the enzyme triokinase (Esmann, 1968).

It is well established that glycolysis is the primary source of energy for leukocytes. It is a curious fact that the rate of glycolysis is different for different types of cells. In general, it can be stated that lymphocytes and related cells utilize less glucose than polymorphonuclear leukocytes and monocytes. This fact has been demonstrated by Beck and Valentine (1952a,b) and confirmed by many other investigators (Hedeskov and Esmann, 1966; Stjernholm, 1967; Stjernholm et al., 1970).

White blood cells established as cell lines in tissue cultures may utilize glucose at a rate which can approach 20–30 μmol/hr/10^8 cells. The classification of these lines by histochemical and immunological techniques or by electron microscopy indicates that they are of lymphoid origin in most cases. Although these cell lines are important research tools, they, nevertheless, hardly resemble the cells occurring in the peripheral blood (Lint, 1973). Macrophages isolated from exudates of guinea pigs also exhibit a high rate of glycolysis approaching 35 μmol/hr/10^8 cells (West *et al.*, 1968).

4. GLYCOGEN METABOLISM

Glycogen is found in all white blood cells, but is most abundant in neutrophils. It probably serves as a store of energy when the supply of glucose or other substrates are limiting.

It was demonstrated by Wagner (1946, 1947, 1950) that the glycogen content of human leukocytes (neutrophils) is about 1% of the wet weight, but that the amount may fluctuate under different conditions. Storage in a glucose-free medium decreases the amount. In conditions such as polycythemia, the glycogen may be 2% of the wet weight, and, in certain cases of glycogen storage disease, the amount may increase to 3%. Lymphocytes contain smaller amounts of glycogen than neutrophils.

Enzymes of glycogen metabolism in white blood cells have been studied quite extensively (Yunis and Arimura, 1964; van der Wende *et al.*, 1964; Luganova and Seits, 1964; Rosell-Perez and Esmann, 1965). Glycogen phosphorylase was assayed in normal leukocytes and in cells from patients with leukemia. Studies were also made of glycogen synthetase. Some of these results indicated that a major portion of phosphorylase activity existed in the active form (phosphorylase a). Glycogen synthetase seems to be present in normal human leukocytes in the D form (absolute dependency on glucose 6-phosphate), and that no conversion of the D to the I form (independent of glucose 6-phosphate) existed under physiological conditions. It was observed earlier by Valentine *et al.* (1953) that chronic myeloid leukemic leukocytes contained lower levels of glycogen than normal leukocytes. This observation would appear not to be related to the maturity of the cells, since leukocytes of leukemoid reactions contain glycogen in concentrations greater than that found in the normal leukocyte. At the present time there are no available data which could explain why leukemic cells contain less glycogen other than lower hexokinase activity (Beck and Valentine, 1952a,b).

Control and regulation of glycogen metabolism are dependent on the activities of the enzymes phosphorylase and glycogen synthetase. The branching and debranching enzymes as well as the α 1→6 glucosidase enzyme are not known to be allosterically modifiable.

Studies of glycogen metabolism using [^{14}C]glucose have been made in order to determine glycogen turnover as well as glycogen pool-size (Stjernholm and Manak, 1970). If the term turnover is defined as the sum of synthesis and degradation, then it can be stated that leukocytes have a glycogen pool which is in a dynamic state or in a state of steady flux. The flux can be changed so that degradation outweighs synthesis if the leukocytes are either starved, stressed, or stimulated to do work such as phagocytosis. Normal human neutrophils exhibit a glucose incorporation into glycogen of about 0.4 μmol/hr/10^8 cells during resting condition. This rate fell to about 0.05 μmol/hr/10^8 cells during phagocytosis (Stjernholm and Manak, 1970).

5. CITRIC ACID CYCLE

The tricarboxylic acid cycle or citric acid cycle has several functions in most animal cells. One important activity is to provide the cell with a catabolic device capable of producing ATP via oxidative phosphorylation. The main metabolic fuel is acetyl CoA which is obtained via oxidative decarboxylation of pyruvate and β-oxidation of fatty acids. However, many other metabolites derived from catabolism of the carbon chains of amino acids are also channeled through the citric acid cycle. Control and regulation of catabolic traffic in the citric acid cycle is intimately connected with the respiratory chain. The flow of acetyl CoA is not only controlled by rate limiting enzymes, but also by the ratios of $NAD^+/NADH$ and ATP/ADP. Many anabolic reactions also begin in the citric acid cycle. Thus, fatty acid synthesis begins with citrate, gluconeogenesis with oxaloacetate, heme synthesis with succinyl CoA, etc. Two nonessential amino acids, aspartate and glutamate, are in equilibrium with two keto acids in the citric acid cycle, oxaloacetate and α-ketoglutarate, respectively. A number of enzyme activities associated with the citric acid cycle have been measured in many different types of leukocytes. An excellent review was published by Blum (1962). It seems logical to assume that neither gluconeogenesis nor fatty acid synthesis are quantitatively important to the leukocytes. Since pyruvate is poorly utilized by the white blood cells, it was also assumed that citric acid cycle activity and oxidative phosphorylation were less important in the overall metabolism of the leukocytes.

Few attempts have been made to show that the intact cycle is operating. In order to prove the existence of a complete cycle, it is necessary to use the techniques outlined by Swim and Krampitz (1954). These investigators, using radioactive tracers, isolated several intermediates of the cycle, determined the radioactive tracer patterns of the intermediates, and then calculated the metabolic flow of the cycle. It is seldom possible to determine the tracer patterns in all intermediates of the cycle, because many of them are present only in catalytic amounts. Two reliable indicators of metabolism can be isolated namely aspartate and glutamate, because these amino acids are in equilibrium with the two keto acids of the cycle by transamination.

Acetyl CoA enters the cycle via a condensation reaction with oxaloacetate to form citrate which is converted via aconitate and isocitrate to α-ketoglutarate (in equilibrium with glutamate). The latter keto acid is decarboxylated and oxidized to succinly CoA, which is metabolized to fumarate and malate. Malate is oxidized to oxaloacetate (in equilibrium with aspartate), and the sequence becomes a true cycle.

If [2-^{14}C]acetate is metabolized by the citric acid cycle, the carbon-4 of α-ketoglutarate is the only ^{14}C-labeled position during the first turn of the cycle. This labeled position becomes randomized equally between the two center carbons of succinate and fumarate, which are symmetrical molecules, to give oxaloacetate equally labeled in the center carbons. When a new molecule of [2-^{14}C]acetate reacts with labeled oxaloacetate during a second turn of the cycle, a different ^{14}C pattern will emerge. From these labeling patterns it is possible to estimate the number of turns of the cycle as well as to measure steady-state metabolism.

The presence of an intact citric acid cycle in human neutrophils was demonstrated by isolation of succinate, fumarate, malate, aspartate, and glutamate after incubation with either [1-^{14}C]acetate, [2-^{14}C]acetate, or [2-^{14}C]palmitate. All intermediates were degraded carbon-by-carbon, and the tracer patterns were shown to obey the equations for determination of metabolism in the citric acid cycle (Stjernholm *et al.*, 1969). These investigators observed that human neutrophils are capable of acetate metabolism to the extent of 0.18 μmol/hr/10^8 cells. No change in citric acid cycle activity was observed during phagocytic conditions. Exudate macrophages of guinea pigs metabolize 1.0 μmol of acetate/hr/10^8 cells (West *et al.*, 1968). Human lymphocytes convert 0.05 μmol of acetate to CO_2 and water/hr/10^8 cells (Stjernholm *et al.*, 1969), and the maximal capacity of the citric acid cycle in human plasma cells is 0.15 μmol/hr/10^8 cells (Stjernholm, 1967).

A number of amino acids such as valine, isoleucine, methionine, and threonine, as well as odd-numbered fatty acids and branched-chain fatty acids, are metabolized via propionyl CoA and methylmalonyl CoA to CO_2 and water by mammalian cells. The vitamins biotin and cobalamin constitute the prosthetic groups of two of the enzymes in the pathway leading to succinyl CoA, which is the link to the citric acid cycle. It has been shown that many types of leukocytes are capable of converting propionate to CO_2 and water via the citric acid cycle (Stjernholm, 1967; West *et al.*, 1968; Stjernholm *et al.*, 1970). These investigators showed that leukocytes contain both propionyl CoA carboxylase and methylmalonyl CoA isomerase using the procedures described by Wood *et al.* (1964). An impaired propionate metabolism by leukocytes of patients with untreated pernicious anemia was also demonstrated. However, leukocytes deficient in cobalamin were still capable of converting propionate to lactate via a pathway which did not involve the citric acid cycle (Stjernholm *et al.*, 1970). The latter pathway, which probably involved acrylyl CoA, was referred to as the "direct pathway," and seemed to be the predominant route for metabolism of propionate to lactate by leukocytes of patients deficient in vitamin B$_{12}$. After treatment with vitamin B$_{12}$, the leukocytes from the same patients converted propionate via the citric acid cycle. Metabolic deviation of propionate utilization was also observed in leukocytes of patients with rheumatoid arthritis (Dimitrov *et al.*, 1969a). Some of the propionate is diverted to the amino acid β-alanine via a pathway involving acrylyl CoA, β-hydroxylpropionyl CoA, and malonic semialdehyde, which is transaminated to β-alanine. No such pathway was observed in leukocytes of normal individuals.

There is no evidence for the presence of leukocytes of a glyoxalate bypass, the modified form of citric acid cycle, which has been found in plants and microorganisms (Kornberg, 1958).

6.   HEXOSE MONOPHOSPHATE PATHWAY (PENTOSE CYCLE)

Oxidation of glucose-6-P to 6-P-gluconate was demonstrated first by Coxon and Robinson (1956) in leukocytes of cats, rabbits, and dogs. Beck (1958b) characterized several enzymes of the oxidative pathway and estimated that 3% or less of the glucose utilized was traversed via the pentose cycle. Sbarra and Karnovsky (1959), in their study of guinea pig neutrophils, noted a significantly higher ratio of carbon-1 to carbon-6 of glucose oxidation to respiratory CO_2 during phagocytosis than during resting conditions. These observations set the stage for an intensive search for more information about phagocytosis and the mechanism of killing bacteria by phagocytes.

The reaction sequence of the hexose monophosphate shunt is glucose-6-P, 6-P-gluconolactone, 6-P-gluconate, ribulose-5-P, xylulose-5-P, and ribose-5-P. The latter two pentose phosphates are then converted to heptulose-7-P and triose phosphate, which are converted to fructose-6-P and erythrose-4-P. The tetrose-P is converted to fructose-6-P in a transketolase reaction. Calculations of the quantitative importance of the hexose monophosphate pathway have been outlined by Katz and Wood (1960). The net result of this pathway is glucose-6-P + 6NADP resulting in $3CO_2$ + 6NADPH + triose phosphate. All calculations of the activity of this cycle are based on this concept.

Most mammalian cells contain a pentose cycle pathway, but for many different reasons. Dietary pentoses are metabolized by the pentose cycle in liver cells. Dividing cells synthesizing DNA and RNA, require a source of pentose phosphates. Adipocytes and cells requiring reducing power in the form of NADPH use the pentose cycle as a synthetic device. Red blood cells utilize the pentose cycle to produce NADPH for reduction of oxidized glutathione, and white blood cells use the NADPH produced in the pentose cycle for production of H_2O_2 (Iyer *et al.*, 1961) and excited states of oxygen (Allen *et al.*, 1972). The intriguing "respiratory burst," observed during phagocytosis by Sbarra and Karnovsky (1959), is connected to NADPH production in the pentose cycle and is the regulatory factor in polymorphonuclear leukocytes which determines the shift from slight to insignificant pentose cycle activity to extremely active flow of metabolites in the cycle.

Accurate calculations of the pentose cycle activity can be made under certain well-defined conditions, and a number of methods are in existence. The most common procedure uses separate incubations with [1-^{14}C]glucose and [6-^{14}C]glucose. The specific yields of CO_2, which is defined as the ^{14}C activities of all metabolic products, are determined. The pentose cycle activity is then calculated and expressed as a fraction of total glucose utilized according to the equation:

$$\frac{Gl(CO_2) - G6(CO_2)}{1 - G6(CO_2)} = \frac{3\ PC}{1 + 2\ PC}$$

The expression 1-PC then becomes the fraction of glucose which was catabolized via the glycolytic pathway. The equation based on specific yields of CO_2 is very useful, and sensitive to small changes in pentose cycle activity.

Two methods of calculating pentose cycle activity involve the randomization of ^{14}C into carbon-1, carbon-2, and carbon-3 of fructose-6-P when either [2-^{14}C]glucose or [3-^{14}C]glucose is the labeled substrate. Since fructose-6-P is a difficult intermediate to isolate and degrade, other metabolites have been used as indicators of metabolism. Glucose units of glycogen, glycerol of triglycerides, alanine, and lactate have been used for these calculations with the additional assumption that the ^{14}C patterns in these metabolites accurately reflect the ^{14}C distribution of the fructose-6-P. The reliability of these methods have been discussed by Stjernholm et al. (1972).

In numerous calculations of the pentose cycle activity during phagocytosis, it was observed that almost 2 μmol of glucose-6-P/hr/10^8 cells may traverse the oxidative pathway. At this point one may ask if the enzymes of the pentose cycle are present in the cytoplasm in amounts which would permit such extensive metabolism. By determination of the V_{max} values for the enzymes of the cycle according to the procedures described by Yaphe et al. (1966), it was demonstrated that all the enzymes of the cycle were present in substantial amounts in the cytoplasm, but 6-P-gluconate dehydrogenase and transketolase could become rate limiting enzymes (Stjernholm, 1968). A depression of transketolase activity in leukocytes of patients treated with diphenylhydantoin has been reported (Markkanen and Peltola, 1971). In this study, the inhibition could reach values from 20–40%. An inhibitory effect of caffein on the pentose cycle activity of polymorphonuclear leukocytes from humans and dogs was reported by Dimitrov et al. (1969b). Impaired pentose cycle activity in lymphocytes of patients with chronic lymphocytic leukemia was observed by Brody et al. (1969).

Pentoses such as ribose, xylose, and arabinose enters the pentose cycle in leukocytes of rabbit via transketolase and transaldolase reactions followed by metabolism via the glycolytic pathway to lactate (Stjernholm and Noble, 1961a, 1963).

7. OTHER PATHWAYS OF CARBOHYDRATE METABOLISM

Galactose can be utilized by human or rabbit leukocytes (Weinberg and Segal, 1960; Stjernholm and Noble, 1961b). There is no reason to assume that galactose is metabolized differently by leukocytes than by other tissues. Thus, galactose is phosphorylated by galactokinase to galactose-1-P, which in a uridyl transferase reaction is converted to UDP-galactose. An epimerase then converts UDP-galactose to UDP-glucose. The latter sugar nucleotide can either be incorporated into glycogen or be converted to glucose-1-P and glucose-6-P for further metabolism by the glycolytic pathway. Incubations with [1-^{14}C]galactose or [2-^{14}C]galactose gave radioactive trace patterns in the glucose units of glycogen, the glycerol moiety of triglycerides, and in the lactate produced by glycolysis, which were consistent with the enzymatic conversions described above (Stjernholm and Noble, 1961b).

Metabolism of fructose by leukocytes from human and rabbit was studied by Martin et al. (1955) and Esmann et al. (1965).

The latter investigators used [2-^{14}C]fructose, [6-^{14}C]fructose, and [1,6-^{14}C]fructose as substrates. Glucose units of glycogen and lactate from glycolysis were degraded to ascertain a metabolic pattern consistent with known pathways for fructose metabolism. The radioactive trace patterns indicated that fructose may have been converted to fructose-1-P as observed in liver, but there was evidence of additional randomizing pathways. One of these pathways may have involved a symmetrical C_3 compound such as proposed by Esmann (1968). In general, the tracer data resembled those produced by radioactive glucose. Thus, it is entirely possible that the main pathway for fructose is via hexokinase leading to fructose-6-P, which is then metabolized via glycolysis.

The conversion of glucosamine to glucose by leukocytes was first observed by Kornfeld and Gregory (1968). Leukocytes from normal human blood were incubated in the presence of [U-^{14}C]glucosamine. After ultrasonic disruption of the cells, 10% of the total radioactivity was located in the particulate fraction. Two-thirds of this ^{14}C activity was found in nonamino sugar derivatives. The supernatant fraction contained ^{14}C-labeled glucose and UDP-*N*-acetyl-glucosamine, acetyl-glucosamine, and free glucosamine. The glucose accounted for 75% of the total radioactivity. In similar experiments, Olsson (1968) isolated glucosaminoglycans. The largest part was identified as chondroitin-4-SO$_4$. The capacity of leukocytes to synthesize glycosaminoglucans was studied by incubation of leukemic myeloid cells with ^{35}SO$_4^{2-}$ and [1-^{14}C]glucosamine. Both the formation of the chondroitin sulfate chain and the sulfation of the latter were demonstrated. Immature myeloid cells had a higher concentration of chondroitin sulfate and a higher rate of biosynthesis than mature myeloid cells. Olsson *et al.* (1968) then demonstrated that human granulocytes are capable of converting glucosamine to galactosamine. In a follow-up study Olsson (1969) showed that incorporation of ^{35}SO$_4^{2-}$, [1-^{14}C]glucosamine, and [^{3}H]serine into chondroitin sulfate was inhibited by puromycin, indicating that the elongation of the polysaccharide chain was dependent on prior synthesis of protein. Gel chromatography of the polysaccharide isolated from subcellular fractions indicated that condroitin sulfate was synthesized as a proteoglycan on microsomal structures.

A significant increase of ^{35}SO$_4^{2-}$ incorporation into glycosaminoglycans by leukocytes of patients with cystic fibrosis was observed by Caudill *et al.* (1974). Although attempts were made to identify obligate heterozygotes by this method it was not uniformly successful. A follow-up of this study would be most helpful.

8. USE OF LEUKOCYTES IN DIAGNOSIS OF INBORN ERRORS OF METABOLISM

Numerous advances has been made to our knowledge of intermediary metabolism during studies of patients with inborn errors of metabolism. A mutation in the gene material often reveals itself as a block of metabolic traffic in a specific pathway. Failure to transcribe and translate an enzyme in a metabolic system leads to accumulation of metabolites prior to the block. Although these metabolites are harmless when present in normal concentrations, they produce aberrations when present in elevated amounts.

We now recognize more than 175 inborn errors of metabolism in man. Almost 40 of these conditions can be diagnosed through the use of leukocytes or homogenates of white blood cells. In some cases mitogen-stimulated lymphocytes have been used for the purpose of identifying an enzymic defect. In many cases it is possible to identify not only the homozygote but also the heterozygote. Short-term cultures of leukocytes in the presence of a particular radioactive substrate followed by isolation of accumulated metabolites are commonly used.

8.1. GLYCOGEN STORAGE DISEASE (GLYCOGENOSIS)

In the following types of glycogen storage disease the leukocytes show decreased activity of the enzyme responsible for a particular type of glycogenosis. Glycogen storage disease type II (Pompe disease), is caused by the absence of α-1,4-glucosidase which is active at an acid pH ("acid maltase," Hers, 1963). A technique was developed for the detection of α-1,4-glucosidase by multiple enzyme analysis of lymphocytes stimulated by phytohemagglutinin. This technique also allowed the detection of heterozygotes (Hirschhorn *et al.*, 1969). The enzyme defect is more expressed in lymphocytes than in polymorphonuclear leukocytes (Koster *et al.*, 1974). In this report it was pointed out that glycogen was a better substrate than maltose for differentiation of heterozygotes from normal controls and patients.

Glycogen storage disease type III is due to the absence of α-1,6-glucosidase. Diminished enzymatic activity was reported by Creveld and Huijing (1965), Steinitz *et al.* (1963), and Huijing (1964).

Glycogen storage disease type IV is caused by a failure of amylo-1,4$\rightarrow$1,6-transglucosylase (branching enzyme). The absence of the enzyme in leukocytes was described by Brown and Brown (1966).

Absence of phosphorylase leads to glycogen storage disease type VI. Diagnosis could be made by determining the enzyme activity in leukocytes (Hulsmann *et al.*, 1961; Schwartz *et al.*, 1970). In some instances the defect may be in the activation of phosphorylase (phosphorylase *b* to phosphorylase *a*), and it is possible that the mutation has affected phosphorylase kinase (Williams and Field, 1961).

Since leukocytes lack glucose-6-phosphatase, they cannot be used in diagnosing glycogen storage disease type I (von Gierke disease). In the type V disease (muscle phosphorylase deficiency) the leukocyte phosphorylase activity is normal (Engel *et al.*, 1963). It has been established that leukocyte phosphorylase resembles liver phosphorylase more than it resembles muscle phosphorylase (Yunis and Arimura, 1964).

8.2. GALACTOSEMIA

Inability to metabolize dietary galactose occurs in galactosemia, an inborn error in which galactose accumulates in the blood and spills over into the urine, when galactose and lactose are ingested. There is also marked accumulation of galactose-1-P in the red blood cells, which would indicate that the mutation did not affect galactokinase. The enzymic defect is located at the galactose-1-P uridyltransferase step. As a result, galactose-1-P is not converted into UDP-galactose. An epimerase is present, however, which converts UDP-glucose to UDP-galactose, thus allowing this metabolite to be produced for other biosynthetic mechanisms important for normal growth and development.

The leukocytes of galactosemic patients share with other tissues the deficiency of galactose-1-P uridyltransferase. A common method utilized by production of $^{14}CO_2$ from [1-^{14}C]galactose and [^{14}C-UDP]glucose as well as the accumulation of galactose-1-P (Weinberg et al., 1960; Weinberg and Segal, 1960; Mellman et al., 1965). The inability to demonstrate the enzyme defect in leukocytes of heterozygous Negroes suggests that the genome for galactosemia in this race is different from that of Caucasians in a manner resembling the genome of glucose-6-P dehydrogenase deficiency (Mellman et al., 1965).

Although it was predicted that the absence of UDP-galactose epimerase would be incompatible with life (Kalckar, 1965), seven individuals have been identified with this deficiency. Elevated levels of galactose-1-P were observed in the patients' erythrocytes. The interesting observation during study of these individuals was that stimulation of their lymphocytes in vitro with phytohemagglutinin led to a reversal of the UDP-galactose epimerase deficiency. It was suggested that lymphocyte stimulation resulted in an increased rate of synthesis of a functioning but mutant enzyme, probably via a derepression of an epimerase locus (Mitchell et al., 1975). This very interesting work would constitute the first example of a reversal of a genetically determined enzyme deficiency using human cells in culture.

8.3. GLUCOSE-6-P DEHYDROGENASE (G6PD) DEFICIENCY

The report on primaquine-induced hemolytic anemia caused by a deficiency of G6PD (Carson et al., 1956) stimulated many new investigations of this enzyme in man (Yoshida, 1973). About 80 variants of G6PD have now been reported (Yoshida et al., 1971). The expression of the deficiency in leukocytes may be dependent on racial and ethnic background (Marks and Gross, 1959). Negroes with the erythrocyte G6PD deficiency have normal levels in their leukocytes, but the enzyme levels in leukocytes of Caucasians are variable. A group of Caucasians in Israel with the erythrocyte deficiency (Ramot et al., 1959) and a Sardinian family (Bonsignore et al., 1966) were found to have levels of leukocyte G6PD below the normal controls. Several reports suggest that the leukocyte G6PD plays a critical role during phagocytosis particularly in the mechanism involved in killing bacteria by neutrophils (Schlegel and Bellanti, 1970; Cooper et al., 1970). Normal bacterial killing by neutrophils has been observed in cases where the G6PD level was about 25% of normal values (Rodey et al., 1970).

Defective bacterial killing by neutrophils may not always be connected with a G6PD deficiency in leukocytes. In chronic granulomatous disease there is a defect in the formation of H_2O_2 by neutrophils (Holmes *et al.*, 1967), and in another case it was a neutrophil pyruvate kinase deficiency which led to recurrent staphylococcal infections (Burge *et al.*, 1976).

8.4. LYSOSOMAL DISEASES

Lysosomes occur in most mammalian tissues. These granules contain hydrolytic enzymes. A decrease or absence of a particular lysosomal hydrolase is the hallmark of lysosomal diseases. The result of such deficiency is an accumulation of the predominant substrate for the missing enzyme.

Mucopolysaccharidosis I has been classified on clinical grounds into three subgroups MPS-IH (Hurler syndrome), MPS-IS (Scheie syndrome) and MPS-IH/S (Hurler–Scheie compound syndrome) according to McKusick *et al.* (1972). The deficient lysosomal enzyme was reported to be α-L-iduronidase (Matalon *et al.*, 1971).

Assay of α-L-iduronidase in leukocytes using phenyl-iduronide has been reported to be a simple and rapid diagnostic tool in screening for mucopolysaccharidosis I. The values in homozygotes are about 3% of normal controls, and obligate heterozygotes were half of normal values (Kelly and Taylor, 1976; Dulaney *et al.*, 1976; Liem and Hooghwinkel, 1975).

Therapeutic attempts to treat a patient suffering from mucopolysaccharidosis type II (Hunter syndrome) was reported by Knudson *et al.* (1971), by infusion of leukocytes obtained from the patient's father and sister. Increased urinary excretions of glucosaminoglycan and products of their degradation was observed.

A simple procedure for the detection of Tay-Sachs disease (G_{m2}-gangliosidosis) by assaying for hexosaminidase A in leukocytes was reported by Klibansky (1971) and by Suzuki *et al.* (1971). The method was based on fluorometric determination of 4-methylumbelliferone liberated from 4-methylumbelliferyl-N-acetyl-glucosamine.

Highly purified hexosaminidase trapped in liposomes subsequently coated with human IgG was observed to be phagocytized by polymorphonuclear leukocytes of patients with Tay–Sachs disease (Cohen *et al.*, 1976). This novel approach represents an attempt to supply enzyme deficient cells with a missing lysosomal enzyme.

Fabry disease is an X-linked inborn error of glycosphingolipid metabolism. The disease is caused by the accumulation of trihexosyl ceramide (galactosyl-galactosyl-glucosylceramide). The primary defect was identified as the absence of the lysosomal enzyme ceramide trihexosidase, which hydrolyzes the terminal galactosyl molecule from the trihexosyl ceramide (Brady *et al.*, 1967). The defective enzyme referred to as α-galactosidase can be assayed using the substrate 4-methyl-umbelliferyl-α-D-galactoside. A sensitive method for diagnosis of Fabry disease using serum, urine, and leukocytes was described by Desnick *et al.* (1973).

A related enzymatic deficiency was observed in a patient with angiokeratoma corporis diffusum characterized by the absence of β-galactosidase (Loonen *et al.*, 1974). However, other less well-defined diseases may be related to a defect in β-galactosidase (Wenger *et al.*, 1974). The source of enzyme was leukocytes and the substrate was 4-methyl-umbelliferyl-β-galactoside. The affected patients had 4% of normal activity.

The skin disease angiokeratoma corporis diffusum may also be caused by a deficiency of α-L-fucosidase causing an accumulation of oligosaccharides and glycoproteins possibly of the following nature, fucose (1→2)-galactose (1→3)-N-acetyl-glucosamine (1→4)-galactose (1→4)-glucose-ceramide. Leukocyte α-fucosidase activity in some patients has been found to be 10% of normal controls (Patel *et al.*, 1972). The problem becomes more vexing, when one learns that some of these patients have normal lysosomal enzyme levels of α-galactosidase, β-galactosidase, and α-L-iduronidase (MacPhee *et al.*, 1975).

Gaucher disease Type I is characterized by the accumulation of glycosylceramide in the reticuloendothelial cells. The disease is caused by the deficiency of the lysosomal enzyme β-D-glucoside glucohydrolase (glucocerebrosidase) as described by Brady *et al.* (1965). When the glycosphingolipids were isolated from the leukocytes of patients with Gaucher disease and normal controls, it was observed that the glucosyl ceramide which is the main glycolipid in the reticuloendothelial cells of patients with Gaucher disease, was present in the same amount in Gaucher leukocytes as in normal controls. Lactosyl ceramide, on the other hand, which is the main glycolipid in leukocytes of normal subjects was significantly increased in the leukocytes of Gaucher patients (Klibansky *et al.*, 1976). This apparent anomaly may be explained if there are different fates for utilization of glucosyl ceramide (e.g., phosphilipid synthesis) by reticuloendothelial cells and leukocytes (Barton and Rosenberg, 1975).

Three diseases are due to a deficiency of arylsulfatase. Sulfatidosis is due to the absence of arylsulfatase A, mucopolysaccharidosis VI is the result of a deficiency of arylsulfatase B and mucosulfatidosis is the result of the deficiency of both arylsulfatase A and B. The two enzymes can be separated from leukocyte lysates by chromatography on DEAE cellulose. A convenient method for measuring the two enzymes using p-nitrocatechol sulfate as the substrate was described by Humbel (1976).

REFERENCES

Abels, J. C., Jones, F. L., Craver, F. L., and Rhoads, C. P., 1944, The metabolism of pyruvate by normal and leukemic white cells, *Cancer Res.* **4**:149.

Allen, R. C., Stjernholm, R. L., and Steele, R. H., 1972, Evidence for the generation of an electronic excitation state(s) in human polymorphonuclear leukocytes and its participation in bactericidal activity, *Biochem. Biophys. Res. Comm.* **47**:679.

Barton, N. W., and Rosenberg, A., 1975, Metabolism of glucosyl-^{3}H-ceramide by human skin fibroblasts from normal and glucosylceramidotic subjects, *J. Biol. Chem.* **250**:3966.

Beck, W. S., 1955, A kinetic analysis of the glycolytic rate and certain glycolytic enzymes in normal and leukaemic leucocytes, *J. Biol. Chem.* **216**:333.

Beck, W. S., 1958a, The control of leukocyte glycolysis, *J. Biol. Chem.* **232**:251.

Beck, W. S., 1958b, Occurrence and control of the phosphogluconate oxidation pathway in normal and leukaemic leucocytes, *J. Biol. Chem.* **232**:271.

Beck, W. S., and Valentine, W. N., 1952a, The aerobic carbohydrate metabolism of leukocytes in health and leukemia. I. Glycolysis and respiration, *Cancer Res.* **12**:818.

Beck, W. S., and Valentine, W. N., 1952b, The aerobic carbohydrate metabolism of leukocytes in health and leukemia. II. The effects of various substrates and coenzymes on glycolysis and respiration, *Cancer Res.* **12**:823.

Blum, K. U., 1962, Enzymepathologie der Blutzellen, *Blut* **8**:239.

Bonsignore, A., Fornaini, G., Leoncini, G., and Fantoni, A., 1966, Electrophoretic heterogeneity of erythrocyte and leukocyte glucose-6-phosphate dehydrogenase in Italians from various ethnic groups, *Nature* **211**:876.

Brady, R. O., Kanfer, J. N., and Shapiro, D., 1965, Metabolism of glucocerebrosides. II. Evidence of an enzymatic deficiency in Gaucher's disease, *Biochem. Biophys. Res. Comm.* **18**:221.

Brady, R. O., Gal, A. E., and Bradley, R. M., 1967, Enzymatic defect in Fabry's disease: Ceramide trihexosidase deficiency, *N. Engl. J. Med.* **275**:1163.

Brody, J. I., Oski, F. A., and Singer, D. E., 1969, Impaired pentose cycle phosphate shunt and decreased glycolytic activity in lymphocytes of chronic lymphocytic leukemia, *Blood* **34**:421.

Brown, B. I., and Brown, D. H., 1966, Lack of an α-1,4-glucan: α-1,4-glucan-6-glucosyl transferase in a case of type IV glycogenosis, *Proc. Natl. Acad. Sci. USA* **56**:725.

Burge, P. S., Johnson, W. C., and Hayward, A. R., 1976, Neutrophil pyruvate kinase deficiency with recurrent staphylococcal infections: First reported case, *Br. Med. J.* **1**:742.

Carson, P. E., Flanagan, C. L., Ickes, C. E., and Alving, A. S., 1956, Enzymatic deficiency in primaquine sensitive erythrocytes, *Science* **124**:484.

Caudill, M., Schafer, I., and Stjernholm, R. L., 1974, Sulfate incorporation of leukocytes from patients with cystic fibrosis, *Lancet* **1**:32.

Cohen, C. H., Weissmann, G., Hoffstein, S., Awasthi, Y. C., and Srivastava, S. K., 1976, Introduction of purified hexosaminidase A into Tay–Sachs leukocytes by means of immunoglobulin-coated liposomes, *Biochemistry* **15**:452.

Cooper, M. R., DeChatelet, L. R., McCall, C. E., LaVia, M. F., Spurr, C. L., and Baehner, R. L., 1970, Leukocyte G6PD deficiency, *Lancet* **2**:110.

Coxon, R. V., and Robinson, R. J., 1956, Carbohydrate metabolism in blood cells studied by means of isotopic carbon, *Proc. Roy. Soc. London Ser. B.* **145**:232.

Desnick, R. J., Allen, K. Y., Desnick, S. J., Raman, M. K., Bernlohr, R. W., and Krivit, W., 1973, Fabry's disease: Enzymatic diagnosis of hemizygotes and heterozygotes, *J. Lab. Clin. Med.* **81**:157.

Dimitrov, N. V., Weir, D., and Stjernholm, R. L., 1969a, Metabolic deviations of polymorphonuclear leukocytes in rheumatoid arthritis, *Blut* **19**:139.

Dimitrov, N. V., Miller, J., and Ziegra, S. R., 1969b, Effects of caffeine on glucose metabolism of polymorphonuclear leukocytes, *J. Pharmacol. Exp. Ther.* **168**:240.

Dulaney, J. T., Milunsky, A., and Moser, H. W., 1976, Detection of the carrier state of Hurler's syndrome by assay of α-L-iduronidase in leukocytes, *Clin. Chim. Acta* **69**:305.

Engel, W. K., Eyerman, E. L., and Williams, H. E., 1963, Late-onset type of skeletal muscle phosphorylase deficiency, *N. Engl. J. Med.* **268**:135.

Esmann, V., 1962, *Carbohydrate Metabolism and Respiration in Leukocytes from Normal and Diabetic Subjects*, Universitetsforlaget, Aarhus, Denmark.

Esmann, V., 1968, Dihydroxyacetone as an intermediate during the metabolism of glycerol and glyceraldehyde in leukocytes from the rat, *Acta Chem. Scand.* **22**:2281.

Esmann, V., Noble, E. P., and Stjernholm, R. L., 1965, Carbohydrate metabolism in leukocytes. VI. The metabolism of mannose and fructose in polymorphonuclear leukocytes of rabbit, *Acta Chem. Scand.* **19**:1672.

Hager, L. P., and Gunsalus, I. C., 1953, Lipoic acid dehydrogenase: The function of *Escherichia coli* Fraction B, *J. Am. Chem. Soc.* **75**:5767.

Hedeskov, C. J., and Esmann, V., 1966, Respiration and glycolysis of normal human leukocytes, *Blood* **28**:163.

Hers, H. G., 1963, α-Glucosidase deficiency in generalized glycogen-storage disease (Pompe's disease), *Biochem. J.* **86**:16.

Hirschhorn, K., Nadler, H. L., Waithe, W. I., Brown, I. B., and Hirschhorn, R., 1969, Pompe's disease: Detection of heterozygotes by lymphocyte stimulation, *Science* **166**:1632.

Holmes, B., Page, A. R., and Good, R. A., 1967, Studies of the metabolic activity of leukocytes from patients with a genetic abnormality of phagocytic function, *J. Clin. Invest.* **46**:1422.

Huijing, F., 1964, Amylo-1,6-glucosidase activity in normal leucocytes and in leucocytes of patients with glycogen-storage disease, *Clin. Chim. Acta* **9**:269.

Hulsmann, W. C., Oei, T. L., and van Creveld, S., 1961, Phosphorylase activity in leukocytes from patients with glycogen-storage disease, *Lancet* **2**:581.

Humbel, R., 1976, Rapid method for measuring arylsulfatase A and B in leukocytes as a diagnosis for sulfatidosis, mucosulfatidosis and mucopolysaccharidosis VI, *Clin. Chim. Acta* **68**:339.

Iyer, G. Y. N., Islam, M. F., and Quastel, J. H., 1961, Biochemical aspects of phagocytosis, *Nature* **192**:535.

Kalckar, H. M., 1965, Galactose metabolism and cell "sociology," *Science* **150**:305.

Katz, J., and Wood, H. G., 1960, The use of glucose-^{14}C for the evaluation of the pathways of glucose metabolism, *J. Biol. Chem.* **235**:2165.

Kelly, E. T., and Taylor, H. A., 1976, Leucocyte values of α-L-iduronidase activity in mucopolysaccharidosis I, *J. Med. Genet.* **13**:149.

Klibansky, C., 1971, Separation of N-acetyl-β-D-hexosaminidase-isoenzymes from human brain and leukocytes by cellulose acetate paper electrophoresis, *Isr. J. Med. Sci.* **7**:1086.

Klibansky, C., Ossimi, Z., Matoth, Y., Pinkhas, J., and DeVries, A., 1976, Accumulation of lactosyl ceramide in leukocytes of patients with adult Gaucher's disease, *Clin. Chim. Acta* **72**:141.

Knudson, A. G., Di Ferrante, N., and Curtis, J. E., 1971, Effect of leukocyte transfusion in a child with type II mucopolysaccharidosis, *Proc. Natl. Acad. Sci. USA* **68**:1738.

Kornberg, H. L., 1958, Synthesis of cell constituents from acetate via the glyoxalate cycle, *Proc. 4th Int. Cong. Biochem.* **13**:251.

Kornfeld, S., and Gregory, W., 1968, The conversion of glucosamine to glucose by white blood cell cultures, *Biochim. Biophys. Acta* **158**:468.

Koster, J. F., Slee, R. G., and Hulsmann, W. C., 1974, The use of leukocytes as an aid in the diagnosis of glycogen storage disease Type II (Pompe's disease), *Clin. Chim. Acta* **51**:319.

Liem, K. O., and Hooghwinkel, G. J. M., 1975, The use of α-L-iduronidase activity determinations in leukocytes for the detection of Hurler and Scheie syndromes, *Clin. Chim. Acta* **60**:259.

Lint, T. F., 1973, Metabolic studies on normal and leukemic established human lymphoblastoid cell lines, Doctoral Dissertation, Tulane University, New Orleans.

Loonen, M. C. B., van den Lugt, L., and Franke, C. L., 1974, Angiokeratoma corporis diffusum and lysosomal enzyme deficiency, *Lancet* **2**:785.

Luganova, I. S., and Seits, I. F., 1959, The metabolic characteristics of human leukocytes in health and leukemia, *Probl. Gematol.* **4**:41 (Pergamon translation).

Luganova, I. S., and Seits, I. F., 1964, Enzymatic activity of the glycogen synthesis system of human leukocytes in leukemia, *Vopr. Onkol.* **10**:38.

MacPhee, G. B., Logan, R. W., and Primrose, D. A. A., 1975, Fucosidosis: How many cases undetected? *Lancet* **2**:462.

Markkanen, T., and Peltola, O., 1971, Pentose phosphate pathway of leukocytes, *Acta Hematol.* **46**:36.

Marks, P. A., and Gross, R. T., 1959, Erythrocyte glucose-6-phosphate dehydrogenase deficiency: Evidence of differences between Negroes and Caucasians with respect to this genetically determined trait, *J. Clin. Invest.* **38**:2253.

Martin, S. P., McKinney, G. R., and Green, R., 1955, The metabolism of human polymorphonuclear leukocytes, *Ann. N.Y. Acad. Sci.* **59**:996.

Matalon, R., Cifonelli, J. A., and Dorfman, A., 1971, α-L-Iduronidase in cultured human fibroblasts and liver, *Biochem. Biophys. Res. Comm.* **42**:340.

McKusick, V. A., Howell, R. R., Hussels, I. E., Neufeld, E. F., and Stevenson, R., 1972, Allelism, nonallelism and genetic compounds among the mucopolysaccharidosis: Hypothesis, *Lancet* **1**:993.

Mellman, W. J., Tedesco, T. A., and Baker, L., 1965, A new genetic abnormality, *Lancet* **1**:1395.

Mitchell, B., Haigis, E., Steinmann, B., and Gitzelmann, R., 1975, Reversal of UDP-galactose 4-epimerase deficiency of human leukocytes in culture, *Proc. Natl. Acad. Sci. USA* **72**:5026.

Noble, E. P., Stjernholm, R. L., and Weissburger, A. S., 1960, Carbohydrate metabolism in the leukocytes I. The pathway of two- and three-carbon compounds in the rabbit polymorphonuclear leukocyte, *J. Biol. Chem.* **235**:1261.

Noble, E. P., Stjernholm, R. L., and Liungdahl, L., 1961, Carbohydrate metabolism in leukocytes. III. Carbon dioxide incorporation in the rabbit polymorphonuclear leukocyte, *Biochim. Biophys. Acta* **49**:593.

Olsson, I., 1968, Biosynthesis of glycosaminoglycans (mucopolysaccharides) in leukemic myeoloid cells, *Biochim. Biophys. Acta* **165**:324.

Olsson, I., 1969, Chondroitin sulfate proteoglycan of human leukocytes, *Biochim. Biophys. Acta* **177**:241.

Olsson, I., Gardell, S., and Thunell, S., 1968, Biosynthesis of glycosaminoglycans in human leukocytes, *Biochim. Biophys. Acta* **165**:309.

Patel, V., Watanabe, I., and Zeman, W., 1972, Deficiency of α-L-fucosidase, *Science* **176**:426.

Perona, G., and Zatti, M., 1959, Sul metabolismo dei carboidrati nei granulociti neutrofili, *Arch. Sci. Biol. (Bologna)* **43**:307.

Ramot, G., Fisher, S., Szeinberg, A., Adam, A., Sheba, C., and Gafni, D., 1959, A study of subjects with erythrocyte G6PD deficiency. II. Investigation of leukocyte enzymes, *J. Clin. Invest.* **38**:2234.

Rodey, G. E., Jacob, H. S., Holmes, B., McArthur, J. R., and Good, R. A., 1970, Leucocyte G6PD levels and bactericidal activity, *Lancet* **1**:355.

Rosell-Perez, M., and Esmann, V., 1965, UDPG-glucan-glycosyl-transferase in human polymorphonuclear leukocytes, *Acta Chem. Scand.* **19**:679.

Sbarra, A. J., and Karnovsky, M. L., 1959, The biochemical basis of phagocytosis. I. Metabolic changes during ingestion of particles by polymorphonuclear leukocytes, *J. Biol. Chem.* **234**:1355.

Schlegel, R. J., and Bellanti, J. A., 1970, Leukocyte G6PD deficiency and bactericidal activity, *Lancet* **1**:677.

Schwartz, D., Savin, M., Drash, A., and Field, J., 1970, Studies in glycogen storage disease. IV. Leukocyte phosphorylase in a family with type VI GSD, *Metabolism* **19**:238.

Steinitz, K., Bodur, H., and Arman, T., 1963, Amylo-1,6-glucosidase activity in leukocytes from patients with glycogen storage disease, *Clin. Chim. Acta* **8**:807.

Stjernholm, R. L., 1967, Carbohydrate metabolism in leukocytes. VII. Metabolism of glucose, acetate and propionate by human plasma cells, *J. Bacteriol.* **93**:1657.

Stjernholm, R. L., 1968, The metabolism of leukocytes, *Congr. Int. Soc. Hematol. Twelfth N.Y. Plenary Session* **5**:175.

Stjernholm, R. L., and Noble, E. P., 1961a, Carbohydrate metabolism in leukocytes. II. The pathway of ribose and xylose metabolism in the rabbit polymorphonuclear leukocyte, *J. Biol. Chem.* **236**:614.

Stjernholm, R. L., and Noble, E. P., 1961b, Carbohydrate metabolism in leukocytes. IV. The metabolism of glucose and galactose in polymorphonuclear leukocytes from rabbit, *J. Biol. Chem.* **236**:3093.

Stjernholm, R. L., and Noble, E. P., 1963, Carbohydrate metabolism in leukocytes. V. The metabolism of five-carbon substrates in polymorphonuclear leukocytes of rabbit, *Arch. Biochem. Biophys.* **100**:200.

Stjernholm, R. L., Dimitrov, N. V., and Pijanowski, L. J., 1969, Carbohydrate metabolism in leukocytes. IX. Citric acid cycle activity in human neutrophils, *J. Reticuloendothel. Soc.* **6**:194.

Stjernholm, R. L., Noble, E. P., Dimitrov, N. W., Morton, D., and Falor, W. H., 1969, Carbohydarte metabolism in leukocytes. XII. Metabolism of the human lymphocyte, *J. Reticuloendothel. Soc.* **6**:590.

Stjernholm, R. L., and Manak, R. C., 1970, Carbohydrate metabolism in leukocytes. XIV. Regulation of pentose cycle activity and glycogen metabolism during phagocytosis, *J. Reticuloendothel. Soc.* **8**:550.

Stjernholm, R. L., Noble, E. P., Dimitrov, N. V., Kellermeyer, R. W., and Falor, W. H., 1970,

Differentiation of normal and leukemic leukocytes into three groups by propionate metabolism, *J. Reticuloendothel. Soc.* **8:**446.

Stjernholm, R. L., Burns, C. P., and Hohnadel, J. H., 1972, Carbohydrate metabolism by leukocytes, *J. Enzyme Physiol. Pathol.* **13:**7.

Suzuki, Y., Berman, P. H., and Suzuki, K., 1971, Detection of Tay-Sachs disease heterozygotes by assay of hexosaminidase A in serum and leukocytes, *J. Pediatr.* **78:**643.

Swim, H. E., and Krampitz, L. O., 1954, Acetic acid oxidation by *Escherichia coli:* Evidence for the occurrence of a tricarboxylic acid cycle, *J. Bacteriol.* **67:**419.

Tibbling, G., 1970, Glycerol uptake in leukocytes and thrombocytes, *Scand. J. Clin. Lab. Invest.* **26:**185.

Valentine, W. N., Folette, J. H., and Lawrence, J. S., 1953, The glycogen content of human leukocytes in health and in various disease states, *J. Clin. Invest.* **32:**251.

van Creveld, S., and Huijing, F., 1965, Glycogen storage disease, *Am. J. Med.* **38:**554.

van der Wende, C., Miller, W. L., and Glass, S., 1964, Glycogen metabolism of normal and leukemic leukocytes, *Life Sci.* **3:**223.

Wagner, R., 1946, The estimation of glycogen in whole blood and white blood cells, *Arch. Biochem.* **11:**249.

Wagner, R., 1947, Studies on the physiology of the white blood cell. The glycogen content of leukocytes in leukemia and polycythemia, *Blood* **2:**235.

Wagner, R., 1950, Enzyme studies on white blood cells, I. Glycogen degradation, *Arch. Biochem.* **26:**123.

Wagner, R., and Yorke, A., 1952, Enzyme Studies on white blood cells. III. Phosphorylating glycogenolysis and phosphorylated intermediates, *Arch. Biochem.* **39:**174.

Weinberg, A. N., and Segal, S., 1960, Effect of galactose-1-phosphate on glucose oxidation by normal and galactosemic leukocytes, *Science* **132:**1015.

Weinberg, A. N., Herring, B., Johnson, P., and Field, J. B., 1960, Galactose metabolism by leukocytes from galactosemic patients, their parents and normals, *Clin. Res.* **8:**27.

Wenger, D. A., Goodman, S. I., and Myers, G. G., 1974, β-Galactosidase deficiency in young adults, *Lancet* **2:**1319.

West, J., Morton, D. J., Esmann, V., and Stjernholm, R. L., 1968, Carbohydrate metabolism in leukocytes. VIII. Metabolic activities of the macrophage, *Arch. Biochem. Biophys.* **124:**85.

Williams, H. E., and Field, J. B., 1961, Low leukocyte phosphorylase in hepatic phosphorylase deficient glycogen storage disease, *J. Clin. Invest.* **40:**1841.

Wood, H. G., Kellermeyer, R. W., Stjernholm, R. L., and Allen, S. H. G., 1964, Metabolism of methylmalonyl-CoA and role of biotin and B_{12} coenzymes, *Ann. N.Y. Acad. Sci.* **112:**661.

Yaphe, W. D., Christensen, D., Biaglow, J. E., Jackson, M. A., and Sable, H. Z., 1966, Enzymes of the pentose cycle, *Can. J. Biochem.* **44:**91.

Yoshida, A., 1973, Hemolytic anemia and G6PD deficiency, *Science* **179:**532.

Yoshida, A., Beutler, E., and Motulsky, A. G., 1971, Human glucose-6-phosphate dehydrogenase variants, *Bull. W. H. O.* **45:**243.

Yunis, A. A., and Arimura, G. K., 1964, Enzyme of glycogen metabolism in white blood cells. I. Glycogen phosphorylase in normal and leukemic human leukocytes, *Cancer Res.* **24:**489.

5

Lipid Metabolism by Phagocytic Cells

PETER ELSBACH and JERROLD WEISS

1. INTRODUCTION

Although the physicochemical, and hence functional, properties of biological membranes result from the interaction of all major membrane constituents, including lipids, proteins, and carbohydrates, lipids are the components that determine most of the properties considered to be characteristic of cell membranes. It is, in fact, the intense preoccupation with membrane structure and function of the past two decades that has led to many of the concomitant advances in the knowledge of lipid chemistry and metabolism.

Endocytosis and the events leading to and following the intracellular sequestration of extracellular material are surface-membrane-associated phenomena. This recognition has prompted investigators interested in membranes and lipids to join the rapidly growing group of students of phagocytosis.

A number of reviews, including very recent ones (Gottfried, 1972; Elsbach, 1972, 1974, 1977a), have dealt with the composition and intracellular distribution of lipids of phagocytic cells and with the enzymatic pathways of synthesis and degradation of cellular lipids in resting and actively engulfing phagocytes. The progress made in these areas, when judged by new data and insights introduced into the recent literature, has slowed down in the last few years. However, there is reason to believe that the pace of progress will quicken again. First, because novel concepts about the behavior of the membrane lipids of both pro- and eukaryotic cells with respect to their distribution and mobility within the membrane, are yet to be applied to studies of the membrane changes that accompany phagocyte–particle interaction. Second, because recent studies of the surface signals that trigger chemotaxis, formation of the phagocytic vacuole, and the multiple metabolic responses of the phagocyte, give strong indications that physicochemical alterations in the membranes and membrane lipids occur upon

PETER ELSBACH and JERROLD WEISS • Department of Medicine, New York University School of Medicine, New York, New York 10016. The literature has been reviewed through March, 1978.

encounter with ingestible material. Exploration of these changes should provide important new insights. Third, particularly during the last decade, knowledge of the chemical anatomy and the functional characteristics of the microbial envelope has vastly increased (Costerton *et al.*, 1974; Braun, 1975; Ghuysen, 1977; Daneo-Moore and Shockman, 1977). The role this envelope plays in the fate of the microorganism *vis à vis* the host defense is now being investigated with increasing intensity (Elsbach, 1977b). Recent findings suggest that the study of the microbial envelope lipids and the enzymes concerned with their metabolism as determinants of phagocyte–microbe interaction will also be quite profitable.

2. LIPID COMPOSITION

Because the literature dealing with the lipid composition of leukocytes has been reviewed recently (Gottfried, 1972), only a brief synopsis of salient findings will be presented in this section.

The relative content of cholesterol, triacylglycerols, and major phospholipid species of polymorphonuclear leukocytes (PMN), lymphocytes, and macrophages from humans and various rodents is quite similar among the different cell types and despite species differences (Elsbach, 1959; Mason *et al.*, 1972; Gottfried, 1972). In all cell types about two thirds of the total phospholipids are lecithin and phosphatidylethanolamine; sphingomyelin represents 10–20% and phosphatidylserine plus phosphatidylinositides 12–15%. The lack of distinctive features in the overall crude lipid composition of this varied group of cells with profoundly different cellular anatomy, presumably reflects properties common to all mammalian cell membranes. However, the degree of similarity among the different cell types and at various stages of differentiation may not be as marked on further scrutiny (Elsbach, 1977a). Thus, the fatty acid composition of the phospholipids of human PMN, monocytes, and lymphocytes is substantially different (Stossel *et al.*, 1974). It should also be noted that glycolipids are unusually abundant in PMN and lymphocytes (up to 18%) (Miras *et al.*, 1966; Gottfried, 1972). The function of these lipids is at present unknown, but the observation that the carbohydrate portion of the glycolipids changes during malignant transformation, thereby contributing to altered cell surface immunogenicity, has aroused great interest (Hakomori and Murakami, 1968; Gahmsberg and Hakomori, 1973). The possibility that altered glycolipids and glycoproteins also modify surface recognition sites important in phagocytosis has not yet been explored.

Furthermore, with advances in cell fractionation, differences are being recognized at the subcellular level. For example, phagocytic vesicles have now been isolated from rabbit alveolar macrophages and PMN of man and guinea pig (Mason *et al.*, 1972), and secondary lysosomes from cultured mouse peritoneal macrophages (Werb and Gordon, 1975a,b; Werb and Cohn, 1972a). The membranes of these structures have higher cholesterol-to-phospholipid and phospholipid-to-protein ratios than do whole cells, and they are also enriched with respect to disaturated lecithins (Werb and Cohn, 1972a; Mason *et al.*, 1972; Smolen and Shohet, 1974). Moreover, the vesicles from different phagocytes exhibit dissimilar phospholipid composition. Thus, compared to whole cells the vesicles isolated from alveolar macrophages possess a higher proportion of sphingomyelin and phosphatidylserine and of an unusual phospholipid lyso-(bis)-phosphatidic acid, but less phosphatidylcholine, phosphatidylethanolamine, and phosphatidylinositides. On the other hand, PMN vesicles are relatively enriched in phosphatidylinositol (Mason *et al.*, 1972).

Although difficulties remain in the collection of adequate yields of reasonably pure plasma membrane preparations (De Pierre and Karnovsky, 1973), it seems rather well established that these membranes from PMN and lymphocytes contain more cholesterol, glycolipids, and sphingomyelin, and proportionately less phosphatidylcholine than whole cells (Marique and Hildebrand, 1973; Allan and Crumpton, 1972; Ferber *et al.*, 1972; Smolen and Shohet, 1974; Hildebrand *et al.*, 1975).

The further exploration of the phagocyte's plasma membrane, where incoming signals are first recognized and translated into ordered responses, should constitute an important target for future research. The application of a number of recent developments in the study of membranes to phagocytes might well be profitable. It would be of interest, for example, to examine the effect of endocytosis on the apparent distribution of lipids between the two monolayers of the lipid bilayer of the phagocyte's plasma membrane. It is conceivable that modification of the asymmetrical distribution of lipid, which allegedly is a feature of all plasma membranes (Bretscher, 1972), is a consequence of the cell surface perturbations that accompany phagocytosis. In this context the role of phospholipid exchange proteins that are thought to play a part in intracellular translocation of specific phospholipids (Wirtz, 1974) could be investigated.

It also has been proposed that alterations in phospholipid methylation by PMN, in response to various stimuli, might affect phospholipid distribution within the membrane and membrane fluidity (Hirata *et al.*, 1979). Since in short-term experiments only a fraction of 1% of total leukocyte phosphatidylcholine derives from transmethylation, it appears that any membrane changes produced by this pathway would have to be restricted to highly discrete areas. It would be of great interest if it could be shown conclusively that such minute changes indeed mediate or transduce surface perturbations and signals. Other approaches that may yield new insights into plasma membrane alterations during phagocytosis include the use of various membrane probes and other physical techniques (Cronan and Gelman, 1975) and experimental modification of the lipid composition of the phagocyte's membranes. Mahoney *et al.* (1977) have shown that the fatty acyl composition of macrophage phospholipids can be altered by incubation with albumin-complexed fatty acids that are not constituents of the normal extracellular environment. Enrichment of the macrophage phospholipids with a saturated fatty acid was accompanied by reduced endocytosis (both pinocytosis and phagocytosis), presumably because of altered membrane fluidity.

3. LIPID BIOSYNTHESIS

The fact that the lipids of mature mononuclear phagocytes and PMN include all species typical of mammalian tissues implies that phagocytes either possess the biochemical apparatus necessary for their synthesis during their entire life span or that they possessed the machinery at some earlier stage of cellular differentiation. As in most cell types, the identification of pathways of lipid biosynthesis in phagocytes, has in the main been based upon the incorporation of a broad range of radioactively labeled precursors into specific products, rather than on the isolation and purification of individual enzymes.

3.1. BIOSYNTHESIS OF FATTY ACIDS

Incorporation of [^{14}C]acetate into fatty acids by mature PMN was first shown for populations of rabbit peritoneal exudate PMN (Elsbach, 1959). The pattern of incorporation suggested acyl chain elongation rather than *de novo* synthesis. Indeed, subsequent studies have demonstrated that mature human mixed peripheral blood leukocytes lack acetyl-CoA carboxylase (Miras *et al.*, 1965; Majerus and Lastra, 1967).

Incorporation of [^{14}C]acetate by alveolar and peritoneal macrophages is much greater than by PMN (Oren *et al.*, 1963; Gottfried, 1972), in part because the macrophages are much larger cells, but, in particular, because these mitochondria-rich cells have a much more vigorous oxidative metabolism. Whether or not [^{14}C]acetate incorporation into fatty acids actually represents *de novo* synthesis or only chain elongation of preexisting fatty acids, as is the case for mature PMN and lymphocytes, has not been determined.

3.2. CHOLESTEROL SYNTHESIS

Mixed human peripheral blood leukocytes (Fogelman *et al.*, 1975) and isolated human lymphocytes (Liljeqvist *et al.*, 1973) incorporate [^{14}C]acetate into cholesterol. Recently Fogelman *et al.* (1977a) have compared the synthesis of cholesterol by highly purified populations of PMN, monocytes, and lymphocytes. Monocytes incorporated far greater amounts of [2-^{14}C]acetate into sterols than lymphocytes. In fact, PMN only incorporated [2-^{14}C]acetate or [2-^{14}C]mevalonate into squalene, and none into sterol. All types of leukocytes showed greater incorporation of label in lipid-depleted medium, presumably because such a medium causes induction of 3-OH-3-methyl-glutaryl-CoA reductase (Fogelman *et al.*, 1975, 1977b).

3.3. SYNTHESIS OF COMPLEX LIPIDS

PMN from various sources also label cellular lipids when incubated with [^{14}C]glucose and inorganic ^{32}P. This indicates the operation of the phosphatidic acid pathway (Kennedy, 1962). Through this pathway, PMN incorporate a major portion of the exogenous albumin-bound [^{14}C]fatty acids that during incubation *in vitro* appear into neutral lipids and phospholipids (Elsbach, 1972). Another portion of free fatty acids incorporated by resting PMN is probably used for acylation of lysophospholipids (monoacylphosphoglycerides) that also circulate in the blood, mainly as albumin complexes (Elsbach, 1968; Switzer and Eder, 1965). Sastri and Hokin (1966) have provided evidence that phosphatidic acid is not only formed by acylation of SN-glycerol-3-P, but under certain experimental conditions can also be synthesized by phosphorylation of diacylglycerols or perhaps of monoacylglycerol followed by acylation of lysophosphatidic acid.

The phosphatidic acid route of complex lipid biosynthesis has been demonstrated in macrophages collected from lungs of normal rabbits and after BCG stimulation (Elsbach, 1972; Gottfried, 1972) and from peritoneal exudates (Day and Fidge, 1962). In BCG-stimulated alveolar macrophages the cytidine diphosphocholine phosphotransferase activity, responsible for the final step in lecithin synthesis, can be demonstrated in plasma membrane preparations as well as endoplasmic reticulum (Wang *et al.*, 1976). Through the phosphatidic acid pathway, the macrophage also actively incorporates extracellular free fatty acids into triacylglycerols and phospholipids (Evans and Mueller, 1963; Elsbach, 1965). Alveolar macrophages can utilize exogenous triacylglycerols as well, presumably because they possess plasma-membrane-associated lipases capable of hydrolyzing the triacylglycerols on the cell surface (Elsbach, 1965b; 1972). PMN are unable to utilize extracellular triglycerides (Elsbach, 1972).

The major phospholipids of PMN and macrophages, namely diacylglycerylphosphocholine and diacylglycerylphosphoethanolamine are not only synthesized via the phosphatidic acid pathway, but also by acylation of the monoacyl derivatives (Elsbach, 1972; Wang *et al.*, 1976, 1977). These products of hydrolysis of phospholipids and of the lecithin–cholesterolacyl transferase reaction (Glomset, 1968) circulate in the blood as albumin complexes and can provide a substantial source of new phospholipid (Elsbach, 1968). The conversion of the monoacyl- to the diacylphospholipid in intact and disrupted PMN and alveolar macrophages is carried out by the type of acylation reaction first described by Lands (1960) in liver. The reaction has been studied in some detail in PMN (Elsbach, 1966, 1967, 1968, 1972) and in BCG stimulated macrophages (Elsbach 1966, 1968, 1972; Wang *et al.*, 1976, 1977). In PMN homogenates, but not in intact PMN, monoacylglycerylphosphocholine (but not monoacylglycerylphosphoethanolamine) is also converted to the diacyl derivative in the reaction 2 monoacylglycerylphosphocholine → diacylglycerylphosphocholine + glycerylphosphocholine. This reaction, in contrast to the Lands pathway, which is associated with the microsomal cell fraction, takes place in the cytosol at acid pH and is independent of added ATP and CoA. No evidence of this reaction is found in macrophage homogenates (Elsbach, 1972).

Miras *et al.* (1964) have reported the incorporation of [L-3-^{14}C]serine by a microsomal preparation from mixed peripheral blood leukocytes into diacylglycerylphosphoserine and diacylglycerylphosphoethanolamine. This incorporation occurs in the presence of ATP, CoA, and Mg^{2+} without participation of cytidine nucleotides. As in other systems, incorporation of serine into serine phospholipids requires Ca^{2+} and also takes place in an energy-independent fashion. The acylation of dihydroxyacetonephosphate as an alternate pathway for phosphatidic acid biosynthesis and the pathway of synthesis of glycerol–ether bonds have not been examined in leukocytes and macrophages.

3.4. LIPID SYNTHESIS DURING AND AFTER PHAGOCYTOSIS

During phagocytosis phagocytes expend more metabolic energy (Karnovsky, 1962), presumably needed for the mechanical work of locomotion and engulfment. The possibility has also been considered that energy is needed for biosynthetic activity, in particular for the production of new membrane to complete formation of the phagocytic vacuole or to replace the internalized plasma membrane. Conceivably, energy might also be used if no new membrane were needed, and if translocation of redundant plasma membrane were the means by which the phagocytic vacuole were formed. Evidence has been presented, however, indicating that newly formed membrane lipids and proteins do accumulate during or following phagocytosis.

Earlier studies had indicated that, while phagocytosis is in progress, human and rabbit PMN and rabbit alveolar macrophages show no appreciably increased incorporation of [^{14}C]acetate into fatty acids, nor of ^{14}C-fatty acids into phospholipids (Elsbach, 1959; Sastry and Hokin, 1966; Wang et al., 1977). The meaning of a modestly increased incorporation of [^{14}C]acetate into triglycerides of rabbit PMN (Elsbach, 1959) and of oleic acid into triglycerides of alveolar macrophages (Wang et al., 1977) is unclear.

The only substantial net addition of membrane phospholipid during phagocytosis that so far has been demonstrated stems from the acylation by PMN of extracellular albumin-bound monoacylglycerylphosphocholine (lysolecithin) and monoacylglycerylphosphoethanolamine to their respective diacyl derivatives according to the accompanying Figure 1 (Elsbach, 1968; Elsbach et al., 1969). In PMN the acylation of exogenous lysolecithin is stimulated up to threefold over resting values during ingestion of various types of particles (Elsbach, 1968). This stimulated single-step conversion of lysolecithin to lecithin indeed appears to represent net addition of phospholipid, by as much as 10% of the total lecithin content of the cell in 30 min, because during this time there is no detectable turnover of phospholipids, nor any loss of lipid into the medium (Elsbach, 1968).

The fatty acid needed for the increased acylation of the exogenous monoacyl compound to the membrane phospholipid seems to derive from hydrolysis of cellular triglycerides rather than from extracellular free fatty acid (Elsbach and Farrow, 1969; Shohet, 1970). Whether resting PMN also use fatty acids from cellular stores for the conversion of monoacylphospholipids to diacylphospholipids has not been determined.

Our observation that most of the radioactively labeled lecithin synthesized

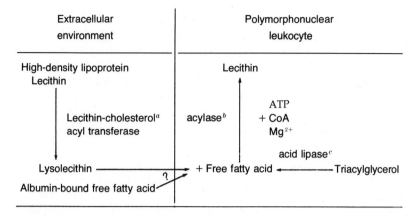

FIGURE 1.   Pathway of lecithin formation from lysolecithin by intact leukocyte. [a]Glomset (1968). [b]Lands (1960); Elsbach (1968, 1972). [c]Elsbach and Farrow (1969); Shohet (1970).

during phagocytosis from medium [^{32}P]lysolecithin is recovered in association with isolated phagosomes, supports the contention that the increased net synthesis reflects formation of new membrane needed for the engulfment process (Elsbach *et al.*, 1972b). Further, inhibition of glycolysis, the main source of energy in PMN, inhibits both phagocytosis and the increase in acylation (Elsbach, 1972). Rabbit alveolar macrophages, obtained after BCG stimulation, exhibit roughly the same increased acylation of albumin-bound lysophospholipids per cell during phagocytosis as do PMN (Elsbach, 1968). However, the actual contribution of phospholipid from this source is rather small, because the much larger macrophage contains approximately eight times more phospholipid per cell than the PMN. Moreover, phospholipid turnover in alveolar macrophages is quite rapid (Franson *et al.*, 1973). It is, therefore, uncertain whether the increased incorporation of monoacylphospholipids during phagocytosis by this type of phagocyte represents net addition of phospholipid or merely increased turnover. Wang *et al.* (1977) also have examined uptake of radiolabeled lysocompounds by alveolar macrophages from rabbits challenged with BCG. These workers found no stimulation of acylation during incubation with autoclaved *Escherichia coli* B. The reason for the divergent results is not clear. One possible cause for failure to achieve stimulation is uptake of too few particles under the experimental conditions chosen.

The studies of Werb and Cohn (1972a,b) on the cultivated mouse peritoneal macrophage further suggest that macrophages during engulfment do not synthesize much new membrane (lipid). In fact, an initial decrease in plasma membrane can be demonstrated. However, both cholesterol and phospholipid accumulate several hours after ingestion. The net increase in lipid mass is proportional to the quantity of particles ingested and also to the amount of plasma membrane interiorized. The reconstitution of plasma membrane following ingestion than permits the macrophages to phagocytize anew. This study also showed for the first time resynthesis of plasma membrane protein as part of membrane replacement: During phagocytosis plasma membrane 5'-nucleotidase activity falls in proportion to the quantity of phagolysosome formed, that is, the amount of plasma membrane internalized. The phagosome-associated enzyme activity begins to reappear some 6 hr after ingestion. At about this time, regeneration of a receptor protein for serum lipoproteins that supply the macrophage with cholesterol also starts (Werb and Cohn, 1972a).

These important studies show not only that at least the mouse peritoneal macrophage can ingest large numbers of particles (more than one hundred 1.1-μm latex particles per cell) by internalizing preexisting plasma membrane, but also that these cells ultimately replace the membrane used. As pointed out by Werb and Cohn (1972b), membrane synthesis by short-lived PMN and by the long-lived macrophage must have different functions. Perhaps this accounts for the fact that the net synthesis of membrane (lipid) by the PMN accompanies its single act of phagocytosis, but follows engulfment in the case of the macrophage which prepares itself for renewed phagosome formation. In these experiments Werb and Cohn used indigestible particles. It is possible that much less new membrane needs to be made if the ingested particle can be degraded. Under such conditions, the internalized plasma membrane may be recycled to be used once again for assembly of outer membrane. This seems to occur in pinocytizing *Acanthamoeba* (Korn *et al.*, 1974). These organisms internalize their whole surface membrane from 5 to 50 times per hour, but exhibit no evidence of increased turnover of plasma membrane phospholipid. The mechanism of reinsertion of internalized membrane into surface membrane might be through fusion of empty vesicles. Another means of delivery of phospholipid (not dependent upon biosynthesis of new lipid) to the plasma membrane from other membrane sites within the phagocyte might be mediated by a class of recently characterized phospholipid exchange proteins (reviews, Wirtz, 1974; Kader, 1977), present in all mammalian tissues examined so far. These proteins can transfer phospholipid between intracellular particles. However, net transfer of phospholipid from one membrane to another has not yet been demonstrated. Finally, as will be discussed in greater detail below, the lipids of ingested bacteria undergo degradation. PMN efficiently incorporate the products of hydrolysis into their own lipids. The amount incorporated is proportional to the extent of hydrolysis and is, therefore, determined by the number of microorganisms taken up and the susceptibility of the microbial lipids to degradative attack during phagocytosis (Elsbach *et al.*, 1972a; Patriarca *et al.*, 1972b). The contribution to membrane assembly of lipid biosynthesis from hydrolysis products of the lipids of microorganisms ingested by phagocytes other than PMN has not yet been investigated to our knowledge.

4. LIPID DEGRADATION

All phagocytic cells so far examined possess an extensive lipid-degradative apparatus. By use of defined substrates triacylglycerol, phospholipid, and

cholesterol-ester-hydrolyzing activities have been identified in polymorphonuclear leukocytes from rabbit and human (Elsbach, 1977a; Braunsteiner *et al.*,
1968), in rabbit alveolar macrophages (Elsbach, 1972; Franson and Waite, 1973;
Franson *et al.*, 1973) and in mouse (Werb and Cohn, 1972a) and rabbit peritoneal
macrophages (Day, 1967). In only a few instances has extensive purification of
single enzymatic activities been achieved (Elsbach *et al.*, 1979; Rindler-Ludwig *et
al.*, 1977).

In addition to functioning in the turnover of lipids as part of normal cellular
metabolism, other roles more specific for the phagocytic cell and the endocytic
process have been investigated. For example, an increase in phospholipid degradation might accompany the membrane alterations that occur during
phagocytosis, such as formation of the phagocytic vacuole and degranulation.
Thus, increased synthesis of membrane lipids in certain regions of the cell, if
required for endocytosis, might depend in part on free fatty acids derived from
endogenous lipid stores. Hydrolysis of lipids, in particular of triacylglycerols,
might also serve to make free fatty acids available as substrate for the increased
energy demands of phagocytosis. Further, to provide the free fatty acid precursors arachidonic acid and di-homo-γ-linolenic acid of prostaglandins and other
biologically active conversion products, known to be synthesized in increased
amounts by engulfing PMN (Higgs *et al.*, 1975; Borgeat *et al.*, 1976), activation of
lipolytic enzymes is thought to be needed (Lands and Rome, 1976). A particularly important function of lipolytic enzymes of phagocytic cells might be the
degradative attack on the lipids of ingested material, including microorganisms.
Evidence in support or against such functions of lipolytic enzymes will be reviewed in this section.

4.1. GENERAL PROPERTIES OF LIPID-HYDROLYZING ENZYMATIC ACTIVITIES IN PHAGOCYTIC CELLS

4.1.1. Cellular Localization

Not unexpectedly, the lipolytic activities are often found in association with
other hydrolytic activities that are typically concentrated in the granules and/or
the lysosomal apparatus of the phagocytes. However, some preliminary evidence has been presented, suggesting that, for example, phospholipase A_2 of
both rabbit and human pmn are not restricted to the granules, but may also
occur in association with other membrane fractions and possibly the plasma
membrane (Franson *et al.*, 1974, 1977). Consistent with a lysosomal localization
of a number of acylglycerol-acyl-hydrolases is the finding that several among the
lipid-hydrolyzing activities exhibit acid pH optima (Elsbach and Rizack, 1963;
Elsbach and Kayden, 1965; Elsbach, 1965b; Franson *et al.*, 1973, 1974). Moreover,
the triacylglycerol-acyl-hydrolases of PMN manifest the latency characteristic of
many lysosomal degradative enzymes (Elsbach and Rizack, 1963; Elsbach and
Kayden, 1965). Others, such as the phospholipase A_2 of rabbit and human PMN

show no latency (Franson *et al.*, 1974, 1977; Weiss *et al.*, 1975), presumably because this activity is so tightly membrane-associated that treatment with weak acids or detergents, or freezing and thawing cause no apparent release.

4.1.2. Triacylglycerol-acyl-hydrolase (E.C. 3.1.1.3)

A number of artificial dispersions of triacylglycerols and the naturally occurring lipoproteins, chylomicrons, serve as substrates. While activity towards all these different substrates is maximal at acid pH, the pH optimum varies from one substrate to another (Elsbach, 1972). The lipase activity of intact rabbit PMN, under certain conditions, is also evident towards endogenous triacylglycerol but the enzyme(s) appear(s) unable to act on extracellular triacylglycerols (Elsbach, 1972; Elsbach and Farrow, 1969). In contrast, indirect evidence has been presented suggesting that the very potent lipase of alveolar macrophages occurs partly in association with the plasma membrane and therefore can hydrolyze triacylglycerols that come in contact with the cell surface (Elsbach, 1965b).

Recently, the partial purification has been reported of an acid triacylglycerol-acyl-hydrolase of human leukocytes (Rindler-Ludwig *et al.*, 1977). Its properties seem rather similar to those of the crude lipase preparation from rabbit PMN.

4.1.3. Phospholipases

Rabbit and human PMN and rabbit alveolar macrophages possess several phospholipid-splitting enzymes (see earlier reviews, Elsbach, 1972, 1977a,b). Figure 2 shows the major phospholipase activities that have been recognized. Phospholipase C and D activity cannot be demonstrated in most animal cells and tissues.

4.1.3a. Phospholipase A (E.C. 3.1.1.4). Rabbit PMN contain prominent phospholipase A_2 activity which is associated with both specific and azurophil granules and also with another membrane-rich cellular fraction obtained by zonal centrifugation (Franson *et al.*, 1974). The granule-associated phospholipase A_2 activity of rabbit peritoneal PMN exhibits two distinct pH optima, one at pH 5.5 and one around pH 7.5 (Franson *et al.*, 1974). Extensive purification of this phospholipase A_2 activity at alkaline pH has been achieved (Elsbach *et al.*, 1979). It is uncertain whether removal during purification of the acid phospholipase A_2 accounts for the single pH optimum of the purified phospholipase A_2, or whether modifiers that seem to be present in the crude preparation (Franson *et al.*, 1974) are responsible for the appearance of a double pH optimum of a single phospholipase A_2. Calcium ions are absolutely required for activity (Elsbach *et al.*, 1972a; Franson *et al.*, 1974). However, the Ca^{2+} concentration needed for maximal activity diminishes with increasing purification. The positional specificity of the rabbit PMN phospholipase A for the 2-acyl position of the major phospholipid species of cell membranes of prokaryotic as well as eukaryotic cells seems well-established (Franson *et al.*, 1974; Weiss and Elsbach, 1977).

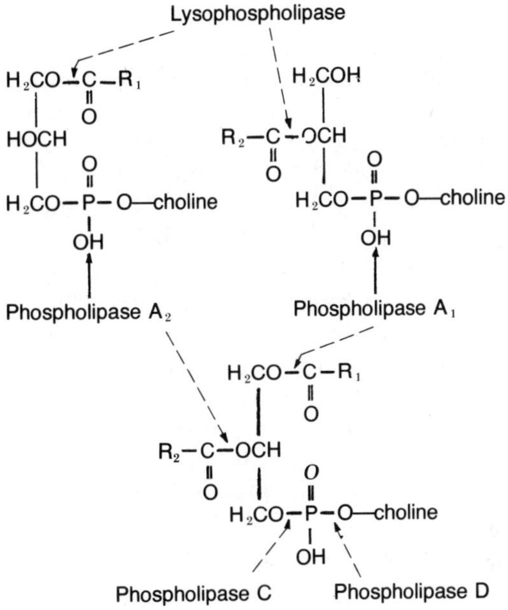

FIGURE 2. Phospholipases acting on glycerolphospholipids (example, lecithin). ‑ ‑ ‑→ Site of action; ——→, product of phospholipase action.

The cell-free ascitic fluid obtained by centrifugation of sterile peritoneal exudates elicited in rabbits to produce large populations of PMN (Hirsch, 1956), contains abundant phospholipase A_2 activity. This activity has been isolated in a highly purified protein fraction of the ascitic fluid (Franson *et al.*, 1978). The properties of the phospholipase A_2 in the fraction are not detectably different from the phospholipase A_2 that we have isolated from rabbit PMN (Franson *et al.*, 1974; Weiss *et al.*, 1975; Elsbach *et al.*, 1979). It is possible, therefore, that leukocytes discharge their phospholipase A_2, perhaps in response to certain (inflammatory?) stimuli. Such release may be akin to the secretion of large amounts of lysozyme (Gordon *et al.*, 1974), and collagenase and elastase (Werb and Gordon, 1975a,b) by mononuclear phagocytes.

A Ca^{2+}-dependent phospholipase A_2 activity has also been identified in human PMN (Franson *et al.*, 1977). Its intracellular distribution, as delineated by rate-zonal centrifugation, among specific and azurophil granules and in a lighter membrane fraction (possibly plasma membrane), is closely similar to that of the phospholipase A_2 of rabbit PMN. However, the phospholipase A_2 in crude preparations of human leukocytes exhibits only a single pH optimum at pH 7.0 and is only about 1/20 as active as the rabbit enzyme (Franson *et al.*, 1977; Weiss *et al.*, in preparation).

In rabbit alveolar macrophages, at least two phospholipases A have been identified; one with acid pH optimum, active in the presence of Ca^{2+} chelators and inhibited by Ca^{2+}, and another with alkaline pH optimum dependent on

Ca^{2+} for activity (Franson *et al.*, 1973). The acid phospholipase appears to be a phospholipase A_2 and occurs predominantly in a lysosome-rich fraction. Determination of the intracellular distribution of the alkaline phospholipase A has been hampered by the fact that fractionation causes loss of activity. Further, since no lysophosphatides accumulate under assay conditions at alkaline pH (Franson *et al.*, 1973), the positional specificity of the phospholipase A activity in this pH range is uncertain. The complete deacylation of diacylphosphoglycerides may be carried out by a phospholipase capable of hydrolyzing both ester positions, or can be the consequence of two types of combined action of two distinct hydrolases: (1) a phospholipase A_2 plus a phospholipase A_1 with limited substrate specificity, and hence capable of degrading lysophosphatides, or (2) a phospholipase A (A_1 or A_2) plus a lysophospholipase. The distinction between these possibilities is usually impossible to determine without purification to homogeneity of the individual enzymes. However, even after this has been achieved, it must be recognized that the strictness of positional specificity often is not absolute and depends on choice of substrate and assay conditions.

4.1.3b. Lysophospholipase. Both rabbit PMN and alveolar macrophages exhibit potent lysophospholipase (E.C. 3.1.1.5) activity at neutral or alkaline pH (Elsbach, 1966, 1967, 1972). Some of this activity can be sedimented by centrifugation, but the bulk of it in the two cell types appears to be nonparticulate and, in PMN, can be dissociated from the granules.

4.1.3c. Other Phospholipid-Hydrolytic Enzymatic Activities. Sphingomyelinase (sphingomyelin phosphodiesterase, E.C. 3.1.4.12) activity has been identified in human peripheral blood leukocytes (Kampine *et al.*, 1967). Its importance lies in its role in the turnover of normally occurring cellular constituents and in the fact that congenital deficiency of this enzymatic activity results in accumulation of sphingolipids causing severe cellular dysfunction, especially in the nervous system (Niemann–Pick disease) (Kampine *et al.*, 1967; Frederickson and Sloan, 1972). There is no known function for sphingomyelinase in phagocytosis and the events that accompany it. So far as we know sphingomyelinase has not been described in other phagocytic cells.

4.1.4. Cholesteryl Esterase (E.C. 3.1.1.13)

Cholesteryl esterases are also ubiquitous lipolytic enzymes. Their lysosome-associated acid hydrolase activity has been found in rabbit peritoneal macrophages (Day, 1967), cultured mouse peritoneal macrophages, and rabbit alveolar macrophages (Werb and Cohn, 1972a), and in human peripheral blood mixed leukocyte preparations (Cortner *et al.*, 1976). We are not aware of studies dealing specifically with cholesteryl-ester hydrolases of relatively homogeneous populations of PMN. In other tissues, it has not been possible to separate acid triacylglycerol-hydrolase and cholesteryl-ester hydrolase activities (Sloan and Frederickson, 1972; Brecher *et al.*, 1977), and it has been suggested that the two activities may reside on a single protein. The coincident deficiency of the two activities in liver, spleen, and lymph nodes of patients with Wolman disease and cholesteryl-ester storage disease (Sloan and Frederickson, 1972) speaks in favor of this possibility.

Cholesteryl esterase activity has been shown to hydrolyze the cholesteryl-esters of low density lipoproteins endocytized by cultured human fibroblasts (Goldstein *et al.*, 1975). The sequential surface receptor binding, endocytosis, and degradation of low-density lipoproteins by cultured fibroblasts and by fresh and cultured lymphocytes (Brown *et al.*, 1976; Ho *et al.*, 1976, 1977) lead to intracellular liberation of free cholesterol and feedback suppression of the rate-limiting enzyme in cholesterol biosynthesis, 3-hydroxy-3-methyl-glutaryl coenzyme A reductase (Brown *et al.*, 1976). The importance of this metabolic sequence is indicated by its interruption in the homozygous form of familial hypercholesterolemia (Brown *et al.*, 1976; Ho *et al.*, 1977; Stein *et al.*, 1976).

Whether cholesteryl-ester hydrolase in phagocytic cells serves a similar function or whether the enzyme(s) fulfill(s) other roles during phagocytosis has not been determined.

4.2. FUNCTION OF LIPOLYTIC ENZYMES IN PHAGOCYTOSIS

4.2.1. Membrane Phospholipid Degradation as Part of the Membrane Modification of Endocytosis

It is unknown what the trigger is that elicits the sequence of multiple metabolic and physical responses following the phagocyte's recognition of signals such as chemotactic factors and contact with particles carrying certain surface properties. We have recently postulated that activation of phospholipase(s) A might be an early event (Elsbach, 1977b). Activation could follow release of Ca^{2+}, known to occur upon stimulation of surface sites by specific agonists, including the chemotactic C_{5a} (Gallin and Rosenthal, 1974; Barthélemy *et al.*, 1977). Hydrolysis of phospholipids by phospholipase A in the plasma membrane would alter the membrane fluidity because of replacement of diacylphospholipids by lysocompounds and free fatty acids (Figure 3). Such a physicochemical change might constitute the signal that leads to activation of the phagocyte's contractile apparatus. It has been suggested, for example, that local anesthetics may transmit their effects on the cytoskeleton of BALB/3T3 cells via a change in the hydrophobic properties of the plasma membrane (Poste *et al.*, 1975; Nicolson *et al.*, 1976, 1977).

It has been postulated repeatedly that the formation of lysophospholipids in discrete domains of outer and inner membranes by membrane-associated phospholipases A might cause sufficient disorganization to facilitate local dissolution and fusion upon intermembrane contact, as occurs during the genesis of the phagocytic vacuole and degranulation (Elsbach and Rizack, 1963; Elsbach *et al.*, 1965; Lucy, 1970). In fact, positive evidence in support of the concept that the production of membrane-labilizing lysophospholipids is involved in fusion is

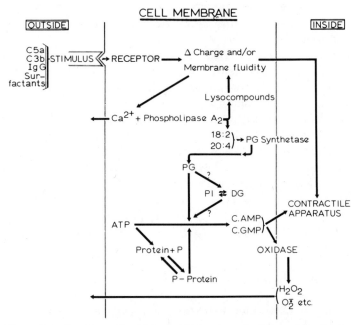

FIGURE 3. Key: 18:2, linoleic acid; 20:4, arachidonic acid; PG, prostaglandin; PI, phosphatidylinositide; DG, diglyceride. Reproduced from P. Elsbach, 1977, Cell surface changes in phagocytosis, in: *The Synthesis, Assembly and Turnover of Cell Surface Components* (G. Poste and G. L. Nicolson, eds.), with permission of Elsevier/North-Holland Biomedical Press.

not available (Korn *et al.*, 1974; Elsbach, 1977b). Thus, net accumulation of lysophospholipids during phagocytosis by PMN and alveolar macrophages has not been detected (Elsbach *et al.*, 1972a). Moreover, both PMN and macrophages possess enzymatic mechanisms to keep the accumulation of lysophospholipids in check, either through reacylation by membrane-associated acylases (acyl-CoA-monoacyl phospholipid-acyltransferase) causing resynthesis of diacyl-phospholipids, or by further deacylation to nonlytic water soluble products by lysophospholipases (Elsbach, 1972). It should be recognized, however, that increases in lysophospholipids in discrete areas of the cell may occur that are not detectable by current methods. Therefore, no final conclusions about the role of lysophospholipids in fusion can yet be drawn.

Partial hydrolysis of the phagocyte's phospholipids and their reconstitution by reacylation of the monoacylphospholipids further provides the cell with a means of changing the fatty acid composition and hence the physical properties of major membrane phospholipids. Mason *et al.* (1972) and Smolen and Shohet (1974) have shown, for example, that during phagocytosis the phospholipid fatty acids of phagocytic vesicles and plasma membranes become more saturated. An increase in saturated fatty acids and a decrease in unsaturated fatty acids is not only caused by replacement in a deacylation–reacylation cycle, but might also be a consequence of peroxidative destruction of polyunsaturated fatty acids. Peroxidation of lipids may be expected to occur during the respiratory burst with concomitant production of H_2O_2 and other highly reactive compounds. Indeed, production of malonyldialdehyde reflecting formation of lipid peroxides has been shown during phagocytosis by rabbit alveolar macrophages, guinea pig PMN (Mason *et al.*, 1972), and by human granulocytes and monocytes from peripheral blood (Stossel *et al.*, 1974). In the case of actively engulfing PMN, peroxidation of endogenous lipids actually has not been demonstrated. Malonyldialdehyde accumulation was detected only during ingestion of particles rich in linolenate (Mason *et al.*, 1972; Stossel *et al.*, 1974).

Another important consequence of the activation of phospholipase A_2 is the release of polyunsaturated fatty acids, substrate for oxygenases that produce prostaglandins. The role of these exceedingly potent chemical messengers in the phagocytic process is still largely conjectural. One possible link they provide in the chain of metabolic reactions triggered by the initial surface perturbation, may be in the regulation of cyclic nucleotide formation (Constantopoulos and Najjar, 1973).

Studies in other cell types suggest that regulatory sequences may operate in the reverse direction. For example, high concentrations of lysolecithin (2×10^{-4} M and higher) have opposite modulatory effects on guanylate cyclase and on adenylate cyclase activities in membrane preparations of 3T3 mouse fibroblasts (Shier *et al.*, 1976), and in platelets Lapetina *et al.* (1977) have observed inhibition of membrane phospholipase by cyclic adenosine 3',5'-monophosphate. For a further speculative discussion of connections between the various events depicted in Figure 3, another recent review may be consulted (Elsbach, 1977b).

4.2.2. The Use of Lipid Stores for the Energy Needs of Phagocytosis

The findings in PMN that glycogenolysis, glycolysis, and lactate production are enhanced very early during phagocytosis, and that inhibition of glycolysis also inhibits phagocytosis, mainly account for the belief that engulfment of particles requires expenditure of metabolic energy (Karnovsky, 1962). If this is so, then phagocytes, in common with muscle and other tissues, may meet increased energy demands not only by increasing glycolysis but also by greater oxidation of fatty acids as a highly efficient means of energy production. As indicated earlier, during phagocytosis PMN hydrolyze part of the substantial triglyceride stores (Elsbach, 1959; Elsbach and Farrow, 1969), presumably because the granule-associated acid lipase(s) is (are) activated by a drop in intracellular pH (Bainton, 1973; Elsbach, 1972). Further, PMN convert ^{14}C-1-free fatty acids to $^{14}CO_2$ (Elsbach and Farrow, 1969). Hence, although the PMN contains few mitochondria, the cell appears capable of fatty acid oxidation. However, $^{14}CO_2$ production by PMN populations whose triglycerides and phospholipids had been labeled during preincubation with ^{14}C-1-fatty acids is no greater during phagocytosis than at rest (Elsbach and Farrow, 1969). Whether macrophages with their more vigorous oxidative metabolism rely on fat as a source of energy during phagocytosis has not been determined to our knowledge.

4.2.3. Conversion of Polyunsaturated Fatty Acids to Prostaglandins, Thromboxanes, and OH-Fatty Acids

It is generally believed that arachidonic acid and other polyunsaturated fatty acid precursors for synthesis of prostaglandins and other biologically active products do not directly stem from the extracellular free fatty acid pool (Lands and Rome, 1976). These diet-derived essential fatty acids first become part of the cellular acyl-esters and must then be released into the exceedingly small cellular free fatty acid pool by acyl-ester hydrolases in order to permit cyclooxygenase and lipoxygenase activity to be expressed (Figure 4). Regulation of the biosynthesis of prostaglandins and other products of conversion of fatty acid precursors depends, therefore, not only on the synthase (oxygenase) activities, but also on the extent of lipolytic activity and its control (Lands and Rome, 1976).

In common with virtually all mammalian cells, PMN and macrophages and, presumably, other phagocytic cells are also capable of biosynthesis of prostaglandins and other biologically active derivatives of polyunsaturated fatty acids such as arachidonic and di-homo-γ-linolenic acid. Moreover, PMN and macrophages share with other cells the ability to respond to certain stimuli with increased formation of these products. Thus, human peripheral blood and rodent peritoneal exudate PMN during phagocytosis of killed bacteria (Bordetella pertussis) or serum treated zymosan particles release at least tenfold more PGE_1 and PGE_2 into the medium than at rest (Higgs and Youlten, 1972; McCall and Youlten, 1973; Bray et al., 1974; Higgs et al., 1975; Tolone et al., 1977). Homogenates of rabbit PMN that has ingested killed bacteria, but not of resting PMN, also produce a thromboxane-A_2-like activity when presented with the endoperoxides PGG_2 or PHG_2 (Higgs et al., 1976). Intact rabbit PMN have further been shown to convert arachidonic acid and di-homo-γ-linolenic acid to monohydroxy acids, presumably by a lipoxygenase system (Borgeat et al., 1976).

Mouse peritoneal macrophages, incubated with serum treated zymosan

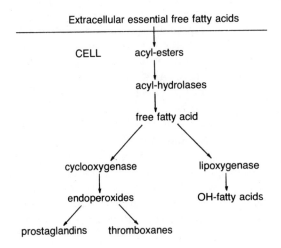

FIGURE 4. Sources of fatty acids for synthesis of prostaglandins and other conversion products.

particles and [³H]arachidonic acid show greatly increased release of [³H]-PGE₂ (Humes *et al.*, 1977). This stimulation is not seen with latex particles. It is surprising that stimulation of prostaglandin formation by zymosan continued linearly for 24 hr.

All these conversion products are believed to be important mediators in several phases of the inflammatory response. Thus, PGE₁ (Kaley and Weiner, 1971), thromboxane B₂ (Boot *et al.*, 1976) and the monohydroxy acid 12-ʟ-OH-5,8,10,14-eicosatetraenoic acid (Hamberg and Samuelsson, 1974; Nugteren, 1975) have been shown to be chemotactic for PMN. The stimulation of synthesis of these agents during phagocytosis would, therefore, promote the migration of other phagocytes to the inflammatory site (Higgs *et al.*, 1975). The observation that various prostaglandins added (in high concentration) to PMN *in vitro* inhibit phagocytosis and lysosomal enzyme release (Zurier *et al.*, 1974; Smith, 1976), may reveal another regulatory function in the inflammatory response.

The inhibitory effect of both steroidal and nonsteroidal antiinflammatory agents such as aspirin, indomethacin, and related compounds on prostaglandin synthesis has often been adduced as supportive evidence for the importance of prostaglandins in modulation of the responses of phagocytic cells.

Of the nonsteroidal drugs the effect of indomethacin on synthesis of prostaglandins and on other conversions of arachidonic acid has been studied most extensively. It must be recognized that indomethacin (and most other inhibitory substances) at higher concentrations (10^{-3}–10^{-4} M) inhibits numerous enzymatic activities (Flower, 1974). However, at relatively low concentrations (10^{-6}–5 × 10^{-5} M), presumed to fall in the pharmacological range (Lands and Rome, 1976), indomethacin inhibits prostaglandin release in all tissues examined, including PMN (Higgs and Youlton, 1972; Higgs *et al.*, 1975) and macrophages (Humes *et al.*, 1977).

Lands and co-workers have shown that indomethacin has two effects on the cyclooxygenase system that converts arachidonate to prostaglandins and thromboxane A₂: (1) a reversible competitive-type inhibition and (2) an irreversible, time-dependent destructive effect (Lands and Rome, 1976).

Indomethacin, at the same minimal concentrations that inhibit prostaglandin synthesis by PMN also inhibits phospholipid hydrolysis by rabbit PMN phospholipase A₂ (Kaplan *et al.*, 1978). In contrast to the effect on cyclooxygenase, indomethacin's effect on PMN phospholipase A₂ is immediate and of the noncompetitive type. Since inhibition of hydrolysis of complex lipids implies reduced availability of precursor polyunsaturated free fatty acids for conversion to biologically active products, the effect of indomethacin on prostaglandin synthesis by PMN may be at more than one site in the enzymatic sequence.

It is not yet known whether other lipolytic enzymes are sensitive to indomethacin and related drugs, nor whether antiinflammatory steroids have a direct effect on the phospholipase A₂ of inflammatory cells.

In transformed mouse fibroblasts hydrocortisone inhibits stimulation by serum of prostaglandin synthesis by inhibition of release of arachidonic acid from cellular lipids (Hong and Levine, 1976).

4.2.4. Microbial Lipid Degradation during Phagocytosis

Early alterations in the microbial envelope accompany killing and digestion of bacteria ingested by phagocytic cells (Elsbach, 1973; Beckerdite et al., 1974). Surprisingly few studies have been carried out, aimed at assessing the importance of degradation of microbial lipid (as major structural components of the envelope) in the fate of ingested microorganisms.

The earliest evidence of phospholipid degradation during and after killing of both gram-negative and gram-positive bacteria by rabbit PMN was reported by Cohn (1963). Not until nearly a decade later were these observations confirmed and extended (Patriarca et al., 1972b). Lipid degradation, although initiated almost as soon as PMN and bacteria make contact, proceeds much slower than killing, reaches a plateau in about 30 min, and does not resume during incubation for several hours. Such a time course does not help clarify whether or not lipid breakdown is an integral part of any bactericidal mechanism.

Further, since many bacteria possess phospholipases (Kleiman and Lands, 1969; Albright et al., 1973; Doi and Nojima, 1973; Kusaka, 1975) that may be activated under conditions adverse to the microorganism (Patriarca et al., 1972a; Kusaka, 1975; Beckerdite-Quagliata et al., 1975), the relative contribution of the PMN phospholipase A_2 and bacterial phospholipases to the degradation measured is uncertain. By heat-inactivating the bacterial phospholipases, it can be shown that rabbit PMN phospholipase A_2 can attack the phospholipids of ingested bacteria (Patriarca et al., 1972b) and therefore this granule-associated hydrolytic enzyme acts within the phagocytic vacuole. It has further been demonstrated (Weiss and Elsbach, 1977) that the PMN phospholipase A_2, present in a highly purified bactericidal fraction, hydrolyzes the bacterial phospholipids during killing of an E. coli mutant devoid of demonstrable phospholipases A (Doi and Nojima, 1973). Similar experiments with the parent strain (possessing the full spectrum of phospholipid hydrolases typical of E. coli) (Doi and Nojima, 1973; Albright et al., 1973) indicate that its rapid loss of viability is accompanied by greater phospholipid hydrolysis, attributable to activation of bacterial phospholipase A (Weiss and Elsbach, 1977). What the relative contribution of the PMN and bacterial phospholipases is to bacterial phospholipid digestion during phagocytosis by PMN should now be amenable to study with these mutant and parent strains.

Normally, the envelope phospholipids of gram-negative and gram-positive bacteria are protected from degradative attack by exogenous as well as endogenous, envelope-associated, phospholipases. For example, pure phospholipases (A_2 and C) added to intact gram-negative wild-type E. coli (Duckworth et al., 1974; van Alphen et al., 1977) and Salmonella typhimurium (Kamio and Nikaido, 1976) or to gram-positive Bacillus subtilis (Op den Kamp et al., 1972) produce no phospholipid hydrolysis. However, after modification of the bacterial envelope, by mutation (van Alphen et al., 1977; Kamio and Nikaido, 1976), antibiotics such as polymixin B (Beckerdite-Quagliata et al., unpublished observations), or heat treatment (Patriarca et al., 1972b), the envelope phospholipids do become susceptible to the action of added phospholipases. Similarly, autolipolysis is only evident after membrane perturbation (Patriarca *et al.*, 1972a; Kusaka, 1975; Beckerdite-Quagliata *et al.*, 1975; Weiss and Elsbach, 1977).

The resistance in general of intact prokaryotic (and eukaryotic) cells to phospholipases has been attributed to several physicochemical properties intrinsic to biological membranes. These include: (1) nonlipid surface constituents (or layers) that shield the envelope phospholipids from access to phospholipases in the extracellular environment (Patriarca *et al.*, 1972b; Kamio and Nikaido, 1976; van Alphen *et al.*, 1977); and (2) tight lipid packing, which prevents insertion of phospholipases into the lipid bilayer (Verger *et al.*, 1973; Demel *et al.*, 1975). The extent of packing is determined by hydrophobic interactions between phospholipid acyl chains (greater with increasing saturation and chain length) and hydrophobic regions of other envelope macromolecules (Cronan and Gelman, 1975). In addition, electrostatic interactions between phospholipid polar headgroups and charged groups of other envelope constituents, often mediated through divalent cations (Mg^{2+}, Ca^{2+}) (Leive, 1974; Nakamura and Mizushima, 1975; Nicolson *et al.*, 1976; Takeuchi *et al.*, 1978) contribute to a tighter envelope structure.

The phospholipids of gram-positive bacteria reside entirely within the cytoplasmic membrane, which is surrounded by a nonlipid cell wall (Salton, 1964). Rapid dissolution of the cell wall peptidoglycan leading to fragmentation of the remaining envelope of the exquisitely lysozyme-sensitive *Micrococcus lysodeikticus* during killing by intact or disrupted PMN is followed by degradation of about 25% of their phospholipids (Patriarca *et al.*, 1972b; Elsbach *et al.*, 1973). The susceptibility of phospholipids of other gram-positive bacteria which possess more complex and lysozyme-resistant cell walls (e.g., *Staphylococcus aureus*) has not yet been examined. That 75% of the membrane phospholipids of *M. lysodeikticus* remain intact clearly demonstrates how effective the envelope protective forces are. More extensive disorganization of the bacterial envelope, produced, for example, by autoclaving, permits greater degradation to occur (Patriarca *et al.*, 1972b).

The gram-negative bacterial envelope contains an additional barrier to attack by exogenous phospholipases on the cytoplasmic membrane phospholipids, namely the highly specialized outer membrane (Glauert and Thornley, 1969; Costerton *et al.*, 1974) which lies peripheral to the thin peptidoglycan layer (corresponding to the cell wall of gram-positive bacteria). In different strains of the same bacterial species a varying portion of the total envelope phospholipids is present in the outer membrane (Smit *et al.*, 1975).

The abundant lipopolysaccharides and highly specific lipoproteins and proteins, characteristic of the outer membrane (Inouye, 1974; Mühlradt and Golecki, 1975; Kamio and Nikaido, 1977), extensively interact among themselves and with the lipid bilayer, markedly restricting lipid mobility (Braun, 1975; Overath *et al.*, 1975; Rottem and Leive, 1977). It is not yet clear what interactions determine the unique properties of this membrane which behaves like a molecular sieve toward hydrophilic molecules, barring only larger ones such as proteins (Leive, 1974; Nakae, 1976). In addition, the presence of this surface layer may account for the inability of exogenous phospholipases, unaided by surface-active agents, to act on the lipids of the outer membrane itself. The outer membrane is also unusual because of its low permeability to hydrophobic substances (Nikaido, 1976).

Killing of several gram-negative bacteria, such as *E. coli*, by intact or disrupted rabbit PMN or by purified PMN proteins is accompanied by hydrolysis of 10–25% of the envelope phospholipids (Elsbach *et al.*, 1972; Weiss *et al.*, 1979). Unlike *M. lysodeikticus*, the cell wall layers of these gram-negative bacteria remain largely intact during killing and, consequently, protect the cytoplasmic membrane from extensive structural disorganization (Elsbach *et al.*, 1973; Beckerdite *et al.*, 1974). Because the PMN phospholipase A_2 is tightly membrane-associated (Franson *et al.*, 1974; Weiss *et al.*, 1975), the enzyme should be unable to penetrate beyond the outer layers of the bacterial envelope. Further, bacterial phospholipase(s), which may become activated, are localized in the outer membrane (Osborn *et al.*, 1972; Albright *et al.*, 1973; Vos *et al.*, 1978). Degradation is, therefore, probably limited to the lipids of the outer membrane (Elsbach, 1973; Weiss *et al.*, 1979).

The effects of PMN on gram-negative bacterial cell wall layers include an almost immediate increase in envelope permeability to actinomycin D. This hydrophobic antibiotic readily traverses cytoplasmic membranes, but not the normal outer membrane, suggesting that it is this layer of the gram-negative envelope where the effect is produced. The permeability-increasing activity resides in a potently bactericidal (toward gram-negative bacteria), noncatalytic cationic protein with an apparent molecular weight of 50,000–60,000 (Weiss *et al.*, 1978; Elsbach *et al.*, 1979). The proteins carrying bactericidal/permeability-increasing (PI) activities and phospholipase A_2 activity are apparently closely associated, and are co-purified as a cationic protein complex with a molecular weight > 50,000. No other catalytic activities (Weiss *et al.*, 1975, 1978; Elsbach *et al.*, 1979) have been detected in this complex. Hydrolysis of bacterial phospholipids during killing of *E. coli* by this highly purified fraction is attributable to both PMN and bacterial phospholipases. Quantitatively, this hydrolysis is about the same as that produced by intact or disrupted PMN (Weiss *et al.*, 1976; Weiss and Elsbach, 1977), suggesting that neither action of the PMN phospholipase on *E. coli* nor activation of *E. coli* phospholipase(s) requires the participation of other constituents (including catalytic activities) of PMN. In contrast, after dissociation of the complex in 1.0 M NaCl, the PMN phospholipase A_2 by itself, like other highly purified phospholipases, is inactive towards intact *E. coli* (Elsbach *et al.*, 1979; Weiss *et al.*, 1979) (Table 1). Other noncatalytic, membrane-active cationic polypeptides have been shown to potentiate phospholipase A_2 activity toward natural (Habermann, 1972) and artificial (Mollay and Kreil, 1974; Yunes *et al.*, 1977) membranes, presumably by altering the physical organization of the lipids within these structures. It is probably in this fashion that the basic peptide antibiotic, polymyxin B, allows several phospholipases to hydrolyze *E. coli* phospholipids (Weiss *et al.*, 1979). The bactericidal/permeability-increasing protein of the rabbit PMN may in the same manner facilitate the action of the PMN phospholipase on intact *E. coli* by virtue of the outer-membrane alterations it produces. However, this potentiation of phospholipase action is not evident with

TABLE 1.   FACILITATION OF EXOGENOUS AND/OR ENDOGENOUS PHOSPHOLIPASE ACTIVITY TOWARD *E. COLI* BY POTENTLY BACTERICIDAL, MEMBRANE-ACTIVE CATIONIC POLYPEPTIDES

| Protein added | Bacterial viability (%) | Permeability to Act D[a] | Susceptibility to added PLA_2[b] | | | Activation of bacterial phospholipase(s)[f] |
			Rabbit PMN[c]	Bee venom[d]	Porcine pancreas[e]	
None	100	−	−	−	−	−
Polymixin B	<1	+	+	+	+	+
Rabbit PMN bactericidal/ PI protein	<1	+	+	±	−	+
Human PMN bactericidal/ PI protein	<1	+	−	−	−	+

[a] Act D, actinomycin D. The procedure used to determine *E. coli* permeability to Act D is described by Beckerdite *et al.* (1974).
[b] PLA_2, phospholipase A_2. Activity of exogenous phospholipases determined toward *E. coli* S_{17}, a phospholipase-A-less mutant strain (Doi and Nojima, 1973).
[c] Purified as described by Weiss *et al.* (1975) and Elsbach *et al.* (1979).
[d] Purified and donated by C. Dutilh and G. de Haas, Laboratory of Biochemistry, University of Utrecht, The Netherlands.
[e] Purified and donated by R.F.A. Zwaal, Laboratory of Biochemistry, University of Utrecht, The Netherlands.
[f] Activation determined in *E. coli* S_{15}, parent strain of S_{17}, equipped with full phospholipid-degradative apparatus (Doi and Nojima, 1973).

several other phospholipases A_2 (Weiss *et al.*, 1979), suggesting that the membrane modifications per se are not sufficient for phospholipase action. This contention is supported by the observation that neither the rabbit PMN phospholipase A_2, nor any other added phospholipase A_2, is active in combination with the human PMN bactericidal/permeability-increasing protein, recently purified to near homogeneity (Weiss *et al.*, 1978) (Table 1). The human PMN protein closely resembles the rabbit PMN protein in its molecular and biological properties, including the envelope alterations it produces. However, unlike the rabbit PMN protein, it exhibits no association with the human PMN phospholipase A_2 during purification (Weiss *et al.*, 1978; Franson *et al.*, 1977). An intriguing possibility is that protein–protein interaction between the phospholipase and bactericidal/permeability-increasing proteins is a requisite for phospholipase action. Alternatively, there may be subtle, presently unrecognized, differences in the envelope alterations produced by the two bactericidal/permeability-increasing proteins that result in action of the rabbit PMN phospholipase in combination with the rabbit bactericidal protein only. In either case, the rabbit PMN phospholipase must possess properties distinct from the other phospholipases A_2 tested to account for the apparently specific cooperativity between the two rabbit PMN proteins.

Both the human and the rabbit bactericidal/permeability-increasing proteins promptly activate bacterial phospholipase(s) (Weiss *et al.*, 1978; Elsbach *et al.*, 1979) (Table 1).

Neither the action of bacterial phospholipase A nor that of the exogenous phospholipases A_2 on the phospholipids of *E. coli* treated with bactericidal proteins requires added Ca^{2+} (Weiss *et al.*, 1979). This is remarkable because all these phospholipases are Ca^{2+}-dependent enzymes when tested against artificial substrates or autoclaved *E. coli*. We have speculated that Ca^{2+} within the bacterial membrane is the source of this cofactor (Weiss *et al.*, 1979). In fact, the addition of high concentrations of either Ca^{2+} or Mg^{2+} to *E. coli* exposed to the PMN proteins causes abrupt cessation of both PMN and bacterial phospholipase activity (Weiss *et al.*, 1976), possibly by squeezing the enzymes out of the bilayer (Weiss *et al.*, 1976; Nicolson *et al.*, 1976). Such an effect would be consistent with the prominent role divalent cations are thought to play in creating a tight outer-membrane structure (Leive, 1974; Nakamura and Mizushima, 1975; Takeuchi *et al.*, 1978).

Net degradation of microbial phospholipids (at least of gram-negative bacteria) during phagocytosis by PMN is also limited by the persistence of microbial biosynthetic activity. For at least 1 hr after *E. coli*, exposed to bactericidal PMN fractions, have lost their ability to multiply, phospholipid synthesis, as judged by incorporation of radioactively labeled precursors, continues at approximately normal rates (Mooney and Elsbach, 1975). In addition, partial reincorporation takes place of products of hydrolysis (Weiss *et al.*, 1976). Furthermore, the synthesis of cardiolipin is increased and that of phosphatidylglycerol decreased (Mooney and Elsbach, 1975), possibly conferring greater envelope stability (Kanemasa *et al.*, 1967) and thereby reducing the susceptibility of envelope phospholipids to further hydrolysis.

Neither the bactericidal nor the permeability-increasing activities appear to require phospholipid degradation. Whether or not phospholipid breakdown facilitates digestion of other microbial macromolecules during phagocytosis is still unknown. The limited lipid degradation measured in the microorganisms studied to date apparently reflects the fate of the microbial envelope as a whole; other envelope macromolecules are also incompletely degraded (Cohn, 1963; Elsbach, 1973; Elsbach *et al.*, 1973).

5. CONCLUDING REMARKS

Where should future studies be directed?

The study of the pathways of lipid biosynthesis and degradation in phagocytic cells is incomplete. However, better understanding of the membrane perturbations that initiate and accompany phagocytosis will require other approaches in addition to conventional biochemical ones. The application of new physicochemical techniques, designed to detect discrete envelope alterations not recognizable by steady state changes in membrane lipids, may be particularly revealing.

We believe that the study of lipid metabolism during the interaction of phagocytes and microorganisms should include further exploration of the microbial envelope as the major determinant of the fate of the microorganism.

Only initial attempts have been made to relate the respiratory burst to peroxidative changes in lipids and other constituents of both phagocyte and microorganism. Insight into the molecular basis of the lethal effect of the O_2-dependent microbicidal system may well derive from dissection of the consequences of such changes for microbial structure and function.

REFERENCES

Albright, F. R., White, D. A., and Lennarz, W. J., 1973, Studies on enzymes involved in the catabolism of phospholipids in *Escherichia coli*, *J. Biol. Chem.* **248**:3968.

Allan, D., and Crumpton, M. J., 1972, Isolation and composition of human thymocyte plasma membrane, *Biochim. Biophys. Acta* **274**:22.

Bainton, D. F., 1973, Sequential degranulation of the two types of polymorphonuclear leukocyte granules during phagocytosis of microorganisms, *J. Cell Biol.* **58**:249.

Barthélemy, A., Paridaens, R., and Schell-Frederick, E., 1977, Phagocytosis-induced [45]Calcium efflux in polymorphonuclear leukocytes, *FEBS Lett.* **82**:283.

Beckerdite, S., Mooney, C., Weiss, J., Franson, R., and Elsbach, P., 1974, Early and discrete changes in permeability of *Escherichia coli* and certain other gram-negative bacteria during killing by granulocytes, *J. Exp. Med.* **140**:396.

Beckerdite-Quagliata, S., Simberkoff, M., and Elsbach, P., 1975, Effects of human and rabbit serum on viability, permeability and envelope lipids of *Serratia marcescens*, *Infect. Immun.* **11**:758.

Boot, J. R., Dawson, S., and Kitchen, E. A., 1976, The chemotactic activity of thromboxane B_2: A possible role in inflammation, *J. Physiol.* **257**:47P.

Borgeat, P., Hamberg, M., and Samuelsson, B., 1976, Transformation of arachidonic acid and homo-γ-linolenic acid by rabbit polymorphonuclear leukocytes. *J. Biol. Chem.* **251**:7816.

Braun, V., 1975, Covalent lipoprotein from the outer membrane of *Escherichia coli*, *Biochim. Biophys. Acta* **15**:335.

Braunsteiner, H., Dienstl, F., Sailer, S., and Sandhofer, F., 1968, Lipase activity in leukocytes and macrophages, *Blood* **24**:607.

Bray, M. A., Gordon, D., and Morley, J., 1974, Role of prostaglandins in reactions of cellular immunity, *Br. J. Pharmacol.* **52**:453.

Brecher, P., Pyun, H. Y., and Chobanian, A. V., 1977, Effect of atherosclerosis on lysosomal cholesterol esterase activity in rabbit aorta, *J. Lipid Res.* **18**:154.

Bretscher, M. S., 1972, Assymetrical lipid bilayer structure for biological membranes, *Nature New Biol.* **236**:11.

Brown, M. S., Ho, Y. K., and Goldstein, J. L., 1976, The low-density lipoprotein pathway in human fibroblasts: Relation between cell surface receptor binding and endocytosis of low-density lipoproteins, *Ann. N.Y. Acad. Sci* **275**:244.

Cohn, Z. A., 1963, The fate of bacteria within phagocytic cells. I. The degradation of isotopically labeled bacteria by polymorphonuclear leukocytes and macrophages, *J. Exp. Med.* **117**:27.

Constantopoulos, A., and Najjar, V. A., 1973, The activation of adenylate cyclase: II. The postulated presence of (A)adenylate cyclase in phospho(inhibited) form(B), a dephospho(activated) form with a cyclic adenylate stimulated membrane protein kinase, *Biochem. Biophys. Res. Commun.* **53**:794.

Cortner, J. A., Coates, P. M., Swoboda, E., and Schnatz, J. D., 1976, Genetic variation of lysosomal acid lipase, *Pediatr. Res.* **10**:927.

Costerton, J. W., Ingram, J. M., and Cheng, K. J., 1974, Structure and function of the cell envelope of gram-negative bacteria, *Bacteriol. Rev.* **38**:87.

Cronan, J. E., and Gelman, E. P., 1975, Physical properties of membrane lipids: Biological relevance and regulation, *Bacteriol. Rev.* **39**:232.

Daneo-Moore, L., and Shockman, G. D., 1977, The bacterial cell surface in growth and division, in: *The Synthesis, Assembly and Turnover of Cell Surface Components* (G. Poste and G. L. Nicolson, eds.), Vol. 4, *Cell Surface Reviews*, pp. 597–716, Elsevier/North Holland, Amsterdam.

Day, A. J., 1967, Lipid metabolism by macrophages and its relationship to atherosclerosis, *Adv. Lipid Res.* **5**:185.

Day, A. J, and Fidge, N. H., 1962, The uptake and metabolism of ^{14}C-labeled fatty acids by macrophages *in vitro, J. Lipid Res.* **3**:33.

Demel, R. A., Geurts van Kessel, W. S. M., Zwaal, R. F. A., Roelofsen, B., and van Deenen, L. L. M., 1975, Relation between various phospholipase actions on human red cell membranes and the interfacial phospholipid pressure in monolayers, *Biochim. Biophys. Acta* **406**:97.

De Pierre, J. W., and Karnovsky, M. L., 1973, Plasma membranes of mammalian cells. A review of methods of their characterization and isolation, *J. Cell Biol.* **56**:275.

Doi, O., and Nojima, S., 1973, Detergent-resistant phospholipase A_1 and A_2 in *Escherichia coli, J. Biochem.* **74**:667.

Duckworth, D. H., Bevers, E. K., Verkleij, A. J., Op den Kamp, J. A. F., and van Deenen, L. L. M., 1974, Action of phospholipase A_2 and phospholipase C on *Escherichia coli, Arch. Biochem. Biophys.* **165**:379.

Elsbach, P., 1959, Composition and synthesis of lipids in resting and phagocytizing leukocytes, *J. Exp. Med.* **110**:969.

Elsbach, P., 1965a, Uptake of fat by phagocytic cells. An examination of the role of phagocytosis. I. Rabbit polymorphonuclear leukocytes, *Biochim. Biophys. Acta* **98**:402.

Elsbach, P., 1965b, Uptake of fat by phagocytic cells. An examination of the role of phagocytosis. II. Rabbit alveolar macrophages, *Biochim. Biophys. Acta* **98**:420.

Elsbach, P., 1966, Phospholipid metabolism by phagocytic cells. I. A comparison of conversion of ^{32}P lysolecithin to lecithin and glycerylphosphorylcholine by homogenates of rabbit polymorphonuclear leukocytes and alveolar macrophages, *Biochim. Biophys. Acta* **125**:510.

Elsbach, P., 1967, Metabolism of lysophosphatidylethanolamine and lysophosphatidylcholine by homogenates of rabbit polymorphonuclear leukocytes and alveolar macrophages, *J. Lipid Res.* **8**:359.

Elsbach, P., 1968, Increased synthesis of phospholipid during phagocytosis, *J. Clin. Invest.* **47**:2217.

Elsbach, P., 1972, Lipid metabolism by phagocytes, *Semin. Hematol.* **9**:227.

Elsbach, P , 1973, On the interaction between phagocytes and microorganisms, *N. Engl. J. Med.* **289**:846.

Elsbach, P., 1974, Phagocytosis, in: *The Inflammatory Process* (B. W. Zweifach, L. Grant, and R. T. McCluskey, eds.), Vol. 1, pp. 363–408, Academic Press, New York.

Elsbach, P., 1977a, White cells, in: *Lipid Metabolism in Mammals* (F. Snyder, ed.), Vol. 1, pp. 259–276, Plenum Press, New York.

Elsbach, P. 1976, Cell surface changes during phagocytosis, in: *The Synthesis, Assembly and Turnover of Cell Surface Components* (G. Poste and G. C. Nicolson, eds.), Vol. 4, *Cell Surface Reviews,* pp. 363–402, Elsevier/North Holland, Amsterdam.

Elsbach, P., and Farrow, S., 1969, Cellular triglyceride as a source of fatty acid for lecithin synthesis during phagocytosis, *Biochim. Biophys. Acta* **176**:438.

Elsbach, P., and Kayden, H. J., 1965, Chylomicron-lipid-splitting activity of rabbit polymorphonuclear leukocytes, *Am. J. Physiol.* **209**:765.

Elsbach, P., and Rizack, M. A., 1963, Acid lipase and phospholipase activity in homogenates of rabbit polymorphonuclear leukocytes, *Am. J. Physiol.* **205**:1154.

Elsbach, P., Van den Berg, J. W. O., van den Bosch, H., and van Deenen, L. L. M., 1965, Metabolism of phospholipids by polymorphonuclear leukocytes, *Biochim. Biophys. Acta* **106**:338.

Elsbach, P., Zucker-Franklin, D., and Sansaricq, C., 1969, Increased lecithin synthesis during phagocytosis by normal leukocytes and by leukocytes of a patient with chronic granulomatous disease, *N. Engl. J. Med.* **280**:1319.

Elsbach, P., Goldman, J., and Patriarca, P., 1972a, Phospholipid metabolism by phagocytic cells. VI. Observations on the fate of phospholipids of granulocytes and ingested *Escherichia coli* during phagocytosis, *Biochim. Biophys. Acta* **280**:33.

Elsbach, P., Patriarca, P., Pettis, P., Stossel, T. P., Mason, R. J., and Vaughan, M., 1972b, The appearance of ^{32}P-lecithin, synthesized from ^{32}P-lysolecithin, in phagosomes from polymorphonuclear leukocytes, *J. Clin. Invest.* **51**:1910.

Elsbach, P., Pettis, P., Beckerdite, S., and Franson, R., 1973, Effect of phagocytosis by rabbit granulocytes on macromolecular synthesis and degradation in different species of bacteria, *J. Bacteriol.* **115**:490.

Elsbach, P., Weiss, J., Franson, R. C., Beckerdite-Quagliata, S., Schneider, A., and Harris, L., 1979, Separation and purification of a potent bactericidal/permeability increasing protein and a closely associated phospholipase A_2 from rabbit polymorphonuclear leukocytes, *J. Biol. Chem.* **254**:11000.

Evans, W. H., and Mueller, P. S., 1963, Effects of palmitate on the metabolism of leukocytes from guinea pig exudate, *J. Lipid Res.* **4**:39.

Ferber, E., Resch, K., Wallach, D. F. H., and Imm, W., 1972, Isolation and characterization of lymphocyte plasma membranes, *Biochim. Biophys. Acta* **266**:494.

Flower, R. J., 1974, Drugs which inhibit prostaglandin biosynthesis, *Pharmacol. Rev.* **26**:33.

Fogelman, A. M., Edmond, E., Seager, J., and Popjak, G., 1975, Abnormal induction of 3-hydroxy-3-methylglutaryl coenzyme A reductase in leukocytes from subjects with heterozygous familial hypercholesterolemia, *Biol. Chem.* **250**:2045.

Fogelman, A. M., Seager, J., Edwards, P. A., Hokom, M., and Popjak, G., 1977a, Cholesterol biosynthesis in human lymphocytes, monocytes and granulocytes, *Biochem. Biophys. Res. Commun.* **76**:167.

Fogelman, A. M., Seager, J., Edwards, P. A., and Popjak, G., 1977b, Mechanism of induction of 3-hydroxy-3-methylglutarylcoenzyme A reductase in human leukocytes, *J. Biol. Chem.* **252**:644.

Franson, R. C., and Waite, M., 1973, Lysosomal phospholipases A_1 and A_2 of normal and Bacillus Calmette Guerin-induced alveolar macrophages, *J. Cell Biol.* **56**:621.

Franson, R., Beckerdite, S., Wang, P., Waite, M., and Elsbach, P., 1973, Some properties of phospholipases of alveolar macrophages, *Biochim. Biophys. Acta* **296**:365.

Franson, R., Patriarca, P., and Elsbach, P., 1974, Phospholipid metabolism by phagocytic cells. Phospholipases A_2 associated with rabbit polymorphonuclear leukocyte granules, *J. Lipid Res.* **15**:380.

Franson, R., Weiss, J., Martin, L., Spitznagel, J. K., and Elsbach, P., 1977, Phospholipase A activity associated with the membranes of human polymorphonuclear leukocytes, *Biochem. J.* **167**:839.

Franson, R., Dobrow, R., Weiss, J., Elsbach, P., and Weglicki, W., 1978, Isolation and characterization of phospholipase A_2 from an inflammatory exudate, *J. Lipid Res.* **19**:18.

Fredrickson, D. S., and Sloan, H. R., 1972, Sphingomyelin lipidoses: Niemann-Pick disease, in: *The Metabolic Basis of Inherited Disease* (J. B. Stanbury, J. B. Wyngaarden, and D. S. Frederickson, eds.), pp. 783–807, McGraw-Hill, New York.

Gahmberg, C. G., and Hakomori, S., 1973, Altered growth behavior of malignant cells associated with changes in externally labeled glycoprotein and glycolipid, *Proc. Natl. Acad. Sci.* **70**:3329.

Gallin, J. T., and Rosenthal, A. S., 1974, The regulatory role of divalent cations in human granulocyte chemotaxis, *J. Cell Biol.* **62**:594.

Ghuysen, J. M., 1977, Biosynthesis and assembly of bacterial cell walls, in: *The Synthesis, Assembly and Turnover of Cell Surface Components* (G. Poste and G. L. Nicolson, eds.), Vol. 4, *Cell Surface Reviews*, pp. 463–596, Elsevier/North Holland, Amsterdam.

Glauert, A. M., and Thornley, M. J., 1969, The topography of the bacterial cell wall, *Annu. Rev. Microbiol.* **23**:159.

Glomset, J. A., 1968, The plasma lecithin-cholesterol acyltransferase reaction, *J. Lipid Res.* **9**:155.

Goldstein, J. L., Dana, S. E., Faust, J. R., Beaudet, A. L., and Brown, M. S., 1975, Role of lysosomal acid lipase in the metabolism of plasma low density lipoprotein: Observations in cultured fibroblasts from a patient with cholesteryl ester storage disease, *J. Biol. Chem.* **250**:8487.

Gordon, S., Todd, J., and Cohn, Z. A., 1974, *In vitro* synthesis and secretion of lysozyme by mononuclear phagocytes. *J. Exp. Med.* **139**:1228.

Gottfried, E. L., 1972, Lipid composition and metabolism of leukocytes, in: *Blood Lipids and Lipoproteins: Quantitation, Composition and Metabolism* (G. J. Nelson, ed.), pp. 387–415, Wiley-Interscience, New York.

Habermann, E., 1972, Bee and wasp venoms, *Science* **177**:314.

Hakomori, S., and Murakami, W. T., 1968, Glycolipids of hamster fibroblasts and derived malignant-transformed cell lines, *Proc. Natl. Acad. Sci. USA* **59**:254.

Hamberg, M., and Samuelsson, B., 1974, Prostaglandin endoperoxides. Novel transformations of arachidonic acid in human platelets, *Proc. Natl. Acad. Sci. USA* **71**:3400.

Higgs, G. A., and Youlton, L. J. F., 1972, Prostaglandin production by rabbit peritoneal polymorphonuclear leukocytes *in vitro*, *Br. J. Pharmacol.* **44**:330P.

Higgs, G. A., McCall, E., and Youlton, L. J. F., 1975, A chemotactic role for prostaglandins released from polymorphonuclear leukocytes during phagocytosis, *Br. J. Pharmacol.* **53**:539.

Higgs, G. A., Bunting, S., Moncada, S., and Vane, J. R., 1976, Polymorphonuclear leukocytes produce thromboxane A_2-like activity during phagocytosis, *Prostaglandins* **12**:749.

Hildebrand, J., Marique, D., and Vanhouche, J., 1975, Lipid composition of plasma membranes from human leukemic lymphocytes, *J. Lipid Res.* **16**:195.

Hirata, F., Corcoran, B. A., Venkatasubramanian, K., Schiffman, E., and Axelrod, J., 1979, Chemoattractants stimulate degradation of methylated phospholipids and release of arachidonic acid in rabbit leukocytes, *Proc. Natl. Acad. Sci. USA* **76**:2640.

Hirsch, J. G., 1956, Phagocytin: A bactericidal substance from polymorphonuclear leukocytes, *J. Exp. Med.* **103**:589.

Ho, Y. K., Brown, M. S., Kayden, H. J., and Goldstein, J. L., 1976, Binding, internalization, and hydrolysis of low density lipoprotein in long-term lymphoid cell lines from a normal subject and a patient with homozygous familial hypercholesterolemia, *J. Exp. Med.* **144**:444.

Ho, Y. K., Fause, J. R., Bilheimer, D. W., Brown, M. S., and Goldstein, J. L., 1977, Regulation of cholesterol synthesis by low density lipoprotein in isolated human lymphocytes, *J. Exp. Med.* **145**:1531.

Hong, S. L., and Levine, L., 1976, Inhibition of arachidonic acid release from cells as the biochemical action of anti-inflammatory corticosteroids, *Proc. Natl. Acad. Sci. USA* **73**:1730.

Humes, J. L., Bonney, R. J., Pelus, L., Dahlgren, M. E., Saldowski, S. J., Kuehl, F. A., Jr., and Davies, P., 1977, Macrophages synthesize and release prostaglandins in response to inflammatory stimuli, *Nature* **269**:149.

Inouye, M., 1974, A three-dimensional molecular assembly model of a lipoprotein from the *Escherichia coli* outer membrane, *Proc. Nat. Acad. Sci. USA* **71**:2396.

Kader, J. C., 1977, Exchange of phospholipids between membranes, in: *Dynamic Aspects of Cell Surface Organization* (G. Poste and G. L. Nicolson, eds.), Vol. 3, pp. 127–204, Elsevier/North-Holland, Amsterdam.

Kaley, G., and Weiner, R., 1971, Prostaglandin E_1: A potential mediator of the inflammatory response, *Ann. N.Y. Acad. Sci.* **180**:338.

Kamio, Y., and Nikaido, H., 1976, Outer membrane of *Salmonella typhimurium*: Accessibility of phospholipid headgroups to phospholipase C and cyanogen bromide activated dextran in the external medium, *Biochemistry* **15**:2561.

Kamio, Y., and Nikaido, H., 1977, Outer membrane of *Salmonella typhimurium*. Identification of proteins exposed on cell surface, *Biochim. Biophys. Acta* **464**:589.

Kampine, J. P., Brady, R. O., Kanfer, J. N., Feld, K., and Shapiro, D., 1967, Diagnosis of Gaucher's disease and Niemann-Pick disease with small samples of venous blood, *Science* **155**:86.

Kanemasa, Y., Okamatsu, Y., and Nojima, S., 1967, Composition and turnover of phospholipids in *Escherichia coli*, *Biochim. Biophys. Acta* **114**:382.

Kaplan, L., Weiss, J., and Elsbach, P., 1978, Low concentrations of indomethacin inhibit phospholipase A_2 of rabbit polymorphonuclear leukocytes, *Proc. Natl. Acad. Sci. USA* **75**:2955.

Karnovsky, M. L., 1962, Metabolic basis of phagocytic activity, *Physiol. Rev.* **42**:143.

Kennedy, E. P., 1962, The metabolism and function of complex lipids, *Harvey Lect.* **57**:143.

Kleiman, J. H., and Lands, W. E. H., 1969, Purification of a phospholipase C from *Bacillus cereus*, *Biochim. Biophys. Acta* **187**:477.

Korn, E. D., Bowers, B., Batzri, S., Simmons, S. R., and Victoria, E. J., 1974, Endocytosis and exocytosis: Role of microfilaments and involvement of phospholipids in membrane fusion, *J. Supramolec. Struct.* **2**:517.

Kusaka, I., 1975, Degradation of phospholipid and release of diglyceride-rich membrane vesicles during protoplast formation in certain gram-positive bacteria, *J. Bacteriol.* **121**:1173.

Lands, W. E. M., 1960, Metabolism of glycerolipids. II. The enzymatic acylation of lysolecithin, *J. Biol. Chem.* **235**:2233.

Lands, W. E. M., and Rome, L. H. 1976, Inhibition of prostaglandin synthesis, in: *Prostaglandins: Chemical and Biochemical Aspects* (S. M. M. Karim, ed.), pp. 87–138, MTP, London.

Lapetina, E. G., Schmitges, C. J., Chandrabose, K., and Cuatrecasas, P., 1977, Cyclic adenosine 3',5'-monophosphate and prostacyclin inhibit membrane phospholipase activity in platelets, *Biochem. Biophys. Res. Commun.* **76:**828.

Leive, L., 1974, The barrier function of the gram-negative envelope, *Ann. N.Y. Acad. Sci.* **235:**109.

Liljeqvist, L., Gürtler, J., and Blomstrand, R., 1973, Sterol and phospholipid biosynthesis in phytohemagglutinin stimulated human lymphocytes, *Acta Chem. Scand.* **27:**197.

Lucy, J. A., 1970, The fusion of biological membranes, *Nature* **227:**815.

Mahoney, E. M., Hamill, A. L., Scott, W. A., and Cohn, Z. A., 1977, Response of endocytosis to altered fatty acyl composition of macrophage phospholipids, *Proc. Natl. Acad. Sci. USA* **74:**4895.

Majerus, P. W., and Lastra, R., 1967, Fatty acid biosynthesis in human leukocytes, *J. Clin. Invest.* **46:**1596.

Marique, D., and Hildebrand, J., 1973, Isolation and characterization of plasma membranes from human leukemic lymphocytes, *Cancer Res.* **33:**2761.

Mason, R. J., Stossel, T. P., and Vaughan, M., 1972, Lipids of alveolar macrophages, polymorphonuclear leukocytes, and their phagocytic vesicles, *J. Clin. Invest.* **51:**2399.

McCall, E., and Youlton, L. J. F., 1973, Prostaglandin E_1 synthesis by phagocytizing rabbit polymorphonuclear leucocytes: Its inhibition by indomethacin and its role in chemotaxis, *J. Physiol.* **234:**98P.

Miras, C. J., Mantzos, J. D., and Levis, G. M., 1964, Incorporation of L-3-^{14}C-serine into microsomal phospholipids of human leukocytes, *Biochim. Biophys. Acta* **84:**101.

Miras, C. J., Mantzos, J. D., and Levis, G. M., 1965, Fatty acid synthesis in human leucocytes, *Biochem. Biophys. Res. Commun.* **19:**79.

Miras, C. J., Mantzos, J. D., and Levis, G. M., 1966, The isolation and partial characterization of glycolipids of normal human leucocytes, *Biochem. J.* **98:**782.

Mollay, C., and Kreil, G., 1974, Enhancement of bee venom phospholipase A_2 activity by melittin, direct lytic factor from cobra venom and polymixin B, *FEBS Lett.* **46:**141.

Mooney, C., and Elsbach, P., 1975, Altered phospholipid metabolism in *Escherichia coli* accompanying killing by disrupted granulocytes, *Infect. Immun.* **11:**1269.

Mühlradt, P. F., and Golecki, J. R., 1975, Asymmetrical distribution and artifactual reorientation of lipopolysaccharide in the outer membrane bilayer of *Salmonella typhimurium*, *Eur. J. Biochem.* **51:**343.

Nakae, T., 1976, Outer membrane of *Salmonella*. Isolation of protein complex that produces transmembrane channels, *J. Biol. Chem.* **251:**2176.

Nakamura, K., and Mizushima, S., 1975, *In vitro* reassembly of the membranous vesicle from *Escherichia coli* outer membrane components. Role of individual components and magnesium ions in reassembly, *Biochim. Biophys. Acta* **413:**371.

Nicolson, G. L., Smith, J. R., and Poste, G., 1976, Effects of local anaesthetics on cell morphology and membrane-associated cytoskeletal organization in BALB/3T3 cells, *J. Cell Biol.* **68:**395.

Nicolson, G. L., Poste, G., and Ji, T. H., 1977, The dynamics of cell membrane organization, in: *Dynamic Aspects of Cell Surface Organization* (G. Poste and G. L. Nicolson, eds.), Vol. 3, *Cell Surface Reviews*, pp. 1–73, Elsevier/North-Holland, Amsterdam.

Nikaido, H., 1976, Outer membrane of *Salmonella typhimurium*. Transmembrane diffusion of some hydrophobic substances, *Biochim. Biophys. Acta* **433:**118.

Nugteren, D. H., 1975, Arachidonate lipoxygenase in blood platelets, *Biochim. Biophys. Acta* **380:**299.

Op den Kamp, J. A. F., Kauerz, M. T., and van Deenen, L. L. M., 1972, Action of phospholipase A_2 and phospholipase C on *Bacillus subtilis* protoplasts, *J. Bacteriol.* **112:**1090.

Oren, R., Farnham, A. E., Saito, K., Milofsky, E., and Karnovsky, M. L., 1963, Metabolic patterns in three types of phagocytic cells, *J. Cell Biol.* **17:**487.

Osborn, M. J., Gander, J. E., Parisi, E., and Carson, J., 1972, Mechanism of assembly of the outer membrane of *Salmonella typhimurium*. Isolation and characterization of cytoplasmic and outer membrane, *J. Biol. Chem.* **247:**3962.

Overath, P., Bremer, M., Gulik-Krzywicki, T., Shecter, E., and Letellier, L., 1975, Lipid phase transitions in cytoplasmic and outer membranes of *Escherichia coli, Biochim. Biophys. Acta* **389**:358.

Patriarca, P., Beckerdite, S., and Elsbach, P., 1972a, Phospholipases and phospholipid turnover in *Escherichia coli* spheroplasts, *Biochim. Biophys. Acta* **260**:593–600.

Patriarca, P., Beckerdite, A., Pettis, P., and Elsbach, P., 1972b, Phospholipid metabolism by phagocytic cells. VII. The degradation and utilization of phospholipids of various microbial species by rabbit granulocytes, *Biochim. Biophys. Acta* **280**:45.

Poste, G., Papahadjopoulos, D., and Nicolson, G. L., 1975, Local anesthetics affect transmembrane cytoskeletal control of mobility and distribution of cell surface receptors, *Proc. Natl. Acad. Sci. USA* **72**:4430.

Rindler-Ludwig, R., Patsch, W., Sailer, S., and Braunsteiner, H., 1977, Characterization and partial purification of acid lipase from human leukocytes, *Biochim. Biophys. Acta* **488**:294.

Rottem, S., and Leive, L., 1977, Effect of variations in lipopolysaccharide on fluidity of the outer membrane of *Escherichia coli, J. Biol. Chem.* **252**:2077.

Salton, M. R. J., 1964, *The Bacterial Cell Wall*, Elsevier, Amsterdam.

Sastry, P. S., and Hokin, L. E., 1966, Studies on the role of phospholipids in phagocytosis, *J. Biol. Chem.* **241**:3354.

Shier, W. T., Baldwin, J. H., Nilsen-Hamilton, M., Hamilton, R. T., and Thanassi, N. M., 1976, Regulation of guanylate and adenylate cyclase activities by lysolecithin, *Proc. Natl. Acad. Sci. USA* **73**:1586.

Shohet, S. B., 1970, Changes in fatty acid metabolism in human leukemic granulocytes during phagocytosis, *J. Lab. Clin. Med.* **75**:659.

Sloan, H. R., and Frederickson, D. S., 1972, Enzyme deficiency in cholesteryl ester storage disease, *J. Clin. Invest.* **51**:1923.

Smit, J., Kamio, Y., and Nikaido, H., 1975, Outer membrane of *Salmonella typhimurium:* Chemical analysis and freeze-fracture studies with lipopolysaccharide mutants, *J. Bacteriol.* **124**:942.

Smith, R. J., 1976, Modulation of phagocytosis by and lysosomal enzyme secretion from guinea-pig neutrophils: Effect of nonsteroid anti-inflammatory agents and prostaglandins, *J. Exp. Pharmacol. Ther.* **200**:647.

Smolen, J. E., and Shohet, S. B., 1974, Remodeling of granulocyte membrane fatty acids during phagocytosis, *J. Clin. Invest.* **53**:726.

Stein, O., Weinstein, D. B., Stein, Y., and Steinberg, D., 1976, Binding, internalization and degradation of low density lipoprotein by normal human fibroblasts and by fibroblasts from a case of homozygous familial hypercholesterolemia, *Proc. Natl. Acad. Sci. USA* **73**:14.

Stossel, T. P., Mason, R. J., and Smith, A. L., 1974, Lipid peroxidation by human blood phagocytes, *J. Clin. Invest.* **54**:638.

Switzer, S., and Eder, H. K., 1965, Transport of lysolecithin by albumin in human and rat plasma, *J. Lipid Res.* **6**:506.

Takeuchi, Y., Ohnishi, S.-I., Ishinaga, M., and Kito, M., 1978, Spin-labeling of *E. coli* membranes by enzymatic synthesis of phosphatidylglycerol: Divalent cation-induced interaction of phosphatidyl glycerol with membrane proteins, *Biochim. Biophys. Acta* **506**:54.

Tolone, G., Bonasera, L., Bray, M., and Tolone, C., 1977, Prostaglandin production by human polymorphonuclear leukocytes during phagocytosis *in vitro, Experientia* **33**:961.

van Alphen, L., Lugtenberg, B., van Boxtel, P., and Verhoef, K., 1977, Architecture of the outer membrane of *Escherichia coli* K12. I. Action of phospholipases A_2 and C on wild-type strains and outer membrane mutants, *Biochim. Biophys. Acta* **466**:257.

Verger, R., Mieras, M. C. E., and de Haas, G. H., 1973, Action of phospholipase A at interfaces, *J. Biol. Chem.* **248**:4023.

Vos, M., Op den Kamp, J. A. F., Beckerdite-Quagliata, S., and Elsbach, P., 1978, Acylation of monoacylphosphorylethanolamine in the inner and outer membranes of the envelope of an *Escherichia coli* K12 strain and its phospholipase-A deficient mutant, *Biochim. Biophys. Acta* **508**:165.

Wang, P., De Chatelet, L. R., and Waite, M., 1976, Enzymes of phospholipid synthesis in Bacillus Calmette-Guerin-induced rabbit alveolar macrophages. Characterization and localization of cytidine diphosphocholine phosphotransferase and monoacylphospholipid acyltransferase, *Biochim. Biophys. Acta* **450**:311.

Wang, P., Waite, M., and De Chatelet, L. R., 1977, Membrane lipid metabolism of Bacillus Calmette Guerin-induced rabbit alveolar macrophages, *Biochim. Biophys. Acta* **487**:163.

Weiss, J., and Elsbach, P., 1977, The use of a phospholipase A-less *Escherichia coli* mutant to establish the action of granulocyte phospholipase A on bacterial phospholipids during killing by a highly purified granulocyte fraction, *Biochim. Biophys. Acta* **466**:23.

Weiss, J., Franson, R., Beckerdite, S., Schmeidler, K., and Elsbach, P., 1975, Partial characterization and purification of a rabbit granulocyte factor that increases permeability of *E. coli*, *J. Clin. Invest.* **55**:33.

Weiss, J., Schmeidler, K., Beckerdite-Quagliata, S., Franson, R., and Elsbach, P., 1976, Reversible envelope effects during and after killing of *E. coli* by a highly purified granulocyte preparation, *Biochim. Biophys. Acta* **436**:154.

Weiss, J., Elsbach, P., Olsson, I., and Odeberg, H., 1978, Purification and characterization of a potent bactericidal and membrane-active protein from the granules of human polymorphonuclear leukocytes, *J. Biol. Chem.* **253**:2664.

Weiss, J., Beckerdite-Quagliata, S., and Elsbach, P., 1979, Determinants of the action of phospholipases A on the envelope phospholipids of *Escherichia coli*, *J. Biol. Chem.* **254**:11010.

Werb, Z., and Cohn, Z. A., 1972a, Cholesterol metabolism in the macrophage. III. Ingestion and intracellular fate of cholesterol and cholesterol esters, *J. Exp. Med.* **135**:21.

Werb, Z., and Cohn, Z. A., 1972b, Plasma membrane synthesis in the macrophage following phagocytosis of polystyrene latex particles, *J. Biol. Chem.* **247**:2439.

Werb, Z., and Gordon, S., 1975a, Secretion of a specific collagenase by stimulated macrophages, *J. Exp. Med.* **142**:346.

Werb, Z., and Gordon, S., 1975b, Elastase secretion by stimulated macrophages. Characterization and regulation, *J. Exp. Med.* **142**:361.

Wirtz, K. W. A., 1974, Transfer of phospholipids between membranes, *Biochim. Biophys. Acta* **344**:95.

Yunes, R., Goldhammer, A. R., Garner, W. K., and Cordes, E. H., 1977, Phospholipases: Melittin facilitation of bee venom phospholipase A_2-catalyzed hydrolysis of unsonicated lecithin liposomes, *Arch. Biochem. Biophys.* **183**:105.

Zurier, R. B., Weissmann, G., Hofstein, S., Kammerman, S., and Tai, H. H., 1974, Mechanisms of lysosomal enzyme release from human leukocytes. II. Effects of cAMP and cGMP, autonomic antagonists and agents which affect microtubule function, *J. Clin. Invest.* **53**:297.

6

Glutathione Metabolism in Leukocytes

R. E. BASFORD

1. INTRODUCTION

Glutathione (L-γ-glutamyl-L-cysteinylglycine) occurs in appreciable concentration in almost all living cells and has been the subject of a great deal of research for more than 75 years. The extensive literature on glutathione has been summarized by a conference and a number of review articles (Colowick *et al.*, 1954; Crook, 1959; Flohé *et al.*, 1974; Meister, 1975; Arias and Jakoby, 1976).

Since glutathione (GSH) contains a free thiol group and is oxidized both enzymatically and nonenzymatically to the disulfide form (GSSG) and reduced enzymatically, much of the early research led to the conclusion that a major role of GSH is to protect cells from oxidative damage. A number of other functions of GSH are also known. These include: detoxification of aromatic and aliphatic compounds in liver by a group of enzymes called glutathione *S*-transferases and excretion as mercapturic acids (Boyland and Chasseaud, 1969); action as a coenzyme in several enzyme-catalyzed reactions including glyoxalase and formaldehyde dehydrogenase (Knox, 1960); and participation in the γ-glutamyl cycle, believed to serve as a mechanism for amino acid transport across cell membranes (Meister, 1975). Only those reactions of glutathione which have been studied in leukocytes* are reviewed in this chapter. A literature review and some personal

---

*Throughout this review, the word *leukocyte* is used rather than *polymorphonuclear leukocyte, PMN, granulocyte,* or *neutrophil* since in much of the cited research, PMNs were not separated from other leukocytes in peritoneal exudates or peripheral blood buffy-coat fractions.

Abbreviations: 5-OP, 5-oxoproline; γGC, γ-glutamylcysteine; LDH, lactate dehydrogenase; NEM, *N*-ethylmaleimide; CGD, Chronic granulomatous disease; BHP, tertiary butyl hydroperoxide; HMPS, hexose monophosphate shunt; Hb-SSG, hemoglobin-glutathione mixed disulfide; Con A, concanavalin A.

R. E. BASFORD • Department of Biochemistry, University of Pittsburgh School of Medicine, Pittsburgh, Pennsylvania 15261. Research reported in this chapter was supported in part by Grant S58 from Health Research and Services Foundation, United Way, and by Training Grant De 00108, National Institute of Dental Research, N.I.H., U.S.P.H.S.

observations on the protective role of glutathione in phagocytic leukocytes has recently been published by Roos *et al.* (1979).

2.   OCCURRENCE AND BIOSYNTHESIS OF GLUTATHIONE IN LEUKOCYTES

2.1.   OCCURRENCE AND CONCENTRATION

The presence of glutathione in leukocytes was first recorded by Platt (1931) who measured by nonspecific methods (iodine titration and nitroprusside reaction) the GSH content of normal human and myelogenous leukemic leukocytes and reported that leukemic leukocytes contain 3–5 times higher levels of GSH than normal leukocytes. Since these original observations, many investigators have determined the concentration of GSH in human leukocytes by specific enzymatic methods, and report concentrations of 10–20 μmol/10^{10} cells, values which are roughly seven times higher than in human erythrocytes (Hardin *et al.*, 1954; Koj, 1962; Burchill *et al.*, 1978).

2.2.   BIOSYNTHESIS

Glutathione is synthesized in two sequential ATP-requiring steps; the first reaction is catalyzed by γ-glutamylcysteinyl synthetase [L-glutamate; L-cysteine γ-ligase (ADP) (E.C. 6.3.2.2)]:

$$\text{L-Glutamate} + \text{L-cysteine} + \text{ATP} \xrightarrow{\text{Mg}^{2+}} \gamma\text{-glutamylcysteine} + \text{ADP} + \text{P}_i \quad (1)$$

and the second reaction by glutathione synthetase [γ-L-glutamyl-L-cysteine: glycine ligase (ADP) (E.C. 6.3.2.3)]:

$$\gamma\text{-L-glutamylcysteine} + \text{glycine} + \text{ATP} \xrightarrow{\text{Mg}^{2+}} \text{glutathione} + \text{ADP} + \text{P}_i \quad (2)$$

Koj (1962) showed that human leukocytes are capable of synthesizing GSH at a rate of 0.38 μmol/10^{10} cells/hr, or about three times the rate of synthesis in human erythrocytes.

2.3.   GLUTATHIONE SYNTHETASE DEFICIENCY

Deficiency of glutathione synthetase activity (5-oxoprolinuria) leads to diminished levels of GSH in erythrocytes, leukocytes, and fibroblasts from affected patients, with a concomitant overproduction of 5-oxoproline (5-OP; pyroglutamic acid) (Hagenfeldt *et al.*, 1974; Larsson *et al.*, 1974; Wellner *et al.*, 1974; Spielberg *et al.*, 1977). γ-Glutamylcysteine synthetase may be subject to feedback inhibition by GSH. A deficiency of GSH synthetase could lead to enhanced production of γ-glutamylcysteine (γ-GC), which may be a substrate for two enzymes of the proposed γ-glutamyl cycle: γ-glutamyl transpeptidase and γ-glutamyl cyclotransferase (Figure 1).

It is probable that in 5-oxoprolinuria, increased production and excretion of 5-OP results from the conversion of γ-GC to 5-OP by γ-glutamylcyclotransferase (Spielberg *et al.*, 1977). The consequences of diminished GSH levels on microtubule assembly and function in leukocytes from a patient with 5-oxoprolinuria have been studied by Oliver *et al.* (1978) and Spielberg *et al.* (1979), and are discussed in Section 3.3.3.

3. THE REDOX ROLE OF GLUTATHIONE IN LEUKOCYTE METABOLISM AND PHAGOCYTOSIS

3.1. BACKGROUND

Perturbation of the membrane of phagocytic cells by a wide variety of agents, including opsonized microorganisms, zymosan granules, polystyrene latex beads, phospholipase c, surfactants, ionophores, C3 and C567 components of complement, or antibody to the phagocytic cell, elicits a burst of metabolic activity. There are rapid increases in oxygen consumption, glucose utilization, hexose monophosphate shunt (HMPS) activity, and the production of superoxide anion (O_2^-) and hydrogen peroxide (H_2O_2). One or both of the latter two compounds are important components of the microbicidal system of phagocytic cells. Two other reduced forms of oxygen, singlet oxygen ($^1O_2$) and hydroxyl radical ($\cdot OH$) may also be important microbicidal agents under some circumstances (Boxer *et al.*, 1979; Badwey *et al.*, 1979; Rosen and Klebanoff, 1979; Torres *et al.*, 1979). For further discussion of these aspects of phagocytic metabolism, the reader is referred to Chapters 4, 7, 8, 12–14, and 16.

The reduction of O_2 to O_2^- and H_2O_2 and the increase in HMPS activity are thought to be causally related. The reducing agent and enzyme system involved in the reduction of O_2 to O_2^- and H_2O_2, and hence the mechanism of HMPS stimulation, have been the subject of controversy for more than a decade. All enzymes of the HMPS pathway are present in substantial amounts in the cytoplasm of normal leukocytes and other phagocytic cells and their kinetic properties do not change during phagocytosis (Stjernholm, 1968; Stjernholm *et al.*, 1972). The availability of NADP, which is required for the oxidation of glucose-

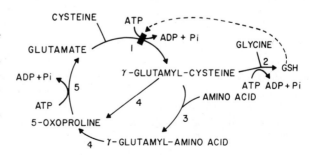

FIGURE 1. Pathway of glutathione formation and degradation. GSH, glutathione; 1, γ-glutamyl-cysteine synthetase; 2, glutathione synthetase; 3, γ-glutamyl transpeptidase; 4, γ-glutamylcyclotransferase; 5, 5-oxoprolinase. Dotted line indicates possible feedback inhibition. Redrawn from Meister (1975).

6-phosphate and 6-phosphogluconate in the first two reactions of the HMPS pathway is, therefore, the rate-limiting factor in HMPS stimulation (Beck, 1958; Iyer and Quastel, 1963).

The most direct mechanism linking HMPS stimulation and increased reduction of O_2 to O_2^- and H_2O_2 is an NADPH oxidase with a pH optimum of 5.5 which is activated by phagocytosis and other membrane perturbations (Iyer and Quastel, 1963; Rossi and Zatti, 1964; Patriarca *et al.*, 1971; Hohn and Lehrer, 1975). An alternate mechanism has been proposed in which an NADH oxidase with optimal activity at pH 7.4 is activated by phagocytosis (Evans and Karnovsky, 1962; Noseworthy and Karnovsky, 1972; Segal and Peters, 1976; Briggs *et al.*, 1977). Those interested in this controversy are referred to Chapter 7 of this volume.

An NADH-linked oxidase would require at least one additional enzymatic reaction to link NADH oxidation, O_2^- and H_2O_2 production, and regeneration of NADP for continued HMPS activity. Evans and Karnovsky (1962) reported a NADPH-linked lactate dehydrogenase (LDH) in guinea pig leukocytes which was active at low pH. They proposed that pyruvate, generated from glycolysis which is stimulated by enhanced NAD produced by the activated NADH oxidase, plus NADPH-linked LDH, regenerates NADP. Patriarca *et al.* (1971) expressed doubts of the validity of this mechanism based on the observed lag between the burst of respiration and the increase in glycolytic rate and on the discrepancy between the amount of lactate produced and the level of NADP required for HMPS stimulation during phagocytosis.

Baehner *et al.* (1970a) found no evidence for an NADPH-linked LDH in alkaline KCl supernatant fractions from human leukocyte homogenates, and Stjernholm and Manak (1970) also found no evidence to support a role for NADPH-linked LDH during phagocytosis in human leukocytes, based on the lack of incorporation of ^3H from [^3H]-NADP into lactate. Reed and Tepperman (1969) found no NADPH-dependent LDH activity in rat leukocytes.

Evans and Kaplan (1966) described a NAD–NADPH transhydrogenase in human leukocytes which could link NADH oxidation to HMPS stimulation but Evans and Karnovsky (1962) found only low levels of this enzyme in guinea pig leukocytes.

A third possibility is the glutathione cycle in which H_2O_2, produced by NADH oxidase, and GSH in the presence of glutathione peroxidase are converted to H_2O and GSSG. GSSG plus NADPH are converted to GSH and NADP, catalyzed by glutathione reductase. These interrelationships are indicated in the scheme of Figure 2, with NADPH rather than NADH oxidase.

3.2. THE GLUTATHIONE CYCLE

In a study of phagocytosis-associated metabolisn in rat leukocytes, Reed and Tepperman (1969) reported that no increase in glucose uptake, lactate production, or $^{14}CO_2$ evolution from [6-^{14}C]glucose could be demonstrated in rat leukocytes presented with latex particles. Particle-stimulated increase in HMPS

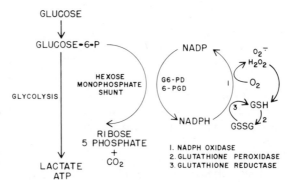

FIGURE 2. Leukocyte metabolism associated with phagocytosis. Reprinted from Basford *et al.* (1977), with permission of Piccin Medical Books.

activity was inhibited by 0.1 mM iodoacetate but was potentiated by 1.0 mM KCN. A cyanide-insensitive NADH oxidase in the cytosol and a cyanide-sensitive NADPH oxidase in granules were demonstrated. As mentioned earlier, no NADPH-dependent LDH could be measured. It was concluded that phagocytosis-associated metabolism in rat leukocytes differed from that in guinea pig and human leukocytes.

Cohen and Hochstein (1963) and Jacobs and Jandl (1966) had established earlier the importance of GSSG reductase and GSH peroxidase (the glutathione cycle) in the regulation of the HMPS pathway in human erythrocytes. Reed (1969) was the first to show that the glutathione cycle is operative in rat leukocytes and that phagocytosis-associated stimulation of HMPS activity is blocked by titration of GSH with N-ethylmaleimide (NEM). Stimulation of HMPS activity was observed by H_2O_2 added directly or generated by the D-amino acid oxidase system as well as by aminotriazole, an inhibitor of catalase-H_2O_2 compound I, and sodium azide, and inhibitor of myeloperoxidase. Thus, the observations of Reed and Tepperman (1969) appeared to be explained by the operation of the glutathione cycle as reported by Reed (1969). Reed also suggested that the glutathione cycle might constitute a protective mechanism in rat leukocytes to prevent inactivation of enzymes by H_2O_2 generated in excess of that required for microbicidal function or produced by ingested microorganisms.

Noseworthy and Karnovsky (1972) presented evidence in support of the findings of Reed (1969) of an enzymatically-mediated glutathione cycle in rat and human leukocytes. However, they concluded that in guinea pig leukocyte extracts, stimulation of HMPS activity by GSH is a result of nonenzymatic oxidation of GSH to GSSG, since GSH peroxidase levels in alkaline KCl extracts of guinea pig leukocytes were only 1% of the levels in rat leukocyte extracts. Extracts from human leukocytes indicated GSH peroxidase levels of 10% of those in rat leukocyte extracts. The authors point out, however, that since enzyme activities were measured in the soluble fraction of leukocyte homogenates, they may not accurately reflect activities *in situ*. Gee and Basford (1969) and Vogt *et al.* (1971) also reported significant levels of GSH peroxidase and GSSG reductase as well as GSSG- and GSH plus H_2O_2-stimulation of HMPS activity in lysates of rabbit alveolar macrophages.

Strauss *et al.* (1969) examined GSSG reductase and NADPH and NADH oxidase activities in intact guinea pig leukocytes under resting and phagocytizing conditions by direct spectrophotometric observations of leukocytes suspended in 70% glycerol to diminish light scattering. GSSG reductase and NADH oxidase activities were measured at pH 7.4 and NADPH oxidase at pH 5.5. As expected, 1 mM KCN stimulated GSSG reductase activity due to catalase and myeloperoxidase inhibition. The activities of both GSSG reductase and NADPH oxidase were increased in leukocytes after phagocytizing latex particles for 15 min. Kinetic studies showed that maximal stimulation of GSSG reductase (2.4-fold) occurred 15 sec after challenge by latex particles while maximal stimulation of NADPH oxidase (2.1-fold) was not observed until 15 min after latex particle addition. All of the GSSG reductase activity measured in resting and phagocytizing cells suspended in 70% glycerol could be accounted for in the 20,000g supernatant fraction of leukocyte homogenates. GSH peroxidase activity was not directly measured, but since KCN caused a marked stimulation of NADPH but not NADH oxidation at pH 7.4, 2 mM NEM caused 75% inhibition of resting and activated GSSG reductase activity, and GSSG was required for NADPH oxidation at pH 7.4, the authors concluded that the glutathione cycle was operative.

In addition, since GSSG reductase appeared to be activated before NADPH oxidase, Strauss *et al.* (1969) suggested that NADP produced by the GSSG reductase reaction serves to enhance HMPS activity and that the H_2O_2 produced by unactivated NADPH oxidase serves both to further increase GSSG reductase activity via GSH peroxidase and to labilize lysosomes resulting in the activation of NADPH oxidase.

A study of leukocytes from two female patients with chronic granulomatous disease which supports this interpretation was published by Holmes *et al.* (1970) and is discussed further in Section 3.2.2.

On the other hand, Rossi *et al.* (1972) measured the kinetics of GSSG reductase spectrophotometrically, in the 20,000g supernatant fraction, and NADPH oxidase polarographically, in the granule fraction of guinea pig leukocytes at 30 and 60 sec after challenge with *Bacillus subtilis*. Their results indicated that activation of NADPH oxidase occurs within a few seconds after the addition of bacteria while the rate of GSSG reduction did not change after challenge by bacteria.

Mandell (1972) studied the effects of NEM on phagocytosis and associated metabolism in human leukocytes. He observed that at a bacteria to leukocyte ratio of 5:1 or less, treatment with 0.1 mM NEM caused no impairment of phagocytosis; however, when the ratio was increased to 100:1, inhibition of bacterial uptake by 0.1 mM NEM was 80–92%. Even at low bacteria to leukocyte ratios, considerable impairment in bacterial killing was observed by treatment with 0.1 mM NEM.

Noseworthy and Karnovsky (1972) also observed inhibition of uptake of [^{14}C]starch granules (5 mg starch/10^7 guinea pig or rat leukocytes) by 0.01–0.10 mM NEM. No impairment of bacterial uptake by rabbit alveolar macrophages (AM) by 0.15 mM NEM was observed by Vogt *et al.* (1971).

In agreement with other investigators, Mandell (1972) showed marked inhibition of HMPS stimulation and diminished intracellular GSH levels by 0.1 mM NEM. However, the formation of H_2O_2 as measured by $[^{14}C]$formate oxidation was unimpaired by 0.1 mM NEM. This result would not be expected if the GSSG reductase is a vital source of H_2O_2 as suggested by Strauss *et al.* (1969) and Holmes *et al.* (1970).

Vogt *et al.* (1971) also observed no inhibition of H_2O_2 production by 0.1 mM NEM as measured by $[^{14}C]$formate oxidation in lysates of rabbit AM. However, as shown by Iyer *et al.* (1961) and amplified by Homan-Müller *et al.* (1975) the assessment of H_2O_2 by $[^{14}C]$formate oxidation has many drawbacks and probably greatly underestimates intracellular H_2O_2 concentration.*

It is, therefore, unclear from the research reviewed to this point whether or not the glutathione cycle is an integral link between O_2^- and H_2O_2 production and HMPS stimulation in phagocytizing cells.

3.2.1.   Glutathione Peroxidase

Glutathione peroxidase [glutathione : H_2O_2 oxidoreductase (E.C. 1.11.1.9)] catalyzes the reaction:

$$R-O-O-H + 2GSH \rightarrow ROH + GSSG + H_2O \qquad (3)$$

R–O–O–H may be hydrogen peroxide or a variety of hydroperoxides including linoleic acid and ethyl linoleate hydroperoxide, cumene hydroperoxide, *t*-butyl hydroperoxide, and most steroid hydroperoxides (Little and O'Brien, 1968). The enzyme is highly specific for GSH; most other thiol-containing compounds will not serve as substrates (Flohé and Günzler, 1974).

Rotruck *et al.* (1973) reported that glutathione failed to protect hemoglobin from oxidative damage when added to hemolysates of erythrocytes of selenium deficient rats pretreated with ascorbate or H_2O_2. The lack of protection by added glutathione was due to the virtual absence of glutathione peroxidase. Purified preparations of the enzyme from erythrocytes of selenium-deficient rats injected with ^{75}Se showed that the ^{75}Se cochromatographed on DEAE-Sephadex A-50 with glutathione peroxidase.

Subsequent studies in three different laboratories (Flohé *et al.*, 1973; Oh *et al.*, 1974; Nakamura *et al.*, 1974) showed that the pure enzyme contains 4 moles of tightly bound selenium per mole of enzyme. Since the enzyme is a tetramer, molecular weight 88,000, there is probably one Se per subunit. Prohaska *et al.* (1977) were able to show that the highly purified enzyme was inhibited by KCN which released Se from the enzyme. GSH or other thiol-containing compounds protected the enzyme from inhibition. The crude enzyme in rat liver cytosol was not inhibited by KCN unless preincubated with cumene hydroperoxide, a cosubstrate for the peroxidase. Although not rigor-

---

*Dr. Dirk Roos (personal communication) has shown that 0.1 mM NEM completely inhibited H_2O_2 production when assayed by the method of Homan-Müller *et al.* (1975).

ously proven, it is probable that Se undergoes oxidation–reduction during the enzymatic reaction (Flohé *et al.*, 1976).

3.2.2. Chronic Granulomatous Disease and Glutathione Peroxidase Deficiency

Chronic granulomatous disease (CGD) is a syndrome of infancy and childhood characterized by recurrent purulent infections. Leukocytes from affected patients phagocytize normally but fail to show the usual stimulation of oxygen uptake, O_2^- and H_2O_2 production, and HMPS activity, and have impaired ability to kill catalase-positive and H_2O_2-negative bacteria (reviewed by Johnston and Newman, 1977). CGD was originally described as a familial syndrome with X-linked transmission affecting only male children (Bridges *et al.*, 1959; Carson *et al.*, 1965).

The molecular defect in this "classical" form of CGD is generally agreed to be a deficiency in the reduction of O_2 to O_2^- and H_2O_2 (Baehner *et al.*, 1970b; Curnette *et al.*, 1974, 1975; Johnston and Newman, 1977). Evidence that the granule fraction of leukocytes from patients with CGD fail to show normal activation of NADPH oxidase has been documented by several groups of investigators (Curnette *et al.*, 1975; DeChatelet *et al.*, 1975; Hohn and Lehrer, 1975; McPhail *et al.*, 1977). On the other hand, Baehner and Karnovsky (1968) and Segal and Peters (1976, 1977) have reported that NADH oxidase is abnormally low in homogenates and plasma membrane fractions of resting leukocytes from CGD patients.

A variant of CGD was first reported by Quie *et al.* (1968) who studied two unrelated females with all of the clinical and pathological features of X-linked CGD. Shortly after this report, a number of other cases of female patients with CGD were described (Azimi *et al.*, 1968; Baehner and Nathan, 1968; Chandra *et al.*, 1969; Douglas *et al.*, 1969). Holmes *et al.* (1970) studied the two female patients reported by Quie *et al.* (1968) as well as seven male patients with the disease and reported that glutathione peroxidase activity in leukocytes from the female patients was significantly lower ($p < 0.001$) than in normal subjects or male CGD patients. Glutathione peroxidase activity in the erythrocytes of the two female patients was normal. Intermediate levels of glutathione peroxidase were demonstrated in the leukocytes of some but not all parents of the CGD patients. Holmes *et al.* (1970) suggested that the scheme proposed by Strauss *et al.* (1969) (discussed in Section 3.2) in which H_2O_2 serves both to further increase GSSG reductase activity via GSH peroxidase and to activate NADPH oxidase would explain why both glutathione peroxidase-deficient and glucose-6-phosphate dehydrogenase-deficient leukocytes show the same metabolic and microbicidal abnormalities.

Malawista and Gifford (1975) later showed that a brother of one of the female patients studied by Holmes *et al.* (1970) had both the clinical features of CGD and glutathione peroxidase deficiency. Both parents of the patients showed normal microbicidal and glutathione peroxidase activity, suggesting an autosomal recessive mode of inheritance.

Matsuda *et al.* (1976) also described a male patient with decreased leukocyte glutathione peroxidase activity. Superoxide anion production and HMPS activity were only slightly stimulated by phagocytosis, whereas NADPH oxidase and glucose-6-phosphate dehydrogenase activities were normal. Both parents of the patient showed intermediate leukocyte glutathione peroxidase activity suggesting an autosomal recessive mode of inheritance with variable penetrance.

In contrast to these studies showing decreased levels of glutathione peroxidase in both female and male patients with an apparent autosomally inherited CGD, DeChatelet *et al.* (1976) studied leukocyte glutathione peroxidase activity from two female and four male patients with autosomal recessive CGD and one male patient with X-linked recessive CGD, and found activity values in the same range (28.6 $\pm$ 3.99 nmol/min/mg protein) as the mean normal control value (23.4 $\pm$ 2.42). Windhorst and Katz (1972) also reported a normal level of glutathione peroxidase activity in one female patient with CGD.

Thus, while some patients with autosomal recessive CGD appear to have diminished leukocyte GSH peroxidase activity, not all such CGD patients exhibit this defect; the defect and syndrome may not be causally related.

3.2.3. Experimental Glutathione Peroxidase Deficiency

The discovery that glutathione peroxidase is a selenoenzyme and that rats are rendered enzyme deficient when fed a Se-free diet, suggested to us the possibility of testing the hypothesis that the enzyme-mediated glutathione cycle is a necessary link between O_2^- and H_2O_2 production and HMPS stimulation in phagocytizing leukocytes (Bartus *et al.*, 1975; Basford *et al.*, 1977).

Fischer CDF 344 rats were obtained at 3 weeks of age, divided into three groups and placed on Se-deficient, sodium selenite-supplemented (Teklad Mills), and standard laboratory chow diets for various periods of time. Leukocytes were obtained from peritoneal exudates after injection of sodium caseinate 12 hr before isolation. The Se content of tissues was determined by measuring the fluorescence of the Se^{4+}–diaminonaphthalene complex according to Olson *et al.* (1973).

Microbicidal activity was assessed at various time periods after the addition of bacteria to a suspension of leukocytes at a bacteria to cell ratio of 2:1, by plating appropriate dilutions of lysed incubation mixtures on Trypticase Soy agar plates, incubation at 37°C overnight, and counting viable colonies.

After 6 weeks on the Se-deficient diet, the Se content of whole blood, heart, kidney, liver, lungs, and spleen dropped to 10% of the control levels. GSH peroxidase activity reflected the tissue Se content. At this time period, GSSG reductase activity, HMPS activity of resting leukocytes, as well as leukocytes challenged with latex, heat-killed *Staphylococcus aureus* and *Streptococcus faecalis*, and microbicidal activity were similar to cells collected from Se-supplemented or chow-fed animals. HMPS activity stimulated by H_2O_2 generated by the D-amino acid oxidase system was the same as latex-stimulated HMPS activity in leukocytes from all three dietary groups.

The time period on experimental diets was then extended from 6 to 20 weeks in order to reduce the Se content and GSH peroxidase levels below 10% of controls. As shown in Figure 3, the Se content of liver (typical of all tissues measured) and GSH peroxidase activity of leukocytes after the rats were fed 7 weeks on the Se-deficient diet, dropped to 5% of control values and remained at that level throughout the 20-week period. These GSH peroxidase levels are near the limit of detection of the assay. Figure 4 shows the HMPS activity of resting and phagocytizing leukocytes from animals at time intervals between 3 and 20 weeks on Se-deficient (SeD) and Se-supplemented (SeS) diets. After 10 weeks on the SeD diet, HMPS activity of leukocytes, stimulated by latex particles, began to decrease and by 14 weeks had fallen to 60% of the value of SeS leukocytes and remained at that level. After 17 weeks, the HMPS activity of resting leukocytes from SeD animals rose above the value measured in SeS leukocytes.

The microbicidal activity of peritoneal leukocytes toward *S. aureus* and *S. faecalis* from rats on SeD diet for 12 weeks decreased slightly (45% inhibition at 30 min to 12% inhibition at 60 min) compared to controls. After 20 weeks on the diets, microbicidal activity remained essentially at the same level (Figure 5).

Serfass and Ganther (1975) also studied the microbicidal activity of

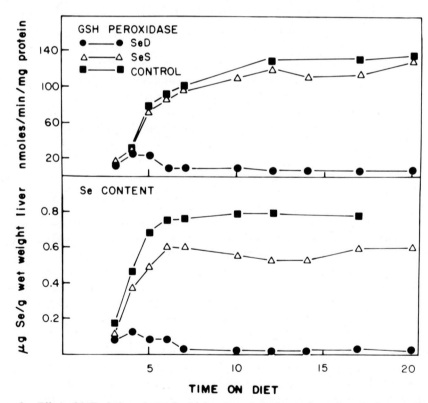

FIGURE 3.   Effect of SeD, SeS, and standard laboratory chow (control) on the selenium content of liver and the glutathione peroxidase activity in peritoneal leukocytes from rats maintained on the respective diets for up to 20 weeks. The points represent the average of 3–8 determinations. Reprinted from Basford *et al.* (1977), with permission of Piccin Medical Books.

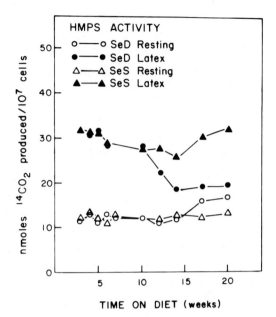

FIGURE 4. HMPS activity in resting and phagocytizing peritoneal leukocytes from SeD and SeS rats at various times after having been placed on their respective diets. Reprinted from Basford *et al.* (1977), with permission of Piccin Medical Books.

peritoneal and peripheral blood leukocytes from SeD and SeS rats toward the yeast *Candida albicans*. Microbicidal ability was assayed by microscopic observations of leukocyte monolayers challenged with *C. albicans* at a ratio of 25–50 yeast cells per leukocyte. Viability of yeast was determined with methylene blue, which stains dead but not living cells.

No statistically significant difference in the ability of peritoneal leukocytes from SeS and SeD rats (17–18 weeks on diets) to ingest or kill *C. albicans* was observed. However, the microbicidal ability of peripheral blood leukocytes from animals on SeD diet for 12–17 weeks was decreased to 6–7% killed yeast cells compared to 21–22% killed cells by leukocytes from SeS rats. This 70% decrease was significant at p values of 0.050–0.001. Peripheral blood leukocytes from both dietary groups ingested an average of 2 yeast cells per leukocyte in 75 min.

We therefore checked the ability of peripheral blood leukocytes from rats on SeD, SeS, and control diets for 20 weeks to kill *S. aureus* by the same method used for peritoneal exudate leukocytes. The results were the same as those obtained with peritoneal leukocytes, i.e., only a slight diminution in microbicidal activity.

Our observations that microbicidal ability was not diminished until animals had been made Se deficient for 12 weeks and that after 17 weeks of Se deficiency the HMPS activity of resting leukocytes rose (Figure 4), suggests that glutathione peroxidase deficiency may be expressed primarily as cellular damage due to peroxidation. Also, consistent with this interpretation, particle uptake as judged by Wright-stained smears appeared to be impaired by 15 weeks on the SeD diet, and ingestion of paraffin emulsions dyed with Oil Red O (Stossel *et al.*, 1972b) was inhibited by about 50% (Bartus, 1978).

Lawrence and Burk (1976) examined liver supernatant fractions from rats fed a Se-deficient diet for 2 weeks for glutathione peroxidase activity using both

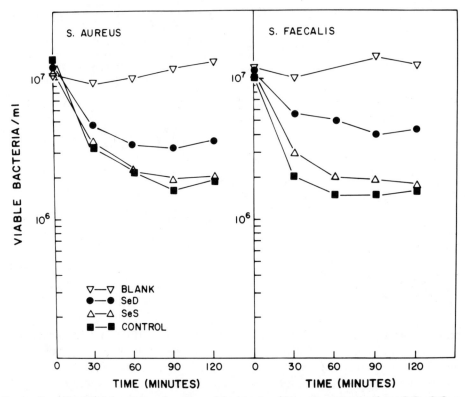

FIGURE 5. Microbicidal activity of peritoneal leukocytes from rats maintained on SeD, SeS, or standard laboratory chow (control) for 20 weeks. The reaction mixture contained 2.1 ml modified Krebs–Ringer phosphate, pH 7.4, 1×10^7 leukocytes, 10% normal sera, 2 μmol glucose and 2×10^7 *S. aureus* or *S. faecalis* cells. Incubation was at 37°C and aliquots removed at the indicated times, diluted, plated on Trypticase Soy agar plates, and colonies counted after incubation overnight at 37°C. Reprinted from Basford *et al.* (1977), with permission of Piccin Medical Books.

H_2O_2 and cumene hydroperoxide as substrates. Enzyme activity dropped to 8% of control when measured with H_2O_2 but retained 42% of control values with cumene hydroperoxide. Sephadex G150 gel chromatography of liver supernatants from Se-sufficient rats separated glutathione peroxidase activity with cumene hydroperoxide into two peaks. Peak I corresponded to the Se-containing peroxidase, which was active with both H_2O_2 and cumene hydroperoxide, while Peak II was active only with cumene hydroperoxide and had a molecular weight of 39,000. Sephadex chromatography of liver supernatant fractions from Se-deficient rats showed only Peak II activity with cumene hydroperoxide.

Prohaska and Ganther (1977) demonstrated that rat liver GSH peroxidase II (estimated molecular weight 45,000), assayed with cumene hydroperoxide, co-chromatographed with glutathione-*S*-transferase activity, assayed with 1-chloro-2,4-dinitrobenzene, on Sephadex G150, DEAE cellulose, and CM-cellulose. Burk *et al.* (1978) evaluated the function of GSH peroxidase I (selenoenzyme) and GSH peroxidase II in intact rat livers. Function was evaluated by measuring the release of GSSG in response to perfusion of H_2O_2

and tertiary butyl hydroperoxide (BHP) with a hemoglobin-free perfusion system. Infusion of H_2O_2 into control livers released GSSG in proportion to the rate of H_2O_2 perfusion. Both Se-deficient and Se-sufficient livers released GSSG in response to BHP infusion. GSSG release from Se-deficient livers was higher than from Se-sufficient livers when perfused with hemoglobin-free medium with no H_2O_2 or BHP present. The authors conclude that liver GSH peroxidase II, which is probably one or more of the glutathione-S-transferases, is operative in Se-deficient rats for the reduction of organic hydroperoxides. They also suggest that the higher release of GSSG from Se-deficient livers without hydroperoxide infusion indicates a higher basal rate of lipid peroxidation in Se-deficient livers compared to controls.

Habig et al. (1974) have examined the substrate specificities and kinetic properties of several of the glutathione-S-transferases and report overlapping substrate specificities which do not conform to the older designations such as aryl-alkyl-, alkene-, and epoxidetransferases. We compared the glutathione-S-transferase activities in the high-speed supernatant fraction of normal rat liver and peritoneal leukocyte and human blood leukocyte homogenates using 1,2-dichloro-4-nitrobenzene and 5 mM GSH at pH 7.5 and 1-chloro-2,4-dinitrobenzene and 1 mM GSH at pH 6.5 as substrates (Bartus, 1978).

Activities with 1-chloro-2,4-dinitrobenzene of 0.081 and 0.050 μmol/min/mg protein were measured in human and rat leukocytes, respectively; no measurable activity was detected with 1,2-dichloro-4-nitrobenzene. Corresponding values for rat liver soluble fractions were 1.44 and 0.040 μmol/min/mg protein for 1-chloro-2,4-dinitro- and 1,2-dichloro-4-nitrobenzene, respectively. It is therefore probable that the glutathione peroxidase activity toward lipid hydroperoxides remaining in our Se-deficient rat leukocytes is due to the glutathione-S-transferase with activity toward 1-chloro-2,4-dinitrobenzene.

Bactericidal activity in leukocytes from normal rats, humans, and guinea pigs (with high, intermediate, and low glutathione peroxidase activity, respectively), from Se-deficient rats, and from a patient with chronic granulomatous disease (X-linked, defective production of O_2^- and H_2O_2) were compared by Bass et al. (1977). No correlation was observed between the level of glutathione peroxidase activity and ability to kill Proteus mirabilis. Only leukocytes from the CGD patient showed significant impairment of bactericidal ability. Glutathione peroxidase was assayed with both H_2O_2 and cumene hydroperoxide as cosubstrates. Activities measured for Se-deficient rat leukocytes ranged from 0 to 7% and 1 to 8% of control values for H_2O_2 and cumene hydroperoxide, respectively.

Bass et al. (1977) also compared HMPS stimulation by latex particles, heat-killed Proteus, and opsonized zymosan in rat leukocytes and found that although latex particles only caused a minimal stimulation of $^{14}CO_2$ production from [1-^{14}C]glucose (8–20%), heat-killed bacteria and opsonized zymosan stimulated HMPS activity over 10-fold. HMPS stimulation by opsonized zymosan in glutathione-peroxidase-deficient leukocytes remained high ($\sim$ 10-fold) but was somewhat lower than in leukocytes from animals fed lab chow or Se-supplemented diets (15- to 17-fold).

The production of O_2^- and the incorporation of ^{131}I into trichloroacetic-acid-precipitable protein by rat leukocytes phagocytizing zymosan was not significantly different from Se-deficient and -sufficient animals. These authors concluded that phagocytic oxidative responses and bactericidal activity were not compromised by glutathione peroxidase deficiency and that the results are not compatible with the hypothesis that CGD can be caused by glutathione peroxidase deficiency.

In a preliminary paper, McCallister *et al.* (1977) reported that rat leukocytes rendered GSH-peroxidase-deficient (6.5% of control) and after 8 weeks on a Se-free diet exhibited normal O_2^- production as judged by nitro-blue tetrazolium reduction, normal HMPS stimulation and ingestion of lipopolysaccharide-coated paraffin oil. However, levels of GSH in GSH-peroxidase-deficient cells were somewhat lower (29%) than normal, capping induced by Concanavalin A (Con A) was increased from $10 \pm 3\%$ to $43 \pm 7\%$ and no migration was observed in the Boyden chamber, indicating enhanced microtubule disassembly (see Section 3.3).

It seems reasonable to conclude from the results obtained by Bass *et al.* (1977), Basford *et al.* (1977), and McCallister *et al.* (1977) on experimentally-induced glutathione peroxidase deficiency that glutathione peroxidase is not an essential link between O_2^- and H_2O_2 production and HMPS stimulation. The paper by Serfass and Ganther (1975) did not include measurements of O_2^- and H_2O_2 or HMPS activity during phagocytosis, but their finding of defective killing of *C. albicans* by rat peripheral leukocytes deserves further investigation with a variety of microorganisms. The results of Holmes *et al.* (1970), Malawista and Gifford (1975), and Matsuda *et al.* (1976) on some male and female patients with the CGD syndrome appear not to support this general conclusion and may reflect host species differences, possibly in the degree of peroxidative damage to the phagocytic cell. Although the results published using experimentally induced leukocyte glutathione peroxidase deficiency do not support an obligatory role for the enzyme as a link between oxygen reduction and HMPS stimulation, an essential role for glutathione reductase, and hence the glutathione cycle cannot be eliminated, since nonenzymatic oxidation of GSH to GSSG occurs at an appreciable rate as shown by Noseworthy and Karnovsky (1972).

3.2.4. Glutathione Reductase

Intracellular concentrations of GSH are very much higher than GSSG, and the enzyme glutathione reductase [NAD(P)H:glutathione oxidoreductase (E.C. 1.6.4.2)] has been detected in all cells examined (Meister, 1975). GSSG reductase catalyzes the reaction:

$$GSSG + NAD(P)H \rightarrow 2GSH + NAD(P) \tag{4}$$

Although the reductant specificity has been somewhat controversial, it now appears that at least the red cell enzyme is active with both nicotinamide nucleotides with a preference for NADPH of about 5 to 1 (Löhe *et al.*, 1974; Icén, 1967).

The specificity of the human erythrocyte enzyme for low-molecular-weight disulfides is nearly complete for GSSG; the only other naturally occurring disulfide substrate, oxidized lipoic acid, is reduced at about 3% of the rate of GSSG (Icén, 1967). The rat liver enzyme reduces cystine at about 5% of the rate of GSSG, but lipoic acid is not reduced (Mize and Langdon, 1962). Somewhat surprisingly, the erythrocyte enzyme was shown by Srivastava and Beutler (1970) to catalyze the cleavage of hemoglobin-glutathione-mixed disulfide (Hb-SSG). Thiol exchange reactions between GSH and Hb-SSG were thought to be ruled out using [^{35}S]-GSH-labeled Hb-SSG. A later study by Birchmeier *et al.* (1973), which used a milder method for preparing Hb-SSG showed that the mixed disulfide was not reduced by GSSG reductase but that it was a substrate for GSH thiol transferase.

The enzyme from human erythrocytes has a molecular weight $\sim$ 120,000 and contains 2 moles of FAD per mole of enzyme. Unlike the yeast enzyme of comparable molecular weight, the red cell enzyme does not appear to be dissociable into monomers (Staal *et al.*, 1969a). Rat liver GSSG reductase was found to have a molecular weight of 44,000 and to contain one active site per mole of enzyme. The liver enzyme appears to be more specific for NADPH than the erythrocyte enzyme; the NADH activity observed was only about 1% of that with NADPH (Mize and Langdon, 1962; Mize *et al.*, 1962).

Although in the majority of cells studied, GSSG reductase is found predominantly in the soluble portion of homogenates, Flohé and Schlegel (1971) showed that 32% of rat liver GSSG reductase is located in the matrix space of mitochondria and similar findings were reported for rat brain by Aldridge and Johnson (1959).

3.2.5. Glutathione Reductase Deficiency

A deficiency of GSSG reductase has been reported in a large number of patients and associated with a variety of hematological disorders (Beutler, 1971). However, Beutler (1969a) has shown that the enzyme in erythrocyte hemolysates from normal human subjects is often only partly saturated with FAD. Enzyme activity can be restored by FAD addition *in vitro* or by riboflavin administration *in vivo* (Beutler, 1969a,b), but the clinical disorders remain unchanged (Löhe *et al.*, 1974). A few cases have been reported in which *in vitro* or *in vivo* restoration of activity was unsuccessful: one group of patients in which genetically variant enzyme with altered electrophoretic mobility is produced (Löhe *et al.*, 1974) and another in which a lowered association constant for FAD was demonstrated (Staal *et al.*, 1969b).

Loos *et al.* (1976) reported clinical and biochemical findings in a family with three children with no demonstrable erythrocyte GSSG reductase in the presence or absence of added FAD or after riboflavin therapy. Erythrocyte levels of glucose-6-phosphate dehydrogenase, 6-P-gluconate dehydrogenase and pyruvate kinase were well within the normal range and GSH peroxidase values were on the low side of normal (21 μmol/min/g Hb for siblings, 19–48 μmol/min/g Hb for 25 normals). Both parents had red cell GSSG reductase levels of about one-half of control levels. One of the children, a 22-year-old woman was studied because of a hemolytic crisis and was shown to have hemolytic anemia probably due to eating fava beans. Her sister and both parents were well and a brother suffered from juvenile cataracts. Leukocyte GSSG reductase in the three siblings was 13–15% of normal and in the parents was reduced by about 45%. Lymphocytes, monocytes, and platelets of the patient also showed greatly reduced enzyme activity. Oxygen uptake by opsonized zymosan-stimulated leukocytes from the patient showed only a brief burst of activity which stopped after 6–7 min. More recent studies on leukocytes from the eldest daughter of this family (Weening et al., 1977; Roos et al., 1979) have clarified the role of GSSG reductase in leukocyte metabolism and strengthened the proposed role of the glutathione cycle as a necessary link between O_2^- and H_2O_2 production and HMPS stimulation.

As indicated earlier, the rate of O_2 uptake by GSSG-reductase-deficient leukocytes was normal for 3–4 min, and tapered off and stopped by 6–7 min after the addition of opsonized zymosan. This phenomenon occurred in the presence or absence of azide, an inhibitor of catalase and peroxidase. A similar pattern was observed for the evolution of $^{14}CO_2$ from [1-^{14}C]glucose, except that the initial rate of HMPS activity was also less than normal. Hydrogen peroxide production, measured by the sensitive method of Homan-Müller et al. (1975), also ceased about 5 min after the addition of zymosan. This result might be interpreted to support the proposal by Strauss et al. (1969) that the glutathione cycle is necessary for H_2O_2 production. However, zymosan-stimulated O_2^- production, measured as cytochrome c reduction, by GSSG-peroxidase-deficient cells was indistinguishable from control cells for the first 15 min, and was only slightly (and probably not statistically significantly) lower after 30 min. Since O_2^- is thought to be the first product of O_2 reduction by NADPH or NADH oxidase, prior to H_2O_2 formation, it was, therefore, concluded that the mechanism for H_2O_2 production is not impaired by GSSG reductase deficiency. It was postulated that O_2 uptake and H_2O_2 production leveled off due to peroxidative damage to the leukocyte. This did not occur in the presence of cytochrome c in the O_2^- assay, since cytochrome c reduction regenerates O_2 from O_2^-, thus preventing the dismutation of O_2^- to H_2O_2. Support for this conclusion was provided by two experiments in which H_2O_2 was generated by glucose and glucose oxidase. The first experiment showed that HMPS activity was stimulated by H_2O_2 generation in normal leukocytes but not in GSSG reductase-deficient cells. The second experiment showed that normal leukocytes exposed to the H_2O_2 generating system for 10 min and then reisolated, showed normal O_2 uptake at 4 min after exposure to opsonized zymosan, whereas GSSG reductase-deficient cells, treated in the same way, showed almost no O_2 uptake.

Thus, the net result of GSSG reductase deficiency is equivalent to depletion of the leukocyte's supply of GSH, which prevents detoxification of H_2O_2 by GSH peroxidase. Leukocyte catalase and myeloperoxidase are apparently of minor importance in the degradation of H_2O_2.

In view of the deleterious effects of GSSG reductase deficiency on leukocyte oxidative metabolism, a surprising finding was that the children of this family had no histories of recurrent infections and no abnormalities were found in the microbicidal functions of the patient's leukocytes (Weening et al., 1977). Chemotaxis towards casein, determined in vitro by the "leading front" method of Zigmond and Hirsch (1973), in vitro uptake of ^{14}C-labeled S. aureus, and in vitro intracellular killing of this organism by the patient's leukocytes were all normal. These results are in contrast to those observed by Holmes *et al.* (1970), Malawista and Gifford (1975), and Matsuda *et al.* (1976) in patients with GSH-peroxidase-deficient leukocytes. The authors (Roos *et al.*, 1979; Weening *et al.*, 1977) speculate that in GSSG reductase deficiency, leukocytes have no impairment in O_2^- and H_2O_2 generation for microbicidal purposes and can detoxify excess H_2O_2 until all leukocyte GSH is oxidized to GSSG. The amount of intracellular GSH is presumed to be sufficient to allow enough time for the O_2-dependent and independent microbicidal mechanisms to kill phagocytized organisms before oxidative damage by O_2^- and/or H_2O_2 prevents further O_2 uptake and HMPS stimulation.

From the foregoing summary there seems little reason to doubt that the glutathione cycle is essential to protect the leukocyte from damage by H_2O_2. Two sets of experimental findings remain puzzling: the lack of strict correlation between the level of GSH peroxidase activity and HMPS activity or microbicidal efficacy as reported by Bass *et al.* (1977), Basford *et al.* (1977), and McCallister *et al.* (1977) and the lack of correlation between GSSG reductase levels and microbicidal efficacy in the face of impaired oxidative metabolism as reported by Weening *et al.* (1977) and Roos *et al.* (1979). The poor correlation in the case of GSH peroxidase may be due to nonenzymatic oxidation of GSH, but this has not yet been adequately documented. Furthermore, even if the explanation is correct for leukocytes from Se-deficient animals, it cannot account for the discrepancy between the results with human (Holmes *et al.*, 1970; Malawista and Gifford, 1975; Matsuda *et al.*, 1976) and rat leukocytes (Bass *et al.*, 1977; Basford *et al.*, 1977). It would appear that either there is a significant species variability in susceptibility to oxidative damage, or that human GSH peroxidase deficiency is not a primary cause of this form of CGD.

The surprising lack of recurrent infections in the siblings of the family with GSSG reductase deficiency studied by Roos *et al.* (1979) as well as the normal *in vitro* bactericidal ability of the patient's leukocytes toward *S. aureus*, in view of the striking impairment of O_2 uptake and HMPS stimulation is difficult to understand under the framework of the postulated central role of the glutathione cycle. The authors, as well as Babior (1978), who has recently reviewed O_2-dependent microbicidal mechanisms, interpret these data to mean that the short period (about 5 min) of normal O_2 uptake and H_2O_2 production by GSSG reductase-deficient leukocytes is sufficient to kill invading organisms, both *in vivo* and *in vitro*. Such a high degree of microbicidal efficiency is unexpected, particularly when compared with the consequences of glutathione synthetase deficiency, discussed in Section 3.3.3.

3.3. ROLE OF GLUTATHIONE IN MICROTUBULE ASSEMBLY

3.3.1. Background

A great deal of direct and indirect evidence has accumulated from research on normal human leukocytes, leukocytes from patients with Chediak–Higashi Syndrome (CHS), and from the beige mouse, an animal model of CHS, to support the concepts that normal microtubule function is necessary for chemotactic response and for the fusion of primary and secondary lysosomes with the phagosome to produce the phagolysosome, where phagocytized organisms are killed and digested (Oliver, 1975; Oliver and Zurier, 1976; Root *et al.*, 1972; Stossel *et al.*, 1972a; Zurier *et al.*, 1974). Microtubule assembly involves the polymerization of tubulin dimer. The binding of concanavalin A to receptor sites on leukocyte plasma membranes causes a rapid increase in centriole-associated microtubules (Hoffstein *et al.*, 1976; Oliver *et al.*, 1976). Kuriyama and Sakai (1974) reported that exposure of isolated tubulin to *p*-chloromercuribenzene sulfonate and other thiol reagents inhibits microtubule assembly and depolymerizes intact microtubules *in vitro*. The distribution of Con A binding sites on normal leukocytes is random as shown by the uniform distribution of fluorescein-isothiocyanate-conjugated Con A. Disassembly of microtubules by treatment with agents such as colchicine and vinblastine which bind to tubulin, causes aggregation of Con A into a surface cap (reviewed by Oliver, 1975). The aggregation of Con A into caps is inhibited by metabolic inhibitors and agents such as cytochalasin B which impair microfilament function (dePetris, 1975; Schreiner and Unanue, 1976). Thus Con A cap formation has been inferred to depend on active contractile movements of microfilaments.

Cytochalasin-B-treated leukocytes or macrophages fail to ingest bound zymosan particles, but fusion of granules with the plasma membrane occurs and soluble enzymes of granules are extruded from the cells (Malawista *et al.*, 1971; Zurier *et al.*, 1973; Allison *et al.*, 1971). Thus it may be concluded that particle engulfment requires microfilament contraction and degranulation requires intact microtubules. Cytocholasin B inhibition of oxidative response to particle but not soluble factor stimulation (Rossi *et al.*, 1976), without impairment of O_2^- production, has been attributed to a decreased number of particle–cell interactions (Roos *et al.*, 1976).

Oliver, Berlin, and their colleagues have published a brilliant series of articles on the relationship between glutathione metabolism and microtubule dissociation and assembly, which are discussed in the following two sections.

3.3.2. Microtubule Assembly and Microtubule-Dependent Surface Properties of Leukocytes

Nath and Rebhun (1976) showed that oxidation of GSH in fertilized sea urchin eggs inhibits mitotic spindle formation and causes dissolution of preformed spindles. Following this observation, Oliver *et al.* (1976) showed that the GSH-oxidizing agents diamide (diazene dicarboxylic acid bis-*N,N*-dimethylamide) and BHP (tertiary butyl hydroperoxide) added to human leukocytes and lymphocytes, promote the movement of surface-bound Con A into caps and inhibit the assembly of microtubules normally induced by Con-A-binding.

Thus oxidation of intracellular GSH had the same effect as colchicine on leukocytes. Although both diamide and BHP oxidize GSH to GSSG, BHP is considered more specific, since diamide reacts nonenzymatically, and can theoretically cause oxidation of other low-molecular-weight thiols or nonsterically hindered protein-SH groups while BHP oxidation of GSH requires GSH-peroxidase (Srivastava *et al.*, 1974). Con A capping and inhibition of microtubule assembly occurred when GSH levels in cell suspensions decreased by 30–70%. Return of GSH to control levels by GSSG reductase and NADPH is accompanied by the appearance of cytoplasmic microtubules and inhibition of the capping response with Con A. The number of microtubules per cell was evaluated by electron microscopy of thin sections through the centriole. Albertini *et al.* (1977) extended these observations by showing that: (1) a dense network of microfilaments underlies Con A caps which occupy a protuberance or constriction in leukocytes, monocytes, and lymphocytes; (2) the accumulation of filaments and formation of protuberances or constrictions are not induced by Con A, but follow microtubule disassembly induced by GSH oxidants; and (3) randomly distributed Con A receptors can move into preformed caps. Burchill *et al.* (1978) then examined the microtubule dynamics during phagocytosis by human leukocytes using the electron microscopic method. Resting leukocytes were shown to have few ($\sim$ 4) centriole-associated microtubules. Phagocytosis of opsonized zymosan induced a cycle of rapid assembly of microtubules followed by disassembly after phagocytosis had ceased. Assembly is initiated by particle contact and was maximal ($\sim$ 16 microtubules/cell) after 3 min. After 9–10 min, phagocytosis had ceased and the number of centriole-associated microtubules was reduced to about 2/cell. Microtubule formation was not induced by 10% zymosan-treated serum, eliminating the possibility of soluble factors generated during opsonization as the stimulus to microtubule assembly. The kinetics of GSH oxidation were then correlated with microtubule dynamics. Microtubule assembly induced by phagocytosis was found to be accomplished by only small decreases in GSH ($-$ 0.23 nmol/10^6 cells) and increases in GSSG (+0.01 nmol/10^6 cells) levels. However, the microtubule disassembly phase is preceded by a gradual increase in GSSG and is coincident with the generation of mixed disulfides of glutathione with protein (protein-SSG). A similar correlation was shown between BHP inhibition of Con-A-induced microtubule assembly and increases in GSSG and protein-SSG in human leukocytes. Protein-SSG was estimated after sodium borohydride reduction of trichloroacetic-acid-precipitated protein. An approximately 50% loss of GSH, added as an internal control, occurred during the reduction step. Thus the estimates of protein-SSG represent minimal values. The recovery of total cellular glutathione after phagocytosis was only 80%, which may have been due to incomplete reduction of protein-SSG as well as borohydride destruction of released GSH. Nevertheless, plots of protein-SSG production vs. time of phagocytosis yielded sigmoid-shaped curves which reached a plateau at 9 min when phagocytosis was complete and assembled microtubules had depolymerized. Previous results with both sea urchin eggs (Mellon and Rebhun, 1976) and human leukocytes (Oliver *et al.*, 1976) had suggested that GSSG inhibits microtubule assembly, but unphysiologically high concentrations of GSSG were required.

Nath and Rebhun (1976) and Rebhun *et al.* (1976) proposed on theoretical grounds that formation and reduction of mixed disulfides between protein-SH and GSSG may regulate microtubule disassembly. The data of Burchill *et al.* (1978) strongly suggest that an equilibrium between GSH, GSSG and protein-SSG is shifted toward protein-SSG when GSSG is elevated during phagocytosis and by GSH oxidants. Other investigators have proposed mixed protein-glutathione disulfide involvement in metabolic regulation. For example, Isaacs and Binkley (1977) showed diurnal variation in disulfide–sulfhydryl ratios in rat liver, with changes in protein-glutathione mixed disulfides of cellular membranes showing the largest variation. Formation of protein-SSG was shown from indirect data to be dependent on thiol transferase and GSSG concentration. The results of Burchill *et al.* (1978) do not show a causal relationship between protein-SSG formation and microtubule inhibition, since it has not been shown that protein-SSG is tubulin-SSG, but studies in progress with radioactive GSH should provide clarification.

The question of whether GSSG or H_2O_2, both of which are generated during phagocytosis- and Con-A-induced assembly of microtubules (DeChatelet, 1975), is the direct physiological oxidant of tubulin was also addressed by Oliver *et al.* (1977). Microtubule disassembly was evaluated in normal human and in CGD leukocytes which are incapable of generating O_2^- and H_2O_2 (Section 3.2.2). Microtubule disassembly was evaluated by Con A capping of leukocytes induced by the GSH oxidants diamide and BHP and by exogenous H_2O_2. Although little or no O_2^- was generated by CGD leukocytes by zymosan, diamide, BHP, or combinations of Con A and diamide or BHP, the CGD leukocytes showed normal HMPS stimulation, normal oxidation of GSH to GSSG, and normal cap formation induced by the GSH oxidizing agents. These data suggest that increased GSSG is a sufficient stimulus to oxidation and depolymerization of cytoplasmic microtubules in human leukocytes.

3.3.3. Glutathione Deficient Leukocytes: Glutathione Synthetase Deficiency

Spielberg *et al.* (1977) reported clinical and biochemical data on an infant with 5-oxoprolinuria showing a deficiency of glutathione synthetase activity in erythrocytes, leukocytes, and cultured skin fibroblasts and a concomitant decrease in glutathione content in these cells. Leukocyte GSH content was 2.1 μg/mg protein compared to 8.4 ± 0.4 μg/mg in normal leukocytes. Leukocyte glutathione synthetase activity was 2 compared to 55 ± 9 nmol/mg protein/hr in normal leukocytes and the apparent K_m for glycine of the synthetase from the patient's fibroblasts was fivefold higher than the synthetase in three normal cell lines. Oliver *et al.* (1978) examined microtubule dynamics in glutathione-deficient leukocytes from this patient. Con A induced a rapid and normal increase in centriole-associated microtubules in these GSH synthetase-deficient cells in spite of GSH levels which were only 20% of normal. Thus normal, high levels of GSH are not required to induce microtubule assembly, and rule out a decrease in cellular GSH as a trigger for microtubule assembly. On the other hand, oxidation of GSH to GSSG by diamide or BHP failed to initiate cap formation in these cells as it does in control leukocytes. Colchicine reduced the number of centriole-associated microtubules in Con-A-treated normal and

GSH-synthetase-deficient cells equally. BHP did not induce microtubule disassembly as it did in normal leukocytes; however, diamide did induce disassembly. It is concluded from this observation that diamide, in the absence of high levels of GSH, oxidizes tubulin-SH groups (Mellon and Rebhun, 1976) and, probably, other protein thiols, while BHP does not. Phagocytosis of lipopolysaccharide-coated oil droplets (Stossel, 1973) induced the release of 10.5 nmol of $H_2O_2/2.5 \times 10^6$ cells/min from normal cells and 17 nmol of H_2O_2 from GSH-synthetase-deficient cells, but failed to induce normal microtubule formation in synthetase-deficient cells. BHP reduced the GSH:GSSG ratio in both normal and GSH-synthetase-deficient cells to a similar level (4:1 in normal and 6:1 in patient cells). It is, therefore, concluded that the control of tubulin polymerization-depolymerization is probably not a function of GSH/GSSG ratio. The evidence suggests that the absolute level of GSSG may inhibit assembly and promote disassembly of microtubules. In normal leukocytes, microtubule assembly is inhibited and preformed tubules depolymerized under conditions where GSSG is increased 15-fold over resting levels. The GSSG level reached after BHP treatment of the GSH-synthetase-deficient leukocytes was only three times normal resting levels. The authors propose that the accumulation of GSSG was inadequate to react with sufficient tubulin to impair microtubule assembly. It is also suggested that the higher-than-normal production of H_2O_2, measured extracellularly by phagocytizing GSH-synthetase-deficient cells, reflects higher intracellular H_2O_2 levels which cannot be reduced by GSH and GSH peroxidase. Thus the intracellular H_2O_2 oxidizes tubulin-SH and prevents microtubule assembly. This hypothesis was supported by a subsequent paper (Spielberg *et al.*, 1979) in which chemotaxis, bactericidal function, and oxidative response were evaluated in leukocytes from the patient with 5-oxoprolinuria. There was no history of skin abscesses, pneumonia, fungal infections, meningitis, or sepsis. However, during the second year of life, the patient had six episodes of acute otitis media and neutropenia was noted during two of these. In the absence of infections the patient maintains a normal white count and differential, however the circulating PMN leukocytes are less mature than normal suggesting increased peripheral cell destruction. Despite the 80% deficiency in GSH content of the patient's leukocytes, the rate of phagocytosis of opsonized lipopolysaccharide-coated oil droplets, rate of reduction of nitroblue tetrazolium (equivalent of O_2^- generation), and chemotactic index, evaluated by the modified Boyden chamber technique of Hill *et al.* (1975), were all normal. Stimulation of HMPS activity by phagocytosis of opsonized zymosan and by diamide and BHP were also normal. As indicated earlier, phagocytosis generated a higher than normal (34-fold as compared to 21-fold) increase in H_2O_2 production. The patient's leukocytes failed to kill *S. aureus* normally after 30, 60, and 100 min of incubation and protein iodination was drastically reduced. Assays for myeloperoxidase showed normal levels of activity. Thus, the decreased iodination and bacterial killing in the presence of elevated levels of H_2O_2 and normal myeloperoxidase activity suggests impairment in lysosomal fusion as a consequence of increased microtubule disassembly. The addition of 25 mM diamide decreased the iodination of protein in normal leukocytes and abolished it in GSH-synthetase-deficient cells, suggesting a vital role for GSH in iodination. In view of the higher specificity of BHP for GSH, it would appear that BHP would have been a better choice than diamide for assessing the role of GSH.

Electron microscopic studies of the patient's leukocytes in the presence of phagocytizing oil droplets showed decreased centriole-associated microtubules and extensive membrane damage. Thus, all of the parameters studied support the hypothesis that GSH synthetase deficiency results in leukoyctes which are unable to reduce metabolically generated H_2O_2, resulting in increased depolymerization of tubulin as well as general membrane damage. All of the data are consistent with the view that such cells have an impaired ability to fuse lysosomes with phagosomes and thus have impaired microbicidal activity. This study also shows that normal glutathione content is not required for the production of NADP via glutathione peroxidase nor for H_2O_2 production. The latter conclusion was also supported by the study of Weening *et al.* (1977) and Roos *et al.* (1978) on a patient with GSSG reductase deficiency (Section 3.2.5).

Thus the studies of Oliver, Berlin, and their collaborators on glutathione-synthetase-deficient leukocytes strengthens the conclusion that an operative glutathione cycle is required for optimal leukocyte function and points to the microtubules as a primary site for control by GSSG levels. An evaluation of the role of glutathione in the regulation of phagocytic functions of neutrophils and monocytes has recently been published by Baehner and Boxer (1979).

Other, less well-defined oxidative damage by O_2^- or H_2O_2 has been suggested in the studies on glutathione peroxidase deficiency (Basford *et al.*, 1977), glutathione reductase deficiency (Weening *et al.*, 1977; Roos *et al.*, 1979), as well as on glutathione synthetase deficiency (Spielberg *et al.*, 1979).

Possible sites of cellular damage due to oxidation by O_2^- or H_2O_2 are double bonds of fatty acids contained in the lipids of membranes and thiol groups of proteins. In addition, cross-linking of proteins may occur as a result of the reaction of the aldehyde break-down product of peroxidized unsaturated fatty acids with amino groups of proteins (Chio and Tappel, 1969). Khandwala and Gee (1973) showed that linoleic acid hydroperoxide added *in vitro* to alveolar macrophages is a potent inhibitor of bacterial uptake and HMPS stimulation.

Since lipid hydroperoxides are substrates for glutathione peroxidase (Little and O'Brien, 1968) it has been suggested that one of the primary functions of GSH peroxidase is the reduction of membrane-bound lipid hydroperoxides to hydroxy derivatives (Christopherson, 1969; Flohé and Zimmerman, 1970). However, McCay *et al.* (1976) have shown in rat liver microsomes and mitochondria that while the glutathione peroxidase system protects membrane-unsaturated phospholipids from oxidative degradation, it does not appear to do so by reduction of lipid peroxides to hydroxy derivatives. Glutathione peroxidase appears to act by preventing oxidative damage by H_2O_2 rather than by reducing lipid hydroperoxides once formed. In fact, glutathione peroxidase, while able to utilize fatty acid hydroperoxides as substrates, appears unable to utilize fatty acyl hydroperoxides acylated to phospholipid.

If one assumes the same is true in leukocytes, in those deficiency states which result in diminished GSH peroxidase activity, membrane damage due to

H_2O_2 production will be irreversible and should lead to increased destruction of leukocytes. Neutropenia was, in fact, seen in the patient with 5-oxoprolinuria studied by Spielberg *et al.* (1977, 1979).

If, as seems likely from the work of Burchill *et al.* (1978), oxidation of tubulin thiol groups by the formation of mixed GSH-protein disulfides occurs concomitantly with microtubule disassembly, it follows that there must be a mechanism for regeneration of tubulin-SH and GSH from tubulin-SSG. Although no information is currently available from research on leukocytes, it seems likely that a thioltransferase as has been described in liver by Isaacs and Binkley (1977) and Mannervik and Axelsson (1975), or one of the glutathione-*S*-transferases (Habig *et al.*, 1974) will be shown to be active in the regeneration of tubulin-SH from tubulin-SSG prior to its repolymerization to microtubules.

4. LEUKOCYTE GLYOXALASE

Although the existence of the glyoxalase system in animal tissues has been known for more than sixty years, the physiological function is still poorly understood. The overall reaction catalyzed by the two enzymes of the glyoxalase system is an intramolecular oxidation–reduction of methyl glyoxal (or other keto aldehyde) to lactic acid (or the corresponding α-hydroxy acid):

$$
\begin{array}{cc}
\text{(R-)} & \text{(R-)} \\
CH_3\text{–}CO\text{–}CHO + H_2O \rightarrow CH_3\ CHOHCOOH & \qquad (5)
\end{array}
$$

Glyoxalase I [*S*-Lactoyl glutathione methylglyoxalase (isomerizing) (E.C. 4.4.1.5)] requires GSH as a coenzyme and catalyzes the formation of *S*-D-lactoyl glutathione from methylglyoxal (pyruvaldehyde) and GSH (Racker, 1951). Glyoxalase II [*S*-2-hydroxy acyl glutathione hydrolase (E.C. 3.1.2.6)] catalyzes the hydrolysis of lactoyl glytathione (and other α-hydroxy acyl glutathiones) to D-lactate or the corresponding D-α-hydroxy acid and free GSH (Racker, 1951; Drummond and Stern, 1961).

The conversion of methylglyoxal to D-lactate was discovered in muscle and liver extracts by Dakin and Dudley (1913) and by Neuberg (1913), and shortly thereafter in dog leukocytes by Levene and Meyer (1913).

For a number of years methylglyoxal was considered to be an intermediate in glycolysis, until Lohmann (1932) showed that lactic acid production from glycogen occurred in the absence of glyoxalase activity. The fact remains that glyoxalase activity is relatively high in many cells including leukocytes and the physiological significance is not known. Methylglyoxal has been shown to be formed in liver mitochondria from amino acetone which is synthesized from acetyl CoA and glycine (Elliot, 1959; Urata and Granick, 1963). So far as I am aware, the existence of amino acetone synthetase has not been demonstrated in leukocytes.

McKinney and Rundles (1956) published evidence that leukemic leukocytes have a lower glyoxalase activity and produce less lactic acid than normal leukocytes; this observation does not seem to have been pursued further. The α-keto aldehydes are generally considered toxic to cells, thus the glyoxalase system may function solely in detoxification (Mannervik, 1974). Szent-Györgi *et al.* (1967) have proposed that glyoxal derivatives may regulate cell division and, thus, might be related to oncology, but the idea has not received wide acceptance.

5.   OTHER POSSIBLE ROLES FOR LEUKOCYTE GLUTATHIONE

A recurrent theme in this review has been the proposed role of the glutathione cycle as an integral link between phagocytosis-associated respiratory bursts, with concomitant O_2^- and H_2O_2 production and the increase in HMPS activity. The nature of the link has been assumed to be mass action. That is, an increased concentration of NADP, produced as a result of an increased NADPH oxidase activity or increased GSSG reductase activity or of both reactions, activates the first enzyme of the HMP pathway, glucose-6-phosphate dehydrogenase. Although apparently not yet examined in leukocytes, a second possibility is suggested by a study of the regulation of the HMPS in rat liver by Krebs and Eggleston (1974; Eggleston and Krebs, 1974).

Glucose-6-phosphate dehydrogenase is inhibited by its reduced product, NADPH. Inhibition approaches 100% when the ratio of NADPH/NADP reaches a value of nine. In rat liver cytoplasm the ratio of free reduced to oxidized nucleotide is close to 100:1 (Veech *et al.*, 1969). Thus, it must be assumed that under normal resting conditions *in vivo*, the rat liver enzyme is completely inhibited and that a specific activator should exist to relieve the inhibition when metabolically necessary. Eggleston and Krebs (1974) showed that GSSG, at physiologic concentration, in the presence of an unidentified cofactor counteracted the inhibition by NADPH. The effect was specific for GSSG, and was not due to the reduction of GSSG by GSSG reductase. The unidentified cofactor, required to demonstrate reversal of NADPH inhibition by GSSG, was present in most rat tissues, was highest in liver, and could not be replaced by a long list of cofactors, coenzymes, and hormones. Further research on the cofactor does not appear to have been published. The ratio of NADPH/NADP in resting guinea pig leukocytes has been determined by Patriarca *et al.* (1971) to be 8.5, and numerous experiments have shown that HMPS activity in resting cells is low. Phagocytizing leukocytes have an NADPH/NADP ratio of about 3.3 (Patriarca *et al.*, 1971). It is possible that increased GSSG levels are responsible for overcoming NADPH inhibition of glucose-6-phosphate dehydrogenase in phagocytizing leukocytes as proposed by Krebs and Eggleston (1974) for liver.

6.   SUMMARY

An evaluation of the rather large literature on the physiologic role of glutathione in phagocytizing leukocytes leads to the conclusion that adequate levels of glutathione, glutathione reductase and, perhaps, glutathione peroxidase are required to protect the leukocyte from damage due to excess

H_2O_2 or O_2^- not utilized in the killing of phagocytized organisms. Glutathione synthetase deficiency and probably glutathione reductase deficiency leads to H_2O_2-induced damage to microtubules resulting in impaired lysosomal fusion and microbicidal function; further oxidative membrane damage due to impaired ability to remove H_2O_2 is likely but not proven.

The glutathione cycle appears to be required neither for O_2^- or H_2O_2 production nor for NADP production in phagocytosis-mediated HMPS stimulation.

ACKNOWLEDGMENTS.   I thank Drs. Janet Oliver, Dirk Roos, and Stephen Spielberg for making available manuscripts prior to publication, and Drs. Barbara Bartus, Sandra Kaplan, Becky Hughey, and Norman Curthoys for helpful discussions and critical reading of the manuscript.

REFERENCES

Albertini, D. F., Berlin, R. D., and Oliver, J. M., 1977, The mechanism of concanavalin A cap formation in leukocytes, *J. Cell Sci.* **26**:57.

Aldridge, W. N., and Johnson, M. K., 1959, Cholinesterase, succinic dehydrogenase, nucleic acids, esterase and glutathione reductase in subcellular fractions of rat brain, *Biochem. J.* **73**:270.

Allison, A. L., Davies, P., and dePetris, S., 1971, Role of contractile microfilaments in macrophage movement and endocytosis, *Nature New Biol.* **232**:153.

Arias, I. M., and Jakoby, W. B. (eds.), 1976, *Glutathione, Metabolism and Function*, Raven Press, New York.

Azimi, P. H., Bodenbender, J. G., Hintz, R. L., and Kontras, S. B., 1968, Chronic granulomatous disease in three female siblings, *J. Am. Med. Assoc.* **206**:2865.

Babior, B. M., 1978, Oxygen-dependent microbial killing by phagocytes (part two), *N. Engl. J. Med.* **298**:721.

Badwey, J. A., Curnette, J. T., and Karnovsky, M. L., 1979, The enzyme of granulocytes that produces superoxide and peroxide. *N. Engl. J. Med.* **300**:1157.

Baehner, R. L., and Boxer, L. A., 1979, Role of membrane vitamin E and cytoplasmic glutathione in the regulation of phagocytic functions of neutrophils and monocytes, *Am. J. Ped. Hematol./Oncol.* **1**:71.

Baehner, R. L., and Karnovsky, M. L., 1968, Deficiency of reduced nicotinamide adenine dinucleotide oxidase in chronic granulomatous disease, *Science* **162**:1277.

Baehner, R. L., and Nathan, D. G., 1968, Quantitative nitroblue tetrazolium test in chronic granulomatous disease, *N. Eng. J. Med.* **278**:971.

Baehner, R. L., Gilman, N., and Karnovsky, M. L., 1970a, Respiration and glucose oxidation in human and guinea pig leukocytes: Comparative studies, *J. Clin. Invest.* **49**:692.

Baehner, R. L., Nathan, D. G., and Karnovsky, M. L., 1970b, Correction of metabolic deficiencies in the leukocytes of patients with chronic granulomatous disease, *J. Clin. Invest.* **49**:865.

Bartus, B. A., 1978, Metabolic and microbicidal activity of glutathione peroxidase deficient polymorphonuclear leukocytes, Ph.D. Thesis, University of Pittsburgh.

Bartus, B. A., Kaplan, S. S., Platt, D., and Basford, R. E., 1975, The role of glutathione peroxidase in phagocytosis, in: *Abstracts of the Annual Meeting of the American Society of Microbiology*, Abstr. B3, p. 12, American Society for Microbiology, Washington, D.C.

Basford, R. E., Bartus, B., Kaplan, S. S., and Platt, D., 1977, Glutathione peroxidase and phagocytosis, in: *Movement, Metabolism and Bactericidal Mechanisms of Phagocytes* (F. Rossi, P. Patriarca, and D. Romeo, eds.), pp. 265–275, Piccin, Padua.

Bass, D. A., DeChatelet, L. R., Burk, R. F., Shirley, P., and Szejda, P., 1977, Polymorphonuclear leukocyte bactericidal activity and oxidative metabolism during glutathione peroxidase deficiency, *Infect. Immun.* **18**:78.

Beck, W. S., 1958, Occurrence and control of the phosphogluconate oxidation pathway in normal and leukemic leukocytes, *J. Biol. Chem.* **232**:271.

Beutler, E., 1969a, Glutathione reductase: Stimulation in normal subjects by riboflavin supplementation, *Science* **165**:613.

Beutler, E., 1969b, Effect of flavin compounds on glutathione reductase activity: *In vivo* and *in vitro* studies, *J. Clin. Invest.* **48**:1957.

Beutler, E., 1971, Abnormalities of the hexose monophosphate shunt, *Semin. Hematol.* **8**:311.

Birchmeier, W., Tuchschmid, P. E., and Winterhalter, K. H., 1973, Comparison of human hemoglobin A carrying glutathione as a mixed disulfide with naturally occurring human hemoglobin A₃, *Biochem. J.* **12**:3667.

Boxer, L. A., Ismail, G., Allen, J. M., and Baehner, R. L., 1979, Oxidative metabolic responses of rabbit pulmonary alveolar macrophages, *Blood* **53**:486.

Boyland, E., and Chasseaud, L. F., 1969, The role of glutathione and glutathione *S*-transferases in mercapturic acid biosynthesis, *Adv. Enzymol.* **32**:173.

Bridges, R. A., Berenedes, H., and Good, R. A., 1959, A fatal granulomatous disease of childhood: The clinical, pathological and laboratory features of a new syndrome, *Am. J. Dis. Child.* **97**:387.

Briggs, R. T., Karnovsky, M. L., and Karnovsky, M. J., 1977, Hydrogen peroxide production in chronic granulomatous disease. A cytochemical study of reduced pyridine nucleotide oxidase, *J. Clin. Invest.* **59**:1088.

Burchill, B. R., Oliver, J. M., Pearson, C. B., Leinbach, E. D., and Berlin, R. D., 1978, Microtubule dynamics and glutathione metabolism in phagocytosing human polymorphonuclear leukocytes, *J. Cell Biol.* **76**:439.

Burk, R. F., Nishiki, K., Lawrence, R. A., and Chance, B., 1978, Peroxide removal by selenium-dependent and selenium-independent glutathione peroxidases in hemoglobin-free perfused liver, *J. Biol. Chem.* **253**:43.

Carson, M. J., Chadwick, D. L., Brubaker, C. A., Cleland, R. S., and Landing, B. H., 1965, Thirteen boys with progressive septic granulomatosis, *Pediatrics* **35**:405.

Chandra, R. K., Cope, W. A., and Soothill, J. F., 1969, Chronic granulomatous disease: Evidence for an autosomal mode of inheritance, *Lancet* **2**:71.

Chio, K. S., and Tappel, A. L., 1969, Inactivation of ribonuclease and other enzymes by peroxidizing lipids and by malonaldehyde, *Biochemistry* **8**:2827.

Christophersen, B. O., 1969, Reduction of linolenic acid hydroperoxide by a glutathione peroxidase, *Biochim. Biophys. Acta* **176**:463.

Cohen, G., and Hochstein, P., 1963, The primary agent for the elimination of hydrogen peroxide in erythrocytes, *Biochemistry* **2**:1420.

Colowick, S., Lazarow, A., Racker, E., Schwartz, D. R., Stadtman, E., and Waelsch, H. (eds.), 1954, *Glutathione*, Academic Press, New York.

Crook, E. M. (ed.), 1959, *Glutathione*, Biochem. Soc. Symp. No. 17, Cambridge University Press, London.

Curnette, J. T., Whitten, D. M., and Babior, B. M., 1974, Defective superoxide production by granulocytes from patients with chronic granulomatous disease, *N. Engl. J. Med.* **290**:593.

Curnette, J. T., Kipnes, R. S., and Babior, B. M., 1975, Defect in pyridine nucleotide-dependent superoxide production by a particular fraction from the granulocytes of patients with chronic granulomatous disease, *N. Engl. J. Med.* **293**:628.

Dakin, H. D., and Dudley, H. W., 1913, On glyoxalase, *J. Biol. Chem.* **14**:423.

DeChatelet, L. R., 1975, Oxidative bactericidal mechanisms of polymorphonuclear leukocytes, *J. Infect. Dis.* **131**:295.

DeChatelet, L. R., McPhail, L. C., Mullikin, D., and McCall, C. E., 1975, An isotopic assay for NADPH oxidase activity and some characteristics of the enzyme from human polymorphonuclear leukocytes, *J. Clin. Invest.* **55**:714.

DeChatelet, L. R., Shirley, P. S., and McPhail, L. C., 1976, Normal leukocyte glutathione peroxidase activity in patients with chronic granulomatous disease, *J. Pediatr.* **89**:598.

dePetris, S., 1975, Concanavalin A receptors, immunoglobulins and θ antigen of the lymphocyte surface. Interactions with Concanavalin A and with cytoplasmic structures, *J. Cell Biol.* **65**:123.

Douglas, S. D., Davis, W. C., and Fudenberg, H. H., 1969, Granulocytopathies: Pleomorphism of neutrophil dysfunction, *Am. J. Med.* **46**:901.

Drummond, G. I., and Stern, J. R., 1961, Enzymic hydrolysis of glutathione thioesters, *Arch. Biochem. Biophys.* **95**:323.

Eggleston, L. V., and Krebs, H. A., 1974, Regulation of the pentose phosphate cycle, *Biochem. J.* **138**:425.

Elliott, W. H., 1959, Amino-acetone: Its isolation and role in metabolism, *Nature* **183**:1051.

Evans, A. E., and Kaplan, N. O., 1966, Pyridine nucleotide transhydrogenase in normal human and leukemic leukocytes, *J. Clin. Invest.* **45**:1268.

Evans, H. W., and Karnovsky, M. L., 1962, The biochemical basis of phagocytosis. IV. Some aspects of carbohydrate metabolism during phagocytosis, *Biochemistry* **1**:159.

Flohé, L., and Günzler, W. A., 1974, Glutathione peroxidase, in: *Glutathione* (L. Flohé, H. C. Benöhr, H. Sies, H. D. Waller, and A. Wendel, eds.), pp. 132–145, Academic Press, New York.

Flohé, L., and Schlegel, W., 1971, Glutathione peroxidase. IV. Intracellular distribution of the glutathione peroxidase system in rat liver, *Hoppe-Seylers Z. Physiol. Chem.* **352**:1401.

Flohé, L., and Zimmermann, R., 1970, The role of GSH peroxidase in protecting the membrane of rat liver mitochondria, *Biochim. Biophys. Acta* **223**:210.

Flohé, L., Günzler, W. A., and Schock, H. H., 1973, Glutathione peroxidase: A selenoenzyme, *FEBS Lett.* **32**:132.

Flohé, L., Benöhr, H. C., Sies, H., Waller, H. D., and Wendel, A. (eds.), 1974, *Glutathione*, Academic Press, New York.

Flohé, L., Günzler, W. A., and Ladenstein, R., 1976, Glutathione peroxidase, in: *Glutathione: Metabolism and Function* (I. M. Arias and W. B. Jakoby, eds.), pp. 115–138, Raven Press, New York.

Gee, J. B. L., and Basford, R. E., 1969, The effects of sulfhydryl inhibitors on phagocytosis by alveolar macrophages, *Fed. Proc.* **28**:445.

Habig, W. A., Pabst, M. J., and Jakoby, W. B., 1974, Glutathione-S-transferases. The first enzymatic step in mercapturic acid formation, *J. Biol. Chem.* **249**:7130.

Hagenfeldt, L., Larsson, A., and Zetterström, R., 1974, Pyroglutamic aciduria. Studies in an infant with chronic metabolic acidosis, *Acta Paediatr. Scand.* **63**:1.

Hardin, E. B., Valentine, W. N., Follette, J. H., and Lawrence, J. S., 1954, Studies on the sulfhydryl content of human leukocytes and erythrocytes, *Am. J. Med. Sci.* **228**:73.

Hill, H. R., Hogan, N. A., Mitchell, T. G., and Quie, P. G., 1975, Evaluation of a cytocentrifuge method for measuring neutrophil granulocyte chemotaxis, *J. Lab. Clin. Med.* **86**:703.

Hoffstein, S., Soberman, R., Goldstein, I., and Weissmann, G., 1976, Concanavalin A induces microtubule assembly and specific granule discharge in human polymorphonuclear leukocytes, *J. Cell Biol.* **68**:781.

Hohn, D. C., and Lehrer, R. I., 1975, NADPH oxidase deficiency in X-linked chronic granulomatous disease, *J. Clin. Invest.* **55**:707.

Holmes, B., Park, B. H., Malawista, S. E., Quie, P. G., Nelson, D. L., and Good, R. A., 1970, Chronic granulomatous disease in females. A deficiency of leukocyte glutathione peroxidase, *N. Engl. J. Med.* **283**:217.

Homan-Müller, J. W. T., Weening, R. S., and Roos, D., 1975, Production of hydrogen peroxide by phagocytizing human granulocytes, *J. Lab. Clin. Med.* **85**:198.

Icén, A., 1967, Glutathione reductase of human erythrocytes. Purification and properties, *Scand. J. Clin. Lab. Invest.* **20**:(Suppl.) 96.

Isaacs, J., and Binkley, F., 1977, Glutathione dependent control of protein disulfide-sulfhydryl content by subcellular fractions of hepatic tissue, *Biochim. Biophys. Acta* **497**:192.

Iyer, G. Y. N., and Quastel, J. A., 1963, NADPH and NADH oxidation by guinea pig polymorphonuclear leukocytes, *Canad. J. Biochem. Biophys.* **41**:427.

Iyer, G. Y. N., Islam, D. M. F., and Quastel, J. H., 1961, Biochemical aspects of phagocytosis, *Nature* **192**:535.

Jacobs, H. S., and Jandl, J. H., 1966, Effects of sulfhydryl inhibition on red cells. III. Glutathione in the regulation of the hexosemonophosphate pathway, *J. Biol. Chem.* **241**:4243.

Johnston, R. B. Jr., and Newman, S. L., 1977, Chronic granulomatous disease, *Pediatr. Clin. North Am.* **24**:365.

Khandwala, A., and Gee, J. B. L., 1973, Linoleic acid hydroperoxide: Impaired bacterial uptake by alveolar macrophages, a mechanism of oxidant lung injury, *Science* **182**:1364.

Knox, W. E., 1960, Glutathione, in: *The Enzymes* (P. D. Boyer, H. A. Lardy, and K. Myrbäck, eds.), 2nd edn., Vol. 2, Part A, pp. 253–294, Academic Press, New York.

Koj, A., 1962, Biosynthesis of glutathione in human blood cells, *Acta Biochim. Pol.* **9**:11.

Krebs, H. A., and Eggleston, L. V., 1974, The regulation of the pentose phosphate cycle in rat liver, *Adv. Enzyme Regul.* **12**:421.

Kuriyama, R., and Sakai, H., 1974, Role of tubulin-SH groups in polymerization to microtubules, *J. Biochem.* **76**:651.

Larsson, A., Zetterström, R., Hagenfeldt, L., Andersson, R., Dreborg, S., and Hornell, H., 1974, Pyroglutamic aciduria (5-oxoprolinuria), an inborn error in glutathione metabolism, *Pediatr. Res.* **8**:852.

Lawrence, R. A., and Burk, R. F., 1976, Glutathione peroxidase activity in selenium-deficient rat liver, *Biochem. Biophys. Res. Commun.* **71**:952.

Levene, P. A., and Meyer, G. M., 1913, On the action of leukocytes on hexoses. IV. On the mechanism of lactic acid formation, *J. Biol. Chem.* **14**:551.

Little, C., and O'Brien, P. J., 1968, An intracellular GSH-peroxidase with a lipid peroxide substrate, *Biochem. Biophys. Res. Commun.* **31**:145.

Löhe, G. W., Blume, K. G., Rüdiger, H. W., and Arnold, H., 1974, Genetic variability in the enzymatic reduction of oxidized glutathione, in: *Glutathione* (L. Flohé, H. C. Benöhr, H. Sies, H. D. Waller, and A. Wendel, eds.), pp. 165–173, Academic Press, New York.

Lohmann, K., 1932, Beitrag zur enzymatischen Umwandlung von synthetischem Methylglyoxal in Milchsäure, *Biochem. Z.* **254**:332.

Loos, H., Roos, D., Weening, R., and Houwerzijl, J., 1976, Familial deficiency of glutathione reductase in human blood cells, *Blood* **48**:53.

Malawista, S. E., and Gifford, R. H., 1975, Chronic granulomatous disease of childhood (CGD) with leukocyte glytathione peroxidase (LPG) deficiency in a brother and sister. A likely autosomal recessive inheritance, *Clin. Res.* **23**:416A.

Malawista, S. E., Gee, J. B. L., and Bensch, K. G., 1971, Cytochalasin B reversibly inhibits phagocytosis: Functional, metabolic, and ultrastructural effects in human blood leukocytes and rabbit alveolar macrophages, *Yale J. Biol. Med.* **44**:286.

Mandell, G. L., 1972, Functional and metabolic derangements in human neutrophils induced by a glutathione antagonist, *J. Reticuloendothel. Soc.* **11**:129.

Mannervik, B., 1974, Glyoxalase I. Kinetic mechanism and molecular properties, in: *Glutathione* (L. Flohé, H. C. Benöhr, H. Sies, H. D. Waller, and A. Wendel, eds.), pp. 78–90, Academic Press, New York.

Mannervik, B., and Axelsson, K., 1975, Reduction of disulphide bonds in proteins and protein mixed disulphides catalyzed by a thiol-transferase in rat liver cytosol, *Biochem J.* **149**:785.

Matsuda, I., Oka, Y., Taniguchi, N., Fuyuyama, M., Kodama, S., Arashima, S., and Mitsuyama, T., 1976, Leukocyte glutathione peroxidase deficiency in a male patient with chronic granulomatous disease, *J. Pediatr.* **88**:581.

McCallister, J., Boxer, L. A., and Baehner, R. L., 1977, Alteration of microtubule function in glutathione peroxidase deficient polymorphonuclear leukocytes, *Clin. Res.* **25**:381A.

McCay, P. B., Gibson, D. D., Fong, K.-Y., and Hornbrook, K. R., 1976, Effect of glutathione peroxidase activity on lipid peroxidation in biological membranes, *Biochim. Biophys. Acta* **431**:459.

McKinney, G. R., and Rundles, R. W., 1956, Lactate formation and glyoxalase activity in normal and leukemic human leukocytes *in vitro*, *Cancer Res.* **16**:67.

McPhail, L. C., DeChatelet, L. R., Shirley, P. S., Wilfert, C., Johnston, R. B., Jr., and McCall, C. E., 1977, Deficiency of NADPH oxidase activity in chronic granulomatous disease, *J. Pediat.* **90**:213.

Meister, A., 1975, Biochemistry of glutathione, in: *Metabolic Pathways*, Vol. 7, *Metabolism of Sulfur Compounds* (D. M. Greenberg, ed.), pp. 101–188, Academic Press, New York.

Mellon, M. G., and Rebhun, L. I., 1976, Sulfhydryls and the *in vitro* polymerization of tubulin, *J. Cell Biol.* **70**:226.

Mize, C. E., and Langdon, R. G., 1962, Hepatic glutathione reductase. I. Purification and general kinetic properties, *J. Biol. Chem.* **237**:1589.

Mize, C. E., Thompson, T. E., and Langdon, R. G., 1962, Hepatic glutathione reductase. II. Physical properties and mechanism of action, *J. Biol. Chem.* **237**:1596.

Nakamura, W., Syun, H., and Hayashi, K., 1974, Purification and properties of rat liver glutathione peroxidase, *Biochim. Biophys. Acta* **358**:251.

Nath, J., and Rebhun, L. I., 1976, Effects of caffeine and other methylxanthines on the development and metabolism of sea urchin eggs. Involvement of $NADP^+$ and glutathione, *J. Cell Biol.* **68**:440.

Neuberg, C., 1913, Weitere Untersuchungen über die biochemische Umwandlung von Methylglyoxal in Milchsäure nebst Bemerkungen über die Entstehung der verschiedenen Milchsäuren in der Natur, *Biochem. Z.* **51**:484.

Noseworthy, J., Jr., and Karnovsky, M. L., 1972, Role of peroxide in the stimulation of the hexose monophosphate shunt during phagocytosis by polymorphonuclear leukocytes, *Enzyme* **13**:110.

Oh, S. H., Ganther, H. E., and Hoeckstra, W. G., 1974, Selenium as a component of glutathione peroxidase isolated from ovine erythrocytes, *Biochemistry* **13**:1825.

Oliver, J. M., 1975, Microtubules, cyclic GMP and control of cell surface topography, in: *Immune Regulation* (A. S. Rosenthal, ed.), pp. 445–471, Academic Press, New York.

Oliver, J. M., and Zurier, R. B., 1976, Correction of characteristic abnormalities of microtubule function and granule morphology in Chediak-Higashi syndrome with cholinergic agonists, *J. Clin. Invest.* **57**:1239.

Oliver, J. M., Albertini, D. F., and Berlin, R. D., 1976, Effects of glutathione-oxidizing agents on microtubule assembly and microtubule-dependent surface properties of human neutrophils, *J. Cell Biol.* **71**:921.

Oliver, J. M., Berlin, R. D., Baehner, R. L., and Boxer, L. A., 1977, Mechanisms of microtubule disassembly *in vivo*: Studies in normal and chronic granulomatous disease leukocytes, *Br. J. Haematol.* **37**:311.

Oliver, J. M., Spielberg, S. P., Pearson, C. B., and Schulman, J. D., 1978, Microtubule assembly and function in normal and glutathione synthetase deficient polymorphonuclear leukocytes, *J. Immunol.* **120**:1181.

Olson, O. E., Palmer, I. S., and Whitehead, E. I., 1973, Determination of selenium in biological materials, in: *Methods of Biochemical Analysis* (D. Glick, ed.), Vol. 21, pp. 39–78, Wiley, New York.

Patriarca, P., Cramer, R., Moncalvo, S., Rossi, F., and Romeo, D., 1971, Enzymatic basis of metabolic stimulation in leukocytes during phagocytosis: The role of activated NADPH oxidase, *Arch. Biochem. Biophys.* **145**:255.

Platt, R., 1931, The blood glutathione in disease, *Br. J. Exp. Pathol.* **12**:139.

Prohaska, J. R., and Ganther, H. E., 1977, Glutathione peroxidase activity of glutathione-*S*-transferases purified from rat liver, *Biochem. Biophys. Res. Commun.* **76**:437.

Prohaska, J. R., Oh, S.-H., Hoeckstra, W. G., and Ganther, H. E., 1977, Glutathione peroxidase: Inhibition by cyanide and release of selenium, *Biochem. Biophys. Res. Commun.* **74**:64.

Quie, P. G., Kaplan, E. L., Page, A. R., Gruskay, F. L., and Malawista, S. E., 1968, Defective polymorphonuclear-leukocyte function and chronic granulomatous disease in two female children, *N. Eng. J. Med.* **278**:976.

Racker, E., 1951, The mechanism of action of glyoxalase, *J. Biol. Chem.* **190**:685.

Rebhun, L. I., Nath, J., and Remillard, S. P., 1976, Sulfhydryls and regulation of cell division in: *Cold Spring Harbor Conferences on Cell Proliferation*, Vol. 3, *Cell Motility*, Book C, *Microtubules and Related Proteins* (R. Goldman, T. Pollard, and J. Rosenbaum, eds.), pp. 1343–1366, Cold Spring Harbor Laboratory, L.I., New York.

Reed, P. W., 1969, Glutathione and the hexose monophosphate shunt in phagocytizing and hydrogen peroxide-treated rat leukocytes, *J. Biol. Chem.* **244**:2459.

Reed, P. W., and Tepperman, J., 1969, Phagocytosis-associated metabolism and enzymes in rat polymorphonuclear leukocytes, *Am. J. Physiol.* **216**:223.

Roos, D., Homan-Müller, J. W. T., and Weening, R. S., 1976, Effect of cytochalasin B on the oxidative metabolism of human peripheral blood granulocytes, *Biochem. Biophys. Res. Commun.* **68**:43.

Roos, D., Weening, R. S., and Loos, J. A., 1979, The protective role of glutathione. The effect of congenital defects in glutathione metabolism on the function of erythrocytes, eye lens cells, and phagocytic leukocytes. A literature review and some personal observations, in: *Inborn Errors of Immunity and Phagocytosis* (F. Guttler, ed.), pp. 261–286, M.T.P. Press, Lancaster, England.

Root, R. K., Rosenthal, A. S., and Balestra, D. J., 1972, Abnormal bactericidal, metabolic, and lysosomal functions of Chediak-Higashi syndrome leukocytes, *J. Clin. Invest.* **51**:649.

Rosen, H., and Klebanoff, S. J., 1979, Bactericidal activity of a superoxide anion-generating system, *J. Exp. Med.* **149**:27.

Rossi, F., and Zatti, M., 1964, Biochemical aspects of phagocytosis in polymorphonuclear leukocytes. NADH and NADPH oxidation by granules of resting and phagocytizing cells, *Experientia* **20**:21.

Rossi, F., Romeo, D., and Patriarca, P., 1972, Mechanism of phagocytosis-associated oxidative metabolism in polymorphonuclear leukocytes and macrophages, *J. Reticuloendothel. Soc.* **12**:127.

Rossi, F., Patriarca, P., Romeo, D., and Zabucchi, G., 1976, The mechanism of control of phagocytic metabolism, in: *The Reticuloendothelial System in Health and Disease*, Part A (S. M. Reichard, M. R. Escobar, and H. Friedman, eds.), pp. 205–223, Plenum Press, New York.

Rotruck, J. T., Pope, A. L., Ganther, H. E., Swanson, H. B., Hafeman, D. G., and Hoeckstra, W. G., 1973, Selenium: Biochemical role as a component of glutathione peroxidase, *Science* **179**:588.

Schreiner, G. F., and Unanue, E. R., 1976, Membrane and cytoplasmic changes in B lymphocytes induced by ligand-surface immunoglobulin interaction, *Adv. Immunol.* **24**:37.

Segal, A. W., and Peters, T. J., 1976, Characterization of the enzyme defect in chronic granulomatous disease, *Lancet* **1**:1363.

Segal, A. W., and Peters, T. J., 1977, Localization of enzymes in neutrophils implicated in microbicidal processes. The enzyme defect in chronic granulomatous disease, in: *Movement, Metabolism and Bactericidal Mechanisms of Phagocytosis* (F. Rossi, P. Patriarca, and D. Romeo, eds.), pp. 175–181, Piccin, Padua.

Serfass, R. E., and Ganther, H. E., 1975, Defective microbicidal activity in glutathione peroxidase-deficient neutrophils of selenium-deficient rats, *Nature* **255**:640.

Spielberg, S. P., Kramer, L. I., Goodman, S. I., Butler, J., Tietze, F., Quinn, P., and Schulman, J. D., 1977, 5-Oxoprolinuria: Biochemical observations and case report, *J. Pediatr.* **91**:237.

Spielberg, S. P., Boxer, L. A., Oliver, J. M., Allen, J. M., and Schulman, J. D., 1979, Oxidative damage to neutrophils in glutathione synthetase deficiency, *Br. J. Haematol.* **42**:215.

Srivastava, S. K., and Beutler, E., 1970, Glutathione metabolism of the erythrocyte. The enzymatic cleavage of glutathione-haemoglobin preparations by glutathione reductase, *Biochem. J.* **119**:353.

Srivastava, S. K., Awasthi, Y. C., and Beutler, E., 1974, Useful agents for the study of glutathione metabolism in erythrocytes. Organic hydroperoxides, *Biochem. J.* **139**:289.

Staal, G. E. J., Visser, J., and Veeger, C., 1969a, Purification and properties of glutathione reductase of human erythrocytes, *Biochim. Biophys. Acta* **185**:39.

Staal, G. E. J., Hellerman, P. W., DeWael, J., and Veeger, C., 1969b, Purification and properties of an abnormal glutathione reductase from human erythrocytes, *Biochim. Biophys. Acta* **185**:63.

Stjernholm, R. L., 1968, The metabolism of human leukocytes, in: *Plenary Session Papers, Twelfth Congress, International Society of Hematology* (E. R. Jaffé, ed.), pp. 175–186, International Society of Hematology, New York.

Stjernholm, R. L., and Manak, R. L., 1970, Carbohydrate metabolism in leukocytes. XIV. Regulation of pentose cycle activity and glycogen metabolism during phagocytosis, *J. Reticuloendothel. Soc.* **8**:550.

Stjernholm, R. L., Burns, C. P., and Hohnadel, J. H., 1972, Carbohydrate metabolism by leukocytes, *Enzyme* **13**:7.

Stossel, T. P., 1973, Evaluation of opsonic and leukocyte function with a spectrophotometric test in patients with infection and with phagocytic disorders, *Blood* **42**:121.

Stossel, T. P., Root, R. K., and Vaughn, M., 1972a, Phagocytosis in chronic granulomatous disease and the Chediak–Higashi syndrome, *N. Eng. J. Med.* **286**:120.

Stossel, T. P., Mason, R. J., Hartwig, J., and Vaughn, M., 1972b, Quantitative studies of phagocytosis by polymorphonuclear leukocytes: Use of emulsions to measure the initial rate of phagocytosis. *J. Clin. Invest.* **51**:615.

Strauss, R. R., Paul, B. B., Jacobs, A. A., and Sbarra, A. J., 1969, The role of the phagocyte in host-parasite interactions. XIX. Leukocyte glutathione reductase and its involvement in phagocytosis, *Arch. Biochem. Biophys.* **135**:265.

Szent-Györgi, A., Együd, L. G., and McLaughlin, J. A., 1967, Ketoaldehydes and cell division. Glyoxal derivatives may be regulators of cell division and open a new approach to cancer, *Science* **155**:539.

Torres, M., Auclair, C., and Hakim, J., 1979, Protein-mediated hydroxyl radical generation—the primary event in NADH oxidation and oxygen reduction by the granule rich fraction of human resting leukocytes, *Biochem. Biophys. Res. Comm.* **88**:1003.

Urata, G., and Granick, S., 1963, Biosynthesis of α-aminoketones and the metabolism of aminoacetone, *J. Biol. Chem.* **238**:811.

Veech, R. L., Eggleston, L. V., and Krebs, H. A., 1969, The redox state of free nicotinamide-adenine dinucleotide phosphate in the cytoplasm of rat liver, *Biochem. J.* **115**:609.

Vogt, M. T., Thomas, C., Vassallo, C. L., Basford, R. E., and Gee, J. B. L., 1971, Glutathione-dependent peroxidative metabolism in the alveolar macrophage, *J. Clin. Invest.* **50**:401.

Weening, R. S., Roos, D., van Schaik, M. L. J., Voetman, A. A., de Boer, M., and Loos, H. A., 1977, The role of glutathione in the oxidative metabolism of phagocytic leukocytes. Studies in a family with glutathione reductase deficiency, in: *Movement, Metabolism and Bacterial Mechanisms of Phagocytosis* (F. Rossi, P. L. Patriarca, and D. Romeo, eds.), pp. 277–285, Piccin, Padua.

Wellner, V. P., Sekura, R., Meister, A., and Larsson, A., 1974, Glutathione synthetase deficiency, an inborn error of metabolism involving the γ-glutamyl cycle in patients with 5-oxoprolinuria (pyroglutamic aciduria), *Proc. Natl. Acad. Sci. USA* **71**:2505.

Windhorst, D. B., and Katz, E. D., 1972, Normal enzyme activities in chronic granulomatous disease leukocytes, *J. Reticuloendothel. Soc.* **11**:400.

Zigmond, S. H., and Hirsch, J. G., 1973, Leukocyte locomotion and chemotaxis. New methods for evaluation, and demonstration of a cell-derived chemotactic factor, *J. Exp. Med.* **137**:387.

Zurier, R. B., Hoffstein, S., and Weissmann, G., 1973, Mechanisms of lysosomal enzyme release from human leukocytes. I. Effect of cyclic nucleotides and colchicine, *J. Cell Biol.* **58**:27.

Zurier, R. B., Weissmann, G., Hofstein, S., Kammerman, S., and Tai, H. H., 1974, Mechanisms of lysosomal enzyme release from human leukocytes. II. Effects of cAMP and cGMP, autonomic agonists and agents which effect microtubule function, *J. Clin. Invest.* **53**:297.

7

Metabolic Changes
Accompanying Phagocytosis

FILIPPO ROSSI, PIERLUIGI PATRIARCA, and DOMENICO ROMEO

1. INTRODUCTION

A vast knowledge of the biochemical properties of phagocytic cells at rest and during phagocytosis has been gained in the past twenty years. During this period, several excellent reviews have been written on selected aspects of this broad field. The review by Karnovsky (1962) is recommended to those who are interested in the historical development of the biochemical studies on phagocytic cells. Among the more recent reviews, those written by the following authors can be mentioned: Cline (1970), for a survey of the various metabolic pathways of leucocytes, including a chapter on nucleic acid and protein metabolism; Sbarra *et al* (1977), for the oxidative changes in phagocytizing polymorphonuclear leukocytes and their relationship to microbicidal activity; Elsbach (1977), for a detailed analysis of the lipid metabolism; Klebanoff (1975), and Babior (1978) for the biochemical basis of the microbicidal mechanisms of polymorphonuclear leukocytes; Karnovsky *et al.* (1975) and Rossi *et al.* (1975) for the metabolism of resting and phagocytizing mononuclear phagocytes. A treatment of the phagocytic engulfment as a surface phenomenon is found in the book edited by Van Oss *et al.* (1975). Finally, a coverage of the various features of the phagocytic process is found in a book by Klebanoff and Clark (1978) and, less extensively, in a recent review by De Chatelet (1979).

The objective of this chapter will be that of reviewing the present knowledge on the changes in oxidative metabolism of polymorphonuclear leukocytes and

FILIPPO ROSSI • Istituto di Patologia Generale, Università di Padova, Sede di Verona, Italy. PIERLUIGI PATRIARCA • Istituto di Patologia Generale, Università di Trieste, Italy. DOMENICO ROMEO • Istituto di Chimica Biologica, Università di Trieste, Italy.

macrophages during phagocytosis. These changes are an increase of oxygen consumption, coupled to an increase of generation of superoxide anion (O_2^-) and hydrogen peroxide (H_2O_2), and an increased glucose catabolism through the hexose monophosphate (HMP) pathway. Altogether these metabolic changes are referred to as the *respiratory burst* of the phagocytic cells. Special *emphasis* will be placed on the enzymatic basis of the respiratory burst. This subject is one of the most controversial in the biochemistry of phagocytic cells but is basic to the understanding of phagocyte functions in health and disease.

2.   THE RESPIRATORY BURST

2.1.   OXYGEN CONSUMPTION AND THE HMP PATHWAY ACTIVITY

As early as 1933, a report appeared showing that white blood cells engaged in phagocytosis undergo a striking increase in oxygen consumption (Baldridge and Gerard, 1933). All types of mammalian phagocytic cells which have since been investigated (neutrophils, eosinophils, monocytes, and macrophages) have been shown to exhibit an increased respiration during phagocytosis. The stimulation of respiration has an intensity dependent on the type of particles used to challenge the cells. It is proportional to the load of particles taken up within a finite range and is influenced by the presence in the assay medium of various factors, such as serum, divalent cations, etc., which favor the phagocytic act. All these variables might account for the different values of cell extra respiration noted in different laboratories. In any case, by using the same type and load of particles, and the same assay conditions, a variability in the extent of the extra respiration is observed among the various types of phagocytic cell, and, for the same type of phagocytic cell, among various species (Rossi *et al.*, 1975). By and large it may be said that phagocytizing polymorphs and peritoneal macrophages exhibit the greatest enhancement in oxygen consumption, whereas the lung macrophages show only a moderate extra respiration (Rossi *et al.*, 1975). Among the polymorphs so far examined, an exceptionally low response to the phagocytic stimulus is noted in peritoneal chicken cells (Penniall and Spitznagel, 1975; Dri *et al.*, 1978).

The respiratory burst is not a postengulfment event as was originally thought. Studies on the early stages of the respiratory burst performed by Rossi and his associates (Rossi and Zatti, 1964; Zatti and Rossi, 1965; Rossi *et al.*, 1972) led to the suggestion that the stimulation of oxygen consumption is triggered by the initial adhesion of the phagocytizable particles to the cell surface. In fact, when the respiration of phagocytic cells is measured by a Clark oxygen electrode—a procedure that was applied to this field for the first time by Rossi and his associates in 1964—an increased rate of respiration can be detected as early as a few seconds after addition of particles to the cells (Rossi and Zatti, 1964, 1966a; Zatti and Rossi, 1965; Patriarca *et al.*, 1971b; Rossi *et al.*, 1972).

We are not better able to define the initial events of the phagocyte–particle interactions which trigger the respiratory burst. By simultaneously evaluating the kinetics of phagocytosis and of the oxygen uptake by human polymorphs, we have shown that the stimulated respiration is associated with formation and motion of pseudopodia around the phagocytizable particle and that it ceases when the phagocytic act is completed (Rossi and Zatti, 1964; Zabucchi et al. 1978).

Evidence has been provided that leukocytes also undergo a metabolic stimulation when they adhere to nonphagocytizable substrates (Morton et al., 1969; Henson and Oades, 1973; Hawkins, 1973; Romeo et al. 1974; Tedesco et al., 1975). At first glance this finding seems to contradict the statement that the respiratory burst is concomitant to the phagocytic act. One should keep in mind however that adhesion of phagocytes to nonphagocytizable substrates is associated with membrane movements along the substrate surface that mimic pseudopodia emission. Therefore this type of cell–substrate interaction is to be regarded as an equivalent of phagocytosis (frustrated phagocytosis, according to Weissmann).

A peculiarity of the phagocytic cells is that they undergo a stimulation of their oxidative metabolism not only during phagocytosis or adhesion to nonphagocytizable substrates, but also when they interact with a variety of agents. These are listed in Table 1. Some of these agents have been used merely as laboratory tools to investigate the mechanism underlying the activation of the metabolism of the phagocytic cells. Most of them, however, such as immune complexes, the chemotactic fragments of complement, endotoxin, fatty acids, kallikrein, and antileukocyte antibodies, are of potential physiological importance, since they may come in contact with the phagocytic cells at the inflammatory site.

The stimulated respiration of polymorphs is not affected by mitochondrial inhibitors such as cyanide, azide, antimycin A, rotenone, amytal, and dinitrophenol, whereas that of some types of macrophages may be depressed to some extent (Karnovsky et al. 1975; Rossi et al., 1975). Although these findings

TABLE 1. AGENTS, OTHER THAN PHAGOCYTIZABLE PARTICLES, KNOWN TO INDUCE A RESPIRATORY BURST IN PHAGOCYTIC CELLS

1. Endotoxin (Strauss and Stetson, 1960)
2. Surface-active agents (Graham et al., 1967; Zatti and Rossi, 1967)
3. Anti-leucocyte antibodies (Rossi et al., 1971b)
4. Phospholipase C (Patriarca et al., 1970, 1971a; Kaplan et al., 1972)
5. Nonphagocytizable mycoplasma (Simberkoff and Elsbach, 1971)
6. Concanavalin A (Romeo et al., 1973b, 1975a)
7. Aggregated immunoglobulin, immune complexes or concanavalin A bound to nonphagocytizable surfaces (Henson and Oades, 1973; Hawkins, 1973; Romeo et al., 1974)
8. Fatty acids (Kakinuma, 1974)
9. Phorbol myristate acetate (Repine et al., 1974)
10. Kallikrein (Goetzl and Austen, 1974)
11. Cytochalasin E (Nakagawara et al., 1974)
12. Complement fragments, C5a (Goetzl and Austen, 1974), C3b and $\overline{C567}$ (Tedesco et al., 1975)
13. Divalent cation ionophores (Schell-Frederick, 1974; Romeo et al., 1975b)
14. Lanthanum ion (Romeo et al., 1975b)

suggest that mitochondria are not involved in the respiratory burst of polymorphs it cannot be concluded that these organelles are at least partially responsible for the increased respiration of macrophages since the inhibitors might affect the phagocytic act per se. A better approach at the identification of the enzymatic basis of the respiratory burst is that of comparing the rate of CO_2 production from glucose labeled in different carbons. [6-^{14}C]Glucose and [1-^{14}C]glucose are usually employed. When glucose is utilized in the glycolytic pathway, followed by oxidation of pyruvate in the Krebs cycle, the same amount of $^{14}CO_2$ is produced from glucose carbon-1 and glucose carbon-6, whereas the $^{14}CO_2$ produced by the HMP pathway derives from glucose carbon-1 (for a more detailed analysis of this issue, see Chapter 4). This methodological approach has permitted several investigators to conclude that in all phagocytic cells the increase of the oxygen consumption is not supported by an enhanced activity of mitochondria but is strictly dependent on the increased activity of another respiratory system which leads to a concomitant stimulation of the HMP pathway activity. In fact, a marked stimulation of the $^{14}CO_2$ yield from [1-^{14}C]glucose is observed upon phagocytosis, whereas the $^{14}CO_2$ derived from [6-^{14}C]glucose increases only slightly (Stahelin *et al.*, 1957; Sbarra and Karnovsky, 1959; Iyer *et al.*, 1961; Karnovsky, 1962; Rossi and Zatti, 1966b; Stjernholm, 1968; West *et al.*, 1968; Stjernholm and Manak, 1970; Rossi *et al.*, 1975; Karnovsky *et al.*, 1975).

It has been calculated that, while in resting polymorphs the glucose flow through the HMP pathway is less than 2%, in phagocytizing polymorphs it increases to 40–80%, depending on the magnitude of the respiratory burst (Rossi and Zatti, 1966b; Stjernholm, 1968; Stjernholm and Manak, 1970). Furthermore, in phagocytizing polymorphs, the amount of glucose catabolized in the HMP pathway may account for as much as 80% of the oxygen consumed (Stjernholm, 1968).

A milestone in the understanding of the significance of the respiratory burst was the finding that the increased oxygen uptake is associated with a production of hydrogen peroxide and that the peroxide is involved in the bactericidal activity of phagocytic cells.

2.2. GENERATION OF HYDROGEN PEROXIDE

The first observation that phagocytizing cells generate hydrogen peroxide during the metabolic burst was made by Iyer *et al.* (1961). The demonstration was based on the oxidation of formate which is known to require H_2O_2 and catalase ($HCOOH + H_2O_2 \rightarrow CO_2 + 2H_2O$). This procedure, which has the drawback of detecting only a small fraction of the peroxide produced, has been employed to detect the phagocytosis-associated stimulation of H_2O_2 production by polymorphs from various sources as well as by mononuclear phagocytes (Baehner *et al.*, 1970; Gee *et al.*, 1970; Baehner and Johnston, 1972; Klebanoff and Hamon, 1972, 1975; Gee and Khandwala, 1976).

Other methods, based on the enzymatic measurement of H_2O_2, have been developed. All these methods allow the determination of only the peroxide released from the cells. In 1968 Zatti *et al.* quantitated the H_2O_2 released from polymorphs by measuring the oxygen generated after the addition of exogenous catalase to a suspension of phagocytizing cells. This technique obviously permits only an endpoint evaluation of H_2O_2 release.

A method proposed by Keston and Brandt (1965), which relies on the increase in fluorescence of leucodyacetyl-2,7-dichlorofluorescein upon oxidation by H_2O_2 in the presence of horseradish peroxidase, was applied by Paul and Sbarra (1968) to the measurement of H_2O_2 in polymorph dialysates. This method has been used by Homan-Müller *et al.* (1975) with intact cells. Root *et al.* (1975) have studied the kinetics of H_2O_2 release from phagocytizing cells by measuring the decrease in fluorescence of the coumarin derivative scopoletin (Andreae, 1955). More recently the fluorimetric assay of H_2O_2 has been further improved in our (Rossi *et al.*, 1980a) and other laboratories (Takeshige *et al.*, 1979) by using the nonfluorescent compound homovanillic acid which is converted to the highly fluorescent 2,2'-dihydroxy-3,3'-dimetoxy diphenyl-5,5'-diacetic acid by horseradish peroxidase in the presence of H_2O_2.

By using these fluorimetric methods, an increased generation of H_2O_2 can be detected a few seconds after addition of particles or other stimulants to polymorphs or macrophages (Root *et al.*, 1975; Nathan and Root, 1977; Rossi *et al.*, 1979). The amount of H_2O_2 measurable with anyone of these methods varies with the type of phagocytic cell employed, and, in all instances, accounts for only part of the oxygen consumed. This was shown by a simultaneous measurement of H_2O_2 release and O_2 consumption carried out by Root *et al.* (1975) with human polymorphs and by Rossi *et al.* (1979) with polymorphs and macrophages.

Table 2 reports the data obtained in our laboratory on the release of H_2O_2 by polymorphs and macrophages from different sources and different species. It can be seen that the highest amount of H_2O_2, relative to the oxygen consumed, is released by the granulocytes and peritoneal macrophages. Several literature data showing a wide variability in H_2O_2 release from different types of mononuclear phagocytes have been summarized in a paper by Nathan and Root (1977).

The low recovery of H_2O_2 with respect to the oxygen consumed suggests that the rate of degradation of the peroxide is higher than that of its formation. Three H_2O_2-degrading enzymes are present in phagocytic cells, i.e., catalase, peroxidase, and glutathione peroxidase. The activity of these enzymes varies largely among different types of phagocyte (Higgins *et al.*, 1978; Rossi *et al.*, 1979), but in general the H_2O_2 degrading machinery of all phagocytic cells is very efficient. This is supported by the following observations: (1) the amount of H_2O_2 measurable in the incubation medium of all types of phagocytes activated under ordinary conditions, i.e., in absence of inhibitors of the H_2O_2-degrading enzymes and of a H_2O_2 trap, is always very low. (2) The amount of H_2O_2 measurable in the incubation medium is higher when a peroxide-trapping system (such as homovanillic acid or scopoletin and horseradish peroxidase) is included in the medium. (3) For some cell types the amount of peroxide is even

TABLE 2. OXYGEN CONSUMPTION AND H_2O_2 PRODUCTION BY
PHAGOCYTIZING GRANULOCYTES AND MACROPHAGES IN THE PRESENCE OF AN
H_2O_2-TRAPPING SYSTEM[a]

Cell		nmol/4'/1.5 × 10⁷ cells[b]	
		O_2	H_2O_2
Granulocytes			
Human blood	(6)	171.2 ± 22.2	35.4 ± 9.0
Guinea pig peritoneal exudate	(4)	173.0 ± 35.1	41.1 ± 10.5
Rabbit peritoneal exudate	(4)	163.8 ± 24.7	22.2 ± 3.8
Chicken peritoneal exudate	(3)	29.7 ± 2.1	5.1 ± 0.3
Macrophages			
Guinea pig peritoneal elicited	(5)	229.5 ± 28.1	81.3 ± 14.7
Rabbit peritoneal elicited	(4)	66.5 ± 16.7	16.6 ± 3.2
Rabbit alveolar resident	(4)	66.1 ± 12.6	0
Rabbit alveolar BCG-activated	(9)	97.7 ± 14.1	6.5 ± 1.2

[a] Oxygen uptake was measured polarographically by a Clark oxygen electrode. The incubation medium contained 0.5 mM $CaCl_2$, 5 mM glucose, 0.8 mM homovanillic acid (HVA), and 20 μg/ml horseradish peroxidase (HRP). The final volume was brought to 2 ml with Krebs-Ringer phosphate buffer (KRP). Temperature, 37°. Opsonized heat-killed *B. mycoides* was used as the phagocytizable particle (cell: bacteria, 1:100). After recording the O_2 consumption for 4 min, the reaction mixture was centrifuged, and H_2O_2 was assayed on the supernatant. The assay was based on the measurement of the compound 2,2'-dihydroxy-3,3'-dimetoxydiphenil-5,5'-diacetic acid which was formed during the incubation from the oxidation of HVA catalyzed by HRP in the presence of the H_2O_2 produced by the phagocytes. The fluorescence was stable during the time required for the experiment. HVA did not affect O_2 consumption by the cells.
[b] The results are expressed as means ± SEM of the difference between the values of phagocytizing cells and those of resting cells. The number of experiments is shown in parenthesis.

higher than in the previous experimental conditions when both a H_2O_2-trapping system and an inhibitor of catalase and peroxidase are present in the incubation medium.

These observations have been made by several groups (Zatti and Rossi, 1968; Root and Metcalf, 1977a,b; Reiss and Ross, 1978; Homan-Müller *et al.*, 1975; Dri *et al.*, 1979; Rossi *et al.*, 1979). Table 3 reports the data obtained in our laboratory. These data also provide indications on the relative importance of the H_2O_2-degradation pathways in different cells. For example rabbit alveolar macrophages and guinea pig peritoneal macrophages degrade H_2O_2 mainly through the glutathione peroxidase pathway. In fact azide, an inhibitor of catalase and peroxidase, which has no effect on glutathione peroxidase, modifies only slightly the accumulation of peroxide in the medium of these cells. Conversely, degradation of H_2O_2 by catalase and peroxidase occurs to a variable extent in rabbit peritoneal macrophages and guinea pig, rabbit, and human granulocytes as indicated by the increased accumulation of H_2O_2 in the medium in the presence of NaN_3. In guinea pig and rabbit granulocytes the ratio between O_2 consumed and H_2O_2 formed is close to 1.0 in the presence of NaN_3, indicating that the peroxide is almost exclusively degraded by catalase and peroxidase.

In summary the present status of the problem of H_2O_2 generation during the respiratory burst is as follows: (1) in all the phagocytic cells so far investigated, including chicken polymorphs, the enhanced oxygen consumption is associated

TABLE 3. EFFECT OF AN H_2O_2-TRAPPING SYSTEM AND OF AZIDE ON THE ACCUMULATION OF H_2O_2 IN THE INCUBATION MEDIUM OF PHAGOCYTIZING GRANULOCYTES AND MACROPHAGES[a]

Cell	H_2O_2 as percent of O_2 consumed		
	None	+Trapping system	+Trapping system +NaN$_3$
Granulocytes			
Guinea pig peritoneal exudate	Trace	23.7 ± 6.0 (4)	103.0 ± 28.0 (4)
Rabbit peritoneal exudate	Trace	13.6 ± 2.3 (4)	92.3 ± 13.2 (4)
Human blood	Trace	20.7 ± 5.2 (6)	57.0 ± 6.3 (3)
Macrophages			
Guinea pig peritoneal elicited	Trace	35.4 ± 6.4 (5)	42.4 ± 7.2 (3)
Rabbit peritoneal elicited	Trace	24.9 ± 4.8 (4)	58.7 ± 14.8 (4)
Rabbit alveolar resident	0	0 (4)	0.6 ± 0.5 (4)
Rabbit alveolar BCG activated	0	6.6 ± 1.2 (9)	8.9 ± 1.4 (9)

[a] H_2O_2 was trapped by the HVA and HRP system (see legend to Table 2) as fast as it appeared in the incubation medium. The peroxide was then quantitated by measuring the fluorescence of 2,2'-dihydroxy-3,3'-dimetoxydiphenil-5,5'-diacetic acid formed (see legend to Table 2).

with H_2O_2 production; (2) the available methods for H_2O_2 assay allow the measurement of only that aliquot of peroxide which is outside of the cell; (3) the amount of H_2O_2 measurable varies among different types of cells and, in physiological conditions, is only a minor part of the peroxide formed; (4) the variable amounts of H_2O_2 measurable depend on the rate of its formation, on whether or not it is released in sites accessible to the assay system, and on the rate of its degradation; (5) the ratio between H_2O_2 produced and O_2 consumed is 1.0 in some types of phagocytes, as can be determined under appropriate experimental conditions, indicating that all the oxygen consumed is finally reduced to H_2O_2.

Other problems concerning H_2O_2 remain to be discussed, such as its role in microbicidal activity, the enzymatic mechanism, and the subcellular site of its generation. The first point is dealt with in other chapters of this book. The other two points will be discussed below.

The discovery that superoxide anion is generated during the respiratory burst has greatly contributed to the elucidation of the mechanisms of H_2O_2 production by phagocytes.

2.3. GENERATION OF SUPEROXIDE ANION

In 1973, Babior *et al.* observed that human polymorphs produce O_2^-, and that the generation of this radical is increased during phagocytosis. This observation has been confirmed with both polymorphonuclear and mononuclear phagocytes from various species stimulated by different agents (Curnutte and Babior, 1974; Salin and McCord, 1974; Johnston *et al.*, 1975, 1976; Weening *et al.*, 1975; Goldstein *et al.*, 1975; Drath and Karnovsky, 1975; Weiss *et al.*, 1977; Rister and Baehner, 1977; Tsan, 1977; Dri *et al.*, 1978).

The method usually employed for the measurement of O_2^- is based on the superoxide-dismutase-inhibitable reduction of exogenous ferricytochrome c by this radical. Since neither ferricytochrome c nor superoxide dismutase enter the cell, the detection of O_2^- is confined to that portion of the radical which is found extracellularly.

The superoxide may act either as an oxidant or as a reductant. Its main fate in the cell is the dismutation according to the reaction $2O_2^- + 2H^+ \rightarrow H_2O_2 + O_2$. This reaction can proceed either spontaneously or catalyzed by the enzyme superoxide dismutase which is contained in phagocytic cells (Beckman *et al.*, 1973; DeChatelet *et al.*, 1974; Patriarca *et al.*, 1974a; Salin and McCord, 1974; Rister and Baehner, 1976; Rest and Spitznagel, 1977). The superoxide can also react with H_2O_2 to form hydroxyl radical ($OH\cdot$) as follows $H_2O_2 + O_2^- \rightarrow O_2 + OH^- + OH\cdot$ (Haber and Weiss, 1934). Formation of $OH\cdot$ by both mononuclear (Weiss *et al.* 1977) and polymorphonuclear leukocytes has been demonstrated (Tauber and Babior, 1977; Rosen and Klebanoff, 1979). Recent studies have indicated that the direct interaction of H_2O_2 and O_2^- is slow (Weinstein and Bielski, 1979), making it unlikely that $OH\cdot$ is generated in biological systems by the above mechanism. Recently it has been proposed that $OH\cdot$ generation may result from the interaction of O_2^- with trace metal, such as iron, which functions as an oxidation–reduction catalyst (McCord and Day, 1978).

The amount of O_2^- found in the extracellular medium of stimulated phagocytes varies among the various cell types examined. By and large, the situation is similar to that already discussed for the release of H_2O_2 (Drath and Karnovsky, 1975; De Chatelet *et al.*, 1975b; Rister and Baehner, 1977; Tsan, 1977). Table 4 reports the data obtained in our laboratory by using different cell types.

The amount of O_2^- recovered in the incubation medium of stimulated phagocytes does not account for the oxygen consumed by the cells. If one assumes that the main fate of O_2^- formed is the dismutation to H_2O_2 and O_2^-, then the ratio between the O_2^- formed and the O_2 consumed depends on the mechanism of H_2O_2 degradation. This problem has been discussed in detail elsewhere (Dri *et al.*, 1979; Rossi *et al.*, 1979). Briefly, if H_2O_2 is degraded by catalase ($H_2O_2 \rightarrow H_2O + \frac{1}{2}O_2$), for 2 mol of O_2^- formed, 0.5 mol of O_2 are consumed (ratio $O_2^-:O_2 = 4.0$), whereas if H_2O_2 is degraded by peroxidase ($H_2O_2 + RH_2 \rightarrow 2H_2O + R$), for 2 mol of O_2^- formed, 1 mol of O_2 is consumed (ratio $O_2^-:O_2 = 2$).

The $O_2^-:O_2$ ratios found by several investigators in stimulated phagocytes are always considerably lower than those theoretically expected. This is also apparent from the data shown in Table 4. Two explanations are possible: (1) not all the oxygen consumed is reduced to O_2^-; and (2) only part of the O_2^- produced is released from the cell.

It is generally assumed that O_2^- formation is an obligate step in the reduction of oxygen by phagocytes, but no experimental evidence has been provided, as far as we know, which supports such an assumption. Recently it has been shown in our laboratory that (1) the oxygen uptake by guinea pig polymorphs

TABLE 4. OXYGEN CONSUMPTION AND O_2^- GENERATION BY PHAGOCYTIZING GRANULOCYTES AND MACROPHAGES[a]

		nmol/4'/1.5 × 10^7 cells[b]	
Cell		O_2	O_2^-
Granulocytes			
Human blood	(6)	171.2 ± 22.2	70.1 ± 15.8
Guinea pig peritoneal exudate	(4)	173.0 ± 35.1	89.9 ± 10.4
Rabbit peritoneal exudate	(4)	163.8 ± 24.7	40.8 ± 6.0
Chicken peritoneal exudate	(3)	29.7 ± 2.1	13.5 ± 1.0
Macrophages			
Guinea pig peritoneal elicited	(5)	229.5 ± 28.1	163.3 ± 23.1
Rabbit peritoneal elicited	(4)	66.5 ± 16.7	6.3 ± 1.2
Rabbit alveolar resident	(4)	66.1 ± 12.6	0.4 ± 0.4
Rabbit alveolar BCG-activated	(9)	97.7 ± 14.1	5.0 ± 1.4

[a] Oxygen consumption was assayed polarographically by a Clark oxygen electrode as indicated in Table 2. O_2^- generation was assayed by the superoxide dismutase inhibitable reduction of cytochrome c.

[b] The results are expressed as means ± SEM of the difference between the values of phagocytizing cells and those of resting cells. The number of experiments is shown in parenthesis.

stimulated by the inophore A23187 is completely inhibited by the O_2^- scavenger cytochrome c; and (2) the addition of superoxide dismutase (SOD) to the cell suspension containing cytochrome c completely restores the oxygen uptake. These observations suggest that at least in this cell type and in these experimental conditions all the O_2 consumed is converted to O_2^-. In this case the ratio $O_2^-:O_2$ was 2.3:1, which is very close to that expected if all the O_2 consumed produced O_2^-, if all the O_2^- produced dismutated to H_2O_2 and O_2, and if the peroxide were predominantly degraded through a peroxidase mechanism (Zabucchi, unpublished.)

The possibility that only part of the superoxide generated is released from the cells is a real one. This is demonstrated by the finding that the recovery of O_2^- in the incubation medium of stimulated phagocytes is considerably increased in the presence of cytochalasin B (Root and Metcalf, 1977a,b; Goldstein *et al.*, 1975; Rossi *et al.*, 1979, 1980a). This finding is generally interpreted as follows: cytochalasin B inhibits vacuole formation so that the superoxide is accessible to the superoxide trap (usually cytochrome c), while in the absence of cytochalasin B part of the O_2^- is sequestered within completely fused vacuoles.

It appears therefore that the poor recovery of O_2^- with respect to the O_2 consumed depends on the inability of the assay system to reach its target either within the cell or within the vacuole. The experiment with the ionophore A23187 and guinea pig polymorphs, which has been briefly described above, has permitted us to overcome this inconvenience and to obtain quantitative relationships between O_2^- and O_2 close to those theoretically expected (Zabucchi, unpublished).

The data shown in Tables 2 and 4 allow us to draw an additional conclusion, that is, that at least in some types of phagocytes (human, guinea pig, and rabbit granulocytes, and guinea pig peritoneal macrophages), only O_2^- but no H_2O_2 is released and that the H_2O_2 found in the incubation medium of these cells originates from O_2^- dismutation. This conclusion is based on the finding that the ratio $O_2^-:H_2O_2$ is close to the theoretical value of 2.0 for the dismutation reaction of O_2^-.

2.4. ANALYSIS OF THE RESPIRATORY BURST

The respiratory burst can be quantitatively monitored by evaluating any one of the following indexes: oxygen uptake, H_2O_2 formation, O_2^- generation, $^{14}CO_2$ production from [1-^{14}C]glucose, OH· production, and chemiluminescence. The measurements should be carried out in both resting and stimulated cells. The difference between the values obtained with stimulated cells and those with resting cells gives the quantitative measure of the respiratory burst. Stimulation can be induced either by the addition of particles such as heat-killed bacteria or zymosan or latex spherules of a definite size or by the addition of any one of the stimulatory agents listed in Table 1.

Oxygen uptake by intact leukocytes can be measured either polarographically with a Clark oxygen electrode or manometrically with a Warburg apparatus. For assays with blood, phagocyte separation of the cells by the Ficoll-Isopaque method or by some other method is required. An important advancement in the clinical evaluation of the respiratory burst has been made with the discovery of Nakamura *et al.* (1978) that oxygen consumption by phagocytizing blood granulocytes can be measured on whole blood after conversion of hemoglobin to carboxyhemoglobin.

Hydrogen peroxide production can be detected with any one of the methods outlined above. The methods employing fluorimetric assays are recommended for their sensitivity and specificity. A spectrophotofluorimetric assay of H_2O_2 production by stimulated leukocytes using a few microliters of whole blood has been recently described by Takeshige *et al.* (1979).

Superoxide anion generation is usually measured by the superoxide-dismutase-inhibitable reduction of cytochrome c or of other electron acceptors such as nitroblue tetrazolium (NBT). Isolated cells are incubated with the stimulatory agent and cytochrome c in the presence or absence of superoxide dismutase. The absorbance of reduced cytochrome c at 550 nm is measured. The difference in absorbance between the assay in the absence of superoxide dismutase and the assay in the presence of superoxide dismutase indicates the O_2^--dependent reduction of cytochrome c. Recently it has been shown that this assay can be carried out also with a few microliters of whole blood (Takeshige *et al.*, 1979; Bellavite *et al.*, 1980a).

The production of $^{14}CO_2$ from [1-^{14}C]glucose is a simple and sensitive measure of the stimulated hexose monophosphate shunt activity in phagocytizing or otherwise stimulated leukocytes. Flasks with two compartments are usually employed. The main compartment of the flask contains the cells suspended in a solution with glucose labeled in carbon-1. The smaller compartment contains a trapping system for CO_2, for example, 0.1–0.2 ml of 20% KOH. After addition of the stimulant to the main compartment, the flask is tightly capped and incubated in a shaking water bath. At the end of the incubation time, the reaction is stopped by the addition of an acid into the main compartment, and the radioactivity of the smaller compartment is counted. This assay can be performed also with whole blood (Keusch *et al.*, 1972). In this case, the specific activity of glucose varies with glucose concentration in the blood. Furthermore, the amount of $^{14}CO_2$ produced will vary with the number of granulocytes in the blood. Thus, absolute values for the test can be given only if the initial glucose concentration and the number of granulocytes in the blood are known. The results can be expressed as the ratio between $^{14}CO_2$ produced by stimulated cells and that produced by resting cells. This ratio is independent of the specific activity of glucose and of the white cell count.

The emission of light (chemiluminescence) by phagocytizing leukocytes has been discovered by Allen *et al.* (1972). Chemiluminescence is usually measured in a liquid scintillation counter operated in an out-of-coincidence mode. The cell suspension is placed in a vial and the background counts are recorded. The stimulatory agent is then added and the increase in counts is followed over a given period of time. All manipulations must be performed in the dark in order to lower background counts.

Production of OH· has been detected by measuring ethylene formation from methional (Weiss *et al.*, 1977; Tauber and Babior, 1977) or using a spin trap (Rosen and Klebanoff, 1979), but these methods are not employed for clinical studies yet.

3. ENZYMATIC BASIS OF THE STIMULATION OF THE OXIDATIVE METABOLISM

3.1. RELATIONSHIPS BETWEEN THE INCREASE OF RESPIRATION AND OF HMP PATHWAY ACTIVITY

There is a little doubt that O_2 consumption and glucose oxidation through the HMP pathway, which are simultaneously activated by phagocytosis, are two events linked one to the other (Rossi *et al.*, 1972; Romeo *et al.*, 1973a).

The controlling mechanisms which determine the flow of glucose via the HMP pathway rely on the activity of the cytosolic enzymes glucose-6-phosphate dehydrogenase, 6-phosphogluconate dehydrogenase, and transketolase, and on the NADPH/NADP$^+$ ratio (Glock and McLean, 1953; Beck, 1958; Stjernholm, 1968; Eggleston and Krebs, 1974). Under normal steady state conditions, the NADPH is much higher than the NADP$^+$, and NADPH exerts a strong inhibition on glucose-6-phosphate dehydrogenase. Among several cell constituents only oxidized glutathione (GSSG) appears to be able to counteract the inhibition by NADPH at physiological concentrations (Eggleston and Krebs, 1974).

In phagocytizing cells, the kinetic properties of the limiting enzymes 6-phospho-gluconate dehydrogenase and transketolase do not change (Stjernholm, 1968). On the contrary, evidence has been provided that in the phagocytizing polymorphs the NADPH/NADP$^+$ ratio shifts from 9 to about 3 (Zatti and Rossi, 1965; Rossi *et al.*, 1971a). It thus appears that the oxidation of NADPH to NADP$^+$ is the event which reverses the inhibition of glucose-6-phosphate dehydrogenase thereby increasing severalfold the activity of the HMP pathway.

The central problem of the metabolic stimulation of the phagocytic cells is, therefore, the identification of the oxidative system(s) generating NADP$^+$ from NADPH. Two main hypotheses have been advanced.

According to one of these, the increased O_2 consumption is supported by activation of a NADH oxidizing system (Evans and Karnovsky, 1961; Karnovsky, 1962; Cagan and Karnovsky, 1964; Mandell and Sullivan, 1971; Segal and Peters, 1976). According to this hypothesis, the linkage between the activated NADH oxidase and the HMP pathway activity would be provided by ancillary reactions which oxidize NADPH to NADP$^+$. Several of such reactions have been postulated: (a) a peroxidation of reduced glutathione (GSH) by H_2O_2, coupled to NADPH oxidation catalzyed by glutathione reductase (Reed, 1969; Baehner *et al.*, 1970); (b) an oxidation of NADPH by glycolytic pyruvate catalyzed by an NADPH-dependent lactate dehydrogenase (Evans and Karnovsky, 1961); (c) an oxidation of NADPH by NAD$^+$ catalyzed by a mitochondrial NADPH/NAD$^+$ transhydrogenase (Evans and Karnovsky, 1962).

The other hypothesis postulates that the activation of a NADPH-oxidizing system is the primary event leading to the enhancement in O_2 consumption and in generation of O_2^-, H_2O_2, and $NADP^+$ (Iyer *et al.*, 1961; Rossi and Zatti, 1964, 1966a,b); Zatti and Rossi, 1965; Selvaraj and Sbarra, 1967; Paul *et al.*, 1970, 1972; Stjernholm and Manak, 1970; Patriarca *et al.*, 1971b; Romeo *et al.*, 1971, 1973a; Rossi *et al.*, 1972; De Chatelet *et al.*, 1975a; Hohn and Lehrer, 1975; Curnutte *et al.*, 1975; Babior *et al.*, 1975, 1976; Iverson *et al.*, 1977). The increased generation of $NADP^+$ by this oxidizing system would *directly* link the increased O_2 consumption to the stimulation of glucose oxidation through the HMP pathway.

3.2. THE NADH-OXIDASE HYPOTHESIS

The involvement of a NADH oxidase in the respiratory burst of phagocytes was first postulated by Evans and Karnovsky (1961). These authors found that "granules" obtained from guinea pig leukocyte homogenates were able to oxidize both NADH and NADPH; the activity was higher with NADH than with NADPH and the optimal pH of the reaction was in the acid range with both nucleotides. Karnovsky and his associates have also reported that the NADH oxidase could be obtained in soluble form when guinea pig and human neutrophils were homogenized in isotonic KCl and 0.34 M sucrose, respectively (Cagan and Karnovsky, 1964; Baehner *et al.*, 1970). In a review on the enzymatic mechanism of the respiratory burst, Karnovsky specifies the properties of the NADH oxidase, i.e., pH optimum 4.5–5.0, no activity with NADPH, K_m of about 1×10^{-3} M, and 80% flarin adenine dinucleotide (FAD) content (Karnovsky, 1973). He also claims that the enzyme is deficient in leukocytes from patients with chronic granulomatous disease, but this view is not shared by other authors.

More recently Briggs *et al.* (1975) have developed a histochemical assay of the oxidase by exploiting the formation of a precipitate, which can be visualized by electron microscopy, when cerous ions react with hydrogen peroxide. With this technique, the authors have shown that a precipitate is formed on the plasma membrane of phagocytizing neutrophils incubated with NADH, but no precipitate was formed with NADPH. When neutrophils from patients with chronic granulomatous disease were used, less precipitate was seen at the cell surface than with cells from control subjects (Briggs *et al.*, 1977). The soluble NADH oxidase obtained from guinea pig resting neutrophils homogenized in alkaline isotonic KCl has been partially purified and characterized by Badwey and Karnovsky (1979). The activity of this preparation, measured at pH 7.0, has a K_m of 0.4 mM, is insensitive to inhibition by cyanide, is inhibited by a variety of nucleotides (ATP, ADP, AMP) and other anions (for example citrate and phosphate), and has a molecular weight of approximately 310,000. The enzyme produces O_2^- and H_2O_2 with a ratio $NADH:O_2^-:H_2O_2$ of about 1:0.25:1. According to the authors, the enzyme is adequate to account for the phenomena (that is, increased oxygen consumption and O_2^-, and H_2O_2 generation) observed in whole cells.

Another group of authors has indicated an NADH oxidase as the enzyme responsible for the respiratory burst (Segal and Peters, 1976, 1977). These authors have reported the presence in the plasma membrane fraction of NBT reductase activity which was active with low concentrations of NADH but not of NADPH.

3.2.1. Inadequacy of the Postulated Ancillary Reactions as Primary Sources of NADP$^+$

3.2.1a. The Glutathione Cycle. A key role for the glutathione cycle in the utilization of H_2O_2 associated with generation of NADP$^+$ from NADPH has been suggested by several authors (Reed, 1969; Baehner *et al.*, 1970; Vogt *et al.*, 1971). The coupling of O_2 consumption to the HMP pathway activity would procede as follows:

$$NADH + H^+ + O_2 \xrightarrow{\quad NADH\ oxidase \quad} NAD^+ + H_2O_2 \tag{1}$$

$$H_2O_2 + 2GSH \xrightarrow{\quad glutathione\ peroxidase \quad} O_2 + GSSG \tag{2}$$

$$GSSG + NADPH + H^+ \xrightarrow{\quad glutathione\ reductase \quad} 29SH + NADP_+ \tag{3}$$

$$NADP_+ + glucose\text{-}6\text{-}P \xrightarrow{\quad HMP\ activity \quad} NADPH + H^+ + \tfrac{1}{2}CO_2 + Ribulose\text{-}5\text{-}P \tag{4}$$

The following experimental evidence indicate that the glutathione cycle is operative in some types of phagocytic cells but has no relevance in others:

1. If H_2O_2 were preferentially utilized in reaction (2), the stoichiometry of the overall process [Reactions (1)–(4)] would be 0.5 mol of CO_2 produced for 1 mol of O_2 consumed. Thus, when the respiratory stimulation is very high and the glucose is predominantly catabolyzed through the HMP pathway, the respiratory quotient CO_2/O_2 should be close to 0.5. On the contrary, in phagocytizing guinea pig polymorphs, we have consistently found a respiratory quotient close to 1.0 (Rossi *et al.*, 1972).

2. The activity of the H_2O_2-degrading enzymes varies in different types of phagocytes. The activity of glutathione peroxidase is very low in rabbit, guinea pig, and human polymorphs, while it is high in macrophages (Rossi *et al.*, 1979). There is an inverse correlation between the activity of glutathione peroxidase and other peroxidases in the phagocytes (Rossi *et al.*, 1979). As an extreme example, chicken polymorphs have no myeloperoxidase (Brune and Spitznagel, 1973) or catalase (Bellavite *et al.*, 1977) activity but do have a high glutathione peroxidase activity (Bellavite *et al.*, 1977).

These findings indicate that different mechanisms are operative in the intracellular degradation of H_2O_2. In fact, experiments performed in our laboratory have shown that when the respiratory burst of guinea pig and rabbit granulocytes takes place in the presence of KCN or NaN$_3$ (which inhibit catalase and myeloperoxidase but not glutathione peroxidase), H_2O_2 accumulates with a ratio of about 1 with respect to oxygen consumed (Zatti *et al.*, 1968; Dri *et al.*, 1979; Rossi *et al.*, 1979). Under these experimental conditions, stimulation of HMP-pathway activity occurs normally although no appreciable H_2O_2 degradation takes place. Conversely, in those types of phagocytes which are rich in glutathione peroxidase activity (macrophages), KCN and azide do not markedly increase H_2O_2 accumulation. In these cells part of the H_2O_2 is degraded by glutathione peroxidase.

3. It has been shown that the initial phagocytic stimulation of the HMP-pathway activity is normal in polymorphs genetically or experimentally deficient in glutathione reductase or glutathione peroxidase (Basford *et al.*, 1977; Weening *et al.*, 1977; Bass *et al.*, 1977).

4. A stimulation of the HMP-pathway activity, unrelated to the activity of glutathione peroxidase and glutathione reductase, has been shown by Mendelson *et al.* (1977) in phagocytizing human granulocytes.

In conclusion, present knowledge indicates that the relevance of the glutathione cycle in the disposal of H_2O_2 and in the supply of $NADP^+$ for the stimulation of the HMP-pathway activity varies in different phagocytic cells. Thus the possibility that the glutathione cycle provides the necessary link between the enhanced oxygen consumption and the HMP-pathway activity cannot be regarded as a general rule. In particular, the glutathione cycle is not operative in guinea pig polymorphs where the NADH oxidase has been claimed to be operative as the primary enzyme of the respiratory burst (Evans and Karnovsky, 1961; Cagan and Karnovsky, 1964; Badwey and Karnovsky, 1979).

3.2.1b. NADPH-Dependent Lactate Dehydrogenase. An oxidation of NADPH by pyruvate catalyzed by an NADPH-dependent lactate dehydrogenase has been proposed by Evans and Karnovsky (1961, 1962) as a possible ancillary reaction linking the increased O_2 consumption to the HMP-pathway activity in phagocytizing polymorphs. Strong evidence exists, however, that this reaction has no physiologic relevance. The evidence is as follows.

1. According to Evans and Karnovsky, an intracellular acidification caused by the increased rate of glycolysis would activate the NADPH-dependent lactate dehydrogenase. Rossi and Zatti (1966a) have proved, however, that the increase in the rate of glycolysis in phagocytizing polymorphs follows the increase in O_2 consumption. Furthermore, when the polymorphs are stimulated by non-phagocytizable agents, such as saponin or phospholipase C, the metabolic burst is accompanied by a depression of glycolysis (Zatti and Rossi, 1967; Patriarca *et al.*, 1970).

2. Were NADPH oxidized by pyruvate, the amount of lactate produced by phagocytizing cells should at least correspond to the amount of $NADP^+$ utilized for the oxidation of glucose through the HMP pathway. By using suitably labeled substrates Rossi and Zatti (1966b) have shown, however, that in phagocytizing guinea pig polymorphs the entire production of lactate is less than that required to account for the regeneration of an adequate amount of $NADP^+$.

3. The activity of the NADPH-dependent lactate dehydrogenase in the polymorphs of some species is either extremely low or even absent (Reed and Tepperman, 1969; Baehner *et al.*, 1970; Noseworthy and Karnovsky, 1972), which casts doubts on its physiological role. By using glucose labeled with tritium in C-1, Stjernholm and Manak (1970) have proved that no such enzyme activity is present in human blood polymorphs, either at rest or during

phagocytosis. In fact, the atoms of tritium, which are transferred from glucose to NADP$^+$ in the first reaction of the HMP pathway, did not appear in the α-position of lactate, as expected if the postulated lactate dehydrogenase had catalyzed the reduction of pyruvate by NADPH. Similar results have been obtained by Roberts and Camacho (1967) by following the fate of tritium labeling of the C-3 of glucose in phagocytizing guinea pig polymorphs.

3.2.1c. NADPH/NAD$^+$ Transhydrogenase. Evans and Karnovsky (1962) have looked for a possible oxidation of NADPH by NAD$^+$, catalyzed by the classical nicotinamide transhydrogenase and have observed that no such enzyme activity is exhibited by a postnuclear extract of guinea pig polymorphs. Subsequently, Evans and Kaplan (1966) and Rossi *et al.* (1972) were able to measure a transhydrogenase activity in the homogenates of human and guinea pig polymorphs, but the activity was too low to account for an NADP$^+$ production sufficient to sustain glucose oxidation through the HMP pathway.

3.2.2. Competition of the NADH Oxidase with Lactate Dehydrogenase and the Fate of Pyruvate

Apart from the doubtful physiological relevance of the ancillary reactions invoked to explain the generation of NADP$^+$ from NADPH, the hypothesis of the primary role of a NADH oxidase in the metabolic burst of the phagocytic cells is marred by other weak points.

If a soluble NADH oxidase were the initiating enzyme of the respiratory burst, it should be able to successfully compete with lactate dehydrogenase for NADH. Zatti and Rossi (1966) have indeed shown that the addition of pyruvate to polymorph postnuclear supernatants strongly inhibits the oxygen consumption in the presence of NADH. This result is consistent with the notion that lactate dehydrogenase has a K_m (NADH) of the order of 10^{-6} M (Rossi *et al.*, 1972; Stjernholm *et al.*, 1972), whereas the NADH oxidase has a K_m (NADH) of the order of 10^{-3} M (Cagan and Karnovsky, 1964; Karnovsky, 1973; Babior *et al.*, 1976; Babior, 1977a,b; Badwey and Karnovsky, 1979). Coupled with the observation that the intracellular concentration of NADH in the polymorphs is 0.8–1.6×10^{-4} M (Selvaraj and Sbarra, 1967; Rossi *et al.*, 1971a; Aellig *et al.*, 1977), this suggests that the oxidation of NADH by pyruvate *in vivo* is highly favored with respect to its oxidation by O$_2$, catalyzed by the oxidase.

Segal and Peters (1976, 1977) have attributed the role of primary oxidase in human polymorphs to a NADH-dependent NBT reductase, with a K_m (NADH) of 1.66×10^{-6} M which would be located in the plasma membrane. Were this enzyme activated in the metabolic burst, one should still find out what the reactions generating NADP$^+$ for the HMP pathway are, as discussed above.

If an oxidase were acting on NADH during the respiratory burst, the ratio NAD$^+$/NADH should increase. Such an increase has not been observed (Selvaray and Sbarra, 1967; Rossi *et al.*, 1971a), suggesting either that NADH oxidase is inactive in the cell or that NADH is continuously regenerated by glyceraldehyde-3-phosphate dehydrogenase, the enzyme providing the most efficient system for NADH generation in the cell. In any case, an amount of pyruvate equivalent to the NADH consumed by the oxidase would escape reduction to lactate (a reduction of pyruvate by NADPH does not occur in phagocytic cells, as discussed in previously). This possibility has been ruled out by Rossi and Zatti (1966b) and by Stjernholm and Manak (1970), who have shown that in phagocytizing cells virtually all the carbon atoms of glucose, which are not oxidized to CO_2, are recovered as lactate.

In conclusion, an overwhelming set of data provides evidence that an NADH oxidase is not the key enzyme responsible for the metabolic burst of the phagocytic cells.

3.3. HYPOTHESIS CONCERNING THE ACTIVATION OF A NADPH-OXIDIZING SYSTEM

The inadequacy of a NADH oxidase as the primary enzyme responsible for the metabolic burst of the phagocytic cells requires the identification of another oxidative system, whose activation could trigger both the reduction of O_2 to O_2^- and H_2O_2, and the decrease of the $NADPH/NADP^+$ ratio with a concomitant increase in the HMP-pathway activity. Both direct and indirect evidence suggest that this system is an NADPH oxidase.

3.3.1. Direct Evidence in Favor of a NADPH Oxidase

Direct evidence for the involvement of a NADPH oxidase is obtained by the measurements of the oxidation of NADPH by subcellular fractions derived from resting and phagocytizing cells. The attention of investigators was drawn toward an NADPH-oxidizing system by the observation of Iyer and Quastel (1963), who showed that homogenates of resting guinea pig polymorphs oxidize NADPH and NADH with generation of H_2O_2, the rate of NADPH oxidation being much higher than that of NADH. By repeatedly freezing and thawing or homogenizing the leukocytes a clear extract was obtained, which retained most of the oxidase activity. The oxidation of NADPH by both the homogenate and the extract was shown to be Mn^{2+}-dependent, to have an optimal pH of about 5, and to be inhibited by 1 mM potassium cyanide. This led Roberts and Quastel (1964) to suggest that the oxidation of NADPH may be accompanied by a reaction involving peroxidase.

In 1964, Rossi and Zatti by subfractionating homogenates of guinea pig polymorphs and by using an assay system similar to that proposed by Iyer and Quastel (1963), i.e., a medium at pH 5.5 with Mn^{2+}, showed that the NADPH- and NADH-oxidizing activities were confined to a 20,000g fraction. A fundamental observation was that the fraction isolated from phagocytizing cells exhibited an increased oxidizing activity toward both the reduced nucleotides, the enhancement in the rate of NADPH oxidation being higher than that of NADH (Rossi and Zatti, 1964). The same authors found that the rate of oxidation of NADH and NADPH by the 20,000g fraction isolated from phagocytizing cells was the same when measured at pH 7.4, without Mn^{2+} and at 2 mM concentration of the coenzymes.

The sensitivity to cyanide of the oxidase described by Quastel and his associates raises some doubts on its physiological relevance, since the metabolic burst of the phagocytic cells is insensitive to this inhibitor. The uncertainty about the presence of a cyanide-insensitive NADPH-oxidase in polymorphs was removed by Rossi and his associates. In fact, by measuring the oxidation of reduced nicotinamide nucleotides (at 0.5–0.7 mM concentration) as oxygen consumption at neutral pH in the absence of manganese and in the presence of KCN, they established that the 20,000*g* fraction of homogenates of guinea pig polymorphs oxidized both nucleotides, while the soluble fraction did not exhibit any NADH or NADPH oxidase activity under the same conditions (Zatti and Rossi, 1966; Rossi and Zatti, 1968). They also demonstrated, in the same experiments, that the cyanide-insensitive oxidation of both NADPH and NADH was markedly activated in the 20,000*g* fraction derived from phagocytizing and from saponine-treated guinea pig polymorphs, the activation in the rate of NADPH oxidation being higher than that of NADH. It was subsequently shown that the NADPH oxidation assayed at neutral pH in the absence of manganese and in the presence of KCN was also activated in polymorphs treated with phospholipase C (Patriarca *et al.*, 1971a) and that the oxidation of NADPH was accompanied by the generation of H_2O_2 (Rossi *et al.*, 1969; Patriarca *et al.*, 1971b).

That the increased oxidative metabolism associated with phagocytosis could be accounted for by a stimulated NADPH oxidase activity was confirmed by the work of Paul *et al.* (1972), who used an acidic medium with Mn^{2+} to measure the activity of the particulate oxidase. It has been subsequently demonstrated in our laboratory that activation of the NADPH oxidase, assayed at pH 5.5 and with Mn^{2+}, also occurs in polymorphs exposed to various stimulants (Patriarca *et al.*, 1971b; Rossi *et al.*, 1972; Romeo *et al.*, 1973b) as well as in phagocytizing lung and peritoneal macrophages (Romeo *et al.*, 1971, 1973a; Rossi *et al.*, 1975).

What emerges from all those studies is that, under any experimental conditions (neutral or acid pH, with or without Mn^{2+}), the oxidation of NADPH and NADH by a 20,000*g* fraction of leukocyte homogenates, measured either as NAD(P)H disappearance or as O_2 consumption, was highly increased by exposing the cells to phagocytizable particles or to other stimulants. In all conditions, the NADPH-oxidizing activity was higher than the NADH-oxidizing activity, except when the assay was carried out at pH 7.4 with high coenzyme concentration (about 2 mM). In this particular case, in fact, the two activities were about the same (Rossi and Zatti, 1964). The observations summarized above, coupled with the finding that the K_m of the NADPH-oxidizing activity (in the order of 10^{-4} M) is close to the NADPH concentration in the cell, whereas the K_m of the NADH-oxidizing activity is about one order of magnitude higher (Patriarca *et al.*, 1971b; Rossi *et al.*, 1972), support the original hypothesis of Rossi and Zatti that a NADPH oxidase is the key enzyme of the respiratory burst of phagocytes.

Additional support to this hypothesis came from the studies of Hohn and Lehrer (1975) who concluded that the primary metabolic defect in polymorphs of patients with the sex-linked chronic granulomatous is a defective activation of the KCN-insensitive NADPH oxidase.

In the last few years interesting data on the properties of the NADPH oxidase and on its role in the respiratory burst have been added.

1. It has been established that O_2^- is an intermediate product in the NADPH oxidase reaction (Patriarca *et al.*, 1975; Curnutte *et al.*, 1975; Babior, 1977a,b). This conclusion is based on the observation that at pH 5.5, and in the presence of Mn^{2+}, SOD inhibits the oxidation of NADPH by a polymorph particulate fraction (Patriarca *et al.*, 1975) and that the oxidation of NADPH is associated with an SOD-inhibitable reduction of ferricytochrome c or NBT (Patriarca *et al.*, 1975; Curnutte *et al.*, 1975; Babior *et al.*, 1975; Patriarca and Dri, 1976). Production of superoxide anion during NADH oxidation has also been demonstrated (Babior *et al.*, 1975, 1976).

2. A new method to measure NADPH and NADH oxidation by leukocyte subcellular particles has been designed by Babior *et al.* (1975). This method is based on the assay of the O_2^- produced during oxidation of the piridine nucleotides. The results obtained by the authors support the role of NADPH oxidation as the primary event in the respiratory burst. The authors in fact have shown that (a) the O_2^--generating activity of a particulate fraction from stimulated human polymorphs is higher with NADPH than with NADH; (b) the pH optimum is 7.0; (c) the affinity of the enzyme for NADPH is greater than for NADH (K_m for NADH of approximately 0.7 mM regardless of pH, while with NADPH the K_m varied from 0.02 mM at pH 6.0 to 0.3 mM at pH 5.0); and (d) the O_2^- generating activity was deficient in cells from patients with chronic granulomatous disease.

3. It has been suggested that in an acid solution NADPH is first oxidized enzymatically with generation of O_2^- and then its oxidation is carried out in a chain reaction, propagated by a NADPH free radical and by O_2^- (Patriarca *et al.*, 1975; Curnutte *et al.*, 1976). This chain reaction, whose rate is increased by the presence of Mn^{2+}, amplifies the consumption of O_2 and the parallel oxidation of NADPH. It must be pointed out that the activity of the nonenzymatic chain reaction does not rule out the physiological relevance of the NADPH-oxidizing system as measured *in vitro*. In fact, activation of the NADPH-oxidizing system in a 20,000g fraction derived from stimulated cells is also detectable at neutral pH which inhibits the chain reaction (Patriarca *et al.*, 1975; Curnutte *et al.*, 1976; Babior *et al.*, 1976). Furthermore no oxidation of NADPH occurs if a heat inactivated preparation of the oxidase is used. However it remains to be established whether a chain reaction is peculiar to the *in vitro* assay condition or may be operative also *in vivo*.

The mechanism of NADPH oxidation has been recently reinvestigated in our laboratory using a 100,000g fraction of homogenates from phagocytizing guinea pig polymorphs (Bellavite *et al.*, 1980b). In this study no Mn^{2+} was used in the assays. The results showed that the NADPH oxidase reaction can be split into at least three components, namely an enzymatic univalent reduction of oxygen with generation of O_2^-, a nonenzymatic chain reaction initiated by O_2^-, and an apparently nonunivalent reduction of oxygen. The rate of each of the three components is greatly influenced by changing pH and/or NADPH concentration. For example, the chain reaction is virtually absent at pH 7.0 with 0.15 mM NADPH while it is greatly favored at acid pH.

It is clear, therefore, that the optimum pH of the oxidase depends on the assay conditions. When the assay is carried out under conditions which allow the measurement of both the enzymatic component and the chain reaction component [this is obtained by measuring oxygen uptake or NAD(P)H disappearance], the pH optimum is acid. Conversely, when the assay is carried out as an O_2^--generating activity, which excludes the chain reaction, the pH optimum is neutral.

4. Two other methodological approaches for the assay of NADPH oxidase have been introduced by De Chatelet and his associates. These investigators have developed an isotopic assay which measures the amount of $NADP^+$ formed in the oxidase reaction from the amount of $^{14}CO_2$ released from [1-^{14}C]-6-phosphogluconate, in the presence of 6-phosphogluconate dehydrogenase (De Chatelet *et al.*, 1975a). The particulate NADPH oxidase, as measured by this assay, is active in the presence of cyanide, is stimulated by Mn^{2+}, has a pH optimum of 5.5, and is much more active in preparations derived from phagocytizing human polymorphs than in preparations from resting cells. The activation was not apparent when a particulate fraction from a patient with chronic granulomatous disease was used. De Chatelet and his associates have also carried out experiments in the absence of Mn^{2+} with granules derived from both neutrophils and eosinophils. These experiments have fully confirmed the activation of the KCN-insensitive NADPH oxidase in human cells of normal donors stimulated by phagocytosis or by phorbol myristate acetate (McPhail *et al.*, 1977; De Chatelet *et al.*, 1977). The same group of investigators has developed another assay method for the oxidation of NADPH and NADH which is based on the fluorimetric assay of $NADP^+$ and NAD^+ formed (Iverson *et al.*, 1977). The oxidation of NADPH by granules isolated from phagocytizing cells, as measured by this assay, was four to five times higher than that of NADH.

5. Kakinuma *et al.* (1977) have developed a method for the assay of nicotinamide coenzymes oxidation based on the measurement of the H_2O_2 with cytochrome c peroxidase. Using this method Kakinuma and Kaneda (1980) have found that the granule-rich fraction isolated from phagocytizing guinea pig PMN oxidizes NADPH [pH 5.8, NAD(P)H 0.1 mM] at a higher rate than NADH and that the K_m of the oxidase for NADPH is lower than for NADH.

3.3.2. Indirect Evidence in Favor of the NADPH Oxidase

At least three lines of indirect evidence support the role of the NADPH oxidase in the metabolic burst. First, Stjernholm and Manak (1970) have shown that in phagocytizing human blood polymorphs the ^{3}H bound to C-1 of glucose exclusively appears as ^3H_2O. This indicates that the isotope is transferred to $NADP^+$ in the reaction catalyzed by glucose-6-phosphate dehydrogenase, the resulting NADP^{3}H being oxidized to produce $^3H_2O_2$, which, in turn, is transformed into ^3H_2O by an exchange with H_2O or in reactions catalyzed by myeloperoxidase or catalase.

Second, Zatti and Rossi (1965) have shown that the addition of bacteria to guinea pig polymorphs induces a very rapid decrease of fluorescence intensity at 420–500 nm (with an excitation wave-length of 366 nm), suggesting a rapid activation of NAD(P)H oxidation during phagocytosis. By determining the cell concentration of reduced and oxidized nicotinamide nucleotides, the same authors have shown that the decrease of fluorescence was associated with a marked decrease in the NADPH/NADP$^+$ ratio (Rossi and Zatti, 1966a; Patriarca *et al.*, 1971b) with no change in the NADH/NAD$^+$ ratio. Furthermore, by comparing the time course of the increased production of $^{14}CO_2$ from [1-^{14}C]glucose and that of the decrease in fluorescence, they could demonstrate that the increased oxidation of glucose in the HMP pathway follows that of NADPH to NADP$^+$.

Finally, the behavior of polymorphs with severe genetic deficiency of glucose-6-phosphate dehydrogenase (Baehner *et al.*, 1972; Cooper *et al.*, 1972) during phagocytosis also supports the contention of the primary role of NADPH oxidase in the metabolic burst. These cells, in which the continuous regeneration of NADPH is impeded, do not undergo the metabolic burst during phagocytosis, as shown by the absence of NBT reduction and H_2O_2 production. In these polymorphs the NADPH oxidase very likely cannot operate because of the deficient supply of NADPH.

3.4. NATURE OF THE OXIDASE

Regardless of whether NADPH or NADH is the physiological substrate used by the phagocytes during the respiratory burst, it is generally believed that all the biochemical events of the respiratory burst depend on the activity of a single enzyme which is able to employ either NADPH or NADH as the electron donor (Babior, 1978).

Initial attempts at purification of the guinea pig enzyme were unsuccessful, since any one of the procedures employed to solubilize it resulted in a marked loss of the activity (Patriarca *et al.*, 1974b).

Gabig *et al.* (1978) were able to prevent loss of activity by adding FAD to the extraction buffer (containing Triton X-100) used to solubilize the enzyme from zymosan-activated human neutrophils. By this procedure, they obtained about 14% of the NAD(P)H-dependent O_2^--forming activity of the starting preparation in soluble form. The soluble enzyme had a molecular weight lower than 300,000. In a subsequent paper, they showed that the soluble enzyme required phosphatidylethanolamine and FAD for maximal activity, had a broad pH optimum in the region of neutrality, and had a K_m for NADPH and NADH of 33 and 930 μM respectively (Gabig and Babior, 1979).

Tauber and Goetzl (1979) have used deoxycholate to solubilize the NAD(P)H-dependent O_2^--generating activity from human neutrophils activated by opsonized zymosan or phorbol myristate acetate. A mean of 12.4% of the particle-associated superoxide-generating activity was solubilized by this procedure. The yield, specific activity, molecular weight, and K_m for NADH and NADPH of the solubilized enzyme were similar to those of the enzyme solubilized by Gabig and co-workers, with the exception that FAD dependence could not be demonstrated.

Segal and Jones (1978) have recently indicated a cytochrome b as a possible component of the oxidase system. This cytochrome has a dual subcellular localization, the major component having a similar distribution to plasma membrane markers, and a denser peak located with the specific granules (Segal and Jones, 1979a). The presence of a cytochrome b in rabbit granulocytes had been already described in 1966 by Shinagawa *et al.* According to Segal and co-workers the evidence for the involvement of cytochrome b in the oxidase reaction is dual. First, the stimulation of neutrophils with phorbol myristate acetate activates the oxidase system and results in the reduction and subsequent oxidation of cytochrome b (Segal and Jones, 1979b). Second, the cytochrome b was not found (Segal *et al.* 1978) or failed to become reduced upon stimulation (Segal and Jones, 1980) in leukocytes from patients with chronic granulomatous disease. Borregaard *et al.* (1979), however, were able to find the cytochrome b in homogenates of neutrophils from patients with both the autosomal recessive and X-linked form of chronic granulomatous disease. The absence of cytochrome b from leukocytes with chronic granulomatous disease could not be confirmed even by Hamers *et al.* (1980). These authors were able to separate the NADPH oxidase from cytochrome b without loss of activity.

3.5. SUBCELLULAR LOCALIZATION OF THE ENZYME RESPONSIBLE FOR THE RESPIRATORY BURST

In 1961, Evans and Karnovsky suggested that a KCN-insensitive NADH oxidase located in the granules was released into the soluble cytoplasm during phagocytosis due to a drop in intracellular pH. In 1964, Cagan and Karnovsky emphasized the importance of a soluble KCN-insensitive NADH oxidase which appeared to contain FAD. In 1964, Rossi and Zatti showed that the NADPH oxidase activity of guinea pig polymorphonuclear leukocytes was associated with the $20,000g$ fraction of the postnuclear supernatant of cell homogenates and proposed that this enzymatic activity was granule-bound. In 1972, Paul *et al.* described two NADPH-oxidizing activities with different subcellular localization in guinea pig polymorphonuclear leukocytes: one was found in the $19,000g$ pellet and was cyanide insensitive, while the other was found in the $19,000g$ supernatant and was inhibited by cyanide. In 1973, Patriarca *et al.* used rate zonal sedimentation techniques to fractionate the postnuclear supernatant of resting rabbit polymorphs and showed that all the NADPH oxidase activity was recovered in the fraction containing azurophilic granules. Since no attempt was made in that study to characterize the plasma membrane fraction the possibility that part of the NADPH-oxidizing activity of the azurophilic granule fraction could be contributed by contamination with plasma membrane fragments cannot be excluded. In a subsequent study with human blood polymorphs, where the membrane fraction was also separated and characterized, Patriarca *et al.* (1977) showed that even in this type of cell the NADPH oxidase was entirely confined to the azurophilic granules. In this case, the peak of the oxidase activity was shifted towards higher sucrose densities as compared to the peak of myeloperoxidase, suggesting that the heavy azurophilic granules are richer in NADPH oxidase than in myeloperoxidase. This result is consistent with the observation of Iverson *et al.* (1978) who found that both NADH and NADPH oxidase of human blood polymorphs at rest or after phagocytosis cosedimented in a linear sucrose gradient to a region with a higher density than that of the azurophilic granules.

In all the studies summarized above the concept of an intracellular location of the initiating enzyme of the respiratory burst, regardless of its nature and substrate specificity, is clearly stated or, at least, surmised.

In 1974, a new trend began to emerge which favored a plasma membrane location of the oxidase. The experimental support to this idea was largely based on indirect evidence, such as the observations that (1) enhanced O_2^- production occurs in cells stimulated by immunoreactants in the presence of cytochalasin B used as an inhibitor of phagocytosis (Goldstein *et al.*, 1975); (2) PMN pretreated with cytochalasin B and then exposed to serum opsonized zymosan have a decreased oxygen consumption, H_2O_2 generation, and HMP activity, as compared with cells without cytochalasin B pretreatment, whereas O_2^- generation is not affected (Roos *et al.*, 1976); and (3) the oxidase activity of whole polymorphonuclear leukocytes subjected to hypotonic shock is inhibited by the nonpenetrating agent *p*-chloromercury benzoate (Takanaka and O'Brien, 1975).

A more direct approach in the study of the subcellular localization of the oxidase has been followed by Briggs *et al.* (1975) and by Segal and Peters (1976). The first group described an NADH-dependent enzyme which produced hydrogen peroxide and was located at the cell surface as judged by a histochemical method based on the formation of an electron-dense precipitate of cerous ions in the presence of H_2O_2. This method however has some inconveniences: (1) a cerium precipitate may be formed even in the absence of H_2O_2 production as indicated by the finding of Briggs *et al.* (1977) that when the method was applied to leukocytes from patients with chronic granulomatous disease some of these cells showed a precipitate on their surface; (2) it has been shown in our laboratory that cerous ions form a complex with NADH and more strongly with NADPH. The formation of cerium-NADPH complex may well explain the failure of Briggs *et al.* to show any precipitate with NADPH since, in their system, there was probably no free nucleotide available for reaction with H_2O_2 (Soranzo, unpublished results).

Segal and Peters (1976) have described an NADH-dependent NBT reductase activity which appeared to localize in a light region of a sucrose-density gradient corresponding to the plasma membrane fraction. It should be pointed out, however, that in this study an NBT reductase activity and not an oxygen-consuming activity was measured. Indeed, when the oxygen consumption assay was used the authors found that the NADPH oxidase activity was localized deep in the gradient in a zone corresponding to that of azurophilic granules.

In 1979, Dewald *et al.* fractionated the postnuclear supernatant of human blood polymorphonuclear leukocytes stimulated by phorbol myristate acetate, using zonal centrifugation techniques in sucrose gradients, and assayed the NADPH oxidase activity in the various fractions as O_2^- generation. The results showed that about 60% of the NADPH-dependent O_2^--generating activity was found in the membrane fraction, about 25% in the region of azurophilic granules, and about 5% in the region of specific granules. These findings are interpreted by the authors as a strong support for the view that the activable O_2^--forming system is localized in the plasma membrane of human neutrophils.

TABLE 5.   SUMMARY OF DIRECT EVIDENCE SO FAR AVAILABLE ON THE SUBCELLULAR LOCALIZATION OF THE ENZYMATIC SYSTEM RESPONSIBLE FOR THE METABOLIC BURST OF NEUTROPHILIC GRANULOCYTES

Authors	Animal species	Fractionation technique	Oxidase assay	Stimulatory agent[a]	Subcellular localization
Patriarca et al. (1973)	Rabbit	Zonal rate sedimentation on discontinuous sucrose density gradient	O_2 uptake	None	Azurophilic granules
Baehner (1975)	Human	Differential centrifugation. Isopycnic equilibration on continuous sucrose density gradient	NAD(P)H-NBT reductase	None	Microsomes or plasma membrane
Rossi and Zatti (1976a,b)	Guinea pig	Centrifugation on discontinuous sucrose density gradient	O_2 uptake	Paraffin oil particles	Granules
Segal and Peters (1977)	Human	Isopycnic equilibration on continuous sucrose density gradient	NAD(P)+ formed NAD(P)H-cyt.c and NBT-reductase	None None	Azurophilic granules Plasma membrane
Iverson et al. (1977)	Human	Isopycnic equilibration on continuous sucrose density gradient	NADP+ formation	STZ	Particles of higher density than azurophilic granules
Patriarca et al. (1977)	Human	Isopycnic equilibration on continuous sucrose density gradient	O_2 uptake	None	Azurophilic granules

Reference	Species	Method	Measurement	Stimulus	Localization
Iverson et al. (1978)	Human	Isopycnic equilibration and zonal rate sedimentation on continuous sucrose density gradient	NADP$^+$ formation	None and STZ	Particles of higher density than azurophilic granules
Cohen et al. (1978)	Guinea pig	Podosomes (differential centrifugation)	O$_2^{\cdot-}$ production	Digitonin	Plasma membrane
Auclair et al. (1978)	Human	Differential centrifugation	NADP$^+$ formation	None and STZ	Cytosol, and both heavy and low density particles
Tauber and Goetzl (1978)	Human	Centrifugation on continuous sucrose density gradient	O$_2^{\cdot-}$ production	STZ	Plasma membrane and azurophilic granules
Dewald et al. (1979)	Human	Rate zonal sedimentation on sucrose density gradient	O$_2^{\cdot-}$ production	PMA	Plasma membrane
Tauber and Goetzl (1979)	Human	Centrifugation on continuous sucrose density gradient	O$_2^{\cdot-}$ production	STZ PMA	Membrane fraction which sediments between azurophilic granules and plasma membrane
Rossi et al. (1980b)	Guinea pig	Rate zonal sedimentation on discontinuous sucrose density gradient. Isopycnic equilibration on continuous sucrose density gradient	O$_2^{\cdot-}$ production and oxygen uptake	None PMA	Azurophilic granules Plasma membrane

[a] STZ, serum–treated zymosan; PMA, phorbol myristate acetate.

A recent report by Tauber and Goetzl (1979) indicates that most of the O_2^--generating activity of the $400g$ postnuclear supernatant of human PMN sediments in a region of a linear sucrose gradient different from both the region of the granule markers and that of the plasma membrane markers.

It is clear from the above discussion that the problem of the subcellular localization of the oxidase is an intriguing one. It is also difficult to establish to what extent the discrepancies in the experimental results may be accounted for by differences in the functional state (resting or stimulated) of the cells, in the stimulatory agents, in the fractionation techniques, and/or in the methods of assay for the oxidase used by the investigators.

Recently, we made a comparative study of the subcellular localization of the NADPH oxidase in guinea peritoneal polymorphs both at rest and stimulated with phorbol myristate acetate using two fractionation techniques (isopycnic zonal equilibration and rate zonal sedimentation) and two assay methods for the oxidase (oxygen consumption and superoxide-anion generation). The results were similar with both fractionation techniques and with both assay methods. The subcellular distribution of the oxidase in resting cells differed from that in stimulated cells. In the former cells the bulk of the NADPH-oxidizing activity was found in a high-density, myeloperoxidase-containing zone of the gradient, while in stimulated cells most of the activity was recovered in a low-density zone containing empty vesicles (Rossi *et al.*, 1980b; Berton *et al.*, 1981). The results of the studies on the localization of the enzyme of the respiratory burst are summarized in Table 5.

3.6.   MECHANISM OF ACTIVATION OF THE OXIDASE

The requirement of intact cells for elicitation of the respiratory burst, the direct relationship existing between the number of particles bound to the cell and the extent of the metabolic stimulation (Romeo *et al.*, 1977; Root and Metcalf, 1977a,b), and finally the stimulation of the respiratory burst by a number of agents active at the cell surface indicate that the initial trigger of the respiratory burst must involve an interaction with the cell membrane. Rossi and Zatti (1968) were the first to advance the hypothesis that the elaboration of a signal for the respiratory burst to occur might be simultaneous with a change in the macromolecular organization of the plasma membrane of phagocytizing cells. Large and rapid surface topographical changes have indeed been observed accompanying phagocytosis. Oliver *et al.* (1974) have reported that in phagocytizing rabbit polymorphs certain lectin receptors are selectively removed from the surface, while transport proteins are selectively spared (Tsan and Berlin, 1971). Berlin and Fera (1977) have also observed that phagocytosis of either oil emulsion or polystyrene beads by polymorphs induced a rapid decrease in plasma membrane viscosity that parallels the extent of endocytosis. The authors have suggested that this is due to a lipid phase separation. This suggestion is consistent with the observation made in our laboratory (Romeo *et al.*, 1970) that the exposure of intact polymorphs to polystyrene beads brings about an immediate increase in fluorescence intensity of the probe 1-anilino-8-naphtalene sulphonate. This rapid change could correspond to initial steps in the development of large regional phase separations. More recently Chau and Karnovsky (1977) have observed a "lacy" pattern of distribution of intramembranous particles on the surface of phagocytizing macrophages and have interpreted this pattern as representing a lipid–protein lateral separation. If this reorganization of the surface membrane provides the trigger for activation of the oxidase, it would be important to establish whether it occurs only along the membrane segment enveloping the particles or is propagated over the whole cell surface.

Also some of the soluble stimulants of the phagocyte oxidative metabolism (see Table 1) are very likely to provoke a rapid topographical rearrangement of plasmalemmal components. In fact, the reaction between antibodies and surface antigens, as well as the binding of lectin molecules to their surface receptors, are known to induce a redistribution of antigens and receptors in the plane of the plasma membrane of various cells (Taylor *et al.*, 1971; Loor *et al.*, 1972; Unanue *et al.*, 1973; Ryan *et al.*, 1974; Nicolson, 1976). The oxidative metabolism of guinea pig polymorphs is stimulated by tetravalent Con A more than by trivalent Con A, suggesting that the stimulation is at least in part accomplished by cross linkage of plasma membrane constituents (Romeo *et al.*, 1978). Furthermore, phospholipid degradation and charge neutralization by appropriate counter ions are known to cause an aggregation of negatively charged macromolecules in the plane of plasma membrane (Nicolson, 1973, 1976). An increase in cytosol Ca^{2+} concentration may also cause sequestration of proteins into specific membrane domains (Nicolson, 1976) by promoting clusters of membrane lipids (Ohnishi and Ito, 1974). Finally, the cholesterol-displacing detergent saponin and the saturated short-chain or unsaturated fatty acids might increase the fluidity of the phagocyte plasma membrane, thereby inducing a lipid–protein phase separation.

The complex series of events occurring at the level of plasma membrane (such as enhanced membrane fluidity, change in membrane potential, loss of calcium from the plasma membrane, ion fluxes, coupling of receptors to adenylate cyclase, etc.) in stimulated leukocytes and their relationships to the respiratory burst, secretion, metabolism of arachidonate, and activation of movement have been recently reviewed (Weissmann *et al.*, 1979; Gallin *et al.*, 1978).

Taken for granted that a perturbation of the plasma membrane is somehow involved in the trigger of the respiratory burst, it remains to be explained how this is translated into activation of the oxidase. At this point, a discussion of the possible mechanisms involved necessarily reflects the uncertainties—which have been brought to light in a previous section—concerning the subcellular localization of the oxidase. There is no doubt that a plasma membrane location of the oxidase would offer a simple explanation for its activation: the enzyme would be in a position to be directly affected by the particle–cell interaction, particularly if it were exposed at the exterior face of the cell membrane. Conversely, if the oxidase were located intracellularly, then some mechanisms or signals must be invoked which transduce the information from the perturbed plasma membrane to the intracellular oxidase. This field is open to speculation and offers abundant material for future experiments.

A new line of research is just emerging which calls attention to a possible involvement of proteases in the initiation of the respiratory burst, as suggested by the inhibition of the metabolic stimulation of both polymorphonuclear or mononuclear leukocytes by a series of compounds which are known to inhibit chymotripsin-like proteases (Goldstein *et al.*, 1979; Kitagawa *et al.*, 1979; Simchowitz *et al.*, 1979).

REFERENCES

Aellig, A., Maillard, M., Phavorin, A., and Frei, J., 1977, The energy metabolism of the leukocyte. IX. Changes in the concentrations of the coenzymes NAD, NADH, NADP, NADPH in polymorphonuclear leucocytes during phagocytosis of *Staphylococcus aureus* and due to the action of phospholipase C, *Enzyme* **22**:207.

Allen, R. C., Stjernholm, R. C., Steele, R. H., 1972, Evidence for the generation of an electron excitation state(s) in human polymorphonuclear leukocytes and its participation in bactericidal activity, *Biochem. Biophys. Res. Commun.* **47**:677.

Andreae, W. A., 1955, A sensitive method for the estimation of hydrogen peroxide in biological material, *Nature (London)* **175**:859.

Auclair, C., Torres, M., Hakim, J., and Troube, H., 1978, NADPH-oxidation activities in subcellular fractions isolated from resting or phagocytosing human polymorphonuclears, *Am. J. Hematol.* **4**:113.

Babior, B. M., 1977a, Recent studies on oxygen metabolism in human neutrophils: Superoxide and chemiluminescence, in: *Superoxide and Superoxide Dismutases* (A. M. Michelson, J. M. McCord, and I. Fridovich, eds.), pp. 271–282, Academic Press, London.

Babior, B. M., 1977b, The superoxide forming enzyme responsible for the respiratory burst of neutrophils, in: *Movement, Metabolism and Bactericidal Mechanisms of Phagocytes* (F. Rossi, P. Patriarca, and D. Romeo, eds.), pp. 235–242, Piccin Medical Books, Padova.

Babior, B. M., 1978, Oxygen-dependent microbial killing by phagocytes, *N. Engl. J. Med.* **298**:659.

Babior, B. M., Kipnes, R. S., and Curnutte, J. T., 1973, Biological defense mechanisms. The production by leukocytes of superoxide, a potential bactericidal agent, *J. Clin. Invest.* **52**:741.

Babior, B. M., Curnutte, J. T., and Kipnes, R. S., 1975, Pyridine nucleotide-dependent superoxide production by a cell-free system from human granulocytes, *J. Clin. Invest.* **56**:1035.

Babior, B. M., Curnutte, J. T., and McMurrich, B. J., 1976, The particulate superoxide-forming system from human neutrophils. Properties of the system and further evidence supporting its participation in the respiratory burst, *J. Clin. Invest.* **58**:989.

Badwey, J. A., and Karnovsky, M. L., 1979, Production of superoxide anion hydrogen peroxide by an NADH-oxidase in guinea pig polymorphonuclear leukocytes, *J. Biol. Chem.* **254**:11530.

Baehner, R. L., 1975, Subcellular distribution of nitroblue tetrazolium reductase (NBT-R) in human polymorphonuclear leukocytes (PMN), *J. Lab. Clin. Med.* **86**:785.

Baehner, R. L., and Johnston, R. B., 1972, Monocyte function in children with neutropenia and chronic infection, *Blood* **40**:31.

Baehner, R. L., Gilman, N., and Karnovsky, M. L., 1970, Respiration and glucose oxydation in human and guinea pig leucocytes: Comparative studies, *J. Clin. Invest.* **49**:692.

Baehner, R. L., Johnston, R. B., and Nathan, G. D., 1972, Comparative study of the metabolic and bactericidal characteristics of severely glucose 6-phosphate dehydrogenase-deficient polymorphonuclear leukocytes and leukocytes from children with granulomatous disease, *J. Reticuloendothel. Soc.* **12**:150.

Baldridge, C. W., Gerard, R. W., 1933, The extrarespiration of phagocytosis, *Am. J. Physiol.* **103**:235.

Basford, R. E., Bartus, B., Kaplan, S. S., and Platt, D., 1977, Glutathione peroxidase and phagocytosis, in: *Movement, Metabolism and Bactericidal Mechanisms of Phagocytes* (F. Rossi, P. Patriarca, and D. Romeo, eds.), pp. 265–275, Piccin Medical Books, Padua.

Bass, D. A., De Chatelet, L. R., Burk, R. F., Shirley, P., and Szejda, P., 1977, Polymorphonuclear leukocyte bactericidal activity and oxidative metabolism during glutathione peroxidase deficiency, *Infect. Immun.* **18**:78.

Beck, W. S., 1958, Occurrence and control of the phosphogluconate oxidation pathway in normal and leukemic leukocytes, *J. Biol. Chem.* **232**:271.

Beckman, G., Lundgren, E., and Tärnvick, A., 1973, Superoxide dismutase isoenzymes in different human tissues. Their genetic control and intracellular localization, *Hum. Hered.* **23**:338.

Bellavite, P., Dri, P., Bisiacchi, B., and Patriarca, P., 1977, Catalase deficiency in myeloperoxidase deficient polymorphonuclear leucocytes from chicken, *FEBS Letters* **81**:73.

Bellavite, P., Dri, P., Berton, G., and Zabucchi, G., 1980a. Un nuovo test di funzionalità fagocitaria basato sulla misura della produzione di superossido anione (O_2^-). I. Principi generali di esecuzione, *LAB* **7**:67.

Bellavite, P., Berton, G., and Dri, P., 1980b, Studies on the NADPH oxidation by subcellular particles from phagocytosing polymorphonuclear leukocytes. Evidence for the involvement of three mechanisms, *Biochim. Biophys. Acta* (in press).

Berlin, R. D., and Fera, J. P., 1977, Changes in membrane microviscosity associated with phagocytosis: Effects of colchicine, *Proc. Natl. Acad. Sci. USA* **74**:1072.

Berton, G., Bellavite, P., Dri, P., Cramer, R., 1981, Subcellular localization of NADPH oxidase in resting and stimulated guinea pig neutrophils (submitted for publication).

Borregaard, N., Johansen, K. S., Tandorff, E., and Wandall, J. H., 1979, Cytochrome b is present in neutrophils from patient with chronic granulomatous disease, *Lancet* **1**:949.

Briggs, R. T., Drath, D. B., Karnovsky, M. L., and Karnovsky, M. J., 1975, Localization of NADH oxidase on the surface of human polymorphonuclear leukocytes by a new cytochemical method, *J. Cell Biol.* **67**:566.

Briggs, R. T., Karnovsky, M. L., and Karnovsky, M. J., 1977, Hydrogen peroxide production in chronic granulomatous disease. A cytochemical study of reduced pyridine nucleotide oxidases, *J. Clin. Invest.* **59**:1088.

Brune, K., and Spitznagel, J. K., 1973, Peroxidaseless chicken leukocytes: Isolation and characterization of antibacterial granules, *J. Infect. Dis.* **127**:84.

Cagan, R. H., and Karnovsky, M. L., 1964, Enzymatic basis of the respiratory stimulation during phagocytosis, *Nature (London)* **204**:255.

Chau, R. M., and Karnovsky, M. J., 1977, Lipid–protein lateral separation in the plasma membrane of phagocytosing macrophages, *J. Cell Biol.* **75**:238a.

Cline, M. J., 1970, Leukocyte metabolism, in: *Regulation of Hematopoiesis* (A. S. Gordon, ed.), Vol. 2, pp. 1045–1079, Appleton-Century-Crofts, New York.

Cohen, H. J., Chovaniec, M. E., and Davies, W., 1978, Subcellular localization and kinetics of activation of the granulocyte superoxide generating system, *Fed. Proc.* **37**:1276.

Cooper, M. R., De Chatelet, L. R., McCall, C. E., LaVia, M. F., Spurr, C. L., and Baehner, L. R., 1972, Complete deficiency of leukocyte glucose-6-phosphate dehydrogenase with defective bactericidal activity, *J. Clin. Invest.* **51**:769.

Curnutte, J. T., and Babior, B. M., 1974, Biological defense mechanisms: The effect of bacteria and serum on superoxide production by granulocytes, *J. Clin. Invest.* **53**:1562.

Curnutte, J. T., Kipnes, R. S., and Babior, B. M., 1975, Defect in pyridine nucleotide dependent superoxide production by a particulate fraction from granulocytes of patient with chronic granulomatous disease, *N. Engl. J. Med.* **293**:628.

Curnutte, J. T., Karnovsky, M. L., and Babior, B. M., 1976, Manganese-dependent NADPH oxidation by granulocyte particles. The role of superoxide and the nonphysiological nature of the manganese requirement *J. Clin. Invest.* **57**:1059.

De Chatelet, L. R., 1979, Phagocytosis by human neutrophils, in: *Phagocytes and Cellular Immunity* (H. Gadebusch, ed.), pp. 1–55, CRC Press, West Palm Beach, Florida.

De Chatelet, L. R., McCall, C. E., McPhail, L. C., and Johnston, R. B., Jr., 1974, Superoxide dismutase activity in leukocytes, *J. Clin. Invest.* **53**:1197.

De Chatelet, L. R., McPhail, L. C., Mullikin, D., and McCall, C. E., 1975a, An isotopic assay for NADPH oxidase activity and some characteristics of the enzyme from human polymorphonuclear leukocytes, *J. Clin. Invest.* **55**:714.

De Chatelet, L. R., Mullikin, D., and McCall, C. E., 1975b, Quantitative aspects of the production of superoxide radicals by phagocytizing human granulocytes, *J. Lab. Clin. Med.* **85**:245.

De Chatelet, L. R., Shirley, P. S., McPhail, L. C., Huntler, C. C., Muss, H. B., and Bass, D. A., 1977, Oxidative metabolism of eosinophil, *Blood* **50**:525.

Dewald, B., Baggiolini, M., Curnutte, J. T., and Babior, B. M., 1979, Subcellular localization of the superoxide-forming enzyme in human neutrophils, *J. Clin. Invest.* **63**:21.

Dri, P., Bisiacchi, B., Cramer, R., Bellavite, P., de Nicola, G., and Patriarca, P., 1978, Oxidative metabolism of chicken polymorphonuclear leucocytes during phagocytosis, *Mol. Cell. Biochem.* **22**:159.

Dri, P., Bellavite, P., Berton, G., and Rossi, F., 1979, Interrelationship between oxygen consumption, superoxide anion and hydrogen peroxide formation in phagocytosing guinea pig polymorphonuclear leukocytes, *Mol. Cell. Biochem.* **23**:109.

Eggleston, L. V., and Krebs, H. A., 1974, Regulation of the pentose phosphate cycle, *Biochem. J.* **138**:425.

Elsbach, P., 1977, White cells, in: *Lipid Metabolism in Mammals* (F. Snyder, ed.), Vol. 1, pp. 259–276, Plenum Press, New York.

Evans, A. E., and Kaplan, N. O., 1966, Pyridine nucleotide transhydrogenase in normal human and leukemic cells, *J. Clin. Invest.* **45**:1268.

Evans, W. H., and Karnovsky, M. L., 1961, A possible mechanism for the stimulation of some metabolic functions during phagocytosis, *J. Biol. Chem.* **236**:Pc30.

Evans, W. H., and Karnovsky, M. L., 1962, The biochemical basis of phagocytosis. IV. Some aspects of carbohydrate metabolism during phagocytosis, *Biochemistry*, **1**:159.

Gabig, T. G., and Babior, B. M., 1979, The O_2^--forming oxidase responsible for the respiratory burst in human neutrophils, *J. Biol. Chem.* **254**:9070.

Gabig, T. G., Kipnes, R. S., and Babior, B. M., 1978, Solubilization of the O_2^--forming activity responsible for the respiratory burst in human neutrophils, *J. Biol. Chem.* **253**:6663.

Gallin, J. I., Gallin, E. K., Malech, H. L., and Cramer, E. B., 1978, Structural and ionic events during leukocyte chemotaxis, in: *Leukocyte Chemotaxis: Methods, Physiology, and Clinical Implications* (J. I. Gallin and P. G. Quie, eds.) pp. 123–141, Raven Press, New York.

Gee, J. B. L., and Khandwala, A. S., 1976, Oxygen metabolism in the alveolar macrophage: Friend and foe? *J. Reticuloendothel. Soc.* **19**:229.

Gee, J. B. L., Vassallo, C. L., Bell, P., Kaskin, J., Basford, R. E., and Field, J., 1970, Catalase-dependent peroxidative metabolism in the alveolar macrophage during phagocytosis, *J. Clin. Invest.* **49**:1280.

Glock, G. E., and McLean, P., 1953, Further studies on the properties and assay of glucose-6-phosphate dehydrogenase and 6-phosphogluconate dehydrogenase of rat liver, *Biochem. J.* **55**:400.

Goetzl, E. J., and Austen, K. F., 1974, Stimulation of human neutrophil leukocyte aerobic glucose metabolism by purified chemotactic factors *J. Clin. Invest.* **53**:591.

Goldstein, B. D., Witz, G., Amoruso, M., and Troll, W., 1979, Protease inhibitors antagonize the activation of polymorphonuclear leukocyte oxygen consumption, *Biochem. Biophys. Res. Commun.* **88**:854.

Goldstein, I. M., Roos D., Kaplan, H. B., and Weissmann, G., 1975, Complement and immunoglobulins stimulate superoxide production by human leukocytes independently of phagocytosis, *J. Clin. Invest.* **56**:1155.

Graham, R. C., Jr., Karnovsky, M. J., Shafer, A. W., Glass, E. A., and Karnovsky, M. L., 1967, Metabolic and morphological observations on the effect of surface active agents on leukocytes, *J. Cell Biol.* **32**:629.

Haber, F., and Weiss, J., 1934, The catalatyc decomposition of hydrogen peroxide by iron salts, *Proc. R. Soc. London Ser. A* **147**:332.

Hamers, M. N., Wever, R., van Shaik, M. L. J., Weening, R. S., and Roos, D., 1980, Cytochrome b and the superoxide-generating system in human neutrophils, in: *The Significance of Superoxide and Superoxide Dismutase*, Vol. 2, *Biological and Clinical Aspects* (W. H. Bannister and J. V. Bannister, eds.), Elsevier, New York.

Hawkins, D., 1973, Neutrophilic leukocytes in immunologic reactions *in vitro:* Effect of cytochalasin B, *J. Immunol.* **110**:294.

Henson, P. M., and Oades, Z. G., 1973, Enhancement of immunologically induced granule exocytosis from neutrophils by cytochalasin B, *J. Immunol.* **110**:290.

Higgins, C. P., Baehner, R. L., McCallister, J., and Boxer, L. A., 1978, Polymorphonuclear leukocyte species differences in the disposal of hydrogen peroxide (H_2O_2), *Proc. Soc. Exp. Biol. Med.* **158**:478.

Hohn, D. C., and Lehrer, R. I., 1975, NADPH oxidase deficiency in X-linked chronic granulomatous disease, *J. Clin. Invest.* **55**:707.

Homan-Müller, J. W. T., Weening, R. S., and Roos, D., 1975, Production of hydrogen peroxide by phagocytizing human granulocytes, *J. Lab. Clin. Med.* **85**:198.

Iverson, D., De Chatelet, L. R., Spitznagel, J. K., and Wang, P., 1977, Comparison of NADH and NADPH oxidase activities in granules isolated from human polymorphonuclear leukocytes with a fluorimetric assay, *J. Clin. Invest.* **59**:282.

Iverson, D. B., Wang-Iverson, P., Spitznagel, J. K., and De Chatelet, L. R., 1978, Subcellular localization of NAD(P)H oxidases in human neutrophilic polymorphonuclear leukocytes, *Biochem. J.* **176**:175.

Iyer, G. Y. N., and Quastel, J. H., 1963, NADPH and NADH oxidation by guinea pig polymorphonuclear leukocytes, *Can. J. Biochem. Physiol.* **41**:427.

Iyer, G. Y. N., Islam, M. F., and Quastel, H. J., 1961, Biochemical aspects of phagocytosis, *Nature (London)* **192**:535.

Johnston, R. B., Jr., Keele, B. B., Misra, H. P., Lehmeyer, J. E., Webb, L. S., Baehner, R. L., and Rajagopalan, K. V., 1975, The role of superoxide anion generation in phagocytic bactericidal activity: Studies with normal and granulomatous disease leukocytes, *J. Clin. Invest.* **55**:1357.

Johnston, R. B., Jr., Lehmeyer, J. E., and Guthrie, L. A., 1976, Generation of superoxide anion and chemiluminescence of human monocytes during phagocytosis and on contact with surface-bound immunoglobulin G, *J. Exp. Med.* **143**:1551.

Kakinuma, K., 1974, Effects of fatty acids on the oxidative metabolism of leukocytes, *Biochim. Biophys. Acta* **348**:76.

Kaplan, S. S., Finch, S. C., and Basford, R. E., 1972, Polymorphonuclear leukocyte activation: Effects of phospholipase c, *Proc. Soc. Exp. Biol. Med.* **140**:540.

Karnovsky, M. L., 1962, Metabolic basis of phagocytic activity, *Physiol. Rev.* **42**:143.

Karnovsky, M. L., 1973, Chronic granulomatous disease-pieces of a cellular and molecular puzzle, *Fed. Proc.* **32**:1527.

Karnovsky, M. L., Lazdins, J., and Simmons, S. R., 1975, Metabolism of activated mononuclear phagocytes at rest and during phagocytosis, in: *Mononuclear Phagocytes in Immunity, Infection and Pathology* (R. van Furth, ed.), pp. 423–439, Blackwell Scientific Publications, Oxford.

Keston, A. S., and Brandt, R., 1965, The fluorimetric analysis of ultramicro quantities of hydrogine peroxide, *Anal. Biochem.* **11**:1.

Keusch, G. T., Douglas, S. D., Mildvan, D., and Hirschman, S. Z., 1972, ^{14}C-Glucose oxidation in whole blood: A clinical assay for phagocytic disfunction, *Infect. Immun.* **5**:414.

Kitagawa, S., Takaku, F., and Sakamoto, S., 1979, Possible involvement of proteases in superoxide production by human polymorphonuclear leukocytes, *FEBS Letters* **99**:275.

Klebanoff, S. J., 1975, Antimicrobial mechanisms in neutrophilic polymorphonuclear leukocytes, *Semin. Hematol.* **12**:117.

Klebanoff, S. J., and Clark, R. A., 1978, *The Neutrophil: Function and Clinical Disorders,* North Holland, Amsterdam.

Klebanoff, S. J., and Hamon, C. B., 1972, Role of myeloperoxidase-mediated antimicrobicidal systems in intact leukocytes, *J. Reticuloendothel. Soc.* **12**:170.

Klebanoff, S. J., and Hamon, C. B., 1975, Antimicrobial system of mononuclear phagocytes, in: *Mononuclear Phagocytes in Immunity, Infection and Pathology* (R. van Furth, ed.), pp. 507–531, Blackwell Scientific Publications, Oxford.

Loor, F., Forni, L., and Pernis, B., 1972, The dynamic state of the lymphocyte membrane. Factors affecting the distribution and turnover of surface immunoglobulins, *Eur. J. Immunol.* **2**:203.

Mandell, G. L., and Sullivan, G. W., 1971, Pyridine nucleotide oxidation by intact human polymorphonuclear neutrophils, *Biochim. Biophys. Acta* **234**:43.

McCord, J. M., and Day, E. D. J., 1978, Superoxide-dependent production of hydroxil radical catalyzed by iron-EDTA complex, *FEBS Letters* **80**:130.

McPhail, L. C., De Chatelet, R. L., Shirley, P. S., Wilfert, C., Johnston, R. B., Jr., and McCall, C. E., 1977, Deficiency of NADPH oxidase activity in chronic granulomatous disease, *J. Pediatr.* **90**:213.

Mendelson, D. S., Metz, E. N., Sagone, A. L., 1977, Effect of phagocytosis on the reduced soluble sulfhydril content of human granulocytes, *Blood* **50**:1023.

Morton, D. J., Moran, J. F., and Stjernholm, R. L., 1969, Carbohydrate metabolism in leukocytes. XI. Stimulation of eosinophils and neutrophils, *J. Reticuloendothel. Soc.* **6**:525.

Nakagawara, A., Takeshige, K., and Minakami, S., 1974, Induction of a phagocytosis-like metabolic pattern in polymorphonuclear leucocytes by cytochalasin E, *Exp. Cell Res.* **87**:392.

Nakamura, M., Nakamura, M. A., Okamura, J., and Kobajashi, Y., 1978, A rapid and quantitative assay of phagocytosis-connected oxygen consumption by leukocytes in whole blood, *J. Lab. Clin. Med.* **91**:568.

Nathan, C. F., and Root, R. K., 1977, Hydrogen peroxide release from mouse peritoneal macrophages. Dependence on sequential activation and triggering, *J. Exp. Med.* **146**:1648.

Nicolson, G. L., 1973, Anionic sites of human erythrocyte membranes. I. Effects of trypsin, phospholipase C, and pH on the topography of bound positively charged colloidal particles, *J. Cell Biol.* **57**:373.

Nicolson, G. L., 1976. Transmembrane control of the receptors on normal and tumor cells. I. Cytoplasmic influence over cell surface components *Biochim. Biophys. Acta* **457**:57.

Noseworthy, J., and Karnovsky, J. L., 1972, Role of peroxide in the stimulation of the hexose monophosphate shunt during phagocytosis by polymorphonuclear leukocytes, *Enzyme* **13**:110.

Ohnishi, S., and Ito, T., 1974, Calcium-induced phase separations in phosphatydil serinephosphatidylcholine membranes, *Biochemistry* **13**:881.

Oliver, J. M., Ukena, T. E. and Berlin, R. D., 1974, Effects of phagocytosis and colchicine on the distribution of lectin-binding sites on cell surfaces, *Proc. Natl. Acad. Sci USA* **71**:394.

Patriarca, P., and Dri, P., 1976, The NADPH oxidase activity of polymorphonuclear neutrophilic leucocytes revisited, in: *The Reticuloendothelial System in Health and Disease* (S. M. Reichard, M. R. Escobar, and H. Friedman, eds.), p. 151, Plenum Press, New York.

Patriarca, P., Zatti, M., Cramer, R., and Rossi, F., 1970, Stimulation of the respiration of polymorphonuclear leukocytes by phospholipase C, *Life Sciences* (Part I) **9**:841.

Patriarca, P., Cramer, R., Marussi, M., Moncalvo, S., and Rossi, F., 1971a, Phospholipid splitting and metabolic stimulation in polymorphonuclear leukocytes, *J. Reticuloendothel. Soc.* **10**:251.

Patriarca, P., Cramer, R., Moncalvo, S., Rossi, F., Romeo, D., 1971b, Enzymatic basis of metabolic stimulation in leukocytes during phagocytosis: The role of activated NADPH oxidase, *Arch. Biochem. Biophys.* **145**:255.

Patriarca, P., Cramer, R., Dri, P., Fant, L., Basford, R. R., and Rossi, F., 1973, NADPH oxidizing activity in rabbit polymorphonuclear leukocytes: Localization in azurophilic granules, *Biochem. Biophys. Res. Commun.* **53**:830.

Patriarca, P., Dri, P., and Rossi, F., 1974a, Superoxide dismutase in leukocytes, *FEBS Letters* **43**:247.

Patriarca, P., Basford, R. E., Cramer, R., Dri, P., and Rossi, F., 1974b, Studies on the NADPH oxidizing activity in polymorphonuclear leukocytes. The mode of association with the granule membrane, the relationship to myeloperoxidase and the interference of hemoglobin with NADPH oxidase determination, *Biochim. Biophys. Acta* **362**:221.

Patriarca, P., Dri, P., Kakinuma, K., Tedesco, F., and Rossi, F., 1975, Studies on the mechanism of metabolic stimulation in polymorphonuclear leukocytes during phagocytosis. I. Evidence for superoxide anion involvement in the oxidation of $NADPH_2$, *Biochim. Biophys. Acta* **385**:380.

Patriarca, P., Cramer, R., and Dri, P., 1977, The present status of the subcellular localization of the NAD(P)H oxidase in polymorphonuclear leucocytes, in: *Movement, Metabolism and Bactericidal Mechanisms of Phagocytes* (F. Rossi, P. Patriarca and D. Romeo, eds.), pp. 167–174, Piccin Medical Books, Padua.

Paul, B. B., and Sbarra, A. J., 1968, The role of the phagocytes in host–parasite interactions. XIII. The direct quantitative estimation of H_2O_2 in phagocytizing cell, *Biochim. Biophys. Acta* **156**:168.

Paul, B. B., Strauss, R. R., Jacobs, A. A., and Sbarra, A. J., 1970, Function of H_2O_2, myeloperoxidase and hexose monophosphate shunt enzymes in phagocytosing cells from different species, *Infect. Immun.* **1**:338.

Paul, B. B., Strauss, R. R., Jacobs, A. A., and Sbarra, A. J., 1972, Direct involvement of NADPH oxidase with the stimulated respiratory and hexose monophosphate shunt activity in phacocytizing leucocytes, *Exp. Cell Res.* **73**:456.

Penniall, R., and Spitznagel, J. K., 1975, Chicken neutrophils: Oxidative metabolism in phagocytic cells devoid of myeloperoxidase, *Proc. Natl. Acad. Sci. USA* **72**:5012.

Reed, P. W., 1969, Glutathione and hexose monophosphate shunt in phagocytizing and hydrogen peroxide-treated rat leukocytes, *J. Biol. Chem.* **244**:2459.

Reed, P. W., and Tepperman, J., 1969, Phagocytosis associated metabolism and enzymes in the rat polymorphonuclear leukocytes, *Am. J. Physiol.* **216**:223.

Reiss, M., and Roos, D., 1978, Difference in oxygen metabolism of phagocytosing monocytes and neutrophils, *J. Clin. Invest.* **61**:480.

Repine, J. E., White, J. G., Clauson, C. C., and Holmes, B. M., 1974, The influence of phorbol myristate acetate on oxygen consumption of polymorphonuclear leukocytes, *J. Lab. Clin. Med.* **83**:911.

Rest, R. F., and Spitznagel, J. K., 1977, Subcellular distribution of superoxide dismutases in human neutrophils. Influence of myeloperoxidase on the measurement of superoxide dismutase activity, *Biochem. J.* **166**:145.

Rister, R. F., and Baehner, R. L., 1976, A comparative study of superoxide dismutase activity in polymorphonuclear leukocytes, monocytes and alveolar macrophages of the guinea pig, *J. Cell Physiol.* **87**:345.

Rister, M., and Baehner, R. L., 1977, Effect of hypoxia on superoxide anion and hydrogen peroxide production by polymorphonuclear leucocytes and alveolar macrophages, *Br. J. Haematol.* **36**:241.

Roberts, J., and Camacho, Z., 1967, Oxidation of NADPH by polymorphonuclear leucocytes during phagocytosis, *Nature (London)* **216**:606.

Roberts, J., and Quastel, J. H., 1964, Oxidation of reduced triphosphopyridine nucleotide by guinea pig polymorphonuclear leucocytes, *Nature (London)* **202**:85.

Romeo, D., Cramer, R., and Rossi, F., 1970, Use of 1-anilino-8-naphthalene sulfonate to study structural transitions in cell membrane of PMN leucocytes, *Biochem Biophys. Res. Commun.* **41**:582.

Romeo, D., Zabucchi, G., Soranzo, M. R., and Rossi, F., 1971, Macrophage metabolism: Activation of NADPH oxidation by phagocytes, *Biochem. Biophys. Res. Commun.* **45**:1056.

Romeo, D., Zabucchi, G., Marzi, T., and Rossi, F., 1973a, Kinetic and enzymatic features of metabolic stimulation of alveolar and peritoneal macrophages challenged with bacteria, *Exp. Cell Res.* **78**:423.

Romeo, D., Zabucchi, G., and Rossi, F., 1973b, Reversible metabolic stimulation of polymorphonuclear leucocytes and macrophages by concanavalin A, *Nature New Biol.* **243**:111.

Romeo, D., Jug, M., Zabucchi, G., and Rossi, F., 1974, Perturbation of leucocyte metabolism by non phagocytosable concanavalin A-coupled beads, *FEBS Lett.* **42**:90.

Romeo, D., Zabucchi, G., Jug, M., Miani, N., and Soranzo, M. R., 1975a, Concanavalin A as a probe for studying the mechanism of metabolic stimulation of leukocytes, in: *Concanavalin A* (T. K. Chowdhury and A. K. Weiss, eds.), pp. 261–271, Plenum Press, New York.

Romeo, D., Zabucchi, G., Miani, N., and Rossi, F., 1975b, Ion movement across leukocyte plasma membrane and excitation of their metabolism, *Nature* **253**:542.

Romeo, D., Zabucchi, G., and Rossi, F., 1977, Surface modulation of oxidative metabolism of polymorphonuclear leucocytes, in: *Movement, Metabolism and Bactericidal Mechanisms of Phagocytes* (F. Rossi, P. Patriarca, and D. Romeo, eds.), pp. 153–165, Piccin, Padua.

Romeo, D., Zabucchi, G., Berton, G., and Schneider, C., 1978, Metabolic stimulation of polymorphonuclear leucocytes: Effects of tetravalent and divalent concanavalin A, *J. Membr. Biol.* **44**:321.

Roos, D., Homan-Müller, J. W. T., and Weening, R. S., 1976, Effect of cytochalasin B on the oxidative metabolism of human peripheral blood granulocytes, *Biochem. Biophys. Res. Commun.* **68**:43.

Root, R. K., and Metcalf, J. A., 1977a, H_2O_2 release from human granulocytes during phagocytosis. Relationship to superoxide anion formation and cellular catabolism of H_2O_2: Studies with normal and cytochalasin B-treated cells, *J. Clin. Invest.* **60**:1266.

Root, R. K., and Metcalf, J. A., 1977b, Superoxide and hydrogen peroxide formation by human granulocytes: Inter-relationship and activation mechanisms, in: *Movement, Metabolism and Bactericidal Mechanisms of Phagocytes* (F. Rossi, P. Patriarca, and D. Romeo, eds.), pp. 185–199, Piccin, Padua.

Root, R. K., Metcalf, J. A., Oshino, N., and Chance, B., 1975, H_2O_2 release from human granulocytes during phagocytosis. I. Documentation, quantitation, and some regulating factors, *J. Clin. Invest.* **55**:945.

Rosen, H., and Klebanoff, S. J., 1979, Hydroxyl radical generation by polymorphonuclear leukocytes measured by electron spin resonance spectroscopy, *J. Clin. Invest.* **64**:1725.

Rossi, F., and Zatti, M., 1964, Changes in the metabolic pattern of polymorphonuclear leukocytes during phagocytosis, *Br. J. Exp. Pathol.* **45**:548.

Rossi, F., and Zatti, M., 1966a, The mechanism of the respiratory stimulation during phagocytosis in polymorphonuclear leukocytes, *Biochim. Biophys. Acta* **113**:395.

Rossi, F., and Zatti, M., 1966b, Effect of phagocytosis on the carbohydrate metabolism of polymorphonuclear leucocytes, *Biochim. Biophys. Acta* **121**:110.

Rossi, F., and Zatti, M., 1968, Mechanism of the respiratory stimulation in saponine-treated leukocytes. The KCN insensitive oxidation of NADPH, *Biochim. Biophys. Acta* **153**:296.

Rossi, F., Zatti, M., and Patriarca, P., 1969, H_2O_2 production during NADPH oxidation by the granule fraction of phagocytosing polymorphonuclear leucocytes, *Biochim. Biophys. Acta* **184**:201.

Rossi, F., Patriarca, P., Romeo, D., 1971a, Sequence of events leading to the metabolic stimulation in PMN leucocytes during phagocytosis, in: *The Reticuloendothelial System and Immune Phenomena* (N. Di Luzio, ed.), pp. 191–208, Plenum Press, New York.

Rossi, F., Zatti, M., Patriarca, P., and Cramer, R., 1971b, Effect of specific antibodies on the metabolism of guinea pig polymorphonuclear leukocytes, *J. Reticuloendothel. Soc.* **9**:67.

Rossi, F., Romeo, D., and Patriarca, P., 1972, Mechanism of phagocytosis-associated oxidative metabolism in polymorphonuclear leucocytes and macrophages, *J. Reticuloendother. Soc.* **12**:127.

Rossi, F., Zabucchi, G., and Romeo, D., 1975, Metabolism of phagocytosing mononuclear phagocytes, in: *Mononuclear Phagocytes in Immunity, Infection and Pathology* (R. van Furth, ed.), pp. 441–462, Blackwell Scientific Publications, Oxford.

Rossi, F., Patriarca, P., Romeo, D., and Zabucchi, G., 1976a, The mechanism of control of phagocytic metabolism, in: *The Reticuloendothelial System in Health and Disease. Function and Characteristics* (S. M. Reichard, M. R. Escobar, and H. Friedman, eds.), pp. 205–223, Plenum Press, New York.

Rossi, F., Romeo, D. and Patriarca, P., 1976b, Metabolic perturbation of the inflammatory cells, *Agents Actions* **6**:50.

Rossi, F., Dri, P., Bellavite, P., and Zabucchi, G., 1979, Oxidative metabolism of inflammatory cells, in: *Advances in Inflammation Research*, Vol. 1 (G. Weissmann, ed.), pp. 139–155, Raven Press, New York.

Rossi, F., Zabucchi, G., Dri, P., Bellavite, P., and Berton, G., 1980a, O_2^- and H_2O_2 production during the respiratory burst in alveolar macrophages, in: *Macrophages and Lymphocytes. Nature, Functions and Interaction*, Part A (M. R. Escobar and H. Friedman, eds.), pp. 53–74, Plenum Press, New York.

Rossi, F., Patriarca, P., Berton, G., and de Nicola, G., 1980b, Subcellular localization of the enzyme responsible for the respiratory burst in resting and PMA-activated leucocytes, in: *The Significance of Superoxide and Superoxide Dismutase*, Vol. 2, *Biological and Clinical Aspects* (W. H. Bannister and J. V. Bannister, eds.), Elsevier, New York.

Ryan, G. B., Borysenko, J. Z., and Karnovsky, M. J., 1974, Factors affecting the redistribution of surface-bound concanavalin A on human polymorphonuclear leukocytes, *J. Cell Biol.* **62**:351.

Salin, M. L., and McCord, J. M., 1974, Superoxide dismutase in polymorphonuclear leukocytes, *J. Clin. Invest.* **53**:1197.

Sbarra, A. J., and Karnovsky, M. L., 1959, The biochemical basis of phagocytosis. I. Metabolic changes during ingestion of particles by polymorphonuclear leukocytes, *J. Biol. Chem.* **234**:1355.

Sbarra, A. J., Selvaraj, R. J., Paul, B. B., Mitchell, G. W., and Louis, F., 1977, Some newer insights of the peroxidase mediated antimicrobial system, in: *Movement, Metabolism and Bactericidal Mechanisms of Phagocytes* (F. Rossi, P. Patriarca, and D. Romeo, eds.), pp. 295–304, Piccin Medical Books, Padua.

Schell-Frederick, E., 1974, Stimulation of the oxidative metabolism of polymorphonuclear leucocytes by the calcium ionophore A23187, *FEBS Letters* **48**:37.

Segal, A. W., and Jones, O. T. G., 1978, Novel cytochrome b system in phagocytic vacuoles of human granulocytes, *Nature* **276**:515.

Segal, A. W., and Jones, O. T. G., 1979a, The subcellular distribution and some properties of the cytochrome b component of the microbicidal oxidase system of human neutrophils, *Biochem. J.* **182**:181.

Segal, A. W., and Jones, O. T. G., 1979b, Reduction and subsequent oxidation of a cytochrome b of human neutrophils after stimulation with phorbol myristate acetate, *Biochem. Biophys. Res. Commun.* **88**:130.

Segal, A. W., and Jones, O. T. G., 1980, Absence of cytochrome b reduction in stimulated neutrophils from both female and male patients with chronic granulomatous disease, *FEBS Letters* **110**:111.

Segal, A. W., and Peters, T. J., 1976, Characterization of the enzyme defect in chronic granulomatous disease, *Lancet* **1**:1363.

Segal, A. W., and Peters, T. J., 1977, Analytical subcellular fractionation of human granulocytes with special reference to the localization of enzymes involved in microbicidal mechanisms, *Clin. Sci. Mol. Med.* **52**:429.

Segal, A. W., Webster, D., Jones, O. T. G. and Allison, A. C., 1978, Absence of a newly described cytochrome b from neutrophils of patients with chronic granulomatous disease, *Lancet* **12:** 446.

Selvaraj, R. J., and Sbarra, A. J., 1967, The role of the phagocyte in host–parasite interactions. VII. Di- and triphosphopyridine nucleotide kinetics during phagocytosis, *Biochim. Biophys. Acta* **141**:243.

Shinagawa, Y., Tanaka, C., Teradka, A., and Shinagawa, Y., 1966, A new cytochrome in neutrophilic granules of rabbit leucocyte, *J. Biochem. (Tokyo)* **59**:622.

Simberloff, M. S., and Elsbach, P., 1971, The interaction *in vitro* between polymorphonuclear leukocytes and mycoplasma, *J. Exp. Med.* **134**:1417.

Simchowitz, L., Metha, J., and Spilberg, I., 1979, Chemotactic factor-induced superoxide radical generation by human neutrophils. Requirement for proteinase (esterase) activity, *J. Lab. Clin. Med.* **94**:403.

Stahelin, H., Karnovsky, M. L., Farnham, A. E., and Smith, E., 1957, Studies on the interaction between phagocytes and tubercle bacilli. III. Some metabolic effect in guinea pig associated with injection with tubercle bacilli, *J. Exp. Med.* **105**:265.

Stjernholm, R. L., 1968, The metabolism of leukocytes, *Congr. Int. Soc. Hematol. 12, New York Plenary Session,* **5**:300.

Stjernholm, R. L., and Manak, R. C., 1970, Carbohydrate metabolism in leukocytes. XIV. Regulation of pentose cycle activity and glycogen metabolism during phagocytosis, *J. Reticuloendothel. Soc.* **8**:550.

Stjernholm, R. L., Burns, C., and Hohnadel., J. H., 1972, Carbohydrate metabolism of leukocytes, *Enzyme* **13**:7.

Strauss, B. S., and Stetson, C. A., 1960, Studies on the effect of certain macromolecular substances on the respiratory activity of the leukocytes of peripheral blood, *J. Exp. Med.* **112**:653.

Takanaka, K., and O'Brien, P. J., 1975, Mechanism of H_2O_2 formation by leukocytes. Evidence for a plasma membrane location, *Arch. Biochem.* **169**:428.

Takeshige, K., Matsumoto, T., Shibata, R., and Minakami, S., 1979, Simple and rapid method for the diagnosis of chronic granulomatous disease, measuring hydrogen peroxide and superoxide anions released from leukocytes in whole blood, *Clin. Chim. Acta* **92**:329.

Tauber, A. I., and Babior, B. M., 1977, Evidence for hydroxyl radical production by human neutrophils, *J. Clin. Invest.* **60**:374.

Tauber, A. I. and Goetzl, E. J., 1978, Subcellular localization and solubilization of the superoxide-generating activities (SGA) of human neutrophils *Blood* **52** (*Suppl.* 1):128.

Tauber, A. I., and Goetzl, E. J., 1979, Structural and catalytic properties of the solubilized superoxide-generating activity of human polymorphonuclear leukocytes. Solubilization, stabilization in solution, and partial characterization, *Biochemistry* **18**:5576.

Taylor, R. B., Duffus, P. H., Raff, M. C., and de Petris, S., 1971, Redistribution and pinocytosis of lymphocyte surface immunoglobulin molecules induced by anti-immunoglobulin antibody, *Nature New Biol.* **233**:225.

Tedesco, F., Trani, S., Soranzo, M. R., and Patriarca, P., 1975, Stimulation of glucose oxidation in human polymorphonuclear leucocytes by C3-sepharose and soluble C567, *FEBS Letters* **51**:232.

Tsan, M. F., 1977, Stimulation of the hexose monophosphate shunt independent of hydrogen peroxyde and superoxide production in rabbit alveolar macrophages during phagocytosis, *Blood* **50**:935.

Tsan, M. F., and Berlin, R. D., 1971, Effect of phagocytosis on membrane transport of nonelectrolytes, *J. Exp. Med.* **134**:1016.

Unanue, E. R., Karnovsky, M. L., and Engers, H. D., 1973, Ligand-induced movement of lymphocyte membrane macromolecules. III. Relationship between the formation and fate of anti-Ig-surface complexes and cell metabolism, *J. Exp. Med.* **137**:675.

Van Oss, C. J., Gillman, C. F., and Newmann, A. W., 1975, *Phagocytic Engulfment and Cell Adhesiveness as Cellular Surface Phenomena*, Marcel Dekker, New York.

Vogt, M. T., Thomas, C., Vassallo, C. L., Basford, R. E., and Gee, J. B. L., 1971, Glutathione-dependent peroxidative metabolism in the alveolar macrophages, *J. Clin. Invest.* **50**:401.

Weening, R. S., Wever, R., and Roos, D., 1975, Quantitative aspects of the production of superoxide radicals by phagocytizing human granulocytes, *J. Lab. Clin. Med.* **85**:245.

Weening, R. S., Roos, D., Van Shaik, M. L. J., Voetman, A. A., de Boer, M., and Loos, H. A., 1977, The role of glutathione in oxidative metabolism of phagocytic leukocytes. Studies in a family with glutathione reductase deficiency, in: *Movement, Metabolism and Bactericidal Mechanisms of Phagocytes* (F. Rossi, P. Patriarca, and D. Romeo, eds.), pp. 277–283, Piccin Medical Books, Padua.

Weinstein, J., and Bielski, B. H. J., 1979, Kinetics of the interaction of HO_2 and O_2^- radical with hydrogen peroxide. The Haber–Weiss reaction, *J. Am. Clin. Soc.* **101**:58.

Weiss, S. J., King, G. W., and Lo Buglio, A. F., 1977, Superoxide generation by human monocytes and macrophages, *J. Clin. Invest.* **60**:370.

Weissmann, G., Korchak, H. M., Perez, H. D., Smolen, J. E., Goldstein, I. M., and Hoffstein, S. T., 1979, The secretory code of neutrophil, *J. Reticuloendothel. Soc.* **26**:687.

West, J., Morton, D. J., Esmann, V., and Stjernholm, L. R., 1968, Carbohydrate metabolism in leukocytes. VIII. Metabolic activities of the macrophages, *Arch. Biochem. Biophys.* **124**:85.

Zabucchi, G., Soranzo, M. R., Berton, G., Romeo, D. and Rossi, F., 1978, The stimulation of the oxidative metabolism of polymorphonuclear leukocytes: effect of colchicine and cytochalasin B, *J. Reticuloendothel. Soc.* **24**:451.

Zatti, M., and Rossi, F., 1965, Early changes of hexose monophosphate pathway activity and of NADPH oxidation in phagocytizing leucocyte, *Biochim. Biophys. Acta* **99**:557.

Zatti, M., and Rossi, F., 1967, Relationship between glycolysis and respiration in surfactant-treated leucocytes, *Biochim. Biophys. Acta* **48**:553.

Zatti, M., Rossi, F., and Patriarca, P., 1968, The H_2O_2 production by polymorphonuclear leucocytes during phagocytosis, *Experientia* **24**:669.

Zurier, R. B., Hoffstein, S., and Weissmann, G., 1973, Mechanisms of lysosomal enzyme release from human leukocytes. I. Effect of cyclic nucleotides and colchicine, *J. Cell. Biol.* **58**:27.

8

The Oxidative Metabolism of Monocytes

D. ROOS and A. J. M. BALM

1. INTRODUCTION

The mononuclear phagocytes comprise a widely distributed and morphologically heterogeneous cell system. Originating from bone marrow stem cells, through the stages of monoblasts and promonocytes, these cells mature into monocytes which are released into the blood. After a few days in the circulation, the monocytes migrate into tissues to form the macrophages of the lungs, liver, spleen, lymphoid organs, pleural and peritoneal cavities, bones, etc. Here, a further differentiation takes place according to their specific functions in the various tissues. In general, macrophages are large cells with a high phagocytic and protein-synthesizing capacity but with a low self-replicating ability.

The monocytes of the blood are thus considered to be an intermediate stage in the development of the tissue macrophages. The morphologic features of the monocyte are characterized by few glycogen granules, moderate amounts of rough-surfaced endoplasmic reticulum (RER), some slim mitochondria and a well-developed Golgi complex. An example of a resting monocyte is shown in Figure 1. In comparison to mature macrophages, circulating monocytes contain less mitochondria and less lysosomes. Macrophages also contain more lipid droplets and phagocytic vacuoles with an array of ingested, not (yet) digested material.

Monocytes take part in digestive as well as in host-defense functions. For the latter function, monocytes cooperate with lymphocytes, both in an afferent role (antigen-processing) and in an efferent role (activation by lymphokines, uptake of opsonized material). Monocytes are capable of chemotaxis,

D. ROOS and A. J. M. BALM • Department of Blood Cell Chemistry, Central Laboratory of the Netherlands Red Cross Blood Transfusion Service, Amsterdam; and Laboratory of Clinical and Experimental Immunology, University of Amsterdam, 1006 AD Amsterdam, The Netherlands.

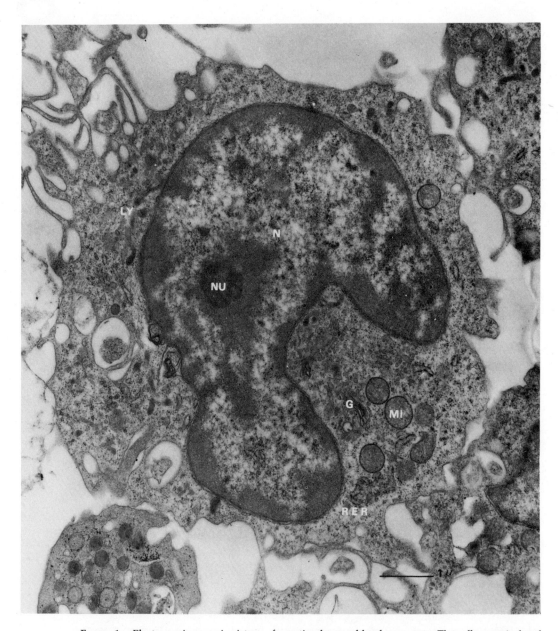

FIGURE 1.   Electron microscopic picture of a resting human blood monocyte. The cells were isolated by buoyant density centrifugation over Ficoll–Isopaque, followed by glass adherence, washing, and scraping with silicone rubber, as described by Reiss and Roos (1978). Magnification: × 15000. Cell preparation: Mr. M. de Boer; electron microscopic photography: Mr. J. Agterberg. N, nucleus; NU, nucleolus; MI, mitochondria; LY, lysosomes; G, Golgi apparatus; RER, rough endoplasmic reticulum.

phagocytosis, and killing of microorganisms, functions shared with polymorphonuclear leukocytes. In general, the neutrophils, in a suicidal way, have a higher killing capacity than monocytes, whereas the latter cells, perhaps as a prelude to their macrophage stage, have also digestive capacity.

This review will focus on the microbicidal properties of the circulating monocyte in comparison to that of the neutrophil and of some macrophages.* Attention will also be paid to the carbohydrate metabolism, in an attempt to define the major pathway of energy conservation in these cells. Neutrophil metabolism is reviewed in Chapter 7 of this volume.

2. METHODS OF ISOLATION AND RECOGNITION

In various species monocytes comprise only 3–7% of the blood leukocytes. For this reason, monocyte research has been confined to large animals and humans. Investigation of monocyte properties *in vitro* is hampered by the difficulty of obtaining relatively pure cell suspensions in sufficient amounts. Therefore, it is relevant to reflect in this review on methods to isolate monocytes and to discriminate these cells from other blood cells. Two methods, sometimes in combination, are usually employed for purification: density centrifugation and surface adherence.

Human monocytes have a density distribution distinctly different from erythrocytes and neutrophils, but partly overlapping with that of lymphocytes (Loos *et al.*, 1976a). Therefore, buoyant-density centrifugation over a layer of albumin (Bennet and Cohn, 1966; Johnson *et al.*, 1977), Ficoll-Hypaque (Johnson *et al.*, 1977) or Ficoll-Isopaque (Böyum, 1968; Loos *et al.*, 1976a) yields mixtures of monocytes and lymphocytes, slightly contaminated with erythrocytes. The red cells can be eliminated by hypotonic (Romeo *et al.*, 1973a) or NH_4Cl-induced (Roos and Loos, 1970) lysis, but lymphocytes have to be removed by other means. This has been performed by three different techniques.

First, using the fact that the density profiles of (human) monocytes and lymphocytes are not *totally* overlapping (Loos *et al.*, 1976a), a second isopycnic centrifugation has been applied with gradient material of lower density than that used in the first step. Albumin (Weston *et al.*, 1975), Ficoll-Isopaque (Loos *et al.*, 1976a) and colloidal silica–polyvinylpyrollidinone (Nathanson *et al.*, 1977) have been used for this purpose. Due to the density overlap, however, yield and purity of the monocyte preparations cannot both be high. Sucrose gradients may be used for enzyme studies, but not for metabolic studies (Zeya *et al.*, 1978).

Second, use has been made of the strong adherence of the monocytes to glass and plastic surfaces. Thus, most of the lymphocytes can be washed away after 90–120 min incubation in a dish at 37°C (Bennet and Cohn, 1966; Cline and Lehrer, 1968; Johnson *et al.*, 1977; Reiss and Roos, 1978). The monocyte monolayers thus obtained can either be used as such, or the cells may be scraped from the dish with a piece of silicone rubber (Brodersen and Burns, 1973).

A third method has recently been described which uses differences in

---

*The literature survey for this chapter was concluded in April, 1978.

shape, size, and density by applying counterflow centrifugation (Sanderson *et al.*, 1977). This technique involves balancing the centrifugal force on the cells with hydrodynamic and buoyancy forces by pumping medium through the rotor during centrifugation. Perhaps this latter method is the most promising for future research: it combines reasonable yields with excellent purity. Moreover, it has none of the disadvantages of the other methods, such as: uptake of gradient material as noted with Ficoll–Isopaque (Splinter *et al.*, 1978), aggregation (Shortman, 1968) and cell-volume changes (Williams and Shortman, 1972) in duced by albumin, adherence of certain lymphocyte subpopulations (Greaves and Brown, 1974), and monocyte activation or damage which may result from adherence and scraping. The counterflow-centrifugation method still has to prove its value, however, for the preparation of monocytes to be used for metabolic studies.

Monocyte recognition is usually based on morphologic criteria such as shape of the nucleus, ability to ingest particles, and/or presence of certain enzymes (lysozyme, peroxidase, nonspecific esterase). Such methods suffer from the unreliability of microscopic recognition, which is due to the statistical error of the limited number of cells that can be visually counted (Rümke *et al.*, 1975) and the variation introduced by staining procedures. Much of this problem has been solved by the introduction of mechanized equipment for cell counting and light-scatter measurements. Thus, it is now possible to estimate the percentage of monocytes by light scattering after esterase staining, even in whole blood (Kaplow *et al.*, 1976).

Another method for monocyte recognition is based on the volume distribution of these cells (Loos *et al.*, 1976b; Kwan *et al.*, 1976; Sanderson *et al.*, 1977). It has been shown that this parameter correlates closely with the prementioned morphologic characteristics (Loos *et al.*, 1976b) and it offers the advantage of measuring up to 100,000 cells in a few seconds. The disadvantage is that red cells (if present in large excess) and granulocytes interfere with the procedure and limit its use to purified mononuclear leukocyte suspensions. Nevertheless, this technique has proven to be of great help in characterizing monocyte (and lymphocyte) suspensions for functional and metabolic studies.

3. GENERATION OF OXIDATIVE MICROBICIDAL PRODUCTS

3.1. MICROBICIDAL FUNCTION

Monocytes, like neutrophils, are capable of moving from the blood to the site of an infection in the tissues, and to ingest and kill invading microorganisms. Neutrophils migrate faster than monocytes: the latter cells lag about 6 hr behind the neutrophils, thereafter replacing these cells at the site of infection (Ward, 1970). Macrophages, many of which are fixed in the tissues, share the property of phagocytosis and killing of microorganisms with the other two phagocytes. Neutrophils appear to be injured more easily during this process than monocytes, resulting in pus formation and increased destruction by the tissue macrophages of the spleen. Also, neutrophils hardly digest the ingested material but merely transport it to the tissue macrophages. Macrophages are equipped with a more powerful lysosomal apparatus which can be refilled with newly synthesized enzymes if needed (Axline and Cohn, 1970). Indigestible material may cause chronic infections by constantly activating the macrophages to secrete inflammatory agents (Ginsburg and Sela, 1976).

In vitro, all three types of phagocytes ingest and kill a great variety of microorganisms, among them numerous strains of bacteria and yeasts (Cline and Lehrer, 1968; Klebanoff and Hamon, 1975; Biggar *et al.,* 1976). In general, neutrophils kill more and faster than monocytes and macrophages, especially at higher ratios of microorganisms to phagocytes (Davis *et al.,* 1968; Rodey *et al.,* 1969; Cline, 1970a; Root *et al.,* 1972; Simmons and Karnovsky, 1973; Steigbigel *et al.,* 1974; Lehrer, 1975). With macrophages, however, one must carefully distinguish between cells directly rinsed from an organ (resident macrophages), cells obtained 3–6 days after injection of an eliciting substance, e.g., casein (elicited macrophages), and cells from antigen-challenged animals (activated macrophages). It has been shown that elicited, and, especially, activated macrophages kill and ingest microorganisms much better than do resident macrophages (Mackaness, 1962; Blanden *et al.,* 1966; Ratzan *et al.,* 1972; Simmons and Karnovsky, 1973; Stubbs *et al.,* 1973; Karnovsky *et al.,* 1975).*

The less avid killing of monocytes as compared to neutrophils and macrophages may be—partly—due to a slower uptake of microorganisms in monocytes (Steigbigel *et al.,* 1974; Reiss and Roos, 1978), although some investigators found neutrophils and monocytes equally effective in this respect (Root *et al.,* 1972; Diamond *et al.,* 1972). In patients with severe and benign congenital neutropenia, the monocytes ingest and kill *Staphylococcus aureus* as rapidly as do normal neutrophils (Bigger *et al.,* 1974; Kay *et al.,* 1975; Greenwood *et al.,* 1978), thus offering a compensatory mechanism for the lack of neutrophils. During monocytosis, however, a decreased monocyte function has been found (Baehner and Johnston, 1972).

3.2. MICROBICIDAL MECHANISM

The most important killing mechanism in neutrophils is the formation of toxic oxygen products which, by themselves and/or in combination with myeloperoxidase, have strong microbicidal properties (for a review see Klebanoff and Hamon, 1975). When neutrophils are incubated with ingestible material, such as immune complexes, microorganisms, or other particles, the oxygen uptake of the neutrophils increases ten- to 15-fold within a few seconds. Several products of this increased respiration have been identified. Hydrogen peroxide is produced in large amounts and can be measured in the

---

*In this review, the term *monocyte* will only be used for the circulating monocyte, and will not include the cells obtained from the peritoneal cavity after short (1–4 days) exposure to an eliciting agent. The latter cells will be considered as elicited macrophages.

medium around the neutrophils (Iyer *et al.*, 1961; Paul and Sbarra, 1968; Root *et al.*, 1975), especially when its enzymatic decomposition is inhibited (Homan-Müller *et al.*, 1975). A few years ago, it was discovered that neutrophils also generate superoxide anion radicals O_2^-, highly reactive products from the one electron reduction of oxygen (Babior *et al.*, 1973). Superoxide probably acts as as intermediate in the formation of H_2O_2 (Weening *et al.*, 1975; Root and Metcalf, 1977).

Indications exist that two more reactive oxygen species may be formed by phagocytizing cells: singlet oxygen and hydroxyl radicals (see also Chapters 12 and 13 of this volume by R. C. Allen and B. M. Babior, respectively). Singlet oxygen molecules ($^1O_2$) are in an energized state; their return to the stable ground state is accompanied by a release of energy that can be measured as chemiluminescence (Kearns, 1971). The occurrence of this reaction in phagocytizing neutrophils has been interpreted as proof for the formation of singlet oxygen by these cells (Allen *et al.*, 1972). Singlet oxygen might be produced by neutrophils in the reaction between hydrogen peroxide and hypochlorite anions (formed in the myeloperoxidase reaction):

$$H_2O_2 + OCl^- \rightarrow H_2O + {}^1O_2 + Cl^-$$

Singlet oxygen may also be derived from the reaction between hydrogen peroxide and superoxide (the Haber–Weiss reaction) (Rosen and Klebanoff, 1976):

$$H_2O_2 + O_2^- \rightarrow {}^1O_2 + \cdot OH + OH^-$$

The emission of light probably occurs via formation of carbonyl radicals (Hodgson and Fridovich, 1976). Although the process of chemiluminescence in phagocytizing neutrophils has been shown to depend to a large extent on the oxidation of particle constituents (Cheson *et al.*, 1976), the same phenomenon has been observed after stimulation of the cells with soluble activators (Hatch *et al.*, 1978). Whatever the correct interpretation may be, we have included the results obtained with monocytes for reasons of comparison.

Hydroxyl radicals ($\cdot OH$) are also formed in the Haber–Weiss reaction, but the indications for the generation in phagocytic leukocytes are even more indirect. Inhibition of bacterial killing by scavengers of superoxide as well as by scavengers of hydrogen peroxide might mean that O_2^- and H_2O_2 react with each other to form the bactericidal hydroxyl radical (Johnston *et al.*, 1975). Moreover, neutrophils also liberate ethylene from methional (Tauber and Babior, 1977; Weiss *et al.*, 1978), a reaction supposed to be mediated by hydroxyl radicals (Beauchamp and Fridovich, 1970). The participation of these reactants in the microbicidal activity of the neutrophils has not been clearly established and is currently debatable.

None of these reactions of the "respiratory burst," initiated by phagocytosis, is present in neutrophils from patients with chronic granulomatous disease (CGD), a syndrome characterized by frequent recurrent infections with life-threatening severity (Johnston and Newman, 1977). Since the capacity

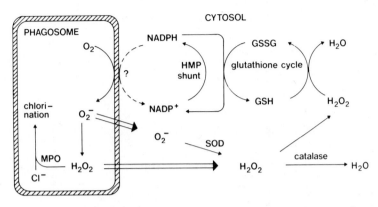

FIGURE 2. Schematic representation of the oxidative metabolism of phagocytic cells. An oxygen-consuming system, presumably located in the plasma membrane, is activated during phagocytosis and releases superoxide into the phagosome. The reducing equivalents are generated in the hexose monophosphate (HMP) shunt and delivered to the system from the cytosolic side of the membrane. The system may either work directly with NADPH as an NADPH oxidase, or indirectly by first reducing NAD as a substrate for an NADH oxidase. In the phagosome, superoxide reacts spontaneously to hydrogen peroxide, which is essential for bacterial killing. One such killing reaction is shown: the myeloperoxidase (MPO)-mediated chlorination.

Superoxide may leak into the cytosol and is converted spontaneously or in a superoxide dismutase (SOD)-catalyzed reaction to hydrogen peroxide. The cytosol is protected against hydrogen peroxide by catalase and by the glutathione redox system. In the latter process, reduced glutathione (GSH) is oxidized by hydrogen peroxide to oxidized glutathione (GSSG), which is reduced again to GSH by NADPH.

in vitro of CGD neutrophils to kill ingested microorganisms is severely depressed, this oxidative killing mechanism is regarded as extremely important for the host-defense of neutrophils.

The reducing equivalents for the formation of these products from molecular oxygen are supposed to be generated from glucose oxidation through the hexose monophosphate shunt. Indeed, the activity of this pathway, measured as formation of $^{14}CO_2$ from $[1-^{14}C]$glucose, is greatly enhanced during phagocytosis (Sbarra and Karnovsky, 1959). The NADPH produced in this reaction sequence may reduce oxygen, either directly by an NADPH oxidase or indirectly by reducing substrates, e.g., NAD, for other oxidases (Figure 2). This complicated matter has been reviewed by Cheson *et al.* (1977).

3.3. OXIDATIVE REACTIONS OF MONOCYTES

Due to the requirement of neutrophil-free suspensions of monocytes for the study of monocyte oxidative metabolism, research in this area has lagged far behind neutrophil studies. A number of recent articles has drastically changed this picture. A summary of the values found in resting and phagocytizing monocytes is given in Table 1.

TABLE 1.   CHANGES IN HUMAN MONOCYTE METABOLISM INDUCED BY PHAGOCYTOSIS[a,b]

Conditions		Metabolism (nmol/10⁷ cells/hr)													Stimulated rate/resting rate	References
Particle	Opsonin	Oxygen uptake		H₂O₂ prod.		O₂⁻ prod.		·OH prod.		Glucose cons.		Lactate prod.		[1-¹⁴C]Gluc → ¹⁴CO₂	Chemiluminescence	
		R	Ø[c]	R	Ø	R	Ø	R	Ø	R	Ø	R	Ø			
Latex	Absent									1330	1170			2.4		Cline and Lehrer (1968)
Latex	Absent	260	1570	1.3[d]	7.2											Baehner and Johnston (1972)
Latex	Absent	317	1230											5.6		Biggar et al. (1974)
Latex	Present													5.9–11.3		Root et al. (1972)
Zymosan	Present					120	350									Weiss et al. (1975)
S. aureus	Present														3	Nelson et al. (1976)
C. albicans	Present														10	Nelson et al. (1976)
Zymosan	Present														10	Nelson et al. (1976)
Zymosan	Present			[d]	0.4									13.6	8.5	Sagone et al. (1976)
Zymosan	Present							0.04	0.1						35	Johnston et al. (1976)
Zymosan	Present					175	1200									Weiss et al. (1977)
Zymosan	Present	60	960	0[e]	480	0	480									Reiss and Roos (1978)
Zymosan	Present											1640	1920	9–25		Roos et al. (1979)

[a] In analogy of Table 1-3, Changes in neutrophil metabolism induced by phagocytosis (Cheson et al., 1977).
[b] For purposes of comparison, data from the references cited were normalized to the units shown here, assuming 10⁷ monocytes to be equal to 2 mg protein (Oren et al., 1963).
[c] R, resting; Ø, phagocytosing.
[d] Formate method.
[e] LDADCF method.

3.3.1. Oxygen Consumption

The stimulated oxygen consumption of purified monocytes from normal donors is lower than that of neutrophils from the same donors, as shown in Figure 3. Monocytes from neutropenic patients consume as much oxygen (Biggar *et al.*, 1974), or even more (Baehner and Johnston, 1972), as do normal neutrophils. A clear increase in monocyte respiration during phagocytosis is found in all studies.

The respiration of resident alveolar macrophages also increases during particle ingestion (Vogt *et al.*, 1971; Romeo *et al.*, 1973b; Gee *et al.*, 1974; Rossi *et al.*, 1975; Rister and Baehner, 1977). With peritoneal macrophages, however, only elicited and activated cells are reported to react with increased oxygen uptake upon phagocytosis (West *et al.*, 1968; Karnovsky *et al.*, 1970).

3.3.2. Superoxide Generation

Monocytes release less superoxide than neutrophils (Johnston *et al.*, 1976; Reiss and Roos, 1978), if the reaction is measured by cytochrome c reduction (Figure 3). Using a less specific reaction, reduction of nitroblue tetrazolium, Weiss *et al.* (1975) found about equal reaction rates in monocytes and neutrophils. Like neutrophils, monocytes also show a positive reaction in the histochemical nitroblue tetrazolium reduction test (Nathan *et al.*, 1969; Douglas, 1970; Park *et al.*, 1972; Cruchaud *et al.*, 1977).

In macrophages, several investigators failed to detect increased superoxide generation during phagocytosis (DeChatelet *et al.*, 1975; Biggar *et al.*, 1976; Gee and Khandwala, 1976; Lowrie and Aber, 1977; Tsan, 1977). Drath and Karnovsky (1975), however, showed that the production of superoxide by macrophages is highly dependent on the origin of the cells (animal species, organ) and the state of cell activation. Moreover, nothing is known about the accessibility of cytochrome c to the site of superoxide production in the various cells.

Superoxide dismutase, the enzyme that protects the cytosol of phagocytes by catalyzing the conversion of superoxide into hydrogen peroxide, is found in neutrophils (DeChatelet *et al.*, 1974; Salin and McCord, 1974; Patriarca *et al.*, 1974a; Johnston *et al.*, 1975) as well as in monocytes (Gee and Khandwala, 1976) and macrophages (DeChatelet *et al.*, 1974; Rister and Baehner, 1977; van Berkel *et al.*, 1977).

3.3.3. Hydrogen Peroxide Production

The measurement of hydrogen peroxide formation by monocytes has met with difficulties. Generation of $^{14}CO_2$ from [^{14}C]formate has generally been employed for the detection of hydrogen peroxide. This method, however, depends on the presence of catalase (Klebanoff and Hamon, 1972), an enzyme more abundantly present in neutrophils than in monocytes (Roos *et al.*, unpublished). Moreover, only a fraction of the consumed oxygen during phagocytosis can be accounted for as hydrogen peroxide when measured by this technique

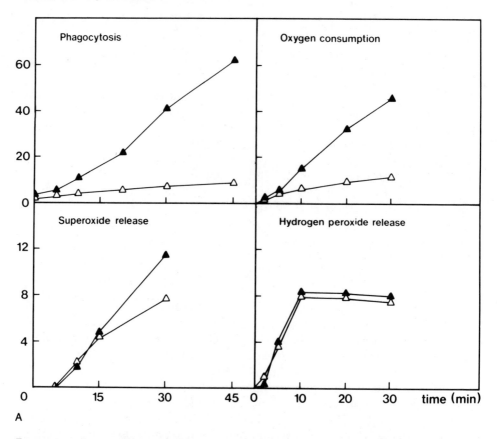

FIGURE 3. Effect of antimycin A on human blood monocyte and neutrophil functions. (A) monocytes; (B) neutrophils. Closed symbols: with DMSO; open symbols: with antimycin in DMSO. Phagocytosis: adherence + uptake of ^{125}I-labeled serum-opsonized zymosan in % of added radioactivity. Oxygen consumption, superoxide release, and hydrogen peroxide release in nmol/10^6 cells during phagocytosis of serum-opsonized zymosan. The data are taken from Reiss and Roos (1978). It

(Klebanoff and Hamon, 1975; Homan-Müller et al., 1975). Stimulation of formate oxidation during particle ingestion is found in neutrophils as well as in monocytes (Baeher and Johnston, 1972; Sagone et al., 1976) and various macrophages (Karnovsky et al., 1970; Gee et al., 1970, 1974; Romeo et al., 1973b; Klebanoff and Hamon, 1975), but quantitative comparison is not possible with this technique. With more specific techniques, H_2O_2 formation by phagocytizing macrophages has been confirmed (Gee et al., 1970; Paul et al., 1970a; Vogt et al., 1971; Romeo et al., 1973b; Rister and Baehner, 1977).

In recent years more sensitive methods have been developed (Homan-Müller et al., 1975; Root et al., 1975). With one of these methods, monocytes were found to release during phagocytosis about one-fifth of the amount of hydrogen peroxide generated by neutrophils from the same donor (see Figure 3). With another of these methods, elicited and activated, but not resident mouse peritoneal macrophages were also shown to generate hydrogen

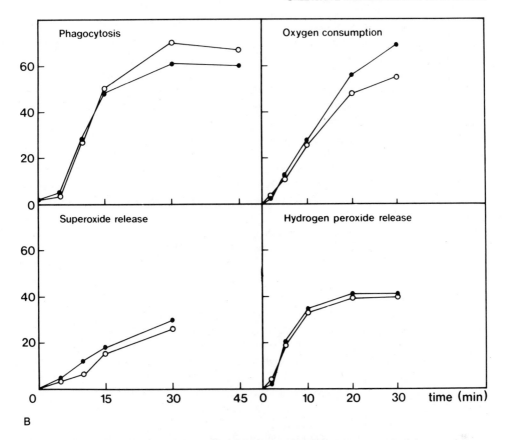

B

can be concluded that (a) monocytes ingest particles more slowly and generate less oxygen products during phagocytosis than neutrophils do; (b) monocytes depend more on mitochondrial respiration for the uptake of particles than neutrophils; (c) monocytes, more than neutrophils, increase mitochondrial respiration during phgagocytosis; and (d) the release of superoxide and hydrogen peroxide from either cell type is unaffected by the presence of antimycin A.

peroxide during particle ingestion (Nathan and Root, 1977). Resident alveolar macrophages from guinea pigs produce about one-fourth the amount of hydrogen peroxide found with neutrophils from the same animals (Rister and Baehner, 1977). Similar studies in the rat failed to detect stimulation of hydrogen peroxide generation by alveolar macrophages during phagocytosis (Biggar and Sturgess, 1978).

It must be emphasized, however, that both methods to determine hydrogen peroxide measure *release* rather than *production* by the cells. Not only will this always result in an underestimation of the true rate of H_2O_2 generation, but—as in the superoxide assay—comparison between different cell types remains hazardous, since it is not known whether the substrate for the detection assay and endogenous substrates compete with each other for the reaction with H_2O_2 with equal efficiency in different cells. The latter problem may also hold the solution for the unexpected report that rabbit alveolar macrophages show increased hexose monophosphate shunt activity during phagocytosis, but not cytochrome c or nitroblue tetrazolium reduction, H_2O_2 release, or chemiluminescence (Tsan, 1977).

The question as to which part of the increment in oxygen consumption during phagocytosis is converted to hydrogen peroxide in monocytes and macrophages is an even more difficult problem than it is in neutrophils. In the latter cells, the degradation of released hydrogen peroxide by catalase can be inhibited by sodium azide. At the same time, the azide addition stimulates the apparent consumption of oxygen because catalase liberates oxygen from hydrogen peroxide. When the neutrophil respiration and hydrogen peroxide formation in the presence of azide are compared, up to 70% of the consumed oxygen is accounted for by hydrogen peroxide formation (Reiss and Roos, 1978).

In monocytes, however, sodium azide *inhibits* the oxygen uptake: several other mitochondrial respiratory inhibitors have the same effect (e.g., antimycin A; see Figure 3). Therefore, it seems very likely that the mitochondrial respiration of monocytes is increased during phagocytosis. Since this metabolic pathway leads to formation of water and not to hydrogen peroxide, it is impossible to use the total increment in oxygen consumption for the calculation of the fraction converted to hydrogen peroxide. If one compares the oxygen consumption in the presence of inhibitors of the mitochondrial respiration with the hydrogen peroxide formation under the same conditions, close to 100% is found for monocytes (no effect of these inhibitors is observed on superoxide or hydrogen peroxide generation by these cells) (Reiss and Roos, 1978). Given the fact that only *release* of hydrogen peroxide is measured and not total production, it is possible that monocytes and neutrophils both convert all oxygen not used in the mitochondria into hydrogen peroxide.

Such calculations have not been made for macrophages, but these cells too, most likely, show an increase in mitochondrial respiration during phagocytosis, because the oxygen consumption of both alveolar and peritoneal macrophages during particle uptake is partly susceptible to inhibition by potassium cyanide (Oren *et al.*, 1963; Romeo *et al.*, 1973b). In contrast, neutrophil respiration is insensitive to KCN (Sbarra and Karnovsky, 1959). The fraction of hydrogen peroxide formed to oxygen consumed during phagocytosis in alveolar macrophages and neutrophils from the same animal is one order of magnitude larger in the neutrophils (Rister and Baehner, 1977). However, as mentioned above, such comparisons are rather hazardous.

3.3.4. Chemiluminescence

Monocytes produce less chemiluminescence during phagocytosis than neutrophils (Johnston *et al.*, 1976; Nelson *et al.*, 1976; Sagone *et al.*, 1976). In both cell types, superoxide dismutase decreases the light emission to about one-fourth of the original signal. Human alveolar macrophages also produce chemiluminescence, with higher resting levels than human neutrophils and with different kinetics during phagocytosis (Beall *et al.*, 1977). The chemiluminescence induced in rabbit alveolar macrophages and elicited peritoneal macrophages by phagocytosis of bacteria is undetectable unless the reaction is amplified by luminol (Allen and Loose, 1976). The reaction is much stronger in neutrophils and shows different kinetics with different cell types and particles studied (Miles *et al.*, 1977). Superoxide dismutase inhibits to different degrees in the various cell types, indicating that perhaps the fractions of singlet oxygen produced in the Haber–Weiss reaction and in the peroxidase-mediated reaction differ in these cells.

3.3.5. Hydroxyl Radical Formation

Monocytes, like neutrophils, liberate ethylene from methional, which may indicate that both cell types produce hydroxyl radicals (Tauber *et al.*, 1977; Weiss *et al.*, 1977, 1978). The ethylene production is inhibited by superoxide dismutase as well as by catalase, thus demonstrating its dependence on superoxide as well as on hydrogen peroxide formation. Moreover, hydroxyl radical scavengers, such as benzoate and ethanol, also strongly depress this reaction. Since only a fraction of the theoretically possible amount of hydroxyl radicals is detected with this method, quantitative comparison between different cell types is impossible. No reports have appeared about the occurrence of this reaction in macrophages.

3.3.6. Hexose Monophosphate Shunt Activity

The stimulation of the hexose monophosphate shunt activity in phagocytizing monocytes has been investigated relatively poorly. Cline and Lehrer (1968) report equal activity in resting neutrophils and monocytes, but a much stronger stimulation upon phagocytosis in neutrophils. Similar values in resting cells are also found by other authors (Baehner and Johnston, 1972; Root *et al.*, 1972; Biggar *et al.*, 1974; Sagone *et al.*, 1976), but the large difference between neutrophils and monocytes during phagocytosis has not been confirmed. Monocytes from neutropenic patients display even more activity than do normal neutrophils during particle ingestion (Baehner and Johnston, 1972). The variation in these results may have been partly due to the problem of obtaining pure cell suspensions, since all contaminating blood cells also show hexose monophosphate shunt activity.

Stimulation of this pathway in peritoneal macrophages during phagocytosis is again restricted to elicited and activated cells (Karnovsky *et al.*, 1970, 1975; Stubbs *et al.*, 1973). Alveolar macrophages, on the other hand, again behave differently: resident cells show increased hexose monophosphate shunt activity during phagocytosis (Oren *et al.*, 1963; Ouchi *et al.*, 1965; Myrvik and Evans, 1967; Gee *et al.*, 1970; Vogt *et al.*, 1971; Romeo *et al.*, 1973b; Rossi *et al.*, 1975; Tsan, 1977) but activated alveolar macrophages react more strongly (Montarosso and Myrvik, 1977; Rossi *et al.*, 1975). This latter effect is in accord with the reported increase in killing capacity and lysosomal enzyme activity of activated alveolar macrophages (Blanden *et al.*, 1966; Myrvik and Evans, 1967; Rossi *et al.*,

1975), but not with the lack of stimulation of nitroblue tetrazolium dye reduction in these cells (Lowrie and Aber, 1977). Possibly, the oxidative killing reactions of alveolar macrophages are already maximally stimulated in the resident cells, whereas the lysosomal enzyme-mediated reactions may be activated by lymphocyte–macrophage interactions (see also Section 6.4).

Monocytes and macrophages are stimulated *in vitro* by lymphokines: products from antigenically activated lymphocytes. A 23,000 molecular weight fraction of supernatants from lymphocyte cultures stimulates the hexose monophosphate shunt activity of guinea pig peritoneal macrophages (Nathan *et al.*, 1971; Fowles *et al.*, 1973) and of human monocytes (Rocklin *et al.*, 1974). Cell mobility, adherence, spreading, phagocytosis, and growth inhibition and killing of *Listeria monocytogenes* are also stimulated, but protein synthesis and DNA synthesis are not. The cells need 2–3 days culturing with the factor(s) to become activated. A lymphokine fraction with a molecular weight of 25,000–75,000 has been reported to induce activation of macrophage NADPH oxidase activity in 10 min (Poulter and Turk, 1975).

The hexose monophosphate shunt is also implicated in phagocyte metabolism in a way other than for the production of reducing equivalents for hydrogen peroxide formation. Reed (1969) has shown that the hexose monophosphate shunt is stimulated by addition of hydrogen peroxide to resting neutrophils, a reaction probably mediated by the glutathione system (see also Chapter 6 of this volume). Indeed, these cells contain glutathione as well as the enzymes glutathione peroxidase and glutathione reductase (Reed, 1969; Strauss *et al.*, 1969; Holmes *et al.*, 1970; Baehner *et al.*, 1970). As shown in Figure 2, the first enzyme catalyzes the reaction between reduced glutathione and hydrogen peroxide, whereas the second enzyme is needed for the reduction of oxidized glutathione with NADPH. As a result of the latter reaction $NADP^+$ is formed, which acts as a stimulant of the hexose monophosphate shunt. Thus, the hexose monophosphate shunt is activated during phagocytosis for two seemingly contradictory reasons: hydrogen peroxide *formation* (in the phagosome) and excess hydrogen peroxide *detoxification* (in the cytosol).

The glutathione system, together with superoxide dismutase and catalase, gives protection to the neutrophil cytosol against oxidation of vital cell components. The glutathione system is also found in the human monocyte (Roos *et al.*, unpublished) and in the alveolar macrophage (Vogt *et al.*, 1971). The importance of this system is illustrated by the damage inflicted on glutathione-reductase-deficient neutrophils and monocytes during phagocytosis (Weening *et al.*, 1977).

3.3.7. Triggering and Localization of the Superoxide-Generating System

A highly interesting aspect of phagocyte metabolism is the "triggering" of the metabolic burst at the start of the phagocytic process. Three lines of evidence exist which indicate that phagocytosis itself is not needed for metabolic stimulation.

1. Several agents are known to inhibit neutrophil phagocytosis without interfering with the stimulation of the oxidative metabolism, among them colchicine (Malawista and Bodel, 1967; DeChatelet *et al.*, 1971; Lehrer, 1973), chloramphenicol (Kaplan *et al.*, 1969; Lehrer, 1973) and cytochalasin B (Goldstein *et al.*, 1975). For monocytes, only one similar indication is known: in the presence of antimycin A, phagocytosis is severely depressed without any effect on the release of superoxide or hydrogen peroxide (Figure 3).

2. The respiratory burst of neutrophils (Goldstein *et al.*, 1976) and monocytes (Johnston *et al.*, 1976) is initiated by the adherence of these cells to nonphagocytosable surfaces.

3. A whole series of soluble agents are known to stimulate the oxidative metabolism of neutrophils. Among these are phorbol myristate acetate (Repine *et al.*, 1974a,b, 1975; Goldstein *et al.*, 1975; DeChatelet *et al.*, 1976), concanavalin A (Romeo *et al.*, 1973c), phospholipase C (Patriarca *et al.*, 1971, 1974a), N-formylmethionyl peptides (Allred and Hill, 1977; Hatch *et al.*, 1978), endotoxin, digitonin, and deoxycholate (Graham *et al.*, 1967). The same agents have been described to stimulate one or more parameters of the oxidative metabolism in monocytes (Johnston *et al.*, 1976) and in macrophages (Graham *et al.*, 1967; Romeo *et al.*, 1973c; Sachs and Gee, 1973; Rossi *et al.*, 1975; Drath and Karnovsky, 1975; Nathan and Root, 1977; Beall *et al.*, 1977; Hatch *et al.*, 1978).

From these and other considerations, many investigators have formed the opinion that the oxygen-consuming, superoxide-generating system in neutrophils must be located in the plasma membrane, inactive in resting cells and activated by particle adherence or membrane perturbation (Stubbs *et al.*, 1973; Salin and McCord, 1974; Takanaka and O'Brien, 1975; Baehner, 1975; Briggs *et al.*, 1975; Johnston *et al.*, 1975; van Berkel and Kruyt, 1977a; Goldstein *et al.*, 1975, 1977; Roos *et al.*, 1977). Whether this system uses NADH, NADPH, or yet another substrate is at present undecided (see Cheson *et al.*, 1977).

The only observation pertinent to the present review is the observation by Romeo *et al.* (1971, 1973b) that alveolar and peritoneal macrophages contain an NADPH oxidase that is stimulated two- to three-fold during phagocytosis but differs from the enzyme found in neutrophils. The macrophage oxidase shows an increased maximal velocity in the phagocytosis-activated state, whereas the neutrophil enzyme also shows an increase in substrate affinity (Patriarca *et al.*, 1971). Moreover, the NADPH oxidase in macrophages is inhibited by 50–80% by KCN, whereas the neutrophil system is unaffected by this agent. The authors took this latter observation to be in accord with a similar inhibition of the macrophage respiration by KCN, but, as mentioned above, this may also be explained by inhibition of mitochondrial respiration. In addition, Romeo and colleagues measured the NADPH oxidase activity in the presence of Mn^{2+} ions, a set-up which favors nonenzymatic, superoxide-mediated free radical chain reactions (Curnutte *et al.*, 1976).

It is at present unknown, therefore, what the significance of the macrophage NADPH oxidase may be in the respiratory burst of these cells during phagocytosis. Similar assays have not been performed in monocytes.

3.4. SIGNIFICANCE OF MICROBICIDAL ACTIVITY

As mentioned in Section 3.2, the importance of the respiratory burst during phagocytosis in the killing of microorganisms by neutrophils is derived from observations on cells from patients with chronic granulomatous disease (CGD). Similar observations on CGD monocytes indicate that in these cells too, the killing of several microorganisms (Davis *et al.*, 1968; Rodey *et al.*, 1969; Lehrer, 1970, 1975) as well as the generation of oxygen products, is severely depressed: NBT reduction (Nathan *et al.*, 1969; Weiss *et al.*, 1975), hydroxyl radical formation (Weiss *et al.*, 1977), hydrogen peroxide release (Fleer *et al.*, 1979), lipid peroxidation (Stossel *et al.*, 1974) and iodination (Lehrer, 1975) are all practically absent in CGD monocytes. In cultured monocytes from CGD patients, only a retardation in NBT reduction is found (Vildé and Vildé, 1976), which may explain the normal killing of *S. aureus* by these cells (Mandell and Hook, 1969). Whether non-oxygen-linked microbicidal mechanisms are also present in cultured CGD monocytes remains to be investigated.

A second indication for the significance of the oxygen products in the microbicidal activity is obtained from experiments performed at low oxygen tensions. Under such conditions bacterial killing is moderately to severely depressed in mouse peritoneal macrophages (Miller, 1971). The problem with such experiments is to separate the effect of anaerobiosis on the uptake of bacteria from the effect on the killing. Miller (1971) solved this problem by examining the anaerobic bacterial killing *in vitro* after bacterial uptake *in vivo*. Other investigators found no effect of low oxygen tensions on phagocytosis by mouse peritoneal macrophages (Thalinger and Mandell, 1971) or cultured human monocytes (Cline, 1970a,b), and performed the whole assay under nitrogen. With this system, no effect of anaerobiosis (Thalinger and Mandell, 1971) or a moderate depression of bacterial killing (Cline, 1970a,b) was found. Since the measurements of phagocytosis do not distinguish between adherence and uptake of bacteria, and since complete absence of oxygen is extremely difficult to obtain, these results must be interpreted with caution. Similar experiments have also been performed with a large number of inhibitors of oxidative metabolism. The results obtained with such systems will not be discussed in this review since dose–response studies were hardly performed, and the specificity of the inhibitors for the oxygen product formation is seldom known.

One more indication for the involvement of oxygen products in bacterial killing is obtained from the effect of oxygen product scavengers on the killing process. Johnston *et al.* (1975) have shown that both superoxide dismutase and catalase inhibit the killing by neutrophils of various strains of bacteria. A similar inhibition was reported with superoxide dismutase in the killing of *S. aureus* by monocytes (Sagone *et al.*, 1976).

Thus, for neutrophils and monocytes, it is clear that the oxygen-dependent killing mechanism is very important in the bactericidal activity of these cells. As will be discussed below, peroxidase plays a major role in this mechanism. For macrophages, the picture is less clear: in these cells, the oxygen-dependent reactions are apparently of less importance (see Section 4). Indications for non-oxygen-dependent and oxygen-dependent, non-peroxidase-dependent mechanisms in the three cell types will be discussed in Section 5.

4. PEROXIDASE-MEDIATED MICROBICIDAL SYSTEMS*

4.1. FUNCTION

The involvement of myeloperoxidase in the killing of microorganisms by neutrophils has first been studied by Klebanoff (1967a, 1968, 1970). Myeloperoxidase, together with H_2O_2 and a halide, is toxic for bacteria (Klebanoff 1967a,b, 1968; McRipley and Sbarra, 1967; Klebanoff and Hamon, 1972; Simmons and Karnovsky, 1973), yeasts (Lehrer, 1969, 1972, 1975; Klebanoff, 1970), and viruses (Belding *et al.*, 1970; Klebanoff and Hamon, 1975). Iodide (Klebanoff, 1967a) as well as chloride and bromide (Klebanoff, 1967b, 1968) are effective electron donors for bactericidal activity. With iodide as substrate, iodine incorporation into bacterial cell wall protein takes place, both in a cell-free peroxidase–H_2O_2–halide system (Klebanoff, 1967b) and in neutrophils (Klebanoff, 1967b; Pincus and Klebanoff, 1971; Root and Stossel, 1974). With chloride as substrate, oxidative decarboxylation of amino acids occurs (Zgliczynski *et al.*, 1968; 1971; Jacobs *et al.*, 1970; Strauss *et al.*, 1970, 1971; Paul *et al.*, 1970b; Stelmaszynska and Zgliczynski, 1974). Since chloride is much more available to the phagocytes than iodide, probably only the latter reaction is physiologically significant. Either the chloramines and aldehydes formed in the latter reaction, or the decarboxylation of amino acids from the bacterial cell wall may cause the death of the ingested microorganisms. Like bacterial killing by neutrophils, this reaction has an acid pH optimum and is inhibited by azide and cyanide. Moreover, chloride, hydrogen peroxide, and peroxidase are present in high concentrations in the phagocytizing neutrophil. Most probably, therefore, this enzyme is important for bacterial killing by neutrophils. In neutrophils, the system is considerably potentiated by elastase (Odeberg and Olsson, 1976a). The question is whether the same system is operative in monocytes and macrophages.

4.2. LOCALIZATION AND DISTRIBUTION

The histochemical localization of peroxidase in neutrophils and monocytes is very similar: in these cells the peroxidase activity is restricted to the azurophil lysosomes. No activity is found in the endoplasmic reticulum, nuclear envelope, or Golgi apparatus (Breton-Gorius and Guichard, 1969; Daems and Brederoo, 1971, 1973; Nichols and Bainton, 1973, 1975).

During maturation of murine promonocytes, the number of peroxidase-positive lysosomes decreases, probably because of cell division without continuous peroxidase synthesis (van Furth *et al.*, 1970). Moreover, when human

*See also Chapters 9, 10, and 11 of this volume.

monocytes are cultured *in vitro,* these cells differentiate within a few days into cells with the morphological appearance of macrophages. Concomitantly, they lose their lysosomal peroxidase activity (Cline, 1970a). Thus, the presence of lysosomal peroxidase activity is clearly determined by the stage of cell differentiation. Rabbit monocytes and promonocytes are an exception in that these cells contain no cytochemically demonstrable peroxidase activity at all (Nichols *et al.,* 1971; Bodel *et al.,* 1977).

Macrophages never show peroxidase activity in the lysosomes. Some macrophages do not stain at all for peroxidase, whereas other macrophages show histochemical staining for peroxidase activity in the RER, nuclear envelope, and some membrane-bound granules. Among the latter cells are guinea pig and rat resident peritoneal macrophages, guinea pig resident alveolar macrophages, and rat Kupffer cells (Cotran and Litt, 1970; Fahimi, 1970; Robbins *et al.,* 1971; Daems and Brederoo, 1971; Widman *et al.,* 1972).

4.3. PROPERTIES

Myeloperoxidase comprises about 5% of the dry weight of human neutrophils (Schultz and Kaminker, 1962). It is a heme protein with a molecular weight of about 120,000 (Felberg and Schultz 1972; Bakkenist *et al.,* 1978) and contains an unusually high number of basic amino acids (Schultz and Shmukler, 1964; Bakkenist *et al.,* 1978). Similar data for the peroxidase in monocytes or macrophages are not available.

Other properties of the peroxidase in the different types of phagocytes *have* been compared, and a number of arguments indicate that the neutrophil enzyme may be identical to the monocyte enzyme. First, in patients with myeloperoxidase deficiency, histochemical staining for peroxidase activity is lacking from neutrophils as well as from monocytes (Undritz, 1966; Lehrer and Cline, 1969), indicating that both enzymes are coded by the same gene. Second, antiserum against human neutrophil myeloperoxidase cross-reacts with peroxidase in human monocytes (Salmon *et al.,* 1970), indicating that both proteins have similar antigenic determinants. In myeloperoxidase deficiency, such antiserum reacts neither with the neutrophils nor with the monocytes (Salmon *et al.,* 1970). Third, physical characteristics of both enzymes, such as optical and electron paramagnetic resonance spectra, are identical (Bos *et al.,* 1978). This indicates the presence of an identical heme group in these enzymes. Fourth, neutrophil and monocyte lysates display similar optimal concentrations for hydrogen peroxide and electron donors, and a similar pH optimum of 5.5 (Bos *et al.,* 1978). This indicates that both peroxidases have the same enzymic properties. Taken together, this is rather strong evidence that monocytes contain the same (myelo) peroxidase as neutrophils do.

From activity measurements, Schultz and Kaminker (1962) have calculated the mean concentration of myeloperoxidase in human neutrophils to be about 5×10^{-5} μmol/10^6 cells. From electron paramagnetic resonance spectra, Wever *et al.* (1976) reach a value of about 2×10^{-5} μmol/10^6 cells. Bos *et al.* (1978), using four different parameters, calculate the concentration to be 3–3.5×10^{-5} μmol/10^6 neutrophils and about 1×10^{-5} μmol/10^6 human monocytes. Thus, in man, neutrophils contain about three times as much peroxidase as monocytes do. A similar conclusion was reached by Baehner and Johnston (1972) from activity measurements.

A number of different arguments indicate that the macrophage peroxidase is not identical with myeloperoxidase. First, there are cytochemical differences: peroxidase in monocytes, measured with the diamino-benzidine method, is less sensitive to aminotriazole than the macrophage enzyme (Daems and Brederoo, 1973). Biochemically, this has been confirmed for the activity with guaiacol (Romeo *et al.*, 1973a). The histochemical substrate specificity of the monocyte and the macrophage peroxidase is also different (van der Ploeg *et al.*, 1974). Second, the amino acid decarboxylation reaction in the presence of chloride ions was shown to occur only with the neutrophil peroxidase, not with the macrophage enzyme (Paul *et al.*, 1973; Romeo *et al.*, 1973a). In accord with this finding, killing of *E. coli* by the neutrophil peroxidase-mediated reaction takes place with iodine as well as with chloride as electron donors, whereas the macrophage enzyme only catalyses the bactericidal reaction with iodide (Paul *et al.*, 1973). Third, with rat Kupffer cell peroxidase, diamino-benzidine is oxidized much faster at pH 8 than at pH 5 (van Berkel and Kruyt, 1977b). In contrast, human myeloperoxidase has a pH optimum of 5.5 (Bos *et al.*, 1978). However, Romeo *et al.* (1973a) found an identical pH optimum for guaiacol oxidation in guinea pig and rabbit neutrophils, resident alveolar macrophages, and elicited peritoneal macrophages. Species differences may play a role in this respect. Together, these results indicate that macrophage peroxidase is different from the enzyme in neutrophils and monocytes.

4.4. IODINATION

As mentioned above, although isolated myeloperoxidase plus hydrogen peroxide plus iodide constitutes a bactericidal system, bacterial killing *in vivo* probably involves chloride rather than iodide. Nevertheless, the incorporation of radioactive iodide into acid-precipitable material by neutrophils during phagocytosis (Klebanoff and Hamon, 1972) is often determined as an easy way to measure both production of adequate amounts of hydrogen peroxide and the release of functionally operative peroxidase into the phagosomes. Thus, in CGD neutrophils and in myeloperoxidase-deficient neutrophils, the iodination reaction is severely depressed (Pincus and Klebanoff, 1971; Simmons and Karnovsky, 1973). Also, in neutrophils from Chediak–Higashi patients, which show impaired release of myeloperoxidase from giant lysosomes (Stossel *et al.*, 1972a), decreased iodination is sometimes found (Weening and Roos, unpublished), but not always (Root *et al.*, 1972).

Monocytes can also iodinate ingested material, albeit less than neutrophils (Baehner and Johnston, 1972; Biggar *et al.*, 1974; Lehrer, 1975; Bos *et al.*, 1978). Although this difference has been ascribed to the difference in peroxidase

activity in these cells (Baehner and Johnston, 1972), other factors, such as particle ingestion, hydrogen peroxide production, and release of peroxidase into the
phagosomes may also be rate-limiting. Monocytes from CGD patients (Lehrer,
1975), myeloperoxidase-deficient patients (Lehrer, 1975), and Chediak–Higashi
patients (Weening and Roos, unpublished) show the same iodination defect as
the neutrophils from these patients.

Macrophages, as could be expected from the low peroxidase activity in these
cells, show hardly any iodinating capacity (Simmons and Karnovsky, 1973; Biggar et al., 1976). Even when the results are expressed per number of ingested
particles, neutrophils are far more capable of fixing iodine (Simmons and Karnovsky, 1973). However, Biggar et al., (1976) have shown that rabbit alveolar
macrophages do not iodinate ingested zymosan particles, even when the particles are coated with horseradish peroxidase. This suggests that hydrogen
peroxide is either not generated by these cells, or not available for iodination.
Unfortunately, no direct comparison between iodinating ability and hydrogen
peroxide release by the various cells was made.

4.5. CORRELATION WITH BACTERICIDAL CAPACITY

The question as to the importance of the peroxidase-mediated reactions for
the microbicidal capacity of a phagocytic cell has two aspects: the nature of the
ingested microorganism and the type of phagocyte. First to be discussed will be
the killing of bacteria. Both myeloperoxidase-deficient neutrophils (Lehrer and
Cline, 1969; Lehrer et al., 1969; Klebanoff, 1970) and normal neutrophils treated
with inhibitors of myeloperoxidase (after ingestion of bacteria) (Klebanoff, 1968,
1970; Klebanoff and Hamon, 1972; Diamond et al., 1972) show a retarded bactericidal capacity. During longer incubations, however, a normal killing of bacteria, both catalase-positive and -negative, is achieved, although the iodination
reaction remains inhibited. In agreement with these findings, MPO-deficient
individuals do not suffer from bacterial infections. A retarded bacterial killing in
vitro has also been reported for Chediak–Higashi neutrophils and monocytes
(Root et al., 1972). Thus, a myeloperoxidase-mediated bactericidal reaction appears to be operative in neutrophils and monocytes, but, as discussed in the next
section, myeloperoxidase-independent mechanisms also contribute to the destruction of bacteria by these cells. Experiments to evaluate the bactericidal
capacity of peroxidase-deficient or peroxidase-inhibited monocytes or macrophages seem not to have been performed.

A direct comparison has been made between the peroxidase activity, the
iodinating capacity, and the bactericidal activity of different phagocytic cells. For
neutrophils and elicited macrophages, these activities closely correlate, especially at higher ratios of bacteria to phagocytes (Cohen and Cline, 1971; Simmons
and Karnovsky, 1973). For resident macrophages, no such correlation is found:
although these cells contain no or very little peroxidase activity, they kill various
strains of bacteria almost as avidly as neutrophils (Cohen and Cline, 1971; Simmons and Karnovsky, 1973; Biggar et al., 1976). Thus, resident macrophages

possess other than peroxidase-dependent bactericidal mechanisms. One warning against this conclusion may be derived from the work of Romeo *et al.* (1973a): extraction of peroxidase from phagocytic cells with the detergent cetyltrimethylammonium bromide results in much higher activities than with other detergents. Therefore, these authors consider the possibility that peroxidase-mediated killing might occur in (some) macrophages. It is not clear, however, how this can happen when killing takes place in the phagosome and the peroxidase activity is located in the endoplasmic reticulum and the nuclear envelope.

A completely different situation exists for the killing of yeasts. *C. albicans* is killed exclusively by the myleoperoxidase-mediated reaction: myleoperoxidase-deficient neutrophils and monocytes kill this organism very poorly (Lehrer, 1970, 1972, 1975). Moreover, resident alveolar macrophages and cultured monocytes, which do not stain for granular peroxidase activity, fail to kill *C. albicans* (Cohen and Cline, 1971). And, in addition, inhibitors of myleoperoxidase also inhibit killing of *C. albicans* by neutrophils (Klebanoff, 1970; Lehrer, 1971) and monocytes (Lehrer, 1970, 1975). Agents have been described, however, which inhibit the iodination of ingested *C. albicans* by human monocytes without any effect on the killing of this yeast (Lehrer, 1975). Thus, iodination is concomitant, rather than a cause of, myeloperoxidase-mediated candidacidal activity against *C. albicans*. Other *Candida* species are killed as effectively or even better by myeloperoxidase-deficient or myeloperoxidase-inhibited neutrophils and monocytes than by normal leukocytes (Lehrer, 1972, 1975).

We may conclude the following:

1. Monocytes and neutrophils contain both peroxidase-dependent and peroxidase-independent microbicidal activity.
2. Resident macrophages contain only peroxidase-independent microbicidal activity.

5. OTHER CIDAL MECHANISMS

5.1. OXYGEN-DEPENDENT, PEROXIDASE-INDEPENDENT MICROBICIDAL ACTIVITY

From the foregoing it is clear that other microbicidal mechanisms than those mediated by peroxidase must exist in phagocytic cells. Some of these depend on oxygen and will be discussed first.

There is some evidence to suggest that catalase, by its peroxidatic activity, might be able to substitute for peroxidase in catalyzing the halogenation reaction. Klebanoff (1969) has shown that catalase, in the presence of a hydrogen-peroxide-generating system and iodide is lethal for bacteria, yeasts, and viruses. Catalase is present in many phagocytic cells (see Klebanoff and Hamon, 1975), and has been shown to be transferred during phagocytosis in rabbit alveolar macrophages from a particulate fraction to the phagosomes (Stossel *et al.*, 1972b). The intraphagosomal pH reaches a value close to the pH optimum of the catalase-mediated microbicidal system (Jensen and Bainton, 1973). According to

Klebanoff and Green (1973), small amounts of iodide may be picked up by phagocytic cells from thyroid hormone degradation. Thus, since hydrogen peroxide is produced by the phagocytes, all the constituents of the system are present. However, actual proof for a physiological role of this system in intact cells is lacking.

In principle, oxygen radicals, such as superoxide, might be microbicidal without the help of any enzyme. The available evidence suggests, however, that superoxide itself is not bactericidal (Gregory *et al.*, 1973; Yost and Fridovich, 1974; Klebanoff, 1974; Babior *et al.*, 1975; DeChatelet *et al.*, 1975; Mandell, 1975), perhaps because virtually every organism that can live in air contains superoxide dismutase (Fridovich, 1974). If killing by a superoxide-producing system occurs, catalase protects, indicating that hydrogen peroxide rather than superoxide is the bactericidal agent. Sometimes both catalase and superoxide dismutase protect, suggesting that the product of a reaction between superoxide and hydrogen peroxide, i.e., singlet oxygen or the hydroxyl radical, is the lethal compound. Similar results have been obtained by Johnston *et al.* (1975) with bacterial killing by neutrophils. The actual proof for the involvement of singlet oxygen and/or hydroxyl radical in the bactericidal process of the phagocyte is scarce. First, a carotenoid-missing mutant of *Sarcina lutea* is killed much more rapidly by neutrophils than the wild-type bacteria, which is consistent with a reaction by singlet oxygen (Krinsky, 1974). Second, degradation of unsaturated fatty acids takes place after ingestion, a process caused by peroxidative cleavage and probably initiated by hydroxyl radicals (Fong *et al.*, 1973; Stossel *et al.*, 1974; Shohet *et al.*, 1974). This reaction is found in monocytes as well as in neutrophils.

Another system which might be operative in phagocytic cells comprises ascorbate plus hydrogen peroxide. It has been found that the bactericidal activity of hydrogen peroxide is highly potentiated by ascorbate, especially in the presence of certain metal ions (Ericsson and Lundbeck, 1955a,b; DeChatelet *et al.*, 1972; Drath and Karnovsky, 1974). Moreover, Miller (1969) has found that a similar treatment renders *Salmonella pullorum* highly sensitive to further lysis by lysozyme. Some of these killing reactions probably proceed through interaction of free radicals with the bacterial cell wall, since they can be inhibited by free radical scavengers (Miller, 1969; Drath and Karnovsky, 1974). Ascorbate and lysozyme are both present in phagocytic leukocytes (Cohn, 1963; Bigley and Stankova, 1974; Drath and Karnovsky, 1974; Lehrer, 1975); therefore, this system may indeed be of physiological importance, especially in macrophages.

5.2. OXYGEN-INDEPENDENT MICROBICIDAL ACTIVITY*

The existence of oxygen-independent antimicrobial systems is illustrated by the normal candidacidal (Lehrer, 1972) and bactericidal (Mandell, 1974) activity of neutrophils at strongly reduced oxygen tension. At least three different systems are known.

*See also Chapter 14 of this volume.

First, the drop in intraphagosomal pH, known to occur in all types of phagocytes (Mandell, 1970; Jensen and Bainton, 1973), may be microbicidal or microbistatic in itself (Looke and Rowley, 1962), and it may also create favorable conditions for the activity of other antimicrobial or digestive systems.

Second lysozyme, also present in all types of phagocytic cells (for a review see Klebanoff and Hamon, 1975), is lytic for a number of peptidoglycan-containing organisms (Salton, 1957). Under extreme pH conditions (Salton, 1957; Brumfitt and Glynn, 1961; Spizizen, 1962), in the presence of chelating agents (Repaske, 1956, 1958), or in the presence of antibody plus complement (Amano *et al.*, 1954; Muschel *et al.*, 1959; Muschel, 1965), organisms which are resistant to lysozyme can be made sensitive to this enzyme. Usually, however, lysozyme acts as a digestive enzyme.

Third, neutrophils also contain cationic bactericidal and fungicidal proteins (Zeya and Spitznagel, 1968; Elsbach, 1973; Lehrer *et al.*, 1974; Odeberg and Olsson, 1976b), whereas monocytes only contain fungicidal proteins not belonging to the class of cathodal esterases (Lehrer, 1975).

Thus, although it is clear that we are far from understanding how exactly phagocytic cells kill microorganisms, we now realize that the different phagocytes contain an array of possibilities to deal with various microbes.

5.3. CYTOTOXIC SYSTEMS

Mononuclear phagocytes, especially macrophages, have also been shown to possess cytotoxic activity against mammalian target cells, including a variety of tumor cells, sensitized erythrocytes, and devitalized bone. These processes will be dealt with below in relation to the oxidative metabolism of the effector cells.

5.3.1. Cytotoxicity against Tumor Cells

Antitumor activity can be raised both *in vivo* and *in vitro*. Peritoneal macrophages from mice injected with syngeneic murine lymphoma cells are specifically cytotoxic against those tumor cells (Evans and Alexander, 1972a). After infection of mice with intracellular growing organisms, such as *Toxoplasma gondii* (Hibbs *et al.*, 1972) and *Corynebacterium parvum* (Olivotto and Bomford, 1974), the peritoneal macrophages become nonspecifically cytotoxic against non-contact-inhibited cells. *In vitro*, macrophages may be rendered specifically cytotoxic by spleen cells from hyperimmunized animals (Evans and Alexander, 1972b), and nonspecifically by double-stranded RNA, endotoxin, lipid A (Alexander and Evans, 1971) and interferon (Huang *et al.*, 1971). Human monocytes are also cytotoxic for tumor cells, even without prior sensitization (Rinehart *et al.*, 1978).

Although contact between effector and target cells seems to be needed, phagocytosis plays no significant role in this process (Keller *et al.*, 1976). Lysosomal enzymes are probably involved (Hibbs, 1974; Bucana *et al.*, 1976), but the importance of oxygen products in the cytotoxic process has hardly been studied. It has been demonstrated that the hydrogen peroxide–myelo-peroxidase–halide system is toxic for mammalian tumor cells (Edelson and Cohn, 1973; Philpott et al., 1973; Clark and Klebanoff, 1975). The products released by zymosan-phagocytizing neutrophils also possess this capacity, in contrast to myeloperoxidase-deficient or chronic granulomatous disease neutrophils (Clark and Klebanoff, 1975).

Human neutrophils are also cytotoxic to certain neoplastic cells. Rat leukemic cells (Gale and Zighelboim, 1975) and mouse ascitic lymphoma cells (Clark and Klebanoff, 1977), but not mouse mastocytoma cells (MacDonald et al., 1975) are rapidly killed in an antibody-dependent, phagocytosis-independent process. The cytotoxic process probably involves oxygen radicals but not myeloperoxidase, since neutrophils from CGD patients fail to kill the target cells, whereas neutrophils from myeloperoxidase-deficient patients or myeloperoxidase-inhibited neutrophils show a normal reaction (Clark and Klebanoff, 1977).

The macrophage-mediated cytotoxicity, on the other hand, seems to be largely oxygen-independent, since neither anaerobiosis nor catalase or superoxide dismutase has much effect on this process (Sorrell et al., 1978). Therefore, although mammalian neoplastic cells are vulnerable to oxygen-dependent killing mechanisms, no evidence is available to suggest this to be a physiologically important process. Most likely, lysosomal enzymes and/or complement components are involved.

5.3.2. Cytotoxicity Against Sensitized Erythrocytes

Red cells sensitized with IgG allo- and autoantibodies can be destroyed either by complement activation or interaction with phagocytes. Since most IgG anti-erythrocyte antibodies are noncomplement binding (Dacie, 1962; Engelfriet et al., 1974), the adherence to, and subsequent cytotoxic lysis by, phagocytes is the most important way of destruction of sensitized erythrocytes (LoBuglio et al., 1967). This process takes place predominantly in the spleen (Lewis et al., 1960).

In vitro, monocytes and macrophages, as well as neutrophils, are capable of lysing red cells sensitized with different allo- and heteroantibodies (Holm, 1972; Holm and Hammarström, 1973; Hersey, 1973; Zeijlemaker et al., 1975; Poplack et al., 1976). The cytotoxic mechanism of human leukocytes against human sensitized red cells strongly depends on the sensitizing antibody. Gill et al. (1977) report red cells coated with blood group A_1 antiserum to be lysed more rapidly by neutrophils than by monocytes in a phagocytosis-dependent process. Fleer et al. (1978a,b) have used anti-D-coated erythrocytes and found these cells to be lysed predominantly extracellularly by monocytes. Probably, the number of antigens per target cell is decisive for the mechanism of lysis. Neutrophils display also phagocytosis-independent cytotoxicity for chicken red blood cells in the presence of leukoagglutinins (Simchowitz and Schur, 1976).

The extracellular cytotoxicity of monocytes appears to be mediated by lysosomal enzymes. Cytochalasin B, which enhances lysosomal enzyme release, also enhances lysis of anti-D-sensitized erythrocytes, whereas colchicine and hydrocortisone, which inhibit lysosomal enzyme release, depress erythrocytotoxicity (Fleer *et al.*, 1978b). On the other hand, monocytes from CGD patients exhibit a normal capacity for red cell lysis, and scavengers of oxygen radicals, such as superoxide dismutase, catalase, and mannitol, do not inhibit this reaction (Fleer, *et al.*, 1979). Thus, although erythrocytes are sensitive to lysis by the isolated hydrogen peroxide–myeloperoxidase–halide system (Klebanoff and Clark, 1975), this mechanism is probably not involved in the monocyte-mediated extracellular process.

5.3.3. Bone Resorption

Osteoclasts, the multinucleated bone macrophages, are involved in bone resorption (Junqueira *et al.*, 1977). There is some evidence that osteoclasts may be derived from circulating monocytes (Walker, 1975; Kahn and Simmons, 1975). Recently, it has been demonstrated that human monocytes resorb killed adult and fetal bone when cocultured for at least 4 days with these tissues (Mundy *et al.*, 1977; Kahn *et al.*, 1978). The monocytes adhere avidly to the bone tissue (Kahn *et al.*, 1978), but contact is not needed for resorption (Mundy *et al.*, 1977). During this process the monocytes acquire some, but not all, characteristics of osteoclasts (Kahn *et al.*, 1978). The process seems to act through lysosomal enzymes, but the involvement of the oxidative metabolism has not been investigated.

6. CARBOHYDRATE METABOLISM AS A SOURCE OF ENERGY*

6.1. RESTING CELLS

Mammalian cells synthetize ATP in aerobic glycolysis. Neutrophils appear to depend heavily on the Embden–Meyerhof pathway for ATP synthesis (Sbarra and Karnovsky, 1959). This conclusion is based on the relatively low oxygen consumption of resting neutrophils, the large proportion of glucose converted to lactate, and the small amount of $^{14}CO_2$ derived from $[6-^{14}C]$glucose in these cells (Sbarra and Karnovsky, 1959; Evans and Karnovsky, 1962; Reed and Tepperman, 1969; Stjernholm and Manak, 1970). Approximately 2% of the consumed glucose is metabolized in the citric acid cycle (Stjernholm *et al.*, 1969) and 1.3% in the hexose monophosphate shunt (Stjernholm and Manak, 1970). Morphologically, a very low number of mitochondria is found in these cells. The ability of the neutrophil to synthesize ATP in the absence of oxygen enables this cell to operate in areas of low oxygen tension.

The carbohydrate metabolism of macrophages differs in cells from different organs. In general, alveolar macrophages display much more oxidative activity (Oren *et al.*, 1963; Romeo *et al.*, 1973b; Simon *et al.*, 1977) and consume more

---

*See also Chapter 4 of this volume.

glucose, especially via the hexose monophosphate shunt and the citric acid cycle, than peritoneal macrophages (Oren *et al.*, 1963; Romeo *et al.*, 1973b). These data correspond to a higher number of mitochondria in alveolar as compared to peritoneal macrophages (Policard *et al.*, 1956; Hirsch and Fedorko, 1970), and seem logical in view of the fact that alveolar macrophages operate in areas with much higher oxygen tensions than those found in the peritoneum.

The regulation of the carbohydrate metabolism in phagocytic cells is shown by the Crabtree effect (depression of respiration by glucose) in elicited peritoneal macrophages and neutrophils, but not in alveolar macrophages (Oren *et al.*, 1963). In the latter cell type, the respiration is *increased* by glucose, indicating that the oxidative phosphorylation is not operating at maximal capacity under normal conditions. In addition, all three cell types have been reported to show a Pasteur effect (increase in lactate production at low oxygen tensions) (Oren *et al.*, 1963; Simon *et al.*, 1977), although this was not confirmed for alveolar macrophages (Ouchi *et al.*, 1965).

Surprisingly little is known about the carbohydrate metabolism of circulating monocytes. Glucose consumption values vary from 260 to 1330 nmol/10^7 cells/hr (Cline and Lehrer, 1968; Para *et al.*, 1972; Brodersen and Burns, 1973; King *et al.*, 1975), and lactate production from 530 to 1640 nmol/10^7 cells/hr (Para *et al.*, 1972; Roos *et al.*, 1979). Lactate production is about twice as high in monocytes as it is in neutrophils (Roos *et al.*, 1979). Nothing is known about the percentage of glucose metabolized in the citric acid cycle or in the hexose monophosphate shunt of these cells. Monocytes contain more mitochondria than neutrophils do (Wintrobe *et al.*, 1974), and may, therefore, be expected to oxidize more glucose in the citric acid cycle.

6.2.　CHANGES DURING PHAGOCYTOSIS

During particle ingestion, the oxygen consumption of phagocytic cells may be increased for two reasons: first, as discussed in Section 3.3.3, oxygen serves as a substrate for hydrogen peroxide formation, and second, oxygen is also used for the oxidation of citric acid cycle intermediates during ATP-generating oxidative phosphorylation.

From the small increase in $^{14}CO_2$ production from [6-^{14}C]glucose and the relative insensitivity of the stimulated oxygen consumption to inhibitors of oxidative phosphorylation, we know that neutrophils do not show a marked increase in citric acid cycle activity during phagocytosis (Sbarra and Karnovsky, 1959; Evans and Karnovsky, 1962; Stjernholm *et al.*, 1969; Stjernholm and Manak, 1970; Reiss and Roos, 1978). Moreover, glycogen consumption, glucose consumption, and lactate production also increase only slightly during phagocytosis (Sbarra and Karnovsky, 1959; Evans and Karnovsky, 1962; Rossi and Zatti, 1966; Zatti and Rossi, 1967; Stjernholm and Manak, 1970), indicating that synthesis of extra ATP for particle uptake by neutrophils is not needed. The ATP levels in these cells do not show a significant change during phagocytosis of

E. coli (Kakinuma, 1970), but uptake of zymosan particles results in about 50% decrease of the ATP (Roos *et al.*, 1979).

Alveolar, as well as peritoneal, macrophages show increased citric acid cycle activity during phagocytosis; the highest increase is found in peritoneal macrophages (Oren *et al.*, 1963; Romeo *et al.*, 1973b). Moreover, comparison of the increase in oxygen consumption and in hexose monophosphate shunt activity (West *et al.*, 1968), or superoxide release (Rister and Baehner, 1977), indicates that part of the respiration increase in both types of phagocytes may be due to enhanced oxidative phosphorylation. Neither the consumption of glucose, nor the production of lactate by macrophages is increased during particle ingestion (Oren *et al.*, 1963; Ouchi *et al.*, 1965; West *et al.*, 1968). During phagocytosis, the level of ATP in alveolar macrophages decreases (Gee *et al.*, 1968).

Again, very little is known about changes in these reactions during phagocytosis by circulating monocytes. Neither glucose consumption nor lactate production are substantially changed by this process (Cline and Lehrer, 1968; Roos *et al.*, 1979). Production of $^{14}CO_2$ from [1-^{14}C]glucose is increased (see Section 3.3.6), and ATP levels decrease about 20% during zymosan phagocytosis (Roos *et al.*, 1979).

Two remarks must be made here. First, it must be realized that ATP generation in the citric acid cycle is so much more efficient than in the Embden–Meyerhof pathway that only a slight increase in the citric acid cycle activity may generate sufficient ATP for particle uptake. Such subtle changes may be very hard to detect. Second, it must also be realized that substrates derived from other substrates than glucose can enter the citric acid cycle, and indeed this has been proven for neutrophils and macrophages (Ouchi *et al.*, 1965; West *et al.*, 1968; Stjernholm *et al.*, 1969). Thus, $^{14}CO_2$ formation from [6-^{14}C]glucose does not give a complete picture of the activity of this pathway.

6.3. EFFECT OF METABOLIC INHIBITORS

Many investigators have studied the effect of metabolic inhibitors on carbohydrate metabolism and the uptake of particles by phagocytic cells. Two problems are eminent: the specificity of the inhibitors used and the methods to study particle uptake.

For the inhibition of the Embden–Meyerhof pathway, sodium fluoride and monoiodoacetic acid have been generally used. Since the first agent reacts with metal ions and the latter with many active thiol groups, inhibition of phagocytosis by these compounds may have been caused for many reasons: inhibition of particle adherence, of microfilament activity, of ATP generation (at *any* point in carbohydrate metabolism), etc. Another inhibitor, potassium cyanide, has been used to inhibit mitochondrial respiration. KCN inhibits also catalase and peroxidases, however, and, thus, affects oxygen uptake and liberation in several ways. Anaerobiosis has also been used, although exclusion of oxygen is notoriously difficult. In addition, proper dose–response studies are seldom reported. It is apparent that results from such experiments must be interpreted with caution.

The quantitation of particle uptake has long been very primitive: either microscopic examination or culturing of noningested bacteria has been used. Although the use of radiolabeled particles has enabled the measurement of reaction *rates*, differentiation between adherence and ingestion is still seldom seen.

The report that monoiodoacetic acid and NaF inhibit particle uptake in neutrophils as well as in alveolar and peritoneal macrophages, whereas KCN, 2,4-dinitrophenol, and anaerobiosis inhibit this function only in alveolar macrophages (Oren *et al.*, 1963), might be interpreted as a dependence of alveolar macrophages on oxidative phosphorylation for ATP synthesis (fed by a pyruvate-dependent citric acid cycle), and a dependence on the Embden–Meyerhof pathway for ATP synthesis in neutrophils and peritoneal macrophages. The effects of monoiodoacetic acid and NaF may have been nonspecific, however, since lower concentrations, which effectively inhibit the metabolic reactions, had little or no effect on phagocytosis in alveolar macrophages (Ouchi *et al.*, 1965; Cohen and Cline, 1971).

More specific inhibitors of mitochondrial electron transport, such as phenazine metasulfate and other artificial electron acceptors, effectively inhibit particle uptake in alveolar macrophages (Ouchi *et al.*, 1965). Inhibitors of oxidative phosphorylation (antimycin, oligomycin) depress the level of ATP in these cells (Gee *et al.*, 1968; Schneider *et al.*, 1978), indicating that ATP synthesis is dependent on this reaction. Antimycin has also been reported to inhibit particle uptake in alveolar macrophages (Ouchi *et al.*, 1965). It seems reasonably safe, therefore, to assume that alveolar macrophages indeed depend on mitochondrial activity for phagocytosis.

On the other hand, the completely normal uptake of particles by neutrophils and peritoneal macrophages at very low oxygen tensions (Sbarra and Karnovsky, 1959; Oren *et al.*, 1963; Cline, 1966; Thalinger and Mandell, 1971), in contrast to alveolar macrophages (Oren *et al.*, 1963; Ouchi *et al.*, 1965; Cohen and Cline, 1971), indicates that the main source of energy in neutrophils and peritoneal macrophages is anaerobic glycolysis. However, direct comparison with the ATP levels in the cells is needed for a correct interpretation, since phagocytosis is directly correlated with this parameter (Waters *et al.*, 1976).

In monocytes, but not in neutrophils, both antimycin and oligomycin have been shown to severely depress the stimulated oxygen consumption and the uptake of particles (Reiss and Roos, 1978). Thus, these cells seem to resemble alveolar macrophages in a dependence on oxidative phosphorylation for the process of phagocytosis.

6.4. EFFECT OF ACTIVATION AND DIFFERENTIATION

Antigenic stimulation of macrophages results in a shift of carbohydrate metabolism in resting cells towards a more glycolytic pattern: less oxygen consumption and more lactate production (Hard, 1970; Para *et al.*, 1972). Moreover, elicited and activated macrophages show a stronger stimulation of glucose consumption and citric acid cycle activity during phagocytosis (Stubbs *et al.*, 1973; Rossi *et al.*, 1975).

The morphologic changes in these cells during the process of activation closely resemble the changes seen in monocytes or macrophages cultured *in vitro*. The cells become larger and acquire more and larger mitochondria, more phase-dense granules and more lipid droplets. The total activity of lysosomal enzymes as well as the capacity to ingest, kill, and digest bacteria increases (Mackaness, 1962; Cohn and Benson, 1965; Bennett and Cohn, 1966; Cline, 1970a; Cohen and Cline, 1971; Hirt and Bonventre, 1973; Simon *et al.*, 1973; Stubbs *et al.*, 1973; Ødegaard *et al.*, 1974; Viken and Ødegaard, 1974).

During cultivation *in vitro*, alveolar and peritoneal macrophages adapt their carbohydrate metabolism to their surroundings: at high oxygen tensions, the cytochrome c oxidase activity increases and the pyruvate kinase activity decreases, whereas the opposite effect is seen at low oxygen tensions (Simon *et al.*, 1973, 1977). If similar changes take place *in vivo*, monocytes may be thought to lose their dependence on mitochondrial activity when they mature into peritoneal macrophages, and maintain or even increase this characteristic when developing into alveolar macrophages (Reiss and Roos, 1978).

7. CONCLUSIONS

This review focuses on three functions of phagocytic cells: oxidative intracellular killing of microorganisms, nonoxidative mechanisms of intracellular microbicidal activity, and extracellular cytotoxicity. Monocytes and neutrophils share at least one intracellular oxidative killing mechanism: a peroxidase-mediated system. Since monocytes ingest microorganisms slower, produce less oxygen products, and contain less myeloperoxidase than neutrophils, the killing of microorganisms is also slower. These differences are only quantitative, not qualitative. Monocytes and neutrophils also contain additional, nonoxidative microbicidal systems, but—as judged from the recurrent infections in patients with chronic granulomatous disease—the oxygen-dependent system is the most important.

Macrophages, on the other hand, cannot be considered merely as mature monocytes in this respect: although macrophages do not contain granular peroxidase, they kill actively with intracellular mechanisms. Therefore, macrophages contain efficient non-peroxidase-mediated microbicidal systems, both oxygen-dependent and oxygen-independent. Since relatively few localized dysfunctions of the mononuclear phagocyte system are known (Territo and Cline, 1976), we must conclude that similar systems are operative in all three cell types. However, the intracellular microbicidal activity of macrophages, and to a lesser extent also of monocytes, can be enhanced by antigenic activation *in vivo* and by lymphokines and culturing *in vitro*.

Extracellular cytotoxicity of the mononuclear phagocyte system does not seem to involve the oxidative killing system of these cells. It is unknown exactly which mechanisms are involved in these processes. Mononuclear phagocytes live longer, therefore, the degradative function of these cells may be impaired in cases of storage disease, infection with intracellular parasites, and during excessive cell turnover (such as myelocytic leukemia).

In general, the neutrophil is a simpler cell, the monocyte is an immature cell, and the macrophage is a highly differentiated, multifunctional cell. Together, these phagocytes comprise a very effective host-defense system.

REFERENCES

Alexander, P., and Evans, R., 1971, Endotoxin and double stranded RNA render macrophages cytotoxic, *Nature New Biol.* **232**:76.

Allen, R. C., and Loose, L. D., 1976, Phagocytic activation of a luminol-dependent chemiluminescence in rabbit alveolar and peritoneal macrophages, *Biochem. Biophys. Res. Commun.* **69**:245.

Allen, R. C., Stjernholm, R. L., and Steele, R. H., 1972, Evidence for the generation of an electronic excitation state(s) in human polymorphonuclear leukocytes and its participation in bactericidal activity, *Biochem. Biophys. Res. Commun.* **47**:679.

Allred, C. D., and Hill, H. R., 1977, Effect of chemoattractants in PMN chemiluminescence, *Clin. Res.* **25**:117A.

Amano, T. S., Inai, Y., Seki, S., Kashiba, K., Fujikawa, K., and Nishimura, S., 1954, Studies on immune bacteriolysis. 1. Accelerating effect on immune bacteriolysis by lysozyme-like substance of leukocytes and egg-white lysozyme, *Med. J. Osaka Univ.* **4**:401.

Axline, S. G., and Cohn, Z. A., 1970, *In vitro* induction of lysosomal enzymes by phagocytosis, *J. Exp. Med.* **131**:1239.

Babior, B. M., Kipnes, R. S., and Curnutte, J. T., 1973, Biological defense mechanisms. The production by leukocytes of superoxide, a potential bactericidal agent, *J. Clin. Invest.* **52**:741.

Babior, B. M., Curnutte, J. T., and Kipnes, R. S., 1975, Biological defense mechanisms. Evidence for the participation of superoxide in bacterial killing by xanthine oxidase, *J. Lab. Clin. Med.* **85**:235.

Baehner, R. L., 1975, Subcellular distribution of nitroblue tetrazolium reductase (NBT-R) in human polymorphonuclear leukocytes (PMN), *J. Lab. Clin. Med.* **86**:785.

Baehner, R. L., and Johnston, R. B. Jr., 1972, Monocyte function in children with neutropenia and chronic infections, *Blood* **40**:31.

Baehner, R. L., Gilman, N., and Karnovsky, M. L., 1970, Respiration and glucose oxidation in human and guinea pig leukocytes: Comparative studies, *J. Clin. Invest.* **49**:692.

Bakkenist, A. R. J., Wever, R., Vulsma, T., Plat, H., and van Gelder, B. F., 1978, Isolation procedure and some properties of myeloperoxidase from human leukocytes, *Biochim. Biophys. Acta,* **524**:45.

Beall, G. D., Repine, J. E., Hoidal, J. R., and Rasp, F. L., 1977, Chemiluminescence by human alveolar macrophages: Stimulation with heat-killed bacteria or phorbol myristate acetate, *Infect. Immun.* **17**:117.

Beauchamp, C., and Fridovich, I., 1970, A mechanism for the production of ethylene from methional. The generation of the hydroxyl radical by xanthine oxidase, *J. Biol. Chem.* **245**:4641.

Belding, M. E., Klebanoff, S. J., and Ray, C. G., 1970, Peroxidase-mediated virucidal systems, *Science* **167**:195.

Bennett, W. E., and Cohn, Z. A., 1966, The isolation and selected properties of blood monocytes, *J. Exp. Med.* **123**:145.

Biggar, W. D., and Sturgess, J. M., 1978, Hydrogen peroxide release by rat alveolar macrophages: Comparison with blood neutrophils, *Infect. Immun.* **19**:621.

Biggar, W. D., Holmes, B., Page, A. R., Deinard, A. S., L'Esperance, P., and Good, R. A., 1974, Metabolic and functional studies of monocytes in congenital neutropenia, *Br. J. Haematol.* **28**:233.

Biggar, W. D., Buron, S., and Holmes, B., 1976, Bactericidal mechanisms in rabbit alveolar macrophages: Evidence against peroxidase and hydrogen peroxide bactericidal mechanisms, *Infect. Immun.* **14**:6.

Bigley, R. H., and Stankova, L., 1974, Uptake and reduction of oxidized and reduced ascorbate by human leukocytes, *J. Exp. Med.* **139**:1084.

Blanden, R. V., Mackaness, G. B., and Collins, F. M., 1966, Mechanisms of acquired resistance in mouse typhoid, *J. Exp. Med.* **124**:585.

Bodel, P. T., Nichols, B. A., and Bainton, D. F., 1977, Appearance of peroxidase reactivity within the rough endoplasmic reticulum of blood monocytes after surface adherence, *J. Exp. Med.* **145**:264.

Bos, A., Wever, R., and Roos, D., 1978, Characterization and quantification of the peroxidase in human monocytes, *Biochim. Biophys. Acta,* **525**:37.

Böyum, A., 1968, Isolation of mononuclear cells and granulocytes from human blood. Isolation of mononuclear cells by one centrifugation, and of granulocytes by combining centrifugation and sedimentation at 1g, *Scand. J. Clin. Lab. Invest.* **21** (Supp. 97):77.

Breton-Gorius, J., and Guichard, J., 1969, Etude au microscope électronique de la localisation des peroxydases dans les cellules de la moelle osseuse humaine, *Nouv. Rev. Franç. Hématol.* **9**:678.

Briggs, R. T., Drath, D. B., Karnovsky, M. L., and Karnovsky, M. J., 1975, Localization of NADH oxidase on the surface of human polymorphonuclear leukocytes by a new cytochemical method, *J. Cell Biol.* **67**:566.

Brodersen, M. P., and Burns, C. P., 1973, The separation of human monocytes from blood including biochemical observations, *Proc. Soc. Exp. Biol. Med.* **144**:941.

Brumfitt, W., and Glynn, A. A., 1961, Intracellular killing of *Micrococcus lysodeikticus* by macrophages and polymorphonuclear leukocytes. A comparative study, *Br. J. Exp. Pathol.* **42**:408.

Bucana, C., Hoyer, L. C., Hobbs, B., Breesman, S., McDaniel, M., and Hanna, M. G., Jr., 1976, Morphological evidence for the translocation of lysosomal organelles from cytotoxic macrophages into the cytoplasm of tumor target cells, *Cancer Res.* **36**:4444.

Cheson, B. D., Christensen, R. L., Sperling, R., Kohler, B. E., and Babior, B. M., 1976, The origin of the chemiluminescence of phagocytosing granulocytes, *J. Clin. Invest.* **58**:789.

Cheson, B. D., Curnutte, J. T., and Babior, B. M., 1977, The oxidative killing mechanisms of the neutrophil, *Progr. Clin. Immunol.* **3**:1.

Clark, R. A., and Klebanoff, S. J., 1975, Neutrophil-mediated tumor cell cytotoxicity: Role of the peroxidase system, *J. Exp. Med.* **141**:1442.

Clark, R. A., and Klebanoff, S. J., 1977, Studies on the mechanism of antibody-dependent polymorphonuclear leukocyte-mediated cytotoxicity, *J. Immunol.* **119**:1413.

Cline, M. J., 1966, Phagocytosis and synthesis of ribonucleic acid in human granulocytes, *Nature* **212**:1431.

Cline, M. J., 1970a, Bactericidal activity of human macrophages: Analysis of factors influencing the killing of *Listeria monocytogenes, Infect. Immun.* **2**:156.

Cline, M. J., 1970b, Drug potentiation of macrophage function, *Infect. Immun.* **2**:601.

Cline, M. J., and Lehrer, R. I., 1968, Phagocytosis by human monocytes, *Blood* **32**:423.

Cohen, A. B., and Cline, M. J., 1971, The human alveolar macrophage: Isolation, cultivation *in vitro,* and studies of morphologic and functional characteristics, *J. Clin. Invest.* **50**:1390.

Cohn, Z. A., 1963, The fate of bacteria within phagocytic cells. 1. The degradation of isotopically labeled bacteria by polymorphonuclear leukocytes and macrophages, *J. Exp. Med.* **117**:27.

Cohn, Z. A., and Benson, B., 1965, The differentiation of mononuclear phagocytes. Morphology, cytochemistry, and biochemistry, *J. Exp. Med.* **121**:153.

Cotran, R. S., and Litt, M., 1970, Ultrastructural localization of horseradish peroxidase and endogenous peroxidase activity in guinea pig peritoneal macrophages, *J. Immunol.* **105**:1536.

Cruchaud, A., Girard, J. -P., and Hitoglou, S., 1977, The functions of human monocytes in normal subjects and in disorders associated with immune deficiency, *Int. Arch. Allergy* **54**:529.

Curnutte, J. T., Karnovsky, M. L., and Babior, B. M., 1976, Manganese-dependent NADPH oxidation by granulocyte particles. The role of O_2^- and the non-physiological nature of the manganese requirement, *J. Clin. Invest.* **57**:1059.

Dacie, J. V., 1962, *The Haemolytic Anaemias,* Part II: *The Autoimmune Haemolytic Anaemias,* 2nd ed., pp. 609–610, Churchill, London.

Daems, W. T., and Brederoo, P., 1971, The fine structure and peroxidase activity of resident and exudate peritoneal macrophages in the guinea pig, in: *The Reticuloendothelial System and Immune Phenomena* (N. R. Diluzio, and K. Fleming, eds.), pp. 19–31, Plenum Press, New York.

Daems, W. T., and Brederoo, P., 1973, Electron microscopical studies on the structure, phagocytic properties, and peroxidatic activity of resident and exudate peritoneal macrophages in the guinea pig, *Z. Zellforsch. Mikrosk. Anat.* **144**:247.

Davis, W. C., Huber, H., Douglas, D., and Fudenberg, H. H., 1968, A defect in circulating mononuclear phagocytes in chronic granulomatous disease of childhood, *J. Immunol.* **5**:1093.

DeChatelet, L. R., Cooper, M. R., and McCall, C. E., 1971, Dissociation by colchicine of the hexose monophosphate shunt activation from the bactericidal activity of the leukocyte, *Infect. Immun.* **3**:66.

DeChatelet, L. R., Cooper, M. R., and McCall, C. E., 1972, Stimulation of the hexose monophosphate shunt in neutrophils by ascorbic acid. Mechanisms of action, *Antimicrob. Agents Chemother.* **1**:12.

DeChatelet, L. R., McCall, C. E., McPhail, L. C., and Johnston, R. B., Jr., 1974, Superoxide dismutase activity in leukocytes, *J. Clin. Invest.* **53**:1197.

DeChatelet, L. R., Mullikin, D., and McCall, C. E., 1975, The generation of superoxide anion by various types of phagocyte, *J. Inf. Dis.* **131**:443.

DeChatelet, L. R., Shirley, P. S., and Johnston, R. B., Jr., 1976, Effect of phorbol myristate acetate on the oxidative metabolism of human polymorphonuclear leukocytes, *Blood* **47**:545.

Diamond, R. D., Root, R. K., and Bennett, J. E., 1972, Factors influencing killing of *Cryptococcus neoformans* by human leukocytes *in vitro*, *J. Inf. Dis.* **125**:367.

Douglas, S. D., 1970, Analytic review: Disorders of phagocyte function, *Blood* **35**:851.

Drath, D. B., and Karnovsky, M. L., 1974, Bactericidal activity of metal-mediated peroxidase-ascorbate systems, *Infect. Immun.* **10**:1077.

Drath, D. B., and Karnovsky, M. L., 1975, Superoxide production by phagocytic leukocytes, *J. Exp. Med.* **141**:257.

Edelson, P. J., and Cohn, Z. A., 1973, Peroxidase-mediated mammalian cell cytotoxicity, *J. Exp. Med.* **138**:318.

Elsbach, P., 1973, On the interaction between phagocytes and micro-organisms, *N. Engl. J. Med.* **289**:846.

Engelfriet, C. P., Kr. von dem Borne, A. E. G., Beckers, D., and van Loghem, J. J., 1974, Autoimmune haemolytic anaemias: Serological and immunochemical characteristics of autoantibodies, *Ser. Haematol.* **7**:328.

Ericsson, Y, and Lundbeck, H., 1955a, Antimicrobial effect *in vitro* of the ascorbic acid oxidation. I. Effect on bacteria, fungi and viruses in pure culture, *Acta Pathol. Microbiol. Scand.* **37**:493.

Ericsson, Y., and Lundbeck, H., 1955b, Antimicrobial effect *in vitro* of the ascorbic acid oxidation. II. Effect of various chemical and physical factors, *Acta Pathol. Microbiol. Scand.* **37**:507.

Evans, R., and Alexander, P., 1972a, Role of macrophages in tumour immunity. 1. Cooperation between macrophages and lymphoid cells in syngeneic tumour immunity, *Immunology* **23**: 615.

Evans, R., and Alexander, P., 1972b, Mechanism of immunologically specific killing of tumour cells by macrophages, *Nature* **236**:168.

Evans, W. H., and Karnovsky, M. L., 1962, The biochemical basis of phagocytosis. IV. Some aspects of carbohydrate metabolism during phagocytosis, *Biochem.* **1**:159.

Fahimi, H. D., 1970, The fine structural localization of endogenous and exogenous peroxidase activity in Kupffer cells of rat liver, *J. Cell Biol.* **47**:247.

Felberg, N. T., and Schultz, J., 1972, Evidence that myeloperoxidase is composed of isozymes, *Arch. Biochem. Biophys.* **148**:407.

Fleer, A., van der Meulen, F. W., Linthout, E., Kr. von dem Borne, A. E. G., and Engelfriet, C. P., 1978a, Destruction of IgG-sensitized erythrocytes by human blood monocytes. Modulation of inhibition by IgG, *Br. J. Haematol.*, **39**:425.

Fleer, A., van Schaik, M. L. J., Kr. von dem Borne, A. E. G., and Engelfriet, C. P., 1978b, Destruction of sensitized erythrocytes by human monocytes *in vitro*. Effects of cytochalasin B, hydrocortisone and colchicine. *Scand. J. Immunol.* **8**:515.

Fleer, A., Roos, D., Kr. von dem Borne, A. E. G., and Engelfriet, C. P., 1979, Cytotoxic activity of human monocytes towards sensitized red cells is not dependent upon the generation of reactive oxygen species, *Blood* **54**:407.

Fong, K. L., McCay, P. B., Poyer, J. L., Keele, B. B., and Misra, H., 1973, Evidence that peroxidation of lysosomal membranes is initiated by hydroxyl free radicals produced during flavin enzyme activity, *J. Biol. Chem.* **248**:7792.

Fowles, R. E., Fajardo, I. M., Leibowitch, J. L., and David, J. R., 1973, The enhancement of macrophage bacteriostasis by products of activated lymphocytes, *J. Exp. Med.* **138**: 952.

Fridovich, I., 1974, Superoxide dismutases, *Adv. Enzymol.* **41**:35.

Gale, R. P., and Zighelboim, J., 1975, Polymorphonuclear leukocytes in antibody-dependent cellular cytotoxicity, *J. Immunol.* **114**:1047.

Gee, J. B. L., and Khandwala, A. S., 1976, Oxygen metabolism in the alveolar macrophage: Friend and foe?, *J. Reticuloendothel. Soc.* **19**:229.

Gee, J. B. L., Robin, E. D., Field, J. B., Smith, J. D., Tanser, A. R., and Kaskin, J., 1968, Roles of ATP and of peroxidative metabolism during phagocytosis in alveolar macrophages (AM), *J. Clin. Invest.* **47**:38A.

Gee, J. B. L., Vassallo, C. L., Bell, P., Kaskin, J., Basford, R. E., and Field, J. B., 1970, Catalase-dependent peroxidative metabolism in the alveolar macrophage during phagocytosis, *J. Clin. Invest.* **49**:1280.

Gee, J. B. L., Kaskin, J., Duncombe, M. P., and Vassallo, C. L., 1974, The effects of ethanol on some metabolic features of phagocytosis in the alveolar macrophage, *J. Reticuloendothel. Soc.* **15**:61.

Gill, P. G., Waller, C. A., and MacLennan, I. C. M., 1977, Relationships between different functional properties of human monocytes, *Immunology* **33**:873.

Ginsburg, I., and Sela, M. N., 1976, The role of leukocytes and their hydrolases in the persistence, degradation and transport of bacterial constituents in tissues: Relation to chronic inflammatory processes in staphylococcal, streptococcal, and mycobacterial infections, and in chronic periodontal disease, in: *Critical Reviews in Microbiology*, Vol. 4, pp. 249–332, CRC Press, Cleveland.

Goldstein, I. M., Roos, D., Kaplan, H. B., and Weissmann, G., 1975, Complement and immunoglobulins stimulate superoxide production by human leukocytes independently of phagocytosis, *J. Clin. Invest.* **56**:1155.

Goldstein, I. M., Kaplan, H. B., Radin, A., and Frosch, M., 1976, Independent effects of IgG and complement upon human polymorphonuclear leukocyte function, *J. Immunol.* **117**:1282.

Goldstein, I. M., Cerquiera, M., Lind, S., and Kaplan, H. B., 1977, Evidence that the superoxide-generating system of human leukocytes is associated with the cell surface, *J. Clin. Invest.* **59**:249.

Graham, R. C., Jr., Karnovsky, M. J., Shafer, A. W., Glass, E. A., and Karnovsky, M. L., 1967, Metabolic and morphological observations on the effect of surface-active agents on leukocytes, *J. Cell Biol.* **32**:629.

Greaves, M. F., and Brown, G., 1974, Purification of human T and B lymphocytes, *J. Immunol.* **112**:420.

Greenwood, M. F., Jones, E. A., Jr., and Holland P., 1978, Monocyte functional capacity in chronic neutropenia, *Am. J. Dis. Child.* **132**:131.

Gregory, E. M., Yost, F. J., and Fridovich, I., 1973, Superoxide dismutases of *Escherichia coli*: Intracellular localization and functions, *J. Bacteriol.* **115**:987.

Hard, G. C., 1970, Some biochemical aspects of the immune macrophage, *Br. J. Exp. Pathol.* **51**:97.

Hatch, G. E., Gardner, D. E., and Menzel, D. B., 1978, Chemiluminescence of phagocytic cells caused by *N*-formylmethionyl peptides, *J. Exp. Med.* **147**:182.

Hersey, P., 1973, Macrophage effector function: An *in vitro* system of assessment, *Transplantation* **15**:282.

Hibbs, J. B., 1974, Heterocytolysis of macrophages activated by *Bacillus* Calmette-Guérin: Lysosome exocytosis into tumor cells, *Science* **184**:468.

Hibbs, J. B., Lambert, L. H., and Remington, J. S., 1972, Possible role of macrophage-mediated non-specific cytotoxicity in tumor resistance, *Nature New Biol.* **235**:48.

Hirsch, J. G., and Fedorko, M. E., 1970, Morphology of mouse mononuclear phagocytes, in: *Mononuclear Phagocytes* (R. van Furth, ed.), pp. 7–28, Blackwell, Oxford.

Hirt, W. E., and Bonventre, P. F., 1973, Cultural, phagocytic, and bactericidal characteristics of peritoneal macrophages, *J. Reticuloendothel. Soc.* **13**:27.

Hodgson, E. K., and Fridovich, I., 1976, The mechanism of the activity-dependent luminescence of xanthine oxidase, *Arch. Biochem. Biophys.* **172**:202.

Holm, G., 1972, Lysis of antibody-treated human erythrocytes by human leukocytes and macrophages in tissue culture, *Int. Arch. Allergy Appl. Immunol.* **43**:671.

Holm, G., and Hammarström, S., 1973, Haemolytic activity of human blood monocytes. Lysis of human erythrocytes treated with anti-A serum, *Clin. Exp. Immunol.* **13**:29.

Holmes, B., Park, B. H., Malawista, S. E., Quie, P. G., Nelson, D. L., and Good, R. A., 1970, Chronic granulomatous disease in females. A deficiency of leukocyte glutathione peroxidase, *N. Engl. J. Med.* **283**:217.

Homan-Müller, J. W. T., Weening, R. S., and Roos, D., 1975, Production of hydrogen peroxide by phagocytizing human granulocytes, *J. Lab. Clin. Med.* **85**:198.

Huang, K. J., Donahoe, R. M., Gordon, F. B., and Dressler, H. R., 1971, Enhancement of phagocytosis by interferon-containing preparations, *Infect. Immun.* **4**:581.

Iyer, G. Y. N., Islam, M. F., and Quastel, J. H., 1961, Biochemical aspects of phagocytosis, *Nature* **192**:535.

Jacobs, A. A., Paul, B. B., Strauss, R. R., and Sbarra, A. J., 1970, The role of the phagocyte in host-parasite interactions. XXIII. Relation of bactericidal activity to peroxidase-associated decarboxylation and deamination, *Biochem. Biophys. Res. Commun.* **39**:284.

Jensen, M. S., and Bainton, D. F., 1973, Temporal changes in pH within the phagocytic vacuole of the polymorphonuclear neutrophilic leukocyte, *J. Cell Biol.* **56**:379.

Johnson, W. D. Jr., Mei, B., and Cohn, Z. A., 1977, The separation, long-term cultivation, and maturation of the human monocyte, *J. Exp. Med.* **146**:1613.

Johnston, R. B., Jr., and Newman, S. L., 1977, Chronic granulomatous disease, *Pediatr. Clin. North Am.* **24**:365.

Johnston, R. B., Jr., Keele, B. B., Jr., Misra, H. P., Lehmeyer, J. E., Webb, L. S., Baehner, R. L., and Rajagopalan, K. V., 1975, The role of superoxide anion generation in phagocytic bactericidal activity. Studies with normal and chronic granulomatous disease leukocytes, *J. Clin. Invest.* **55**:1357.

Johnston, R. B., Jr., Lehmeyer, J. E., and Guthrie, L. A., 1976, Generation of superoxide anion and chemiluminescence by human monocytes during phagocytosis and on contact with surface-bound immunoglobulin G, *J. Exp. Med.* **143**:1551.

Junqueira, L. C., Carneiro, J., and Contopoulos, A., 1977, Bone cells, in: *Basic Histology*, 2nd., pp. 120–122, Lange, Los Altos, Ca.

Kahn, A. J., and Simmons, D. J., 1975, Investigation of cell lineage in bone using a chimaera of chick and quail embryonic tissue, *Nature* **258**:325.

Kahn, A. J., Stewart, C. C., and Teitelbaum, S. L., 1978, Contact-mediated bone resorption by human monocytes *in vitro*, *Science* **199**:988.

Kakinuma, K., 1970, Metabolic control and intracellular pH during phagocytosis by polymorphonuclear leukocytes, *J. Biochem. (Tokyo)* **68**:177.

Kaplan, S. S., Perillie, P. E., and Finch, S. C., 1969, The effect of chloramphenicol on human leukocyte phagocytosis and respiration, *Proc. Soc. Exp. Biol. Med.* **130**:839.

Kaplow, L. S., Dauber, H., and Lerner, E., 1976, Assessment of monocyte esterase activity by flow cytophotometry, *J. Histochem. Cytochem.* **24**:363.

Karnovsky, M. L., Simmons, S., Glass, E. A., Shafer, A. W., and D'Arcy Hart, P., 1970, Metabolism of macrophages, in: *Mononuclear Phagocytes*, (R. van Furth, ed.), pp. 103–120, Blackwell, Oxford.

Karnovsky, M. L., Lazdins, J., and Simmons, S. R., 1975, Metabolism of activated mononuclear phagocytes at rest and during phagocytosis, in: *Mononuclear Phagocytes in Infection, Immunity and Pathology* (R. van Furth, ed.), pp. 424–438, Blackwell, Oxford.

Kay, A. B., White, A. G., Barclay, G. R., Darg, C., Raeburn, J. A., Uttley, W. S., McCrae, W. M., and Innes, E. M., 1975, Monocyte function in chronic benign neutropenia, *Lancet* **1**:391

Kearns, D. R., 1971, Physical and chemical properties of singlet molecular oxygen, *Chem. Rev.* **71**:395.

Keller, R., 1976, Cytostatic and cytocidal effects of activated macrophages, in: *Immunobiology of the Macrophage* (D. S. Nelson, ed.), pp. 487–508, Academic Press, New York.

King, G. W., Para, M. F., LoBuglio, A. F., and Sagone, A. L., Jr., 1975, Human monocyte metabolism: Male vs. female, *J. Reticuloendothel. Soc.* **17**:282.

Klebanoff, S. J., 1967a, A peroxidase-mediated antimicrobial system in leukocytes, *J. Clin. Invest.* **46**:1078.

Klebanoff, S. J., 1967b, Iodination of bacteria: A bactericidal mechanism, *J. Exp. Med.* **126**:1063.

Klebanoff, S. J., 1968, Myeloperoxidase-halide-hydrogen peroxide antibacterial system, *J. Bacteriol.* **95**:2131.

Klebanoff, S. J., 1969, Antimicrobial activity of catalase at acid pH, *Proc. Soc. Exp. Biol. Med.* **132**:571.

Klebanoff, S. J., 1970, Myeloperoxidase: Contribution to the microbicidal activity of intact leuko-cytes, *Science* **169**:1095.

Klebanoff, S. J., 1974, Role of the superoxide anion in the myeloperoxidase-mediated antimicrobial system, *J. Biol. Chem.* **249**:3724.

Klebanoff, S. J., and Clark, R. A., 1975, Hemolysis and iodination of erythrocyte components by a myeloperoxidase-mediated system, *Blood* **45**:699.

Klebanoff, S. J., and Green, W. L., 1973, Degradation of thyroid hormones by phagocytosing human leukocytes, *J. Clin. Invest.* **52**:60.

Klebanoff, S. J., and Hamon, C. B., 1972, Role of myeloperoxidase-mediated antimicrobial systems in intact leukocytes, *J. Reticuloendothel. Soc.* **12**:170.

Klebanoff, S. J., and Hamon, C. B., 1975, Antimicrobial systems of mononuclear phagocytes, in: *Mononuclear Phagocytes in Infection, Immunity and Pathology,* (R. van Furth, ed.), pp. 507–529, Blackwell, Oxford.

Krinsky, N. I., 1974, Singlet excited oxygen as a mediator of the antibacterial action of leukocytes, *Science* **186**:363.

Kwan, D., Epstein, M. B., and Norman, A., 1976, Studies on human monocytes with a multiparame-ter cell sorter, *J. Histochem. Cytochem.* **24**:355.

Lehrer, R. I., 1969, Antifungal effects of peroxidase systems, *J. Bacteriol.* **99**:361.

Lehrer, R. I., 1970, The fungicidal activity of human monocytes: A myeloperoxidase-linked mechanism, *Clin. Res.* **18**:408A.

Lehrer, R. I., 1971, Inhibition by sulfonamides of the candidacidal activity of human neutrophils, *J. Clin. Invest.* **50**:2498.

Lehrer, R. I., 1972, Functional aspects of a second mechanism of candidacidal activity by human neutrophils, *J. Clin. Invest.* **5**:2566.

Lehrer, R. I., 1973, Effects of colchicine and chloramphenicol on the oxidative metabolism and phagocytic activity of human neutrophils, *J. Infect. Dis.* **127**:40.

Lehrer, R. I., 1975, The fungicidal mechanisms of human monocytes. I. Evidence for myeloperoxidase-linked and myeloperoxidase-independent candicacidal mechanisms, *J. Clin. Invest.* **55**:338.

Lehrer, R. I., and Cline, M. J., 1969, Leukocyte myeloperoxidase deficiency and disseminated can-didiasis: The role of myeloperoxidase in resistance to *Candida* infection, *J. Clin. Invest.* **48**:1478.

Lehrer, R. I., Hanifin, J., and Cline, M. J., 1969, Defective bactericidal activity in myeloperoxidase-deficient human neutrophils, *Nature* **223**:78.

Lehrer, R. I., Mitchell, K. I., and Hake, R. B., 1974, The fungicidal proteins of mammalian neu-trophils, *J. Clin. Invest.* **53**:44A.

Lewis, S. M., Szur, L., and Dacie, J. V., 1960, The pattern of erythrocyte destruction in haemolytic anaemia, as studied with radio-active chromium, *Br. J. Haematol.* **6**:122.

LoBuglio, A. F., Cotran, R. S., and Jandl, J. H., 1967, Red cells coated with immunoglobulin G: Binding and sphering by mononuclear cells in man, *Science* **158**:1582.

Looke, E., and Rowley, D., 1962, The lack of correlation between sensitivity of bacteria to killing by macrophages or acidic conditions, *Aus. J. Exp. Biol. Med. Sci.* **40**:315.

Loos, J. A., Blok-Schut, B., van Doorn, R., Hoksbergen, R., Brutel de la Rivière, A., and Meerhof, L., 1976a, A method for the recognition and separation of human blood monocytes on density gradients, *Blood* **48**:731.

Loos, J. A., Blok-Schut, B., Kipp, B., van Doorn, R., and Meerhof, L., 1976b, Size distribution, electronic recognition, and counting of human blood monocytes, *Blood* **48**:743.

Lowrie, D. B., and Aber, V. R., 1977, Superoxide production by rabbit pulmonary alveolar macrophages, *Life Sci.* **21**:1575.

MacDonald, H. R., Bonnard, G. D., Sordat, B., and Zawodnik, S. A., 1975, Antibody-dependent cell-mediated cytotoxicity: Heterogeneity of effector cells in human peripheral blood, *Scand. J. Immunol.* **4**:487.

Mackaness, G. B., 1962, Cellular resistance to infection, *J. Exp. Med.* **116**:381.

Malawista, S. E., and Bodel, P. T., 1967, The dissociation by colchicine of phagocytosis from increased oxygen consumption in human leukocytes, *J. Clin. Invest.* **46**:786.

Mandell, G. L., 1970, Intraphagosomal pH of human polymorphonuclear neutrophils, *Proc. Soc. Exp. Biol. Med.* **134**:447.

Mandell, G. L., 1974, Bactericidal activity of aerobic and anaerobic polymorphonuclear neutrophils, *Infect. Immun.* **9**:337.

Mandell, G. L., 1975, Catalase, superoxide dismutase, and virulence of *Staphylococcus aureus. In vitro* and *in vivo* studies with emphasis on staphylococcal-leukocyte interactions, *J. Clin. Invest.* **55**:561.

Mandell, G. L., and Hook, E. W., 1969, Leukocyte function in chronic granulomatous disease of childhood, *Am. J. Med.* **47**:473.

McRipley, R. J., and Sbarra, A. J., 1967, Role of the phagocyte in host-parasite interactions. XII. Hydrogen peroxide myeloperoxidase bactericidal system in the phagocyte, *J. Bacteriol.* **94**:1425.

Miles, P. R., Lee, P., Trush, M. A., and van Dyke, K., 1977, Chemiluminescence associated with phagocytosis of foreign particles in rabbit alveolar macrophages, *Life Sci.* **20**:165.

Miller, T. E., 1969, Killing and lysis of gram-negative bacteria through the synergistic effect of hydrogen peroxide, ascorbic acid, and lysozyme, *J. Bacteriol.* **98**:949.

Miller, T. E., 1971, Metabolic event involved in the bactericidal activity of normal mouse macrophages, *Infect. Immun.* **3**:390.

Montarosso, A. M., and Myrvik, Q. N., 1977, HMS response in normal and BCG immunized rabbit alveolar macrophages, *J. Reticuloendothel. Soc.* **22**:9A.

Mundy, G. R., Altman, A. J., Gondek, M. D., and Bandelin, J. G., 1977, Direct resorption of bone by human monocytes, *Science* **196**:1109.

Muschel, L. H., 1965, Immune bactericidal and bacteriolytic reactions, in: *Ciba Foundation Symposium on Complement* (G. E. W. Wolstenholme and J. Knight, eds.), pp. 155–169, Little, Brown, Boston.

Muschel, L. H., Carey, W. F., and Baron, L. S., 1959, Formation of bacterial protoplasts by serum components, *J. Immunol.* **82**:38.

Myrvik, Q. N., and Evans, D. G., 1967, Metabolic and immunologic activities of alveolar macrophages, *Arch. Environ. Health* **14**:92.

Nathan, C. F., and Root, R. K., 1977, Hydrogen peroxide release from mouse peritoneal macrophages. Dependence on sequential activation and triggering, *J. Exp. Med.* **146**:1648.

Nathan, C. F., Karnovsky, M. L., and David, J. R., 1971, Alterations of macrophage functions by mediators from lymphocytes, *J. Exp. Med.* **133**:1356.

Nathan, D. G., Baehner, R. L., and Weaver, D. K., 1969, Failure of nitro blue tetrazolium reduction in the phagocytic vacuoles of leukocytes in chronic granulomatous disease, *J. Clin. Invest.* **48**:1895.

Nathanson, S. D., Zamfirescu, P. L., Drew, S. I., and Wilbur, S., 1977, Two-step separation of human peripheral blood monocytes on discontinuous density gradients of colloidal silica-polyvinylpyrrolidinone, *J. Immunol. Meth.* **18**:225.

Nelson, R. D., Mills, E. L., Simmons, R. L., and Quie, P. G., 1976, Chemiluminescence response of phagocytizing human monocytes, *Infect. Immun.* **14**:129.

Nichols, B. A., and Bainton, D. F., 1973, Differentiation of human monocytes in bone marrow and blood: Sequential formation of two granule populations, *Lab. Invest.* **29**:27.

Nichols, B. A., and Bainton, D. F., 1975, Ultrastructure and cytochemistry of mononuclear phagocytes, in: *Mononuclear Phagocytes in Immunity, Infection and Pathology*, (R. van Furth, ed.), pp. 15–55, Blackwell, Oxford.

Nichols, B. A., Bainton, D. F., and Farquhar, M. G., 1971, Differentiation of monocytes. Origin, nature, and fate of their azurophilic granules, *J. Cell Biol.* **50**:498.

Odeberg, H., and Olsson, I., 1976a, Microbicidal mechanisms of human granulocytes: Synergistic effects of granulocyte elastase and myeloperoxidase or chymotrypsin-like cationic protein, *Infect. Immun.* **14**:1276.

Odeberg, H., and Olsson, I., 1976b, Mechanisms for the microbicidal activity of cationic proteins of human granulocytes, *Infect. Immun.* **14**:1269.

Ødegaard, A., Viken, K. E., and Lamvik, J., 1974, Structural and functional properties of blood monocytes cultured *in vitro, Acta Pathol. Microbiol. Scand. Sect. B* **82**:223.

Olivotto, M., and Bomford, R., 1974, *In vitro* inhibition of tumour cell growth and DNA synthesis by peritoneal and lung macrophages from mice infected with *Corynebacterium parvum, Int. J. Cancer* **13**:478.

Oren, R., Farnham, A. E., Saito, K., Milofsky, E., and Karnovsky, M. L., 1963, Metabolic patterns in three types of phagocytizing cells, *J. Cell Biol.* **17**:487.

Ouchi, E., Selvaraj, R. J., and Sbarra, A. J., 1965, The biochemical activities of rabbit alveolar macrophages during phagocytosis, *Exp. Cell Res.* **40**:456.

Para, M., Sagone, A., Balcerzak, S., and LoBuglio, A., 1972, Metabolism of normal and activated monocytes, *Clin. Res.* **20**:742A.

Park, B. H., Biggar, W. D., L'Esperance, P., and Good, R. A., 1972, N. B. T. test on monocytes of neutropenic patients, *Lancet* **1**:1064.

Patriarca, P., Cramer, R., Marussi, M., Moncalvo, S., and Rossi, F., 1971, Phospholipid splitting and metabolic stimulation in polymorphonuclear leukocytes, *J. Reticuloendothel. Soc.* **10**:251.

Patriarca, P., Dri, P., and Rossi, F., 1974a, Superoxide dismutase in leukocytes, *FEBS Lett.* **43**:247.

Patriarca, P., Basford, R. E., Cramer, R., Dri, P., and Rossi, F., 1974b, Studies on the NADPH oxidizing activity in polymorphonuclear leukocytes. The mode of association with the granule membrane, the relationship to myeloperoxidase and the interference of hemoglobin with NADPH oxidase determination, *Biochim. Biophys. Acta* **362**:221.

Paul, B. B., and Sbarra, A. J., 1968, The role of the phagocyte in host-parasite interactions. XIII. The direct quantitative estimation of H_2O_2 in phagocytizing cells, *Biochim. Biophys. Acta* **156**:168.

Paul, B. B., Strauss, R. R., Jacobs, A. A., and Sbarra, A. J., 1970a, Function of H_2O_2, myeloperoxidase, and hexose monophosphate shunt enzymes in phagocytizing cells from different species, *Infect. Immun.* **1**:338.

Paul, B. B., Jacobs, A. A., Strauss, R. R., and Sbarra, A. J., 1970b, The role of the phagocyte in host-parasite interactions. XXIV. Aldehyde generation by the myeloperoxidase-H_2O_2-chloride antimicrobial system: A possible *in vivo* mechanism of action, *Infect. Immun.* **2**:414.

Paul, B. B., Strauss, R. R., Selvaraj, R. J., and Sbarra, A. J., 1973, Peroxidase-mediated antimicrobial activities of alveolar macrophage granules, *Science* **181**:849.

Philpott, G. W., Bower, R. J., and Parker, C. W., 1973, Selective iodination and cytotoxicity of tumor cells with an antibody-enzyme conjugate, *Surgery (St. Louis)* **74**:51.

Pincus, S. H., and Klebanoff, S. J., 1971, Quantitative leukocyte iodination, *New Eng. J. Med.* **284**:744.

Policard, A., Collet, A., and Pregermain, S., 1956, Electron microscope studies on alveolar cells from mammals, in: *Electron Microscopy, Proc. Stockholm Conf.* (F. S. Sjöstrand and J. Rodin, eds.), pp. 244–256, Academic Press, New York.

Poplack, D. G., Bonnard, G. D., Holiman, B. J., and Blaese, R. M., 1976, Monocyte-mediated antibody-dependent cellular cytotoxicity: A clinical test of monocyte function, *Blood* **48**:809.

Poulter, L. W., and Turk, J. L., 1975, Studies on the effect of soluble lymphocyte products (lymphokines) on macrophage physiology. I. Early changes in enzyme activity and permeability, *Cell. Immunol.* **20**:12.

Ratzan, K. R., Musher, D. M., Keusch, G. T., and Weinstein, L., 1972, Correlation of increased metabolic activity, resistance to infection, enhanced phagocytosis, and inhibition of bacterial growth by macrophages from *Listeria*- and BCG-infected mice, *Infect. Immun.* **5**:499.

Reed, P. W., 1969, Glutathione and the hexose monophosphate shunt in phagotizing and hydrogen peroxide-treated rat leukocytes, *J. Biol. Chem.* **244**:2459.

Reed, P. W., and Tepperman, J., 1969, Phagocytosis-associated metabolism and enzymes in the rat polymorphonuclear leukocyte, *Am. J. Physiol.* **216**:223.

Reiss, M., and Roos, D., 1978, Differences in oxygen metabolism of phagocytosing monocytes and neutrophils, *J. Clin. Invest.* **61**:480.

Repaske, R., 1956, Lysis of gram-negative bacteria by lysozyme, *Biochim. Biophys. Acta* **22**:189.

Repaske, R., 1958, Lysis of gram-negative organisms and the role of versene, *Biochim. Biophys. Acta* **30**:225.

Repine, J. E., White, J. G., Clawson, C. C., and Holmes, B. M., 1974a, The influence of phorbol myristate acetate on oxygen consumption by polymorphonuclear leukocytes, *J. Lab. Clin. Med.* **83**:911.

Repine, J. E., White, J. G., Clawson, C. C., and Holmes, B. M., 1974b, Effects of phorbol myristate acetate on the metabolism and ultrastructure of neutrophils in chronic granulomatous disease, *J. Clin. Invest.* **54**:83.

Repine, J. E., White, J. G., Clawson, C. C., and Holmes, B. M., 1975, The influence of phorbol myristate acetate on the metabolism of neutrophils from carriers of sex-linked chronic granulomatous disease, *J. Lab. Clin. Med.* **85**:82.

Rinehart, J., Lange, P., Gormus, B. J., and Kaplan, M. E., 1978, Human monocyte-induced tumor cell cytotoxicity, *Blood* **52**:211.

Rister, M., and Baehner, R. L., 1977, Effect of hyperoxia on superoxide anion and hydrogen peroxide production of polymorphonuclear leucocytes and alveolar macrophages, *Br. J. Haematol.* **36**:241.

Robbins, D., Fahimi, H. D., and Cotran, R. S., 1971, Fine structural cytochemical localization of peroxidase activity in rat peritoneal cells: Mononuclear cells, eosinophils and mast cells, *J. Histochem. Cytochem.* **19**:571.

Rocklin, R. E., Winston, C. T., and David, J. R., 1974, Activation of human blood monocytes by products of sensitized lymphocytes, *J. Clin. Invest.* **53**:559.

Rodey, G. E., Park, B. H., Windhorst, D. B., and Good, R. A., 1969, Defective bactericidal activity of monocytes in fatal granulomatous disease, *Blood* **33**:813.

Romeo, D., Zabucchi, G., Soranzo, M. R., and Rossi, F., 1971, Macrophage metabolism: Activation of NADPH oxidation by phagocytosis, *Biochem. Biophys. Res. Commun.* **45**:1056.

Romeo, D., Cramer, R., Marzi, T., Soranzo, M. R., Zabucchi, G., and Rossi, F., 1973a, Peroxidase activity of alveolar and peritoneal macrophages, *J. Reticuloendothel. Soc.* **13**:399.

Romeo, D., Zabucchi, G., Marzi, T., and Rossi, F., 1973b, Kinetic and enzymatic features of metabolic stimulation of alveolar and peritoneal macrophages challenged with bacteria, *Exp. Cell Res.* **78**:423.

Romeo, D., Zabucchi, G., and Rossi, F., 1973c, Reversible metabolic stimulation of polymorphonuclear leukocytes and macrophages by Concanavalin A, *Nature* **243**:111.

Roos, D., and Loos, J. A., 1970, Changes in the carbohydrate metabolism of mitogenically stimulated human peripheral lymphocytes. I. Stimulation by phytohaemagglutinin, *Biochim. Biophys. Acta* **222**:565.

Roos, D., van Schaik, M. L. J., Weening, R. S., and Wever, R., 1977, Superoxide generation in relation to other oxidative reactions in human polymorphonuclear leukocytes, in: *Superoxide and Superoxide Dismutases* (A. M. Michelson, J. M. McCord, and I. Fridovich, eds.), pp. 307–316, Academic Press, New York.

Roos, D., Reiss, M., Balm. A. J. M., Palache, A. M., Cambier, P. H., and Van der Stijl-Neijenhuis, J. S., 1979, A metabolic comparison between human blood monocytes and neutrophils, in: *Macrophages and Lymphocytes: Nature, Functions and Interaction* (M. R. Escobar and H. Friedman, eds.), pp. 29–36, Plenum Press, New York.

Root, R. K., and Metcalf, J. A., 1977, H_2O_2 release from human granulocytes during phagocytosis. Relationship to superoxide anion formation and cellular catabolism of H_2O_2: Studies with normal and cytochalasin B-treated cells, *J. Clin. Invest.* **60**:1266.

Root, R. K., and Stossel, T. P., 1974, Myeloperoxidase-mediated iodination by granulocytes. Intracellular site of operation and some regulating factors, *J. Clin. Invest.* **53**:1207.

Root, R. K., Rosenthal, A. S., and Balestra, D. J., 1972, Abnormal bactericidal, metabolic, and lysosomal functions of Chediak–Higashi syndrome leukocytes, *J. Clin. Invest.* **51**:649.

Root, R. K., Metcalf, J., Oshino, N., and Chance, B., 1975, H_2O_2 release from human granulocytes during phagocytosis. I. Documentation, quantitation, and some regulating factors, *J. Clin. Invest.* **55**:945.

Rosen, H., and Klebanoff, S. J., 1976, Chemiluminescence and superoxide production by myeloperoxidase-deficient leukocytes, *J. Clin. Invest.* **58**:50.

Rossi, F., and Zatti, M., 1966, Effect of phagocytosis on the carbohydrate metabolism of polymorphonuclear leukocytes, *Biochim. Biophys. Acta* **121**:110.

Rossi, F., Zabucchi, G., and Romeo, D., 1975, Metabolism of phagocytosing mononuclear phagocytes, in: *Mononuclear Phagocytes in Infection, Immunity and Pathology* (R. van Furth, ed.), pp. 441–460, Blackwell, Oxford.

Rümke, C. L., Bezemer, P. D., and Kuik, D. J., 1975, Normal values and least significant differences for differential leukocyte counts, *J. Chron. Dis.* **28**:661.

Sachs, F. L., and Gee, J. B. L., 1973, Comparison of the effects of phagocytosis and phospholipase C on metabolism and lysozyme release in rabbit alveolar macrophages, *J. Reticuloendothel. Soc.* **14**:52.

Sagone, A. L., Jr., King, G. W., and Metz, E. N., 1976, A comparison of the metabolic response to phagocytosis in human granulocytes and monocytes, *J. Clin. Invest.* **57**:1352.

Salin, M. L., and McCord, J. M., 1974, Superoxide dismutases in polymorphonuclear leukocytes, *J. Clin. Invest.* **54**:1005.

Salmon, S. E., Cline, M. J., Schultz, J., and Lehrer, R. I., 1970, Myeloperoxidase deficiency. Immunologic study of a genetic leukocyte defect, *New Eng. J. Med.* **282**:250.

Salton, M. R. J., 1957, The properties of lysozyme and its action on microorganisms, *Bacteriol. Rev.* **21**:82.

Sanderson, R. J., Shepperdson, F. T., Vatter, A. E., and Talmage, D. W., 1977, Isolation and enumeration of peripheral blood monocytes, *J. Immunol.* **118**:1409.

Sbarra, A. J., and Karnovsky, M. L., 1959, The biochemical basis of phagocytosis. I. Metabolic changes during the ingestion of particles by polymorphonuclear leukocytes, *J. Biol. Chem.* **234**:1355.

Schneider, C., Gennaro, R., de Nicola, G., and Romeo, D., 1978, Secretion of granule enzymes from alveolar macrophages. Regulation by intracellular Ca^{2+}-buffering capacity, *Exp. Cell Res.*, **112**:249.

Schultz, J., and Kaminker, K., 1962, Myeloperoxidase of the leucocyte of normal human blood. I. Content and localization, *Arch. Biochem. Biophys.* **96**:465.

Schultz, J., and Shmukler, H. W., 1964, Myeloperoxidase of the leucocyte of normal human blood. II. Isolation, spectrophotometry and amino acid analysis, *Biochemistry* **3**:1234.

Shohet, S. B., Pitt, J., Baehner, R. L., and Poplack, D. G., 1974, Lipid peroxidation in the killing of phagocytized pneumococci, *Infect. Immun.* **19**:1321.

Shortman, K., 1968, The separation of different cell classes from lymphoid organs. II. The purification and analysis of lymphocyte populations by equilibrium density gradient centrifugation, *Austr. J. Exp. Biol. Med. Sci.* **46**:375.

Simchowitz, L., and Schur, P. H., 1976, Lectin-dependent neutrophil-mediated cytotoxicity. I. Characteristics, *Immunology* **31**:303.

Simmons, S. R., and Karnovsky, M. L., 1973, Iodinating ability of various leukocytes and their bactericidal activity, *J. Exp. Med.* **138**:44.

Simon, L. M., Axline, S. G., Horn, B. R., and Robin, E. D., 1973, Adaptations of energy metabolism in the cultivated macrophage, *J. Exp. Med.* **138**:1413.

Simon, L. M., Robin, E. D., Phillips, J. R., Acevedo, J., Axline, S. G., and Theodore, J., 1977, Enzymatic basis for bioenergetic differences of alveolar versus peritoneal macrophages and enzyme regulation by molecular O_2, *J. Clin. Invest.* **59**:443.

Sorrell, T. C., Lehrer, R. I., and Cline, M. J., 1978, Mechanism of non-specific macrophage-mediated cytotoxicity: Evidence for lack of dependence upon oxygen, *J. Immunol.* **120**:347.

Spizizen, J., 1962, Preparation and use of protoplasts, in: *Methods in Enzymology* (S. R. Colowick and N. O. Kaplan, eds.), Vol. 5, pp. 122–134, Academic Press, New York.

Splinter, T. A. W., Beudeker, M., and van Beek, A., 1978, Changes in cell density induced by Isopaque, *Exp. Cell Res.* **111**:245.

Steigbigel, R. T., Lambert, L. H., Jr., and Remington, J. S., 1974, Phagocytic and bactericidal properties of normal human monocytes, *J. Clin. Invest.* **53**:131.

Stelmaszynska, T., and Zgliczynski, J. M., 1974, Myeloperoxidase of human neutrophilic granulocytes as chlorinating enzyme, *Eur. J. Biochem.* **45**:305.

Stjernholm, R. L., and Manak, R. C., 1970, Carbohydrate metabolism in leukocytes. XIV. Regulation of pentose cycle activity and glycogen metabolism during phagocytosis, *J. Reticuloendothel. Soc.* **8:**550.

Stjernholm, R. L., Dimitrov, N. V., and Pijanowski, L. J., 1969, Carbohydrate metabolism in leukocytes. IX. Citric acid cycle activity in human neutrophils, *J. Reticuloendothel. Soc.* **6:**194.

Stossel, T. P., Root, R. K., and Vaughan, M., 1972a, Phagocytosis in chronic granulomatous disease and the Chediak-Higashi syndrome, *N. Engl. J. Med.* **286:**120.

Stossel, T. P., Mason, R. J., Pollard, T. D., and Vaughan, M., 1972b, Isolation and properties of phagocytic vesicles. II. Alveolar macrophages, *J. Clin. Invest.* **51:**604.

Stossel, T. P., Mason, R. J., and Smith, A. L., 1974, Lipid peroxidation by human blood phagocytes, *J. Clin. Invest.* **54:**638.

Strauss, R. R., Paul, B. B., Jacobs, A. A., and Sbarra, A. J., 1969, The role of the phagocyte in host-parasite interactions. XIX. Leukocytic glutathione reductase and its involvement in phagocytosis, *Arch. Biochem. Biophys.* **135:**265.

Strauss, R. R., Paul, B. B., Jacobs, A. A., and Sbarra, A. J., 1970, Role of the phagocyte in host-parasite interactions. XXII. H_2O_2-dependent decarboxylation and deamination by myeloperoxidase and its relationship to antimicrobial activity, *J. Reticuloendothel. Soc.* **7:**745.

Strauss, R. R., Paul, B. B., Jacobs, A. A., and Sbarra, A. J., 1971, Role of the phagocyte in host-parasite interactions. XXVII. Myeloperoxidase $H_2O_2 - Cl^-$-mediated aldehyde formation and its relationship to antimicrobial activity, *Infect. Immun.* **3:**595.

Stubbs, M., Vreeland Kühner, A., Glass, E. A., David, J. R., and Karnovsky, M. L., 1973, Metabolic and functional studies on activated mouse macrophages, *J. Exp. Med.* **137:**537.

Takanaka, K., and O'Brien, P. J., 1975, Mechanisms of H_2O_2 formation by leukocytes. Evidence for a plasma membrane location, *Arch. Biochem. Biophys.* **169:**428.

Tauber, A. I., and Babior, B. M., 1977, Evidence for hydroxyl radical production by human neutrophils, *J. Clin. Invest.* **60:**374.

Territo, M., and Cline, M. J., 1976, Macrophages and their disorders in man, in: *Immunobiology of the Macrophage* (D. S. Nelson, ed.), pp. 593–616, Academic Press, New York.

Thalinger, K. K., and Mandell, G. L., 1971, Bactericidal activity of macrophages in an anaerobic environment, *J. Reticuloendothel. Soc.* **9:**393.

Tsan, M. F., 1977, Stimulation of the hexose monophosphate shunt independent of hydrogen peroxide and superoxide production in rabbit alveolar macrophages during phagocytosis, *Blood* **50:**935.

Undritz, E., 1966, Die Alius-Grignaschi-Anomalie: der erblich-konstitutionelle Peroxydase Defekt der Neutrophilen und Monozyten, *Blut* **14:**129.

van Berkel, T. J. C., and Kruyt, J. K., 1977a, Distribution and some properties of NADPH and NADH oxidase in parenchymal and non-parenchymal liver cells, *Arch. Biochem. Biophys.* **179:**8.

van Berkel, T. J. C., and Kruyt, J. K., 1977b, Identity of peroxidatic activities in non-parenchymal rat liver cells in relation to parenchymal liver cells, in: *Kupffer Cells and Other Liver Sinusoidal Cells* (E. Wisse, and D. L. Knook, eds.), pp. 307–314, Elsevier, Amsterdam.

van Berkel, T. J. C., Kruyt, J. K., Slee, R. G., and Koster, J. F., 1977, Identity and activities of superoxide dismutase in parenchymal and nonparenchymal cells from rat liver, *Arch. Biochem. Biophys.* **179:**1.

van der Ploeg, M., Streefkerk, J. G., Daems, W. T., and Brederoo, P., 1974, Quantitative aspects of cytochemical peroxidase reactions with 3,3'-diaminobenzidine and 5,6-dihydroxyindole as substrates, in: *Electron Microscopy and Cytochemistry* (E. Wisse, W. T. Daems, I. Molenaar, and P. van Duijn, eds.), pp. 123–126, North Holland, Amsterdam.

van Furth, R., Hirsch, J. G., and Fedorko, M. E., 1970, Morphology and peroxidase cytochemistry of mouse promonocytes, monocytes and macrophages, *J. Exp. Med.* **132:**794.

Viken, K. E., and Ødegaard, A., 1974, Phagocytosis of heat-killed radio-labelled *Candida albicans* by human blood monocytes cultured *in vitro*, *Acta Pathol. Microbiol. Scand. Sect. B*, **82:**235.

Vildé, J. L., and Vildé, F., 1976, Nitroblue tetrazolium reduction by human macrophages: Studies in chronic granulomatous disease, in: *The Reticuloendothelial System in Health and Disease* (S. M. Reichard, M. R. Escobar, and H. Friedman, eds.), pp. 139–144, Plenum Press, New York.

Vogt, M. T., Thomas, C., Vassallo, C. L., Basford, R. E., and Gee, J. B. L., 1971, Glutathione-dependent peroxidative metabolism in the alveolar macrophage, *J. Clin. Invest.* **50:**401.

Walker, D. G., 1975, Bone resorption restored in osteopetrotic mice by transplants of normal bone marrow and spleen cells, *Science* **190**:784.

Ward, P. A., 1970, Comparisons and contrasts in the chemotactic behavior of mononuclear cells and neutrophils, in: *Cellular Interactions in the Immune Response*, Second International Convention on Immunology, Buffalo, N.Y. (S. Cohen, G. Gudkowicz, and R. T. McCluskey, eds.), p. 191, Karger, Basel.

Waters, M. D., Vaughan, T. O., Campbell, J. A., Stead, A. G., and Coffin, D. L., 1975, Adenosine triphosphate concentration and phagocytic activity in rabbit alveolar macrophages exposed to divalent cations, *J. Reticuloendothel. Soc.* **18**:29A.

Weening, R. S., Wever, R., and Roos, D., 1975, Quantitative aspects of the production of superoxide radicals by phagocytizing human granulocytes, *J. Lab. Clin. Med.* **85**:245.

Weening, R. S., Roos, D., van Schaik, M. L. J., Voetman, A. A., de Boer, M., and Loos, J. A., 1977, The role of glutathione in the oxidative metabolism of phagocytic leukocytes. Studies in a family with glutathione reductase deficiency, in: *Movement, Metabolism and Bactericidal Mechanisms of Phagocytes* (F. Rossi, P. L. Patriarca, and D. Romeo, eds.), pp. 277–283, Piccin, Padua.

Weiss, S. J., King, G. W., and LoBuglio, A. F., 1975, Quantitative NBT reduction and superoxide (O_2^-) generation by monocytes: The effect of cyclic AMP, *Clin. Res.* **23**:497A.

Weiss, S. J., King, G. W., and LoBuglio, A. F., 1977, Evidence for hydroxyl radical generation by human monocytes, *J. Clin. Invest.* **60**:370.

Weiss, S. J., Rustagi, P. K., and LoBuglio, A. F., 1978, Human granulocyte generation of hydroxyl radical, *J. Exp. Med.* **147**:316.

West, J., Morton, D. J., Esmann, V., and Stjernholm, R. L., 1968, Carbohydrate metabolism in leukocytes. VIII. Metabolic activities of the macrophage, *Arch. Biochem. Biophys.* **124**:85.

Weston, W. L., Dustin. R. D., and Hecht, S. K., 1975, Quantitative assays of human monocyte-macrophage function, *J. Immunol. Meth.* **8**:213.

Wever, R., Roos, D., Weening, R. S., Vulsma, T., and van Gelder, B. F., 1976, An EPR study of myeloperoxidase in human granulocytes, *Biochim. Biophys. Acta* **421**:328.

Widman, J.-J., Cotran, R. S., and Fahimi, H. D., 1972, Mononuclear phagocytes (Kupffer cells) and endothelial cells. Identification of two functional cell types in rat liver sinusoids by endogenous peroxidase activity, *J. Cell Biol.* **52**:159.

Williams, N., and Shortman, K., 1972, The separation of different cell classes from lymphoid organs. The effect of pH on the buoyant density of lymphocytes and erythrocytes, *Austr. J. Exp. Biol. Med. Sci.* **50**:133.

Wintrobe, M. M., Lee, G. R., Boggs, D. R., Bithell, T. C., Athens, J. W., and Foerster, J., 1974, Granulocytes and monocytes. Morphology and chemical properties, in: *Clinical Hematology*, 7th ed., pp. 221–285, Lea and Febiger, Philadelphia.

Yost, F. J., and Fridovich, I., 1974, Superoxide radicals and phagocytosis, *Arch. Biochem. Biophys.* **161**:395.

Zatti, M., and Rossi, F., 1967, Relationship between glycolysis and respiration in surfactant-treated leukocytes, *Biochim. Biophys. Acta* **148**:553.

Zeijlemaker, W. P., Roos, M. T. L., Schellekens, P. T. A., and Eijsvoogel, V. P., 1975, Antibody-dependent human lymphocytotoxicity: A microassay system, *Eur. J. Immunol.* **5**:579.

Zeya, H. I., and Spitznagel, J. K., 1968, Arginine-rich proteins of polymorphonuclear leukocyte lysosomes. Antimicrobial specificity and biochemical heterogeneity, *J. Exp. Med.* **127**:927.

Zeya, H. I., Keku, E., DeChatelet, L. R., Cooper, M. R., and Spurr, C. L., 1978, Isolation of enzymatically homogeneous populations of human lymphocytes, monocytes, and granulocytes by zonal centrifugation, *Am. J. Pathol.* **90**:33.

Zgliczynski, J. M., Stelmaszynska, T., Ostrowski, W., Naskalski, J., and Sznajd, J., 1968, Myeloperoxidase of human leukaemic leukocytes. Oxidation of amino acids in the presence of hydrogen peroxide, *Eur. J. Biochem.* **4**:540.

Zgliczynski, J. M., Stelmaszynska, T., Domanski, J., and Ostrowski, W., 1971, Chloramines as intermediates of oxidation reactions of amino acids by myeloperoxidase, *Biochim. Biophys. Acta* **235**:419.

9

Myeloperoxidase

JULIUS SCHULTZ

1. HISTORICAL INTRODUCTION

Dr. K. Agner sought to prepare catalase from sources other than liver because preparations of catalase from liver by the methods of Sumner and Dounce retained a blue fluorescence throughout the purification procedure. There was a question as to the relation of the fluorescence to catalase activity. Using tubercular empyema, principally containing neutrophils known to have catalactic activity, he discovered a green heme enzyme he called verdoperoxidase. In an extensive doctoral thesis Agner described its chemical and physical properties, its absorbtion spectra in the oxidized and reduced states, and its stability (Agner, 1971). The fact that the heme could not be separated from the apoprotein by acid, acetone, or silver acetate showed that its structure differed from hemoglobin, catalase, and myoglobin; but that like cytochrome c and lactoperoxidase the heme was covalently bound. Unlike cytochrome c the linkage was not through a thioether. Thus the new enzyme was unique in nature as to its heme and its linkage to the peptide chain. Agner's continued contributions led to the demonstration that verdoperoxidase [later changed to myeloperoxidase (MPO), to indicate its origin, since other enzymes were also green], was able to detoxify diphtheria toxin, thus indicating a possible biological function (Agner, 1950).

Many problems arose in regard to the structure and function of the new enzyme, but the paucity of available material limited the rate of research to resolve them. Agner later used the pus of infected dog uteri to modify his procedure, and eventually prepared the enzyme in a crystalline state by procedures difficult to reproduce. He also found on solution in pyridine that the enzyme separated into a soluble and insoluble fraction with different absorption spectra on the addition of methanol (Agner, 1958).

About ten years after Agner's discovery Schultz and co-workers, in another laboratory, initiated an investigation of the peroxidase of experimental rat chloroma which had become readily available (Schultz *et al.*, 1954). Myeloperoxidase was prepared using 1RC-50 for the first time in good yield

JULIUS SCHULTZ • Papanicolaou Cancer Research Institute, Miami, Florida 33101; and Department of Biochemistry, University of Miami Medical School, Coral Gables, Florida 33124. This work was supported by NIH Grant CA10904.

(Schultz et al., 1957). The spectra obtained by Agner's procedures were identical to that reported by him, while the use of 1RC-50 resulted in crystallizable material; the crystals still retained the red fluorescence characteristic of experimental chloroma (Schultz, 1958). When this source became unavailable, Schultz obtained from Merck Sharpe and Dohme weekly shipments of one pint of buffy coat from which adequate quantities of the enzyme were prepared. From these preparations the amino acid composition, the nature of the porphyrin of the heme, the linkage to the peptide chain, and its relation to cytochrome oxidase were reported (Schultz and Schmukler, 1964). The subunit structure (Schultz et al., 1972), the presence of isoenzymes (Felberg and Schultz, 1972), the possible presence of two hemes (Schultz et al., 1972), the isolation of the peroxidase-containing granules (Schultz and Kaminker, 1962), and the inhibition of tumor growth in methyl-cholanthrene-induced mammary adenocarcinoma were reported (Schultz et al., 1976). When human enzyme became unavailable, the dog enzyme became the principal source in this laboratory, and, with the development of new procedures, readily crystallizable material in good yield became available (Harrison et al., 1977).

The development of interest in the function of myeloperoxidase grew rapidly after the discoveries from Klebanoff's laboratory (see chapter 11 of this volume) that the killing of bacteria, fungi, and, eventually, cancer cells occurred by halogenation through the system $MPO-H_2O_2-halogen$ (Klebanoff, 1971). These findings helped explain the metabolic defects associated with the newly discovered chronic granulomatous disease of children (Baehner and Karnovsky, 1968) and patients lacking MPO in their neutrophils (Lehrer et al., 1972). Finally, from the laboratories at Karnovsky and of Sbarra (see the other chapters of this volume) extensive research on the phagocytic process was carried out. The stage was, therefore, set for more detailed study of the chemical structure, and its relation to biological activity, of myeloperoxidases, contributions being made by a number of laboratories. The current state of the chemistry of MPO will be reviewed here.

2. PREPARATION OF MYELOPEROXIDASE

The first preparations for myeloperoxidase by Agner were made by homogenization with ammonium sulfate in the presence of ether. The cake produced was then redissolved and followed by repeated precipitation with ammonium sulfate after removal of the sulfate with barium acetate (Agner, 1941). He later modified this by using hydroxyapatite, which resulted in material that was crystallized (Agner, 1950). This latter was from infected dog uteri. Schultz and his co-workers followed this procedure in the first preparations from experimental chloromas but later modified it by introducing CG-50 ion exchange resin (Schultz et al., 1957). This was done by redissolving the ammonium sulfate precipitate in 0.1 molar phosphate and eluting from the column stepwise up to 0.3 M phosphate. (From these preparations the crystalline MPO was prepared.) In subsequent years, with human myeloperoxidase, trypsin digestion was followed by initial ammonium sulfate precipitation and column chromatography using IRC-50 (Schultz and Schmukler, 1964). For many years, the final step of ammonium sulfate precipitation was replaced by using free flow electrophoresis on a Brinkman apparatus; this provided excellent material since the value of A_{430}/A_{280} was greater than 0.8.

The trypsinized material has often been criticized as giving rise to material that may have nicked peptide chains. However, this was reexamined, and it was repeatedly shown that the physical characteristics of the enzyme were not different from those obtained by procedures without trypsin (Felberg *et al.*, 1969). Acrylamide gel analysis of preparations made with or without trypsin showed the same distribution of isoenzymes.

New interest in MPO resulted in a variety of methods of preparation from the laboratories of Rohrer *et al.* (1966), Zgliczynski *et al.* (1968), Olsson *et al.* (1972), and Himmelhoch *et al.* (1969). None of the latter led to readily crystallized material. The introduction of the use of cetyltrimethylammonium bromide as a detergent (Desser *et al.*, 1972) increased the yields and provided products of high purity (Harrison *et al.*, 1977). Bakkenist *et al.* (1978) utilized the same detergent and normal human leukocytes, while Harrison *et al.* (1977) earlier reported the preparation of crystalline enzyme from infected dog uteri. Because of the ease of crystallization of these preparations and because the identical procedure has been found suitable for human leukocytes it is described in detail here.

2.1. PURIFICATION OF CANINE MYELOPEROXIDASE*

The peroxidase was solubilized by extraction of 1 liter of canine pus with 1% cetyltrimethylammonium bromide in 0.1 M K_2HPO_4 in a total volume of 2 liters at 4°C. Following centrifugation, the peroxidase was precipitated by the addition of ammonium sulfate to 55% saturation (4°C), dialyzed against 0.05 M K_2HPO_4, and subject to ion exchange (11 columns) on Amberlite CG-50 (200–400 mesh, Mallinkrodt). The column was washed with 0.1 M K_2HPO_4, and the peroxidase eluted with 0.2 M K_2HPO_4. The peroxidase ($\sim$ 12 μmol) was around 75% pure as evidenced by the ratio of absorbance at 472 nm (Soret max. ferrous form) to that at 280 nm. (This ratio is 0.82 in the pure protein at pH 8.6.) The final step consisted of adsorption chromatography on hydroxyapatite (BioRad Laboratories). The preparation was adsorbed on a 20 × 5 cm column in 0.05 M K_2HPO_4, washed with 0.1 M buffer, and eluted with a 1.0 liter linear gradient of K_2HPO_4 (0.1–0.5 M). Enzyme exhibiting a spectral ratio (A_{472}/A_{280}) greater than 0.76 was eluted at around 0.25 M K_2HPO_4. The yield of this material was 7–9 μmol ($\sim$ 60% yield overall) when the starting material was of good quality.

2.1.1. Crystallization

The purification procedure above yielded peroxidase which was pure by electrophoretic, immunological, and sedimentation criteria. The ratio of the absorbance at the Soret maximum of the ferrous form (472 nm) to that of the ferric form (428 nm) was often less than that (1.02) reported by Agner (1941). Upon crystallization, this unexplained difference was lost, and the spectra of both

*Harrison *et al.* (1977).

oxidation states were as reported by this author. The purity of the protein, during the purification procedure, was best estimated by the ratio A_{472}/A_{280} (0.83 in the crystalline enzyme).

Myeloperoxidase in the effluent of the hydroxyapatite column fully crystallized (in 3 days) after the slow addition of ethyl alcohol to 30% (v/v) at 4°C, following dialysis against 0.2 M K_2HPO_4. At high protein concentrations, (10 mg/ml), turbidity or gross precipitation first occurred, the plate-like crystals (up to 0.5 mm in two dimensions) growing in the precipitate. When crystallization was performed by dialysis against 0.2 M K_2HPO_4, 20% ethyl alcohol, a higher proportion of the needle form (Figure 1) was evident, and this form was observed exclusively when the dialysis was performed at 0°C.

3.   ANALYTICAL PROCEDURES

There are as many analytical methods for determining peroxidase as there are investigators. However, the most fundamental description of the kinetics of the myeloperoxidase is the reaction with hydrogen peroxide and guaiacol as donor described by Maehly (1955). The basis of this method is the reaction time measured at 470 nm following the addition of buffer, myeloperoxidase, and hydrogen peroxide. The time it takes to reach an optical density increase of 0.05 is the basic parameter measured. Of particular significance in this procedure is the fact that hydrogen peroxide level versus a given amount of myeloperoxidase is critical. There is a maximum concentration of H_2O_2, which is required for a given amount of enzyme present. Therefore, less than a maximum will give a

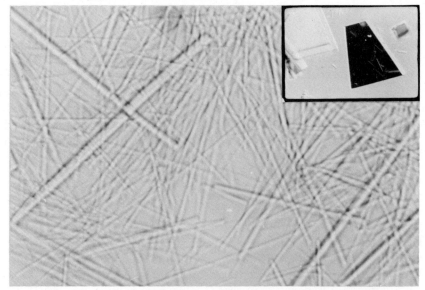

FIGURE 1.   Crystalline canine myeloperoxidase. The protein was crystallized by dialysis against 0.2 M K_2HOP_4/20% ethyl alcohol. Needles (×1300). Inset: typical plate form exhibiting dichroism. Phase contrast (×250). From Harrison *et al.* (1977).

slower reaction rate and greater than a maximum will also give a slower reaction rate. It is, therefore, important that the amount of hydrogen peroxide used is that required for maximum velocity.

In the laboratories of Schultz and his co-workers, the square root of $1/t$ (t being the time to reach 0.05) OD increase was the basic measurement. The method described here has been used for the past 25 years by a large number of researchers including graduate students, technicians, and postdoctoral trainees. Of particular interest is that values obtained are so highly reproducible. The value of K per milligram of rat chloroma MPO was 300 and that of the enzyme from man was 1000. Even over 20 years later with an entirely new team of investigators the same values were obtained on new preparations from both sources. Also the application of this method to determine enzyme content per poly have stood the test of time in this laboratory (Schultz *et al.*, 1954).

The important fact again is the concentration of hydrogen peroxide at a given myeloperoxidase concentration. This is achieved by utilizing the amount of myeloperoxidase that requires 12–15 sec for time t. Therefore, a constant amount of hydrogen peroxide can be used without any fear of having too much or too little. A very recent procedure using K_1 values at pH 10 offers a more sensitive procedure claiming to save a great deal of material. But no indication as to the control of the amount of H_2O_2 in respect to MPO was indicated (Bernay *et al.*, 1978).

3.1. PROCEDURE FOR DETERMINING MPO CONTENT PER POLYMORPHONUCLEAR (PMN) CELL OR PER MILLILITER OF BLOOD

Reagents 1.5% EDTA in saline: Dissolve 15 g EDTA in less than 1 liter of deionized water by adding 6 g NaOH to bring pH to 7.0; add 9 g of NaC1; add deionized H_2O up to 1 liter.

1. Add 9.0–9.5 ml of cold 0.9% saline to a centrifuge tube and place tube(s) and saponin (30 mg/ml) in ice to keep cold.

2. Place $100\,\mu l$ of EDTA in saline solution into 1 ml beaker using micropipette.

3. Cut tail of rat approximately 1–2 mm from end, milk blood into beaker, moving beaker in a circular motion on the table, to mix blood and EDTA-in-saline.

4. For every 100 μl of EDTA use 200 μl whole blood (beaker should be scratched at 300 μl mark).

5. Draw sample from tail up to 0.5 mark in blood pipette and dilute to 11 mark for whole blood white blood cells (WBC).

6. Make whole blood smear from tail.

7. Take another WBC pipette and aspirate up to 0.5 mark from the EDTA-blood mixture, stirring it as above.

8. Transfer 200 μl from the EDTA–blood mixture with 200 μl pipette to centrifuge tube. The mixture tends to adhere to the pipette wall; so carefully clean it out by aspirating the 0.9% saline and blowing out.

9. Shake the tube by placing the thumb over the end—tilting back and forth until all the mixture is in solution.

10. With capillary tube remove some of the remaining liquid in beaker and blow out on slide to make smear.

11. Add 100 μl saponin to centrifuge tube; shake tube 3 or 4 times. If more than 1 rat is being used, do this step to all tubes at the same time.

12. Allow to hemolyze (about 30–60 sec) or until the graduations can be seen clearly through the tube.

13. Spin down at 1000 rpm at 4°C in refrigerated centrifuge for 20 min.

14. Decant supernatant carefully, taking out last few drops by placing tissue at tube's mouth.

15. Dissolve pellet in 100 μl 0.014 N HCl. The resulting solution should be pale yellow.

16. Add deionized H_2O with dropper up to 0.5 ml mark.

17. Shake well and run K test (see Procedure for Assay of MPO above).

18. While the centrifuge is spinning, shake WBC pipettes 3 min.

19. After running K test, determine WBC count and stain smears for differentiation count.

3.1.1. Finger Puncture Blood: Samples are Taken for WBC Count and Differential

For MPO assay on human for finger or venipuncture blood.

1. Add 0.1 ml blood into each of two conical centrifuge tubes containing 10 ml cold saline.

2. 100 μl Saponin (30 mg/ml) to each—shake until solution clears—50 seconds.

3. Centrifuge at 1000 rpm for 10 min in PR2 refrigerant centrifuge.

4. Decant supernatant, wipe inside tube dry, and store in ice bath.

5. Add 100 μl 0.014 N HCl to pellets, shake for 60 sec—dilute to 0.5 ml with ice cold H_2O.

6. Calculation to obtain K/ml blood.
   Original blood sample 0.1 ml.
   K/aliq $\times$ 1/aliq $\times$ 5 = K/ml blood.
   To calculate K/PMN: $\dfrac{K/ml\ blood}{PMN/ml\ blood}$ = K/PMN
   When expressed per 10^7 cells K/PMN is multiplied by 10^7.

3.2. RECOMMENDED PROCEDURE FOR ASSAY OF MYELOPEROXIDASE

Reagents: Guaiacol*; 0.22 ml per 100 ml solution of 0.1 M Na/K phosphate buffer pH 7.0, H_2O_2: 1 ml supernatant diluted to 100 ml with cold distilled H_2O; 0.1 M Na/K phosphate buffer pH 7.0.

Procedure: 2 ml guaiacol solution* is added to each assay tube. (Tubes are

---

*In some cases the guaiacol solution is made up so that the second solution of 0.01 M phosphate buffer is not necessary. However, when aliquots of MPO solutions requiring more than 0.05 ml are necessary the 0.1 M phosphate can be adjusted to make the final volume 3.0 ml. In many cases, however, the single solution of 3 ml guaiacol (0.22 ml per 100 ml buffer) is quite satisfactory as long as the amount of sample does not cause dilutions that would give erroneous results.

TABLE 1. PEROXIDASE CONTENT OF LEUKOCYTES FROM A
NUMBER OF SPECIES[a]

Species	(K/WBC) $\times 10^7$	(K/PN) $\times 10^7$
Man	14	21
Rat	7.6	25
Goat	5.7	30
Turtle	9.6	25
Horse	7.4	24
Calf	9.2	23
Cow	5.2	22
Dog	18	21
Rabbit	9.9	20
Pony	4.9	20
Guinea pig	8.7	19
Sheep	8.4	19

[a] The data illustrate the dependence of peroxidase activity of white blood cells in nature on the neutrophil content. Each blood sample was treated according to the procedure of Gold and Cole (Bachner and Karnovsky, 1968) to separate the white cells. These were counted for total cells (WBC) and polymorphonuclear (PMN) cells, and the peroxidase activity (K) was determined on the repeatedly frozen and thawed samples. These data are representative values of individuals from each species wherein the individual variation can be greater than the species difference. Table from Schultz *et al.* (1965).

previously calibrated for use in Spectronic 20.) Add 1 ml of 0.01 M phosphate buffer and allow tubes to cool to room temperature. Add enzyme samples and adjust spectrometer to zero, or at 470 nm. Add 10 μl H$_2$O$_2$, start stopwatch and stop when needle reaches 0.05. This should require 12–15 sec. If time is beyond these limits enzyme solution should be readjusted. The extent to which this is necessary depends on the particular conditions of the experiment.

TABLE 2. PEROXIDASE CONTENT OF LEUKOCYTES OBTAINED
FROM A NUMBER OF SUBJECTS[a]

Donor (date)		MPO/10^7 cells (μg)	MPO/ml blood (μg)
J.F.	(7/16)	11.2	3.7
S.J.	(7/16)	20.4	4.5
N.B.	(7/16)	17.4	5.0
B.T.	(7/16)	28.5	4.0
S.W.	(7/16)	18.5	4.0
N.B.	(7/20)	22	5.6
N.B.	(7/22)	17.9	5.8
S.W.*	(7/23)	18.8	4.2
S.W.**	(7/23)	32.8	7.2

[a] The value of K is calculated as micrograms of enzyme in humans where 1000K = 1 mg. All values are from finger puncture, assayed as described under procedures. Blood values are taken from differential counts of PMN/ml blood. Dates of sampling are indicated in parentheses, showing the reproducibility of the procedure. In each the first sample (*) was taken by venipuncture and the second** by finger puncture indicating a possible response to injury.

The value $K = \sqrt{1t}$ is calculated. For example, when time t is 15 sec $\sqrt{1/15}$ = 0.26. If 10 μl of enzyme solution was the aliquot, the value is $26K$ per ml of solution assayed. Inasmuch as it has been shown here for many years in hundreds of preparations of purity greater than 0.80 in cases of canine and human MPO, that the value of 1000K equates 1 mg MPO, a K value of $1 = 1 \mu$g enzyme. In the case of rat it is $300K = 1$ mg. Values obtained on a number of species and individuals are presented in Table 1 and 2.

4. PHYSICAL AND CHEMICAL PROPERTIES OF MPO

4.1. IRON CONTENT

Agner (1941) originally reported the value of Fe for the preparations from human empyema as 0.1%. The crystalline enzyme from the dog was 0.078%, which he thought might be due to species difference or to more accurate Fe analysis. The human enzyme was initially reported as 0.093 (Schultz and Schmukler, 1964), but more recently, to be 0.077% (Schultz et al., 1972). A value of 0.08% was reported for the dog enzyme (Harrison et al., 1977). In terms of enzyme per g-atom of Fe, the value is 70×10^3 g, thus indicating a dimeric structure when the molecular weight obtained by other means is considered. Variations in values for Fe were probably due to iron contamination as seen from EPR studies (see below).

4.2. MOLECULAR WEIGHT

There is some confusion as to the molecular weight of MPO when the physical measurement on the ultracentrifuge is compared to the gel filtration values. The two most recent reports find agreement on the human (Bakkenist et al., 1978) and dog enzyme (Harrison et al., 1977) of 144,000 and 142,000, respectively. This is within the values of those reported earlier (Zgliczynski et al., 1968; Desser et al., 1972; Ehrenberg and Agner, 1958; Odajima and Yamazaki, 1972).

These values do not agree with those found by gel filtration. Felberg and Schultz (1972) reported a value of 122,000 using Sephadex 200 for human MPO, and when Biogel A was used Harrison et al. (1977) found a value of 120,000 for the dog enzyme.

The difference between the two procedures was explained on the basis of the assymmetry of the globular conformation of the enzyme as follows (Harrison et al., 1977):

On gel filtration anomalous molecular weight was obtained using Biogel A. This was not explained on the basis of adsorption or dissociation of the molecule into monomers. The difference in elution volume between myeloperoxidase and γ-globulin, a molecule of similar size and chain content, is striking. The possibility exists that the gel filtration properties of myeloperoxidase are a result of a degree of asymmetry, which has recently been shown to produce anomalous retardaion on gel filtration even when the equivalent Stokes' radius is used as calibrating parameter (Nozaki et al., 1976). Myeloperoxidase partitions similarly to serum albumin ($R_2 = 35$Å), while its Stokes' radius, calculated from the diffusion coefficient, is 45Å. This discrepancy is similar in magnitude to those reported by Nozaki *et al.* (1976) for fibrinogen and the sodium dodecyl sulfate complexes of ovalbumin and the heavy chain of γ-globulin. In terms of molecular weight, myeloperoxidase behaves as a globular protein of molecular weight slightly higher than the weight of each heavy subunit (≈ 63,000).

It might be noted here that the above values for molecular weight depend on the properties of the enzyme in solution. In an experiment which is independent of these properties and requiring methods of highest precision, the following results were obtained (Schultz *et al.*, 1972). A sample of 0.83 RZ of enzyme was dissolved and aliquots measured for 1) iron content; 2) dry weight; 3) amino acid analysis. With an iron content of 0.077 of dry wt., grams of enzyme/g-atom of iron was 72,550. Weight of amino acids per g-atom Fe was 67,000 g. If we added to this the weight of the porphyrin and carbohydrate, the agreement is within 2%. The average is 70,000, or 140,000 for the dimer (Schultz *et al.*, 1972). It thus appears that the gel filtration values are anomalous.

4.3. HEME INEQUIVALENCE

A number of experiments suggested the possible presence of at least two kinds of heme in MPO. Agner noted that when MPO was treated with pyridine-methanol for several hours, a precipitate formed. When redissolved, it showed a spectral difference from the remaining supernatant. He did not, however, pursue this further. The following experiments indicate more definitively the presence of more than a single heme species.

1. When MPO was heated at 75°C at pH 6.8, there was a 50% loss of activity within 2 hr, with the formation of a precipitate. The supernatant retained its activity for greater than 8 hr (Schultz *et al.*, 1972).

2. Denaturation at 28°C in 0.5 N NaOH showed two distinct inactivation curves when plotted on a log scale (Schultz *et al.*, 1972).

3. When the heat stable component and heat labile component were treated with hydrazine-acetic acid two distinct cleavage rates were obtained whose average rate was that of the native enzyme (Schultz *et al.*, 1972).

4. Tryptic digestion of the denatured enzyme resulted in appearance of two heme peptide species each appearing at different times.

5. It was difficult to isolate a single heme peptide following conventional procedures. The digest yielded two electrophoretic components of different amino acid composition (Schultz *et al.*, 1972).

6. Odajima and Yamazaki (1972) reported spectral evidence for two cyanide complexes.

7. Finally, most recent studies showed one noncovalently linked heme, while the other was firmly linked to the heavy chain (Harrison and Schultz, 1978), providing the most definitive evidence to this day of the inequivalence of the two hemes.

The above findings helped explain the multiple bands formed on acrylamide gel electrophoresis (Felberg and Schultz, 1972) and free flow electrophoresis in 6

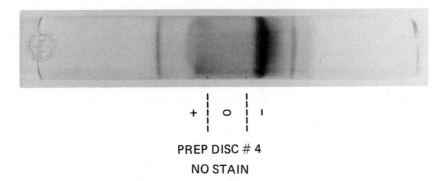

PREP DISC # 4

NO STAIN

FIGURE 2.   Two-directional acrylamide gel electrophoresis of human MPO showing six isoenzymes. From Felberg and Schultz (1972).

M urea. Evidence of the subunit structure determined by cleavage with mercaptoethanol or ethylene imine described below, provided most definitive evidence of the source of the heterogeneity of MPO (Schultz *et al.*, 1972; Harrison *et al.*, 1977; Harrison and Schultz, 1978).

4.4.   ISOENZYMES

The original demonstration of isoenzymes of MPO (Felberg and Schultz, 1972) (Figure 2) was difficult to accept because the source material represented samples from several thousand donors. The possibility of slight individual differences might have accounted for the six isoenzymes seen. However, recently a report confirming this observation demonstrated at least four isoenzymes from single donors (Strauven, 1978). The proportion of each differed with the age of the donor; and more remarkably, there was a distinct difference in temperature sensitivity confirming the earlier observations (Schultz *et al.*, 1972; Bonner and Schultz, 1967). It might be noted that the extraction procedure used by the above authors, referred to as Method I and Method II, yielded different results. This difference, they claimed, was due to the use of "cetrimide" in one case and "HTAB" (hexadecyltrimethyl ammonium bromide) in the other. Actually both are chemically identical detergents, that is, cetyltrimethyl ammonium bromide and hexadecyltrimethyl ammonium bromide are both the bromides of 16 straight chain alkyl trimethyl amines. The difference is most likely due to other factors, since the solutions used for extraction were not the same. Nevertheless, the data obtained demonstrated that the isoenzyme patterns varied with the clinical condition and age of the patient.

4.5.   AMINO ACID COMPOSITION

Analyses of the amino acid composition of MPO obtained from humans, dogs, and pigs revealed only slight differences. The most notable difference is that the enzymes from the dog and pig contain more arginine (Table 3).

TABLE 3.   AMINO ACID COMPOSITION OF MYELOPEROXIDASE OF MAN, PIG, AND DOG[a]

Amino acid	Man[c]	Man[b,d]	Pig[b,e]	Dog[f]
Lys	39	33	41	24
His	14	12	16	14
Arg	107	105	139	122
Asp	136	153	145	152
Thr	58	69	52	61
Ser	61	63	52	74
Glu	129	111	99	108
Gly	80	78	81	100
Ala	75	78	78	68
Cys/2	—	30	26	21
Val	71	51	74	48
Met	31	36	22	22
Ile	48	48	37	48
Leu	121	129	136	128
Tyr	27	24	24	34
Phe	55	51	59	56

[a] The values in the table are in terms of residues per 140,000 molecular weight.
[b] Recalculated to a molecular weight of 140,000.
[c] Bakkenist *et al.* (1978).
[d] Schultz and Schmukler (1964).
[e] Wu and Schultz (1975).
[f] Harrison *et al.* (1977).

4.6.   SPECTROPHOTOMETRY

The absorbance spectrum of MPO has now been reported in numerous publications and all agree with that initially found by Agner (1941). Of all heme proteins in nature, the shift of the Soret maximum at 430 to 472 nm on reduction

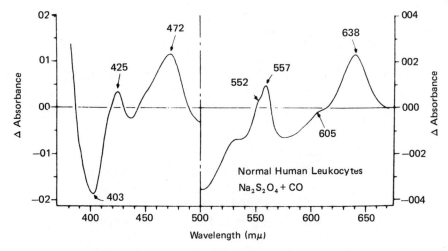

FIGURE 3.   Reduced difference spectrum of whole polymorphonuclear cell frozen in liquid nitrogen. CO was introduced into the reference and sample cuvettes; only the sample cuvette was reduced. From Schultz *et al.* (1972).

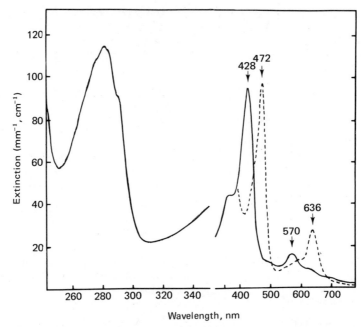

FIGURE 4.   Reduced and oxidized spectrum of canine MPO (R2 = 0.85). Note that reduced Soret 5 has higher extension than oxidized, as originally observed by Agner. Prepared in author's laboratory by J. E. Harrison. Solid line, oxidized; broken line, reduced.

with dithionite is the largest. There is also a shift from 560 to 638 nm in the visible region of the spectrum.

The spectrum of the enzyme in the intact cell has been published and is reproduced here (Schultz *et al.* 1972). It shows the same reduced characteristics of purified MPO, but also the presence of heme enzymes of the electron transport system (Figure 3).

Extensive studies of spectrophotometry of MPO have been reported in great detail by a number of investigators which included the CN, F, NH_2OH, azide, etc. (Agner, 1941; Schultz and Schmukler, 1964; Bakkenist *et al.*, 1978; Maehly and Chance, 1955; Odajima and Yamazaki, 1972); only the oxidized-reduced spectra will be reproduced here (Figure 4). Studies are continuing on the microenvironment of the Fe using NMR and various ligands by Block and Harrison of the author's laboratory.

4.7.   ELECTRON SPIN RESONANCE (EPR) SPECTROSCOPY OF MPO

Several laboratories have reported measurements of EPR spectra of MPO after the first report of Ehrenberg (1962). This was confirmed by Schultz *et al.* (1972). In each case, the signals indicated that the two subunits of the enzyme gave identical signals. Further studies by other laboratories (Bakkenist *et al.*, 1978; Blum *et al.*, 1978) have since appeared.

TABLE 4.   EPR SIGNALS FROM LEUKOCYTES AND
HEME PROTEINS (g values)[a]

Source	Rhombic		Axial
	g_x	g_y	g
Leukocytes			
Guinea pig	6.59	4.88	5.92
	6.31	5.19	
Pig	6.59	5.04	—
	6.37	5.25	
Horse	6.80	5.00	—
	6.50	5.30	
Dog	6.77	5.00	5.92
Sheep	6.68	5.21	—
Cow and calf	6.78	5.00	—
Man	6.90	5.07	—
Chicken	—	—	6.0
Myeloperoxidase			
Dog uterus	6.77	5.02	
	6.33	5.52	—
	6.3	5.3	—
Alveolar macrophages			
Rhesus monkey	6.08	5.86	
Dog	6.77	5.00	
	6.41	5.32	

[a] Table from Blum *et al.* (1978).

The principal signals are at $g = 6.7$–6.9 and g 5.0–5.5 in the high spin region and at about 1.95 in the low spin region. Interpretation of these signals attribute $g = 6$ to axial symmetry of the heme plane. When this is distorted by conformational changes of the enzyme, rhombic distortion values deviating from 6.0 are seen, but $g = 2$ remains the same. The physiological significance of this was proposed as a result of studies on human granulocytes (Weber *et al.*, 1976) and guinea pig PMN leukocytes (Rotilio *et al.*, 1977), which claimed that during phagocytosis signals indicating axial symmetry, at 6.0 appeared which were attributed to MPO. Values of 5.89 indicated more axial symmetry at pH 2.0 (Bakkenist *et al.*, 1978). Since the phagocytic vacuole is acidic in nature this might suggest the change seen. But more recently, Blum *et al.* (1978) demonstrated that this was due to the presence of methemoglobin, either free or from erythrocytes. A summary of the g value of leukocytes and heme proteins reported by the authors is shown in Table 4. The tendency toward axial symmetry at acid pH has been attributed to denaturation of the protein (Blum *et al.*, 1978).

The signal at $g = 4.3$ seen in many preparations of MPO has been related to nonspecific iron; to establish that this is not related to the structure of MPO, an EPR spectrum of highly purified dog enzyme is shown in Figure 5 (Harrison *et al.*, unpublished).

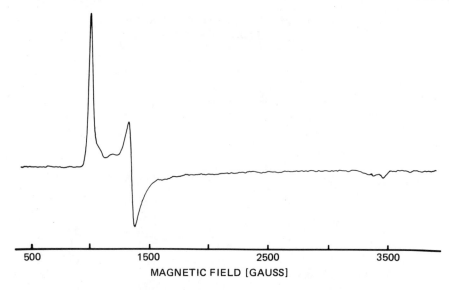

FIGURE 5.   EPR of highly purified MPO shows no signal at 4.3 (at 1505 G). From Harrison *et al.* (unpublished).

4.8.   CIRCULAR DICHROISM (CD)

The optical rotation of a protein measured at each wavelength of light has provided insight into structure. Such spectra are compared with the conventional absorption spectrum. In the case of MPO, the bands of oxidized and reduced enzyme correlate well with the absorption maxima (Bakkenist *et al.*, 1978). These authors report that some weak absorption bands have a large transition dipole moment and thus show intense CD, whereas the strong absorption bands generate little or no optical activity. Aromatic amino acids are attributed to bands at 282, 278, 260, and 252 nm and those at 350, 412, 450, 560, and 620 nm to the heme (see Figure 6).

5.   CATALYTIC ACTIVITY OF MPO

Catalytic action of MPO has been under study since its discovery by Agner. There are, however, two kinds of catalysis that must be kept in mind: one is the peroxidation reaction and the other the chlorination process. Chance and Maehly (1955) reviewed the first process using hydrogen donors and postulated two steps resulting in a complex of MPO and H_2O_2 called compound II, also described by Agner (1963). Compound II was shown by Odajima (1972) to contain two oxidizing equivalents more than the native enzyme. The first compound formed (compound I) is very unstable and difficult to detect, but its rate of formation has been measured by both Chance (personal communication) and Newton *et al.* (1965).

The above peroxidative oxidative reaction takes place in neutral or alkaline

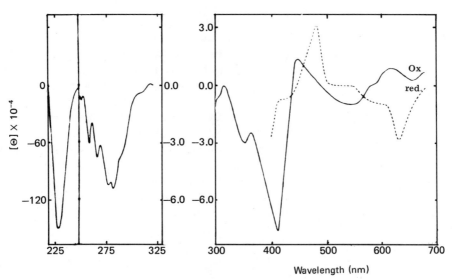

FIGURE 6. Circular dichroic spectrum of MPO of man. From Bakkennist *et al.* (1978).

medium. It could only occur outside the cell. On the other hand, the chlorination reaction can take place in the cell because under physiological conditions, the vacuole of the PMN cell is at a pH in the acid range, and chloride is present. Hence, more attention is now directed toward peroxidative chlorination catalysis by MPO.

It was first believed that compound II ($MPO \cdot H_2O_2$) containing two extra electrons reacted with Cl^- to carry out the chlorination reaction. However, it has been reported that $MPO \cdot H_2O_2$ could not be reduced by chloride (Newton *et al.*, 1965). A chloride form of the enzyme had been suggested to play a role in the reaction (Agner, 1970; Stelmaszynska and Zgliczynski, 1974). But it was later shown that such chloride complexes are also formed with inactive proteins (Morrison and Bayse, 1973). Others showed that the chloride complex was pH dependent (Stelmaszynska and Zgliczynski, 1974).

Any interpretation of the catalytic process requires unequivocal demonstration of the presence of intermediates. It had been assumed, but not proven, that Cl_2 or HOCl was formed as the first oxidation product of Cl^-. Although free Cl_2 can be detected by its odor when MPO, H_2O_2, and Cl^- are mixed, this was not suitable for quantitative studies. Even when the radioactivity of $^{56}Cl_2$ was measured following $^{56}Cl^-$ peroxidation, suitable data could not be collected because the released gas was in equilibrium with nonvolatile HOCl (Harrison and Schultz, 1976). By carrying out the reaction in the presence of immobilized MPO on a membrane and rapidly measuring the product in the filtrate, adequate data could be obtained (Harrison and Schultz, 1976).

Evidence that ClO_2 was not an intermediate was shown by Harrison and Schultz (1976) using the procedures of Hager *et al.* (1966). In this way MPO differs from horseradish peroxidase (HPR) which can use chlorite (ClO_2) as a chlorinating agent; however, HPR cannot use Cl^-.

The above demonstrated that the ultimate chlorinating agent is HOCl. To complete the picture of the catalytic process information was needed as to what catalytic form of the enzyme carries out the oxidation of Cl^- to $HOCl/Cl_2$, and also the relation of the interpretation observed by spectrophotometry to these complexes and the role of pH and Cl^- in its biological function.

In experiments to provide that information Harrison and Schultz (1976) found that at constant pH there is a maximum Cl^- concentration for the peroxidation of Cl^-, and that there is a pH maximum at constant Cl^- concentration. Thus, both Cl^- and H^+ are involved in the reaction with the enzyme. Spectrophotometrically, using stop flow methods, the rate of decomposition of the hydrogen peroxide complex was too slow to account for the rapid rate of catalysis of the oxidation of Cl^-. Thus compound II could not be involved directly in the chlorination reaction. The following scheme of the catalytic process was suggested (Harrison, 1976). Note the dual role of H^+ and Cl^-.

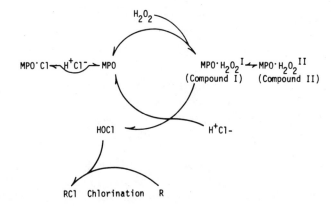

In this scheme the chloride complex of the enzyme as well as compound II are inactive forms at acid pH as long as Cl acceptor R is present to drain off the HOCl formed. In the absence of Cl^-, compound II would be formed, and in the presence of large excess of Cl^-, chloride complex of MPO would be formed, both of which are inactive. Thus, this may be a control mechanism in keeping the enzyme inactive until chlorination products are available. When MPO is extruded outside the cell at pH of 7.4, direct peroxidative oxidation of hydrogen donors takes place.

5.1. ROLE OF SINGLET OXYGEN

When hydrogen peroxide is added to hypochlorite (e.g., Superoxol to Chlorox) one sees a red chemiluminescence. Such chemiluminescence can be measured in a scintillation counter. Allen (1973) found that on phagocytosis such chemiluminescence took place, and postulated that singlet oxygen was the "toxic agent" in the killing involved in the $MPO-H_2O_2-Cl^-$ system. This was widely accepted (Roos, 1977) and, apparently, confirmed (Allen, 1975; Rosen and Klebanoff, 1977; Piatt *et al.*, 1977). From the above scheme, it is apparent that for HOCl and H_2O_2 to be present at the same time it would be required that HOCl accumulate in absence of Cl_2 acceptors.

Two independent reports appeared which called for reevaluation of the role of singlet oxygen in the killing mechanism of the MPO–H_2O_2–Cl system (Harrison *et al.*, 1978; Held and Hurst, 1978). Both depended on the highly reactive HOCl, which could compete with singlet oxygen for the same reagents used to trap and identify singlet oxygen. Thus, tests for singlet oxygen in the presence of trapping agents were positive only because the HOCl formed reacted with the trapping agents. Measurements of chemiluminescence by a more definitive technique showed that the light emission continued long after any singlet oxygen was present (Harrison *et al.*, 1978). The factors involved in the continued fluorescence are most likely what is measured in the phagocytic process. That, however bears further investigation.

6. STRUCTURE OF MPO

Heme proteins in nature are iron–cyclic tetrapyrroles bound to proteins which function in the oxidative processes either in electron transport, with oxygen reduction products such as H_2O_2, or directly with oxygen. They differ mostly in regard to the nature of the protein, inasmuch as, in many cases, for example, myoglobin, hemoglobin, horseradish peroxidase, and catalase, they are Fe–protoporphyrin–proteins; their functions are entirely determined by the apoprotein. These heme proteins are also similar in that the prosthetic group is readily separated from the apoprotein by acid-acetone, indicating an electrostatic linkage. While cytochrome c, and lactoperoxidase not only have iron–tetrapyrroles different from those above, their linkage to the apoprotein is by thioether in the case of cytochrome c, but as ester linkage in case of lactoperoxidase. Cytochrome oxidase, on the other hand, has a distinctly different porphyrin, a long alkyl chain on one ring and a formyl group on another; however, its heme is not covalently bound to the protein.

Myeloperoxidase is quite unique among the heme proteins inasmuch as its heme is related to cytochrome oxidase, it has two hemes, one bound covalently (a peptide bond) and the other noncovalently, and is made up of two subunits, each of which demonstrates charge heterogeneity.

6.1. NATURE OF THE HEME

The first indication of the nature of the heme of MPO was the fact that the pyridine hemochromogen was the same as cytochrome oxidase. Since the latter was known to have a polar side chain, an aldehyde, MPO heme was assumed also to have a polar group (Schultz and Schmukler, 1964). From Morell's laboratory, however, the polar side chain was claimed to be a ketoaldehyde (Nichol *et al.*, 1969). The procedures used in those experiments, however, with present day knowledge, were quite drastic, utilizing hot alkaline treatment for 1 hr. The heme isolated was assumed to be a product resulting from the action of alkali in

(Wu and Schultz, 1975) where the aldehyde was on ring 4 as in cytochrome oxidase. This was firmly established by a relatively mild procedure, in agreement with the presence of the formyl group on ring 4 by Harrison and Schultz (1978), as 2,4-divinyl-8-formyl deuteroporphyrin in agreement with the former projection (Wu and Schultz, 1975).

Although this was determined in the noncovalently bound heme, there is no evidence at present that the second heme is structurally different.

6.2. SUBUNIT STRUCTURE

The first indications of the subunit structure of myeloperoxidase were found as a result of treatment of the human enzyme with mercaptoethanol in guanidine (Schultz *et al.*, 1972). In earlier work on this procedure—dialysis to remove the guanidine and the mercaptoethanol—resulted in the loss of enzyme, which could not be accounted for at that time. However, with the introduction of SDS techniques, it was possible to show that two bands appeared—each of which could be further prepared on Sephadex 200. A light chain and a heavy chain were isolated and examined. The heavy chain was green in color, contained the heme, and was found to have a molecular weight of 60,000. The molecular weight of the light chain was 10,000, and, inasmuch as the total molecular weight was 140,000, and known to be a dimer, it was assumed that two heavy chains and two light chains made up the entire molecule. Using the dog enzyme and the preparations obtained by new procedures, Harrison *et al.*, (1977) reported molecular weights for heavy and light chains of 57,500 and 10,500, respectively (Figure 7). The amino acid composition is shown in Table 5.

More recently, it was demonstrated that linkage of the light and heavy chains were not covalently bound, but could be separated by ethylene imine. More significant, it was also found that the second heme could also be removed with ethylene imine, and that the enzyme, therefore, would consist of one heme covalently bound to the peptide chain and the other heme not so bound. These remarkable findings put an entirely new light on the structure of the enzyme, which is, therefore, made up of two light chains, two heavy chains, and one heme located in a firmly bound area and the other heme much less firmly bound.

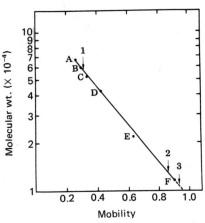

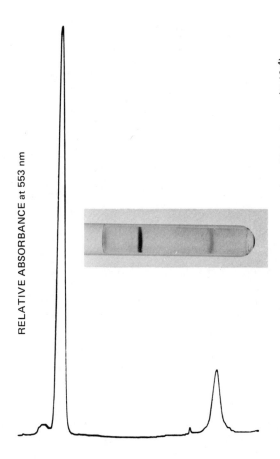

FIGURE 7. (a) Scan of the components of myeloperoxidase separated by polyacrylamide gel electrophoresis in the presence of dodecyl sulfate following denaturation in urea–SDS in the presence of β-mercaptoethanol. The inset shows the actual gel. (b) Estimation of the molecular weights of the component polypeptides of myeloperoxidses on an 8% gel. The molecular weight markers were serum albumin (A); catalase (B); γ-globin heavy chain (C); ovalbumin (D); γ-globin light chain (E); and cytochrome c (F).

However, it must be stated that the second heme can be removed only after the light chain is separated, therefore, indicating a location associated with the light chains.

While these findings provided the first insight into the structure of this unique enzyme, it became apparent that even greater complications existed. The fact that both the light chains and the heavy chains showed charge heterogeneity (Figure 8), recalled the case of 7S gamma globulins, wherein the reduced-alkylated protein could be resolved into light and heavy chains. These are now recognized as consisting of a spectrum of molecular species resulting from variable and constant regions of the L and H chains. Is the same true of MPO?

Inasmuch as both gamma globulin and MPO originate from the RES, the possibility of multiclone origin of the enzyme is suggested. From a description of the origin of the mature PMN cell (Broxmeyer and Moore, 1978), it might be possible that stem cells may be sufficiently variable in expression to provide a spectrum of molecular species. Since more than one gene can be involved in making a polypeptide chain such variability is quite possible.

MPO may be related to gamma globulin in terms of its apoprotein structure

TABLE 5.   AMINO ACID COMPOSITION OF CANINE MYELOPEROXIDASE
LIGHT AND HEAVY CHAIN COMPONENTS

Amino acid	Light chain[a] (residues/10,500 mol. wt.)[c]	Heavy chain[a] (residues/57,500 mol. wt.)	Light plus heavy chain (mol. wt.)	Native[b] (residues/68,000 mol. wt.)
Asx	11	59	70	76
Thr	5	24	29	32
Ser	9	29	38	37
Glx	11	43	54	54
Pro	10	35	45	50
Gly	8	29	37	34
Ala	9	22	31	35
Cys/2	1	9[d]	10	10
Val	3	27	30	24
Met	1	12	13	11
Ile	1	25	26	24
Leu	9	54	63	64
Tyr	1	15	16	17
Phe	5	26	31	28
His	2	5	7	7
Lys	3	9	12	12
Arg	9	52	61	61
Trp	N.D.[e]	N.D.	N.D.	17
Total	98	475	573	593

[a] Values are the nearest integral numbers obtained from a single analysis of a 24 hr acid hydrolysis of S-carboxymethylated proteins.
[b] Except for tryptophan, the values are to the nearest integral number and represent the average results from 24 and 72 hr hydrolysis; threonine and serine were determined by extrapolation to zero time of hydrolysis; tryptophan was determined spectrophotometrically.
[c] Molecular weights were estimated by SDS gel electrophoresis.
[d] Determined as S-carboxymethylcysteine.
[e] Not determined.

and to cytochrome oxidase in the structure of its heme. MPO may be a unique molecule in the scheme of evolution reflecting earlier more primitive species with multiple functions.

7.   BIOLOGICAL IMPLICATIONS OF MPO CONTENT OF THE PMN CELL

The amount of MPO in the PMN cell is fairly constant in nature. It has been shown among a number of species that the enzyme activity is definitely correlated with the neutrophils or heterophils (Table 1). The correlation is not constant when the enzyme activity is related to all the white cells in a given species (Schultz et al., 1965). The location of the enzyme in the granules has been established by gradient separation techniques (Schultz et al., 1965; John et al., 1967; Bretz and Baggiolini, 1974) and by electron microscopy (Gainton and Farquar, 1968).

The variation of the enzyme content in health and disease has not yet been

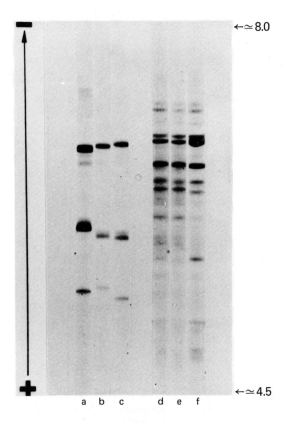

← ≃ 8.0

← ≃ 4.5

a   b   c   d   e   f

FIGURE 8.   (a) Isoelectric heterogeneity of S-carboxymethylated heavy and light subunits of canine myeloperoxidase. 20–40 μg of protein were focused in 4% polyacrylamide gels which were 2% in pH 3.5–10 ampholytes. Focusing was carried out at 1°C for 16 hr at 400 V. Gels were stained in Coomassie blue and only the portions of the gels containing protein bands are shown; approximate pH values are indicated, and the arrow indicates the direction of protein migration. Gels a–c represent the patterns obtained for S-carboxymethylated light chain and gels d–f for S-carboxymethylated heavy chain. Myeloperoxidase used for gels (a) and (d) was isolated from a single animal. The other gels represent protein obtained from two different preparations of myeloperoxidase, using pooled pus from several animals. Gels (b) and (e) constitute one preparation; (c) and (f) represent the other. From Harrison *et al.* (1977).

widely explored. The methods described in this review have been shown to be suitable for such studies. The values obtained by the procedures described are shown in Tables 1 and 2. Those of a patient with cyclic neutropenia following lipexol treatment are presented in Figure 9. Of particular interest is that, although the enzyme content per cell may vary among individuals, the more constant value in the MPO per ml of blood indicates that the total circulating cells have approximately the same value, regardless of the total number of cells. Thus, one would predict a number of clinical states of the PMN cell similar to the anemias, such as hypochromic or hyperchromic depending on the number of cells amongst which the available MPO produced can be distributed. The effect of MPO on mammary adenocarcinoma induced by methylcholanthrene when

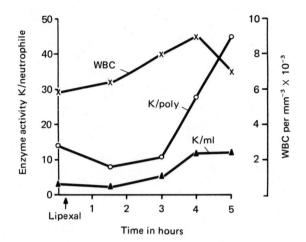

FIGURE 9.   Change in MPO content of neutrophil of patient with cyclic neutropenia following injection with lipexal.

the native heme. Later, it was shown that the heme was a formyl compound administered in conjunction with thiotepa, has been reported as well as the fate of peritoneally injected MPO (Schultz *et al.*, 1976).

REFERENCES

Agner, K., 1941, Verdoperoxidase, ferment isolated from leukocytes, *Acta Physiol. Scand.* **2**, Suppl. 8:1.

Agner, K., 1950, Studies on peroxidative detoxification of purified diphtheria toxin, *J. Exp. Med.* **92**:337.

Agner, K., 1958, Crystalline myeloperoxidase, *Acta Chem. Scand.* **12**:89

Agner, K., 1963, Studies on myeloperoxidase activity. I. Spectrophotometry of the MPO·H_2O_2 compound, *Acta Chem. Scand.* **17**:332.

Agner, K., 1970, Biological effects of hypochlorous acid formed by "MPO"-peroxidation in the presence of chloride ions, in: *Structure and Function of Oxidation-Reduction Enzymes* (A. Akeson and A. Ehrenberg, eds.), pp. 329–335, Pergamon Press, Elmsford, N.Y.

Allen, R. C., 1973, Dissertation, Tulane Univ., New Orleans, University Microfilms, pp. 74–291.

Allen, R. C., 1975, Halide dependence of the myeloperoxidase-mediated antimicrobial system of the polymorphonuclear leukocyte in the phenomenon of electronic excitation, *Biochem. Biophys. Res. Commun.* **63**:675.

Baehner, R. L., and Karnovsky, M. L., 1968, Deficiency of reduced nicotinamide-adenine dinucleotide oxidase in chronic granulomatous disease, *Science* **162**:1277.

Bainton, D. F., and Farquar, M. G., 1968, Differences in enzyme content of azurophil and specific granules of polymorphonuclear leukocytes. II. Cytochemistry and electron microscopy of bone marrow cells, *J. Cell Biol.* **39**:299.

Bakkennist, A. R. J., Wever, R., Vulsma, T., Plat, H., and van Gelder, B. F., 1978, Isolation procedure and some properties of myeloperoxidase from human leukocytes, *Biochim. Biophys. Acta* **524**:45.

Benoy, B., Ratnan, P., Selvarag, J., and Sbarra, A. J., 1978, A sensitive assay method for peroxidases from various sources, *J. Reticuloendothel. Soc.* **23**:407.

Block. R. E., and Harrison, J. E., 1978, Nuclear magnetic resonance studies of myeloperoxidase, 176th American Chemical Society Meeting (Bio-chemistry).

Blum, H., Chance, B., Gunson, D. E., and Lichtfield, W. J., 1978, Electron paramagnetic resonance spectra of myeloperoxidase in polymorphonuclear leukocytes, *FEBS Lett.* **86**:37.

Bonner, M. J., and Schultz, J., 1967, Labile iron and heat sensitive compounds of myeloperoxidase, *Fed. Proc.* **26**:821.

Bretz, U., and Baggiolini, M., 1974, Biochemical and morphological characterization of azurophil and specific granules of human neutrophilic polymorphonuclear leukocytes, *J. Cell Biol.* **63**:251.

Broxmeyer, H. E., and Moore, M., 1978, Communication between white cells and the abnormalities of this in leukemia, *Biochim. Biophys. Acta* **516**:129.

Chance, B., and Maehly, 1955, Assay of catalases and peroxidases. 1. Catalase assay by disappearance of peroxide, in: *Methods in Enzymology*, Vol. 2 (S. P. Colwick and N. O. Kaplan, eds.), pp. 764–775. Academic Press, New York.

Desser, R. K., Himmelhoch, S. R., Evans, W. H., Januska, M., Mage, M., and Shelton, E., (1972), Guinea pig heterophil and eosinophil peroxidase, *Arch. Biochem. Biophys.* **148**:452.

Ehrenberg, A., and Agner, K., 1958, Molecular weight of myeloperoxidase, *Acta Chem. Scand.* **12**:95.

Ehrenberg, A., 1962, Electron spin resonance absorption by some hemoproteins, *Ark. Kemi* **19**: 119–128.

Felberg, N. T., and Schultz, J., 1972, Evidence that myeloperoxidase is composed of isoenzymes, *Arch. Biochem. Biophys.* **148**:407.

Felberg, N. T., Putterman, G. J., and Schultz, J., 1969, Myeloperoxidase X: Comparison of normal human leukocyte myeloperoxidase prepared with and without the use of trypsin, *Biochem. Biophys. Res. Commun.* **37**:213.

Hager, L. P., Morris, D. R., Brown, F. S., and Eberwein, H., 1966, Chloroperoxidase. II. Utilization of halogen atoms, *J. Biol. Chem.* **241**:1769.

Harrison, J., 1976, The functional mechanism of myeloperoxidase, in: *Cancer Enzymology* (J. Schultz and F. Ahmad, eds.), pp. 305–314, Academic Press, New York.

Harrison, J. E., and Schultz, J., 1974, Peroxidation of halide by myeloperoxidase of the rat, *Am. Chem. Soc. Abstr.* **99.**

Harrison, J. E., and Schultz, J., 1976, Studies on the chlorinating activity of myeloperoxidase, *J. Biol. Chem.* **251**:1371.

Harrison, J. E., and Schultz, J., 1978, Myeloperoxidase: Confirmation and nature of heme binding inequivalence, *Biochim. Biophys. Acta* **536**:341.

Harrison, J. E., Pabalan, S., and Schultz, J., 1977, The subunit structure of crystalline canine myeloperoxidase, *Biochim. Biophys. Acta* **493**:247.

Harrison, J. E., Watson, B. D., and Schultz, J., 1978, Myeloperoxidase and singlet oxygen: A reappraisal, *FEBS Lett.* **92**:327.

Held, A. M., and Hurst, J. K., 1978, Ambiguity associated with use of singlet oxygen trapping agents in myeloperoxidase-catalyzed oxidations, *Biochem. Biophys. Res. Comm.* **81**:878.

Himmelhoch, S. R., Evans, W. H., Mage, M. G., and Peterson, E. A., 1969, Purification of myeloperoxidases from the bone marrow of the guinea pig, *Biochemistry* **8**:914.

John, S., Berger, M., Bonner, M. J., and Schultz, J., 1967, Localization of lactic dehydrogenase isozymes in lysosomal fraction of the neutrophil of normal blood, *Nature* **215**:1483.

Klebanoff, S. J., 1971, Intraleukocytic microbicidal defects, *Annu. Rev. Med.* **22**:39.

Lehrer, R. I., Goldberg, L. S., Apple, M. A., and Rosenthal, N. P., 1972, Refractory megaloblastic anemia with myeloperoxidase deficient neutrophils, *Ann. Internal Med.* 76, 447.

Maehly, A. C., 1955, Myeloperoxidase, *Methods Enzymol.* **2**:794.

Morrison, M., and Bayse, G., 1973, Stereospecificity in peroxidase-catalyzed reactions, in: *Oxidases and Related Redox Systems* (T. F. King, H. S. Mason, and M. Morrison, eds.), pp. 375–388, University Park Press, Baltimore.

Newton, N., Morrell, D. B., and Clarke, L., 1965, Myeloperoxidase and associated porphyrins in rat chloroma tissue, *Biochim. Biophys. Acta* **96**:463.

Nichol, A. W., Morrell, D. B., and Thomson, J., 1969, Porphyrins derived from the prosthetic group of myeloperoxidase, *Biochem. Biophys. Res. Comm.* **36**:576.

Nozaki, Y., Scheckter, N. M., Reynolds, J. A., and Tanford, C., 1976, Use of gel chromatography for the determination of the Stokes radii of proteins in the presence and absence of detergents. A reexamination, *Biochemistry* **15**:3884.

Odajima, T., and Yamazaki, I., 1972, Myeloperoxidase of the leukocyte of normal blood. IV. Some physicochemical properties, *Biochim. Biophys. Acta* **284**:360.

Olsson, I., Olofsson, T., and Odeberg, H., 1972, I. Myeloperoxidase-mediated iodination in granulocytes, *Scan. J. Haematol.* **9**:483.

Piatt, J. F., Cheema, A. S., and O'Brien, P. J., 1977, Peroxidase catalyzed singlet oxygen formation from hydrogen peroxide, *FEBS Lett.* **74**:251.

Rohrer, G. F., von Wartburg, J. P., and Aebi, H., 1966, Myeloperoxidase aus menschlichen leukozyten. II. Untersuchunzen zur kinetik und substratspezifität, *Biochem. Z.* **344**:478.

Roos, D., 1977, Oxidative killing of microorganisms by phagocytic cells, *Trends Biochem. Sci.* **2**:61.

Rosen, H., and Klebanoff, S. J., 1977 Formation of singlet oxygen by the myeloperoxidase-mediated antimicrobial system, *J. Biol. Chem.* **252**:4803.

Rotilio, G., Brunori, M., Concetti, A., Dri, P., and Patriarca, P., 1977, Involvement of myeloperoxidase in the metabolic activation of phagocytes, EPR studies, *FEBS Lett.* **73**:181.

Schultz, J., 1958, Myeloperoxidase, *Ann. N. Y. Acad. Sci.* **75**:22.

Schultz, J., and Kaminker, K., 1962, Myeloperoxidase of the leukocyte of normal human blood, *Arch. Biochem. Biophys.* **96**:465.

Schultz, J., and Schmukler, H. W., 1964, Myeloperoxidase of the leukocyte of normal human blood. II. Isolation, spectrophotometry, and amino acid analysis, *Biochemistry* **3**:1234.

Schultz, J., Shay, H., Gruenstein, M., 1954, The chemistry of experimental chloroma. 1. Porphyrins and peroxidases, *Cancer Res.* **14**:157.

Schultz, J., Gordon, A., and Shay, H., 1957, Chemistry of experimental chloroma. IV. Column chromatographic purification of verdoperoxidase, *J. Am. Chem. Soc.* **79**:1632.

Schultz, J., Corlin, R., Oddi, F., Kaminker, K., and Jones, W., 1965, Myeloperoxidase of the leukocyte of normal human blood. III. Isolation of the peroxidase granule, *Archiv. Biochem. Biophys.* **3**:73.

Schultz, J., Felberg, N., and John, S., 1967, Myeloperoxidase, VIII. Separation into ten components by free-flow electrophoresis, *Biochem. Biophys. Res. Comm.* **28**:543.

Schultz, J., Snyder, H., Wu, N. C., Berger, N., and Bonner, M. J., 1972, Chemical nature and biological activity of myeloperoxidase, in: *The Molecular Basis of Electron Transfer* (J. Schultz and B. F. Cameron, eds.), pp. 301–321, Academic Press, New York.

Schultz, J., Baker, A., and Tucker, E., 1976, Myeloperoxidase enzyme therapy on rat mammary tumor, in: *Miami Winter Symposia, Cancer Enzymology* (J. Schultz and F. Ahmad, eds.), p. 319, Academic Press, New York.

Stelmaszynska, T., and Zgliczynski, J. M., 1974, Myeloperoxidase of human neutrophilic granulocytes as chlorinating enzyme, *Eur. J. Biochem.* **45**:305.

Strauven, T. A., Armstrong, D., James, G. T., and Austin, J. H., 1978, Separation of leukocyte peroxidase isoenzymes by agarose-acrylamide disc electrophoresis, *Age* **1**:111.

Wever, R., Roos, D., Weening, R. S., Uvlsma, T., and van Gelder, B. F., 1976, An EPR study of myeloperoxidase in human granulocytes, *Biochim. Biophys. Acta* **421**:328.

Wu, N. C., and Schultz, J., 1975, The prosthetic group of myeloperoxidase, *FEBS Lett.* **60**:141.

Zgliczynski, J. M., Stelmaszynska, T., Ostrowski, W., Naskalski, J., and Sznajd, J., 1968, Myeloperoxidase of human leukemic leukocytes: Oxidation of amino acids in the presence of hydrogen peroxide, *Eur. J. Biochem.* **4**:540.

10

Characteristics of Myeloperoxidase from Neutrophils and Other Peroxidases from Different Cell Types

JAN MACIEJ ZGLICZYNSKI

1. GENERAL PROPERTIES OF MYELOPEROXIDASE FROM NEUTROPHILS

1.1. INTRODUCTION AND SOME STRUCTURAL CONSIDERATIONS

Myeloperoxidase (MPO) belongs to the class of oxidoreductases (EC 1.11.1.7). The electron acceptor substrate for this enzyme is solely hydrogen peroxide, while the electron donor substrates are different reductants including halides (with the exception of fluoride).

Among the electron donor substrates chloride seems to be the most important. This is due to the fact that myeloperoxidase is the only mammalian peroxidase known so far which can oxidize chloride (Agner, 1958, Zgliczyński *et al.*, 1971; Harrison and Schultz, 1976). Moreover, chloride, commonly present in tissues, is now considered as a necessary substrate for the bactericidal process mediated by myeloperoxidase.

The enzyme is a hemoprotein with a molecular weight of 149,000 (Ehrenberg and Agner, 1958), and has an isoelectric point above pH 10 (Agner, 1941). The myeloperoxidase molecule contains two iron porphyrin groups (Ehren-

JAN MACIEJ ZGLICZYŃSKI • Institute of Medical Biochemistry, Nicolaus Copernicus Academy of Medicine, 31-034 Krakow, Poland. Research was supported by Grant II.1.2.11 from the Polish Academy of Sciences.

TABLE 1. LIGHT ABSORPTION MAXIMA OF PEROXIDASE AND THEIR DERIVATIVES

Peroxidase	Oxidized (nm)	Reduced (nm)	CN oxidized (nm)	Pyridine hemochrome (nm)
Myeloperoxidase (Schultz and Schmukler, 1964; Newton et al., 1965a)	430 570 625 690	475 590 637	458 634	438 590
Lactoperoxidase (Carlström, 1969; Morel and Clezy, 1963)	412 501 541 589 631	442 565 600	430 555 595	425 530 571 (558)
Eosinophil peroxidase (Archer et al., 1965)	415 505 640	445 566 595		525 567
Intestine mucosa peroxidase (Stelmaszyńska and Zgliczyński, 1971)	417 490 543 596 642	448 565 596	429 555 596	423 530 568

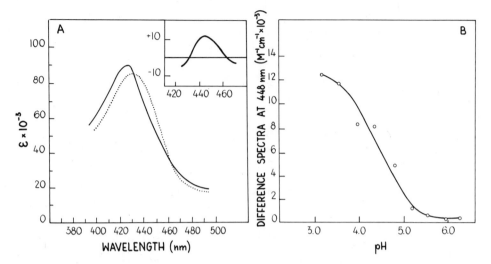

FIGURE 1. The pH-dependent spectral changes of MPO at Soret region. (A) MPO-neutral form, pH 6.28 (——) and MPO-acidic form, pH 3.17 (·····); insert—the pH difference spectrum of the acidic minus neutral form of MPO; (B) The pH dependence of the spectral difference at 448 nm between the acidic and neutral form of MPO (Stelmaszyńska and Zgliczyński, 1974).

berg and Agner, 1958), whose structure has not yet been definitely elucidated. These hematin groups differ from prosthethic groups of other peroxidases as can be recognized in the characteristic spectral features of myeloperoxidase (Table 1), (Schultz and Schmukler, 1964; Newton *et al.*, 1965a; Chang *et al.*, 1970; Nichol *et al.*, 1969). It has been suggested that MPO heme may have a chlorine type conjugation (Chang *et al.*, 1970) and unusual conjugated electrophilic substituents of the heme group (Nichol *et al.*, 1969).

The absorption bands of ferrimyeloperoxidase appear at 280 nm (protein) and 430, 570, 625, and 690 nm (hematin of enzyme). The positions of the bands shift slightly depending on the pH of the medium. This spectral difference which has been measured mainly within the Soret region (430 nm) (Stelmaszyńska and Zgliczyński, 1974) may reflect the ionization state of a group of a protein moiety which is in close proximity to the heme iron. When this group is protonated, the acidic form of the enzyme exists (λmax = 432 nm) and at higher pH values the neutral form of MPO occurs (λmax = 426 nm) (Figure 1a). The pK of 4.4–4.6 (Figure 1b) for the reversible interconversion of both forms suggests that the aspartic or glutamyl carboxyl may participate in this process (Figure 2). The neutral form of MPO changes to the acidic form with the uptake of one proton per hematin group of enzyme, which can be established according to the method of Boeri *et al.* (1953).

A carboxyl ligand at the sixth coordinate position of the heme iron has been proposed previously for chloroperoxidase (Thomas *et al.*, 1970) and for lactoperoxidase (Morrison *et al.*, 1970).

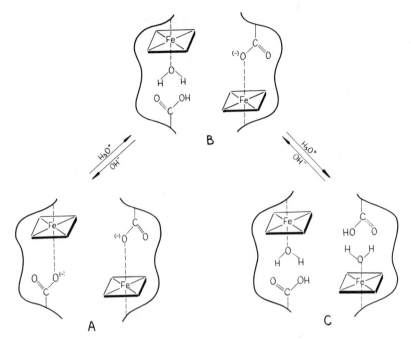

FIGURE 2.   Proposed model of interconversion of the neutral to acidic form of MPO. (A) The neutral form of MPO; (B) the semi-acidic form; (C) the acidic form. (The presence of two hematin groups per mole of MPO is considered.)

The existence of MPO in the acidic and neutral form determines several catalytic properties of this enzyme, which will be discussed later.

1.2. SOURCES FOR THE ISOLATION OF MPO

MPO was obtained from many different sources such as: tubercular empyema (Agner, 1941), pus from infected dog uteri (Agner, 1958), rat chloroma (Schultz, 1958; Newton *et al.*, 1965b), normal human blood (Schultz, 1964; Rohrer *et al.*, 1966), blood of patients with chronic granulocytic leukemia (Zgliczyński *et al.*, 1968; Naskalski, 1977), and guinea pig bone marrow (Desser *et al.*, 1972; Himmelhoch *et al.*, 1969).

All the sources have the common feature of being rich in neutrophils, this means that MPO occurs mainly in this type of cell. The high concentration of MPO in neutrophils (2–5% of the dry weight of the cell) (Agner, 1941; Schultz *et al.*, 1965) indicates a close relationship between the catalytic function of MPO and the biological role of neutrophils. Additional insight into interdependence between the function of the enzyme and its parent cell can be gained by tracing the fate of MPO during functional changes in neutrophils concomitant with their transformation from a resting to an active state.

1.3. DISTRIBUTION OF MPO IN RESTING AND PHAGOCYTIZING NEUTROPHILS

In the resting neutrophils MPO is found only in azurophil or primary granules (Baggiolini *et al.*, 1970; Bainton *et al.*, 1968; Dunn *et al.*, 1968). Phagocytosis results in the formation of phagocytic vacuoles. Around the vacuoles one may observe degranulation resulting from the discharge of the contents of the granules into the vacuole. In effect, the MPO coats the surface of the engulfed bacteria (Klebanoff, 1970). This makes possible the direct action of the enzyme on the bacteria.

During phagocytosis a striking burst of oxidative metabolism occurs (Sbarra and Karnovsky, 1959), and the oxygen consumed is converted in part to hydrogen peroxide (Iyer and Islam, 1961; Paul and Sbarra, 1968; Zatti *et al.*, 1968). This triggers the peroxidative action of MPO. Since the main goal of the neutrophil is to kill the engulfed bacteria, it is clear that MPO may participate in this process by formation of several products with bactericidal properties. Some important data supporting this conclusion were obtained from experiments with purified preparations of MPO.

2. CATALYTIC PROPERTIES OF MPO

2.1. MPO AS A BACTERICIDAL ENZYME

MPO together with hydrogen peroxide and iodide and/or chloride forms a very potent bactericidal system, effective against different bacteria (Klebanoff, 1967 and 1968; McRipley and Sbarra, 1967), fungi (Klebanoff, 1970; Lehrer, 1969), and viruses (Belding *et al.*, 1970).

Although it was known that all halides (except F^-) could participate in the MPO bactericidal system, only iodide was extensively investigated in this respect (for review see Klebanoff, 1975a). MPO in the presence of H_2O_2 mediates the oxidation of iodide by H_2O_2 to form a reactive iodinating species, which can iodinate and/or oxidize several compounds including proteins and lipids of bacterial membranes (Klebanoff, 1975). These chemical events may be associated with bacterial death.

The course of iodination has also been demonstrated in phagocytizing neutrophils in a medium containing $^{131}I^-$ (Klebanoff, 1967, 1970). The incorporation of labeled iodine into a trichloroacetic-acid-precipitable fraction was observed and the fixed iodine was localized near the ingested bacteria in the phagocytic vacuole (Klebanoff, 1970). It has been shown that the MPO system is responsible for this process (Pincus and Klebanoff, 1971; Root and Stossel, 1974).

Further studies on the action of MPO in neutrophils revealed chloride as a participating agent (Zgliczyński and Stelmaszyńska, 1975). When phagocytizing neutrophils were incubated with labeled chloride ($^{36}Cl^-$), the radioactivity was incorporated into the insoluble fraction of the cells. This process was quite dis-

tinct after only 5 min of phagocytosis, whereas it did not occur in resting neutrophils. Azide, a potent inhibitor of peroxidases, retarded the amount of fixed
chlorine incorporated, which strongly suggested that MPO had been involved in
this process. Additional evidence for the chlorinating activity of the MPO system
was obtained from experiments with purified MPO preparations. This chlorinating ability of the MPO system will be the main subject of further considerations.

2.2. MPO AS A CHLORINATING OR SUICIDAL ENZYME

The chlorinating activity of MPO is due to the ability of the enzyme to
oxidize the chloride by H_2O_2 (Agner, 1958; Zgliczyński et $al.$, 1971; Harrison and
Schultz, 1976), according to the equation:

$$Cl^- + H_2O_2 \xrightarrow{MPO} OCl^- + H_2O \tag{1}$$

The amount of OCl^- ions becomes appreciable above pH 6 according to a HOCl
pK value of 7.5. When the pH is lowered HOCl molecules predominate:

$$OCl + H_3O^+ \leftrightarrow HOCl + H_2O \tag{2}$$

HOCl in the presence of an excess of chloride can be converted as follows:

$$HOCl + Cl^- \leftrightarrow Cl_2 + OH^- \tag{3}$$

From the considerations presented one can deduce that MPO may produce a
mixture of OCl^-, HOCl, and Cl_2, with the ratio of constituents depending on
the pH.

Preliminary data regarding the chlorinating ability of MPO were collected
from the experiments on sulfanilamide chlorination (Agner, 1972). This enzyme
system yielded a reaction product identical to that yielded by the chlorination of
sulfanilamide by HOCl. Further evidence was obtained from the MPO-
mediated chlorination of amines and amino acids (Zgliczyński et $al.$, 1971). In
these experiments proof of the formation of chlorinating species by MPO was
deduced indirectly from an estimation of the final chlorinated product.

If the substance which could accept the chlorinating species was absent in
the reaction medium, the enzyme would destroy itself. This means that for the
action of MPO to be permanent the chlorinating species must be removed. This
phenomenon can be demonstrated spectrophotometrically by monitoring the
Soret band during the MPO-mediated reaction (Zgliczyński et $al.$, 1968). When
only MPO, H_2O_2, and Cl^- are present, an irreversible disappearance of the Soret
band can be observed. The addition of cysteic acid, as the chlorine acceptor,
prior to the beginning of the reaction results in the accumulation of the chlorinated product and the Soret band remains the same as in the native enzyme
(Figure 3). This suicidal tendency is so distinct that it may be used as a sensitive
criterion to check whether a compound can be chlorinated by the MPO system.
If a particular substance can prevent the changes in the MPO Soret band, it may
simply be stated that the chlorinated derivative of this compound is at least a
transient, if not the final, product of the reaction.

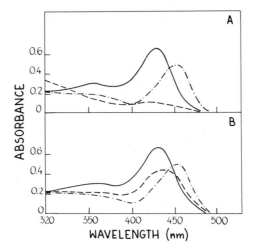

FIGURE 3.   The influence of chlorine accepting substrate (cysteic acid) on the spectrum at Soret region of MPO–H_2O_2–Cl^- system. The spectral changes of MPO–H_2O_2–Cl^- system in the absence (A) and in the presence of cysteic acid (B), pH 5.8. (——) MPO; (–·–·–) MPO –H_2O_2 (after 20 sec); (------) MPO–H_2O_2–Cl^- (after 60 sec) (Zgliczyński et al., 1968).

Recently new evidence supporting the formation of HOCl in the MPO–H_2O_2–Cl system has become available (Harrison and Schultz, 1976). It was directly demonstrated that HOCl is formed as a result of the activity of this system; however this took place only when the chlorinating species (HOCl) had been quickly removed from contact with MPO. This was achieved by using MPO immobilized on a Millipore membrane and forcing through a solution of substrates, H_2O_2 and Cl^-. Under these conditions the initial efficiency of HOCl was 40% of that theoretically made possible by the amount of H_2O_2 used, and decreased in spite of the considerable amount of enzyme and the relatively short duration of its contact with the product. This suggests that, even under these circumstances, the suicidal inactivation of MPO may occur. In conclusion it may be stated that the MPO–H_2O_2–Cl^- system hardly operates at all in the absence of a chlorine-accepting substrate. This simply means that for the full activity of MPO three substrates, i.e., H_2O_2, Cl^-, and a chlorine-acceptor are necessary.

2.3.   SUBSTRATES CHLORINATED BY THE MPO–H_2O_2–Cl^- SYSTEM

A list of substrates which can be easily chlorinated by the MPO–H_2O_2–Cl^- system consists of many compounds with different chemical structures. Among them the amino compounds have been investigated most extensively. All substances with amino groups yielded chloramino compounds as the products of enzymatic chlorination. Compounds such as α-amino acids and secondary amines (diethanolamine) yield only monochloramines when chlorinated by the enzymatic system. Under the conditions of enzymatic chlorination the introduction of the first chlorine atom is easier than that of the second.

Monochloramines of α-amino acids decompose into ammonia, aldehydes, and carbon dioxide (Zgliczyński et al., 1968). The measurement of $^{14}CO_2$ as a product of labeled amino acid degradation has been employed as a test for MPO chlorinating activity (Paul *et al.*, 1970). This method is especially useful for intact cells, granules, and crude preparations of MPO, on condition that only the production of H_2O_2-dependent $^{14}CO_2$ is taken into consideration.

Those amino compounds which form fairly stable monochloramines, for example, ethanolamine, β-amino-carboxylic acid (β-alanine), aminosulfonic acid (taurine) (Stelmaszyńska and Zgliczyński, 1974) and peptides (Stelmaszyńska and Zgliczyński, 1978), usually also yield dichloramino derivatives. The introduction of a second chlorine atom due to the MPO system occurs only when the formation of monochloramine has been completed and, in general, this reaction is not efficient. This is true above pH 5, but below this value a second mechanism operates by which a dichloramino derivative can be formed. This is the nonenzymatic dismutation of mono- to dichloramino derivatives, as shown in equation (4) (Stelmaszyńska and Zgliczyński, 1978).

$$2\ R-NH-Cl + H^+ \leftrightarrow R-NH_3^+ + R-NCl_2 \qquad (4)$$

The lower the pH of the reaction medium the faster the conversion of mono- into dichloramino compounds. In effect, if enzymatic chlorination proceeds within the pH range 4–5, the yield of dichloroamino derivative is higher and it is formed even when the amino substrate is still present.

The chloramines can be detected and estimated due to their chemical and spectral properties (Weil and Morris, 1949; Neale and Walsh, 1965). The chloramines oxidize iodide (as does HOCl) and bromide. The latter is oxidized by chloramine only up to pH 6, which makes it possible to distinguish chloramino compounds from HOCl (Zgliczyński *et al.*, 1971). Both types of chloramine absorb in the UV region with λmax at 250–255 nm and 300–305 nm for the mono- and dichloramino groups, respectively. This UV absorbance proved to be useful for quantitative determination of chloramino compounds and thus the kinetics of chlorination mediated by the MPO system have been investigated by this method. Diethanolamine, a secondary amine, was especially useful in this respect since it yields only monochloramino derivatives of a convenient stability at any pH value (Stelmaszyńska and Zgliczyński, 1974).

Besides amino compounds, the phenolic ring of tyrosine, pyrimidine bases such as cytosine and uracil, as well as their nucleosides and nucleotides act as chlorine-accepting substrates in the MPO system (Stelmaszyńska and Zgliczyński, unpublished). The products of chlorination of these substrates contain a chlorine–carbon bond. Cytosine and uracil are chlorinated in position 5 (Lis and Zgliczyński, unpublished). There is no doubt that such products may be of high biological importance and may, for instance, act as potent cytostatic agents.

The third chlorine-accepting group of substrates for the MPO system is macromolecules. It has been reported that proteins and heat-killed or live bacteria can be efficiently chlorinated (Zgliczyński and Stelmaszyńska, 1975). In the case of proteins, chlorine reacts partly with free amino groups, for example, N-terminal (Stelmaszyńska and Zgliczyński, 1978) or ε-amino groups of lysine. Oxidative peptide cleavage as a consequence of peptide bond chlorination has been also suggested (Selvaraj *et al.*, 1974). The phenolic rings of tyrosyl residues may be a target of chlorination. Enzymatically chlorinated bovine serum albumin subjected to pronase hydrolysis and to amino acid analysis yields 3-chloro- and 3,5-dichlorotyrosine (Stelmaszyńska, unpublished).

Two interesting observations were made when bacteria (*Staphylococcus epidermidis*) were used as the chlorine-accepting substrate (Zgliczyński and Stelmaszyńska, 1975). First, the live bacteria were less efficiently chlorinated by the MPO system than those killed by heat, which indicates that some particular antichlorinating mechanisms exist in live bacteria (catalase?). Secondly, chlorination of bacteria was markedly promoted by preincubation of live microorganisms with MPO prior to the start of chlorination. This preincubation abolished the difference in the extent of chlorination between living and heat-killed bacteria. This phenomenon can be explained, assuming that the preincubation of myeloperoxidase with bacteria results in specific binding of the enzyme with the membrane of the microorganism. In such close proximity, the reactive chlorinating species produced by MPO can directly attack the bacterial surface.

The promoting influence of preincubation on the extent of bacterial chlorination provoked attempts to establish the bactericidal efficiency of myeloperoxidase preincubated with bacteria prior to addition of H_2O_2 and Cl^-. Selvaraj *et al.* (1978) have, in fact, found that such preincubation results in absorption of myeloperoxidase onto the surface of the microorganism leading to enhancement of its bactericidal activity.

2.4. CHLORIDE AS A SUBSTRATE FOR MPO IN THE CHLORINATION REACTION

Chloride seems to be a true substrate for MPO. This is deduced from the kinetic data of the chlorination reaction as well as from the spectral evidence of the formation of the MPO–chloride complex (Stelmaszyńska and Zgliczyński, 1974). The saturation effect of chloride concentration on the rate of chlorination is typical of a true substrate. However, an excess of chloride results in the inhibition of chlorination, but this inhibitory effect is apparent and can be reversed by an increase in the pH of the reaction medium. This indicates that there is a different optimal chloride concentration for every pH value. The higher this value, the higher the chloride concentration required for the optimal reaction course. A reasonable explanation for this dependence would be the assumption that the apparent affinity between Cl^- and MPO varies depending on the pH. Such an assumption is acceptable in view of the effect of pH on the properties of MPO (interconversion of acidic and neutral form) (Figure 1).

There are certain indications that chloride binds exclusively to the acidic form of MPO, since this form may have one free coordinate position of iron (Figure 2). On this assumption the following reaction may occur:

$$MPO + H^+ \xrightarrow{K_H} MPO \cdot H^+ + Cl^- \xrightarrow{K_{Cl}} MPO \cdot HCl \qquad (5)$$

Several data support equation (5) and in turn all of them together can be consistently accounted for by this equation. Since chloride shifts the equilibrium of these reactions (5) to the right and of OH⁻ ions to the left, in an acidic medium only two forms can occur, MPO·H⁺ and MPO·HCl. Indeed, when the conversion of myeloperoxidase to MPO·HC1 was traced spectrophotometrically, within the pH range where the MPO·H⁺ form had predominated, an isobestic point at 450 nm was obtained, which proves the existence of only two forms in equilibrium (Figure 4). An increase in the OH⁻ ion concentration results in the formation of three forms of myeloperoxidase, MPO, MPO·H⁺ and MPO·HCl; thus their inter-conversion occurs without the isobestic point (Figure 4b). Furthermore, within the pH range where a mixture of MPO and MPO·H⁺ exists (e.g., at pH 5.2, Figure 4b) the addition of chloride causes a decrease in the amount of MPO present in the neutral form. This is observed as a shift of the Soret band to longer wavelengths.

Equation (5) also implies the competitive inhibition of Cl⁻ binding to myeloperoxidase by OH⁻ ions. In fact, the higher the pH, the higher the required Cl⁻ concentration to obtain 50% saturation of the enzyme (Fig. 5). This changing affinity of Cl⁻ to myeloperoxidase is only apparent and is brought about by changes in the MPO to MPO·H⁺ ratio. Thus the actual affinity between Cl⁻ and myeloperoxidase, described by the dissociation constant, K_{Cl}, can only be measured on condition that there is a substantial excess of MPO·H⁺. This value was established under such conditions and was found to be 5.2×10^{-3} M, (Figure 5).

The affinity between myeloperoxidase and C1⁻ can also be measured kineti-cally using the chlorination reaction. The data obtained indicates that the appar-ent affinity between MPO and Cl⁻ is dependent upon the pH. However, the presence of H_2O_2 in the medium additionally complicates the interpretation of

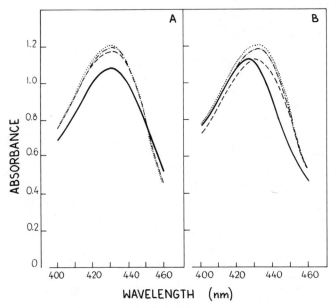

FIGURE 4.   The effect of Cl⁻ on the spectral properties of MPO at pH 3.28 (A) and 4.23 (B). ——, 6.3 μM MPO in phosphate-citrate buffer; (------) + 10 mM Cl⁻; (-·-·-) + 50 mM Cl⁻; (·····) + 400 mM Cl⁻ (Stelmaszyńska and Zgliczyński, 1974).

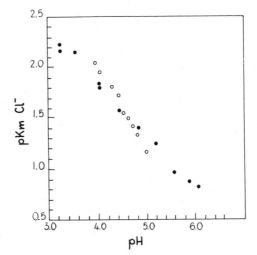

FIGURE 5. The effect of pH on the apparent affinity between MPO and Cl⁻ ions calculated from kinetic data of diethanolamine chlorination by MPO–H₂O₂–Cl⁻ system (O) and from spectral difference at 430 nm resulting from the formation of the enzyme–Cl⁻ complex (●) (Stelmaszyńska and Zgliczyński, 1974).

this data; therefore, it will be discussed after having presented the relationship between H_2O_2 and MPO.

2.5. H_2O_2 AS A SUBSTRATE FOR MPO IN THE CHLORINATION REACTION

Agner (1963) has found that MPO forms a spectrally distinct, dissociable H_2O_2 complex with an apparent dissociation constant $K_{H_2O_2} = 1 \times 10^{-4}$ M. He suggested that below an H_2O_2 concentration of 1×10^{-4} M, only one of the two hematins of the enzyme reacts with H_2O_2, but at a higher concentration two iron atoms become occupied by H_2O_2 and myeloperoxidase acquires the properties of a catalase. Further increase of the H_2O_2 concentration, above 2×10^{-3} M, results in the destruction of the enzyme.

In fact, the MPO-mediated chlorination reaction is inhibited by an excess of H_2O_2 (Agner, 1958b; Zgliczyński *et al.*, 1968; Naskalski, 1977). This inhibitory effect of an excess of H_2O_2 can be reversed by lowering the pH (Zgliczyński *et al.*, 1977). It was found that the optimal pH of the MPO-mediated chlorination changes depended on the H_2O_2 concentration. The lower the H_2O_2 concentration the higher the optimal pH of the MPO-mediated chlorination. The apparent affinity between MPO and H_2O_2 was also found to be pH-dependent, and decreased with a lowering of the pH (Figure 6). Thus, H^+ inhibits the binding of H_2O_2 to myeloperoxidase.

As mentioned above, MPO exists in both a neutral and an acidic form. Since the H^+ ion inhibits H_2O_2 binding to myeloperoxidase it can be assumed that H_2O_2 is bound to the neutral form of MPO, and the reaction may be expressed:

$$\text{MPO} + \text{H}_2\text{O}_2 \leftrightarrow \text{MPO} \cdot \text{H}_2\text{O}_2$$
$$+$$
$$\text{H}^+ \qquad\qquad (6)$$
$$\updownarrow$$
$$\text{MPO} \cdot \text{H}^+$$

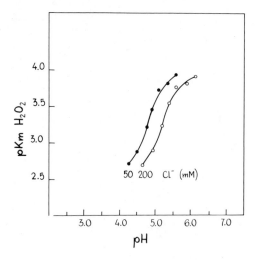

FIGURE 6. The effect of pH on the apparent affinity between MPO and H_2O_2 calculated from kinetic data of diethanolamine chlorination by MPO–H_2O_2–Cl$^-$ system (Zgliczyński *et al.*, 1977).

As can be seen from this reaction, the binding of H_2O_2 to MPO is dependent on the percentage content of the neutral form of the enzyme. The amount of this form is in turn dependent on the level of H$^+$ ions and, additionally, on the Cl$^-$ concentration, as described in the previous section. Thus, the equation summarizing the reactions between MPO and its substrates, H_2O_2 and Cl$^-$, can be presented as follows:

$$MPO + H_2O_2 \leftrightarrow MPO \cdot H_2O_2$$
$$+$$
$$H^+ \tag{7}$$
$$\updownarrow$$
$$MPO \cdot H^+ + Cl^- \leftrightarrow MPO \cdot HCl$$

Such an interdependent reaction system enables one to gain an insight into the relationship between MPO, H_2O_2, Cl$^-$, and pH.

2.6. H_2O_2, Cl$^-$, AND pH INTERDEPENDENCE IN THE CHLORINATION REACTION

It is obvious that during the course of chlorination (i.e., in order to oxidize Cl$^-$ by H_2O_2) two MPO complexes have to occur simultaneously: MPO$\cdot H_2O_2$ and MPO$\cdot$HCl. Taking into account the fact that the myeloperoxidase molecule contains two prosthetic groups, this situation could occur on one enzyme molecule.

If a given myeloperoxidase molecule contains exclusively Cl$^-$ or H_2O_2 on both active sites it is unable to catalyze the chlorination reaction. This may happen if an excess of H_2O_2 in relation to Cl$^-$, or vice versa, occurs in the reaction medium. When, during the course of chlorination, some inhibition is observed due to an excess of one of the substrates, this may be reversed by increasing the concentration of the other. Kinetic data indicate that there exists a mutual competition between Cl^- and H_2O_2 for binding to the actives sites of myeloperoxidase. Thus, only one particular $[H_2O_2]/[Cl^-]$ ratio is required for the simultaneous binding of both H_2O_2 and Cl^- to one molecule of the enzyme.

However, as mentioned in previous sections the binding of both Cl^- and H_2O_2 is pH dependent, but in opposite directions. This means that the optimal $[H_2O_2]/[Cl^-]$ ratio is different for every pH value.

The optimal $[H_2O_2]/[Cl^-]$ ratio for the rate of chlorination has been determined for several H^+ ion concentrations, and a linear relationship between the logarithm of $[H_2O_2]/[Cl^-]$ and the pH value has been obtained (Figure 7) (Zgliczyński et al., 1977).

The relationship presented in Figure 7 was experimentally established within a pH range of 4–6. Extrapolation up to pH 7.4 yielded optimal conditions which were then fully proved experimentally, and confirmed the validity of such an extrapolation. Finally, taking into account the three parameters $[H_2O_2]$, $[Cl^-]$, and $[H^+]$ (Figure 7), the optimal conditions for chlorination can be established from the equation:

$$\frac{[H_2O_2]}{[H^+] \times [Cl^-]} = \text{const.} = B \tag{8}$$

(where B is the constant value for optimal chlorination).

It now becomes clear that optimal chlorination can be carried out over a wide range of pH values providing that there is a proper adjustment between H_2O_2 and Cl^- concentrations. A B value equal to 560 in equation (8) was found experimentally when diethanolamine was used as a chlorine-accepting substrate (Zgliczyński et al., 1977). In all cases when the rate of chlorine acceptance exceeds the rate of HOCl formation by the MPO–H_2O_2–Cl^- system, the B values should be the same.

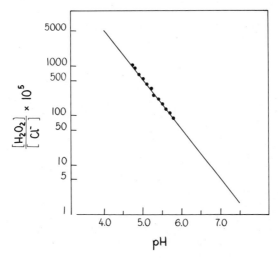

FIGURE 7. The optimal pH for MPO-medicated chlorination at different $[H_2O_2]/[Cl^-]$ ratios (Zgliczyński et al., 1977).

2.7. SOME THEORETICAL CONSIDERATIONS CONCERNING THE MODE OF ACTION OF MPO

When one attempts to establish a general model for the optimal action of the MPO–H_2O_2–Cl^- chlorination system, it is useful to summarize concisely all the relationships already presented:

1. The lower the pH, the more effective the binding of Cl^- with MPO. The OH^- ions competitively inhibit the MPO·HCl complex formation, i.e., Cl^- can react only with the acidic form of MPO.
2. The higher the pH the more effective the binding between H_2O_2 and MPO. This binding is competitively inhibited by H^+ ions, i.e., H_2O_2 can react only with the neutral form of MPO.
3. There is competition between Cl^- and H_2O_2 for active sites in myeloperoxidase molecules. For optimal chlorinating activity of the enzyme there is one particular optimal value for the $[H_2O_2]/[Cl^-]$ ratio. This value decreases when the pH increases, which means that myeloperoxidase can act optimally over a wide range of pH values.
4. A chlorine-accepting substrate is absolutely necessary to maintain the course of Cl^- oxidation by H_2O_2. This means that any accumulation of HOCl, or any species formed during the peroxidation of Cl^-, causes the destruction of myeloperoxidase and results in the interruption of the catalyzed process.

All the relationships summarized here can be accounted for by a system of mutually dependent reactions.

Now we must reconsider all these reactions as one system and determine whether this system is consistent with all known experimental data.

$$MPO + H_2O_2 \leftrightarrow MPO \cdot H_2O_2 \qquad MPO + A \qquad \text{chlorinated}$$
$$+ \qquad\qquad\qquad\qquad\qquad\qquad\qquad\qquad \text{product A–Cl}$$
$$H^+ \qquad\qquad\qquad\qquad\qquad + H_2O + HOCl \qquad (9)$$
$$\updownarrow \qquad\qquad\qquad\qquad\qquad\qquad\qquad\qquad\qquad \text{destroyed}$$
$$MPO \cdot H^+ + Cl^- \leftrightarrow MPO \cdot HCl \qquad MPO \qquad\qquad MPO$$

(A is the chlorine-accepting substrate.)

Each reaction in this system has its own particular equilibrium constant, which can be expressed as follows:

$$K_H = \frac{[MPO] \times [H^+]}{[MPO \cdot H^+]} \tag{10}$$

$$K_{H_2O_2} = \frac{[MPO] \times [H_2O_2]}{[MPO \cdot H_2O_2]} \tag{11}$$

$$K_{Cl} = \frac{[MPOH^+] \times [Cl^-]}{[MPO \cdot HCl]} \tag{12}$$

Since all of these reactions are mutually dependent, one can solve the equations (10), (11), and (12) together. Thus, solving equation (11) for [MPO], equation (12) for [MPOH$^+$], substituting these values into equation (10), and rearranging them, one can obtain:

$$\frac{K_{H_2O_2}}{K_H \times K_{Cl}} = \frac{[MPO \cdot HCl] \times [H_2O_2]}{[MPO \cdot H_2O_2] \times [H^+] \times [Cl^-]} \tag{13}$$

For an optimal rate of chlorination, the [MPO·HCl] should be equal to [MPO·H$_2$O$_2$] (for discussion see Section 2.6), and equation (13) becomes simplified:

$$\frac{K_{H_2O_2}}{K_H \times K_{Cl}} = \frac{[H_2O_2]}{[H^+] \times [Cl^-]} \tag{14}$$

It should be especially stressed that exactly the same equation for optimal chlorination was postulated from purely experimental determinations (see section 2.6). This strongly suggested that the theoretically postulated, mutually dependent reactions system [equation (9)] is correct.

As can be seen in equation (8), the expression $[H_2O_2]/[H^+] \times [Cl^-]$ is equal to 560, so it can be expressed as:

$$560 - \frac{K_{H_2O_2}}{K_H \times K_{Cl}} \tag{15}$$

Since K_H and K_{Cl} values are known to be 3.16×10^{-5} M and 5.2×10^{-3} M, respectively (see Sections 1.1 and 2.4), one can obtain from equation (15) the $K_{H_2O_2}$ value as 0.92×10^{-4} M. It is again striking that similar value for the apparent dissociation constant of the MPO·H$_2$O$_2$ complex has been determined by other investigators from kinetic and spectrophotometric data (Agner, 1963; Newton et al., 1965b; Stelmaszyńska, 1970).

The model of the action of the MPO–H$_2$O$_2$–Cl$^-$ system was based on the assumption that Cl$^-$ can bind only to the acidic while H$_2$O$_2$ binds to the neutral form of myeloperoxidase. However, it is interesting to speculate why this might be the case.

The binding of Cl$^-$ solely with the acidic form is relatively easy to explain, taking into account that MPO·H$^+$ has one free coordinate position of 6 iron, whereas in the neutral form of MPO this position is occupied by carboxylate anion. Moreover, the negative charge of carboxylate anion repels the negatively charged Cl$^-$ ion.

There is, however, insufficient data regarding the structure of MPO to make possible an explanation for the exclusive binding of H$_2$O$_2$ to the neutral form of MPO. But one can imagine that the linking of H$_2$O$_2$ to the neutral form of MPO could be due to mutual interactions of H$_2$O$_2$, carboxylate anion, and hematin

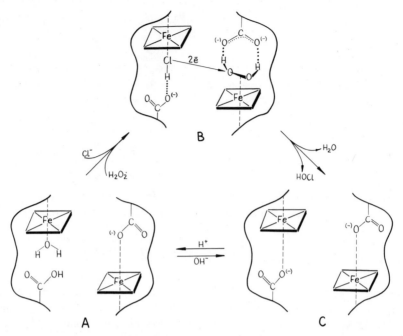

FIGURE 8.   Proposed model for H_2O_2 and Cl^- binding with neutral and acidic subunits of MPO, respectively. (A) Semiacidic form of MPO; (B) catalytically active complex of MPO (Cl^- and H_2O_2 bound to the same molecule of MPO); (C) neutral form of MPO released after HOCl and H_2O formation.

iron. The negatively charged carboxylate anion may attract the positive pole of H_2O_2 with the possibility of the formation of hydrogen bonds, whereas the negative pole of H_2O_2 is, in turn, attached to the hematin iron. This results in the formation of a dipole chain (Figure 8).

The binding of H_2O_2 molecules to the acidic form of MPO could be prevented by the presence of hydrogen on the carboxyl group, i.e., the formation of the postulated dipole chain would be less possible.

The binding of H_2O_2 solely with the neutral form of MPO is additionally supported by studies of the optimum pH for the MPO-mediated oxidation of guaiacol. It has been found that the rate of guaiacol oxidation by $MPO \cdot H_2O_2$ sharply decreases below pH 5, where the neutral form of MPO undergoes conversion to the acidic form (Stelmaszyńska, 1970).

2.8.   SOME PHYSIOLOGICAL ASPECTS OF THE MODE OF ACTION OF MPO

Bearing in mind the relationship, $B = [H_2O_2]/[H^+] \times [Cl^-]$, and its consequences in the proper adjustment of the conditions for the chlorination reaction, the optimum course of this reaction can be obtained over a wide range of substrate concentrations and pH values. It has been found that MPO-mediated chlorination can occur at the same rate over a range of substrate concentrations, i.e., 0.167 mM–1.67 mM and 50 mM–400 mM for H_2O_2 and Cl^-, respectively (Zgliczyński et al., 1977). Moreover, not only can the rate of chlorination be stable over such a wide range of conditions, but the bactericidal potential of the MPO system should also be equally stable. Recently, it has been found that the MPO–H_2O_2–Cl^- system exerts its bactericidal activity at a neutral pH (Sbarra et al., 1976). This finding contradicts the previously held view that the bactericidal activity of this system can be attained mainly within the acidic pH range. It may be that those opinions arose from the fact that the bactericidal experiments were carried out at the same H_2O_2 and Cl^- concentrations for every pH value tested.

Through this ability to act over such a broad spectrum of conditions, MPO seems to be perfectly adapted to function in neutrophils during phagocytosis. The engulfment of bacteria triggers the simultaneous formation of H_2O_2 and lactic acid (Kakinuma, 1970; Mandell, 1970). It is obvious that at the very beginning of phagocytosis both H^+ and H_2O_2 concentrations are low, but even this may constitute the optimal conditions for the action of the MPO system, as can be predicted from equation nine. During the subsequent stages of phagocytosis the level of H_2O_2 and intravacuolar $[H^+]$ increases (Jensen and Bainton, 1973), and in this way the ratio of $[H_2O_2]/[H^+]$ may still be stable. This means that the optimal conditions for MPO activity may also be constant.

The change in Cl^- concentrations opens an additional way in which conditions can be properly adjusted for MPO activity. So far, there is insufficient data which might indicate whether the Cl^- concentration within the neutrophil is actively adjusted. However, there is one indication that such a mechanism for the regulation of MPO activity is possible. In a study using rabbit neutrophils the intracellular chloride concentration was found to vary with that of the extracellular fluid (Wilson and Manery, 1949).

The necessity for flexibility of the conditions required for MPO activity is clear, taking into account the fact that in the neutrophils a wide variety of bacteria are effectively killed. It is known that some strains can produce H_2O_2 and others possess a catalase which decreases the H_2O_2 concentration, and some produce acid, thereby lowering the pH of the surrounding medium. Such strains will themselves alter the conditions for optimal MPO activity, and it is clear that regulation by changing other substrate concentrations or pH is necessary to assure optimal MPO activity. There are some indications that an active adjustment of H_2O_2 production in phagocytizing neutrophils takes place, depending on the kind of particles ingested (Mandell, 1971).

2.9. PRODUCTION OF SINGLET MOLECULAR OXYGEN BY THE MPO SYSTEM

In the last few years, attention has been focused on an interesting event occuring during phagocytosis, namely the emission of light (Allen et al., 1972). This chemiluminescence has not been observed in resting leukocytes (Allen et al., 1972) and in granulocytes with impaired ability to produce H_2O_2 (Stjernholm *et al.*, 1973). It has also been found that a crude preparation of MPO in the presence of H_2O_2 and Cl^- or Br^- can emit photons (Allen and Steele, 1973; Allen, 1975a,b).

It is currently thought that the chemiluminescence observed in phagocytizing neutrophils as well as that originating from the MPO–H_2O_2–Cl^- or Br^- system is due to the formation of singlet oxygen ($^1O_2$) and its relaxation to the ground triplet state. Formation of singlet oxygen in phagocytizing neutrophils may occur through the dismutation of the superoxide anion O_2^- (Allen *et al.*, 1974) or, as has been more frequently suggested (Allen and Steele, 1973; Allen, 1975a,b; Klebanoff, 1975b), from the reaction of HOCl with H_2O_2 (Kasha and Khan, 1970). This latter reaction is associated with MPO activity in two ways. Firstly, it would be competitive with MPO-mediated oxidation of Cl^- for H_2O_2, and, secondly, it would be MPO-dependent since HOCl can only be formed through MPO catalytic action.

Moreover, the process of $^1O_2$ formation from H_2O_2 and HOCl would be competitive with the chlorination due to the consumption of HOCl. In view of this competition, the question arises which of the two processes will predominate: the utilization of HOCl by amino compounds to form chloramines or the reaction of HOCl with H_2O_2 to form singlet oxygen. It will be dependent on the concentration of substrates reacting with HOCl, and on the HOCl affinity for them. Considering the high concentration of amino compounds (e.g., taurine, amino acids, proteins) in neutrophils and their high affinity for HOCl, it seems probable that the chlorination reaction may predominate. Apart from direct HOCl utilization for $^1O_2$ formation, one might imagine that the singlet oxygen is produced as a result of the reaction between chloramines and H_2O_2.

3.   OTHER PEROXIDASES FROM DIFFERENT CELL TYPES

3.1.   ISOLATED AND PURIFIED PEROXIDASES

Peroxidase activity has already been detected in almost all kinds of animal tissues (Neufeld *et al.*, 1958). However, true peroxidases in a purified state have only been obtained from certain types of cells or tissues. Only these enzymes have been studied in detail, and our knowledge concerning their structure and activity is relatively well established. Even though a number of facts are known about their physical and chemical properties, the biological function of only two of them is properly understood. These are myeloperoxidase from neutrophils and thyroid peroxidase from the thyroid follicular cells. Myeloperoxidase has already been described, and, therefore, only some data about thyroid peroxidase will be given.

Thyroid peroxidase is a hemoprotein, tightly bound to microsomal membranes of cells (Neary *et al.*, 1976). The molecular weight of this enzyme has been proposed to be 200,000 with three identical subunits, each containing a hematin prosthetic group (Morrison *et al.*, 1970; Danner and Morrison, 1971). Other authors (Ljunggren and Åkeson, 1968; Taurog *et al.*, 1970) suggest a monomeric form of the enzyme with a molecular weight of 50,000–64,000. The structure of the thyroid peroxidase prosthetic group is also controversial: protohematin has been detected by Danner and Morrison (1971) and Ljunggren and Åkeson (1968), but also other types of hematin groups have been suggested (Taurog *et al.*, 1970). Thus, that more than one peroxidase may exist in the thyroid gland has been expected (Danner and Morrison, 1971).

Thyroid peroxidase is involved in the biological function of the thyroid gland. It participates in thyroxine formation by catalyzing steps such as the oxidation of iodide by H_2O_2, iodination of tyrosyl residues, and even coupling of diiodotyrosine (for review see Hosoya *et al.*, 1971).

The following peroxidases belong to the group of purified peroxidases of unprecisely established biological function: lactoperoxidase, salivary gland peroxidase, eosinophil, and intestinal mucosa peroxidase. All these enzymes have some properties in common, such as similar spectral features (Table 1), the oxidation of iodide, and an iodination ability. These similarities suggest that this group of enzymes will also have other similar properties, for example, molecular weight and content of iron per molecule. However, there is no such data available for all the mentioned enzymes. Nevertheless, the identity of the peroxidases from mammary, salivary, and lacrimal glands was confirmed by proving their antigenic identity (Morrison *et al.*, 1965; Morrison *et al.*, 1970).

As may be seen from Table 1 the peroxidase isolated from eosinophils is distinctly different from myeloperoxidase (Archer *et al.*, 1965). Moreover, it was found that these two enzymes in spite of their presence in cells of common origin are under separate genetic control (Salmon *et al.*, 1970).

Rytömaa (1960) has shown that many tissues containing considerable peroxidase activity are substantially infiltrated by eosinophils, and so he has suggested that these tissue enzymes may in fact originate from eosinophils. This was partially supported by the isolation and purification of intestinal mucosa peroxidase (Stelmaszyńska and Zgliczyński, 1971) which appeared to be spectrally similar to eosinophil peroxidase (Table 1).

Without necessarily taking this hypothesis as a fact it will be convenient in further considerations to call this group of peroxidases the eosinophil-type peroxidase group.

It is striking that the eosinophil-type peroxidases have distinct iodination ability. Lactoperoxidase exhibit a 40 times higher iodination ability than myeloperoxidase and is an even better iodinating enzyme than thyroid peroxidase (Morrison *et al.*, 1970). Also eosinophils which contain *o*-dianisidine peroxidase activity comparable to that of neutrophils can iodinate considerably more efficiently than neutrophils (DeChatelet *et al.*, 1977). Therefore, it may be suggested that the main catalytic function of the eosinophil-type peroxidases is to iodinate organic compounds. There are no data, however, that this group of enzymes can catalyze oxidation of chloride (i.e., can catalyze chlorination), a property which is characteristic for myeloperoxidase from neutrophils.

Both, iodination and chlorination processes are associated with cidal activity of the peroxidase systems. There is no doubt that chlorination is a more drastic process than iodination, thus, it is interesting to speculate that the chlorination process physiologically can operate mainly in a separated, closed compartment such as the phagocytic vacuole of neutrophils. The cidal activity of iodination can be exerted outside the phagocytizing cell. It has also been suggested, that eosinophils are physiologically designated to kill large nonphagocytized objects (DeChatelet *et al.*, 1977), while the main goal of neutrophils is to phagocytize and kill engulfed objects intracellularly.

3.2. PEROXIDASES DETECTED BY THEIR ACTIVITY

Peroxidases are of interest especially in respect to host–parasite interactions, since they display bactericidal properties. From this point of view, the peroxidase activity in cells involved in host-defense has been widely studied. Since neutrophils and eosinophils were already considered in this respect, cells such as lymphocytes, monocytes, tissue macrophages, and Kupffer cells will be discussed in terms of their peroxidase content and peroxidase type.

Data are available regarding the peroxidase activity of mouse spleen cells, predominantly lymphocytes by morphology (Paul *et al.*, 1970; Strauss *et al.*, 1972a,b). The peroxidase present in these cells can form a bactericidal system with a halide and H_2O_2. This enzyme can catalyze the decarboxylation of amino acid in the presence of chloride, which indicates that it is a myeloperoxidase-type enzyme. It has been shown that only myeloperoxidase can chlorinate amino acids to chloramines which in turn are decomposed with decarboxylation (see Section 2.3). However, certain differences have been observed between myeloperoxidase and the peroxidase of spleen lymphocytes in respect to guaiacol oxidation (Paul *et al.*, 1976). Even in spite of this difference in guaiacol oxidation, it is most likely that mouse spleen lymphocyte peroxidase is in fact a myeloperoxidase-type. This does not, however, exclude the possibility that, in addition to myeloperoxidase, another peroxidase-like protein may occur in spleen lymphocytes which accounts for a different pattern of guaiacol oxidation.

The peroxidase activity has also been observed in monocytes from different species (for review see Klebanoff and Hamon, 1975), and as in mouse spleen lymphocytes this peroxidase is most likely a myeloperoxidase. The following data indicate that this is the case. It has been found that monocyte peroxidase cross-reacts with antiserum against neutrophil myeloperoxidase (Salmon *et al.*, 1970). If there is a genetic absence of myeloperoxidase in neutrophils, then monocytes also display biochemical and antigenic lack of peroxidase. This suggests that both enzymes—in neutrophils and monocytes—are coded by the same gene and thus are identical proteins (Salmon *et al.*, 1970). The identity of these enzymes has recently become more evident by the finding of Bos *et al.* (1978) that the EPR and visible light absorption spectra for both enzymes are the same.

The data concerning peroxidase activity in tissue macrophages are contradictory (for review see Klebanoff and Hamon, 1975). Whether this activity is found or not, depended on the source of macrophages and the experimental method for revealing the peroxidase activity.

Romeo *et al.* (1973) have shown the presence of peroxidase activity in both alveolar and peritoneal macrophages of rabbit and guinea pig by using a suitable biochemical method employing an appropriate concentration of H_2O_2 and of cationic detergent. They have further demonstrated that this enzyme lacks the ability to catalyze the decarboxylation of amino acids, which indicates that it is not a myeloperoxidase-type enzyme. Moreover, it has been shown that this enzyme forms a bactericidal system exclusively with iodide (Paul *et al.*, 1973), whereas the MPO-mediated bactericidal system utilizes either Cl^- or I^-.

In the last cell type to be considered, i.e., Kupffer cells, peroxidase activity has been detected (Fahimi *et al.*, 1970; Widmann *et al.*, 1972), however, due to a lack of adequate data it is impossible to say which type of peroxidase is present in these cells.

In summary, it can be said that peroxidase activity is detectable in almost all types of host-defense cells. This indicates that these enzymes play an important role in the host–parasite interaction. However, their degree of participation in the bactericidal systems of different cells varies considerably. It seems that this peroxidase-bactericidal system is most developed in the neutrophils.

ACKNOWLEDGMENT.   I am indebted to Dr. Teresa Stelmaszyńska for co-work in preparation of this chapter.

REFERENCES

Agner, K., 1941, Verdoperoxidase. A ferment isolated from leucocytes, *Acta Physiol. Scand.* **2** (Suppl. 8):1.

Agner, K., 1958a, Crystalline myeloperoxidase, *Acta Chem. Scand.* **12**:89.

Agner, K., 1958b, Peroxidative oxidation of chloride ions, *Proc. 4th Intl. Congr. Biochem. Vienna* **15**:64.

Agner, K., 1963, Studies on myeloperoxidase activity, *Acta Chem. Scand.* **17**:S332.

Agner, K., 1972, Biological effects of hypochlorous acid formed by "MPO"—peroxidation in the presence of chloride ions, in: *Structure and Function of Oxidation-Reduction Enzymes* (Å. Åkeson and A. Ehrenberg, eds.), pp. 329–335, Pergamon Press, Elmsford, N.Y.

Allen, R. C., 1975a, Halide dependence of the myeloperoxidase-mediated antimicrobial system of the polymorphonuclear leucocyte in the phenomenon of electronic excitation, *Biochem. Biophys. Res. Comm.* **63**:675.

Allen, R. C., 1975b, The role of pH in the chemiluminescent response of the myeloperoxidase-halide-HOOH antimicrobial system, *Biochem. Biophys. Res. Comm.* **63**:684.

Allen, R. C., and Steele, R. H., 1973, The functional generation of electronic excitation states by myeloperoxidase, *Fed. Proc.* **32**:478.

Allen, R. C., Stjernholm, R. L., and Steele, R. H., 1972, Evidence for the generation of an electronic excitation state(s) in human polymorphonuclear leucocytes and its participation in bactericidal activity, *Biochem. Biophys. Res. Comm.* **47**:679.

Allen, R. C., Yevitch, S. J., Orth, R. W., and Steele, R. H., 1974, The superoxide anion and singlet molecular oxygen: Their role in the microbicidal activity of the polymorphonuclear leucocyte, *Biochem. Biophys. Res. Comm.* **60**:909.

Archer, G. T., Air, G., Jackas, M., and Morell, D. B., 1965, Studies on rat eosinophil peroxidase, *Biochim. Biophys. Acta* **99**:96.

Baggiolini, M., Hirsch, J. G., and deDuve, D., 1969, Resolution of granules from rabbit heterophil leucocytes into distinct populations by zonal sedimentation, *J. Cell Biol.* **40**:529.

Bainton, D. F., and Farquhar, M. G., 1968, Differences in enzyme content and specific granules of polymorphonuclear leucocytes. II. Cytochemistry and electron microscopy of bone marrow cells, *J. Cell Biol.* **39**:299.

Belding, M. E., Klebanoff, S. J., and Ray, C. G., 1970, Peroxidase-mediated virucidal systems, *Science* **167**:195.

Boeri, E., Ehrenberg, A., Paul, K. G., and Theorell, H., 1953, On the compounds of ferricytochrome c appearing in acid solution, *Biochim. Biophys. Acta* **12**:273.

Bos, A., Wever, R., and Roos, D., 1978, Characterization and qualification of the peroxidase in human monocytes, *Biochim. Biophys. Acta,* **525**:37.

Carlström, A., 1969, Lactoperoxidase, some spectral properties of a hemoprotein with a prosthetic group of unknown structure, *Acta Chem. Scand.* **23**:203.

Chang, V., Morell, D. B., Nichol, A. W., and Clezy, P. S., 1970, Substituted iron chlorins and their complexes with globin: Spectral comparison with myeloperoxidase, *Biochim. Biophys. Acta* **215**:88.

Danner, D. J., and Morrison, M., 1971, Isolation of the thyroid peroxidase complex, *Biochim. Biophys. Acta* **235**:44.

DeChatelet, L. R., Shirley, P. S., McPhail, L. C., Huntley, C. C., Muss, H. B., and Bass, D. A., 1977, Oxidative metabolism of the human eosinophil, *Blood* **50**:3.

Desser, R. K., Himmelhoch, S. R., Evans, W. H., Januska, M., Mage, M., and Shelton, E., 1972, Guinea pig heterophil and eosinophil peroxidase, *Arch. Biochem. Biophys.* **148**:452.

Dunn, W. B., Hardin, J. H., and Spicer, S. S., 1968, Ultrastructural localization of myeloperoxidase in human neutrophil and rabbit heterophil and eosinophil leucocytes, *Blood* **32**:935.

Ehrenberg, A., and Agner, K., 1958, The molecular weight of myeloperoxidase, *Acta Chem. Scand.* **12**:95.

Fahimi, H. D., Cotran, R. S., and Litt, M., 1970, Localization of endogenous and exogenous peroxidase activity in Kupffer cells and peritoneal macrophages, *Fed. Proc.* **29**:493a.

Harrison, J. E., and Schultz, J., 1976, Studies on the chlorinating activity of myeloperoxidase, *J. Biol. Chem.* **251**:1371.

Himmelhoch, S. R., Evans, W. H., Mage, M., and Peterson, E. A., 1969, Purification of myeloperoxidases from the bone marrow of the guinea pig, *Biochemistry* **8**:914.

Hosoya, T., Matsukawa, S., and Nagai, Y., 1971, Localization of peroxidase and other microsomal enzymes in thyroid cells, *Biochemistry* **10**:3086.

Iyer, G. Y. N., Islam, D. M. F., and Quastel, J. H., 1961, Biochemical aspects of phagocytosis, *Nature* **192**:535.

Kakinuma, K., 1970, Metabolic control and intracellular pH during phagocytosis by polymorphonuclear leucocytes, *J. Biochem.* **68**:177.

Kasha, M., and Khan, A. U., 1970, The physics, chemistry and biology of singlet molecular oxygen, *Ann. N.Y. Acad. Sci.* **171**:5.

Klebanoff, S. J., 1967, Iodination of bacteria: A bactericidal mechanism, *J. Exp. Med.* **126**:1063.

Klebanoff, S. J., 1968, Myeloperoxidase-halide-hydrogen peroxidase antibacterial system, *J. Bacteriol.* **95**:2131.

Klebanoff, S. J., 1970, Myeloperoxidase-mediated antimicrobial systems and their role in leucocyte function, in: *Biochemistry of the Phagocytic Process* (J. Schultz, ed.), pp. 89–110., North-Holland, Amsterdam.

Klebanoff, S. J., 1975a, Antimicrobial mechanisms in neutrophilic polymorphonuclear leucocytes, *Sem. in Hematol.* XII:117.

Klebanoff, S. J., 1975b, Antimicrobial system of the polymorphonuclear leucocyte, in: *The Phagocytic Cell in Host Resistance* (J. A. Bellanti and D. H. Dayton, eds.), pp. 45–60, Raven, New York.

Klebanoff, S. J., Hamon, C. B., 1975, Antimicrobial systems of mononuclear phagocytes, in: *Mononuclear Phagocytes in Immunity, Infection and Pathology* (R. van Furth, ed.), pp. 507–531, Blackwell, Oxford.

Lehrer, R. I., 1969, Antifungal effects of peroxidase systems, *J. Bacteriol.* **99**:361.

Ljunggren, J. G., and Åkeson, Å., 1968, Solubilization, isolation and identification of a peroxidase from the microsomal fraction of beef thyroid, *Arch. Biochem. Biophys.* **127**:346.

Mandell, G. L., 1971, Influence of type of ingested particle on human leucocyte metabolism, *Proc. Soc. Exp. Biol. Med.,* **137**:1228.

McRipley, R. J., and Sbarra, A. J., 1967, Role of the phagocyte in host-parasite interactions. XII. Hydrogen peroxide-myeloperoxidase bactericidal system in phagocytes, *J. Bacteriol.* **94**:125.

Morell, D. B., and Clezy, P. S., 1963, The haematin prosthetic groups of some animal peroxidases. I. The preparation and properties of an ether-soluble haematin from milk peroxidase, *Biochim. Biophys. Acta* **71**:157.

Morrison, M., Allen, P. Z., Bright, J., and Jayasinghe, W., 1965, Identification and isolation of lactoperoxidase from salivary gland, *Arch. Biochem. Biophys.* **111**:126.

Morrison, M., Bayse, G., and Danner, G. J., 1970, The role of mammalian peroxidase in iodination reactions, in: *Biochemistry of the Phagocytic Process* (J. Schultz, ed.), pp. 51–66, North-Holland, Amsterdam.

Naskalski, J., 1977, Myeloperoxidase inactivation in the course of catalysis of chlorination of taurine, *Biochim. Biophys. Acta* **485**:291.

Neale, R. S., and Walsh, M. R., 1965, New aspects of the Hoffman-Loeffler *N*-chloramine rearrangement in acetic acid, *J. Am. Chem. Soc.* **87**:1255.

Neary, J. T., Davidson, B., Armstrong, A., Strout, H. V., and Maloof, F., 1976, Solubilization of thyroid peroxidase by nonionic detergents, *J. Biol. Chem.* **251**:2525.

Neufeld, H. A., Levay, A. N., Lucas, F. V., Martin, A. P., and Stotz, E., 1958, Peroxidase and cytochrome oxidase in rat tissues, *J. Biol. Chem.* **233**:209.

Newton, N., Morell, D. B., Clarke, L., and Clezy, P. S., 1965a, The haem prosthetic groups of some animal peroxidases. II. Myeloperoxidase, *Biochim. Biophys. Acta* **96**:476.

Newton, N., Morell, D. B., and Clarke, L., 1965b, Myeloperoxidase and associated porphyrins in rat chloroma tissue. I. *Biochim. Biophys. Acta* **96**:463.

Nichol, A. W., Morell, D. B., and Thomson, J., 1969, Porphyrins derived from the prosthetic groups of myeloperoxidase, *Biochem. Biophys. Res. Commun.* **36**:576.

Paul, B. B., and Sbarra, A. J., 1968, The role of the phagocyte in host-parasite interactions. XIII. The direct quantitative estimation of H_2O_2 in phagocytizing cells, *Biochim. Biophys. Acta* **156**:168.

Paul, B. B., Strauss, R. R., Jacobs, A. A., and Sbarra, A. J., 1970, Function of H_2O_2, myeloperoxidase and hexose monophosphate shunt enzymes in phagocytizing cells from different species, *Infect. Immunity* **1**:338.

Paul, B. B., Strauss, R. R., Selvaraj, R. J., and Sbarra, A. J., 1973, Peroxidase mediated antimicrobial activities of alveolar macrophage granules, *Science* **181**:849.

Paul, B. B., Poskitt, P. K. F., Selvaraj, R. J., Zgliczyński, J. M., and Sbarra, A. J., 1976, Mouse spleen lymphocyte bactericidal and peroxidase activities: Enhancement by whole body X-irradiation, *Proc. Soc. Exp. Biol. Med.* **152**:151.

Pincus, S. H., and Klebanoff, S. J., 1971, Quantitative leucocyte iodination, *N. Engl. J. Med.* **284**:744.

Rohrer, G. F., von Wartburg, J. P., and Aebi, H., 1966, Myeloperoxidase aus menschlichen leukozyten, *Biochem. Z.* **344**:478.

Romeo, D., Cramer, R., Marzi, T., Soranzo, M. R., Zabucchi, G., and Rossi, F., 1973, Peroxidase activity of alveolar and peritoneal macrophages, *J. Reticuloendothel. Soc.* **13**:399.

Root, R. K., and Stossel, T. P., 1974, Myeloperoxidase-mediated iodination by granulocytes. Intracellular site of operation and some regulatory factors, *J. Clin. Invest.* **53**:1207.

Rytömaa, T., 1960, Organ distribution and histochemical properties of eosinophil granulocytes in rat, *Acta Pathol. Microbiol. Scand.* **50**(suppl. 140):1.

Salmon, S. E., Cline, M. J., Schultz, J., and Lehrer, R. I., 1970, Myeloperoxidase deficiency. Immunologic study of a genetic leucocyte defect, *N. Engl. J. Med.* **282**:250.

Sbarra, A. J., and Karnovsky, M. L., 1959, The biochemical basis of phagocytosis. I. Metabolic changes during the ingestion of particles by polymorphonuclear leucocytes, *J. Biol. Chem.* **234**:1355.

Sbarra, A. J., Selvaraj, R. J., Paul, B. B. Zgliczyński, J. M., Poskitt, P. K. F., Mitchell, G. W., and

Louis, F., 1976, Chlorination, decarboxylation, and bactericidal activity mediated by the MPO-H_2O_2-Cl-system. *Adv. Exp. Med. Biol.* **73**:191.

Schultz, J., 1958, Myeloperoxidase, *Ann. N.Y. Acad. Sci.* **75**:22.

Schultz, J., and Schmukler, H. W., 1964, Myeloperoxidase of the leucocytes of normal human blood. II. Isolation, spectrophotometry and amino acid analysis, *Biochemistry* **3**:1234.

Schultz, J., Corlin, R., Oddi, F., Kaminker, K., and Jones, W., 1965, Myeloperoxidase of the leucocyte of normal human blood. I. Content and localization, *Arch. Biochem. Biophys.* **111**:73.

Selvaraj, R. J., Paul, B. B., Strauss, R. R., Jacobs, A. A., and Sbarra, A. J., 1974, Oxidative peptide cleavage and decarboxylation by the myeloperoxidase-H_2O_2-Cl$^-$ antimicrobial system, *Infect. Immun.* **9**:255.

Selvaraj, R. J., Zgliczyński, J. M., Paul, B. B., and Sbarra, A. J., 1978, Enhanced killing of myeloperoxidase coated bacteria in the MPO-H_2O_2-Cl$^-$ system, *J. Inf. Dis.* **137**:481.

Stelmaszyńska, T., 1970, Studies on hog intestinal mucosa peroxidase, Ph.D. Thesis, Jagiellonian University, Krakow.

Stelmaszyńska, T., Zgliczyński, J. M., 1971, Studies on hog intestinal mucosa peroxidase, *Eur. J. Biochem.* **19**:56.

Stelmaszyńska, T., and Zgliczyński, J. M., 1974, Myeloperoxidase of human neutrophilic granulocytes as chlorinating enzyme, *Eur. J. Biochem.* **45**:305.

Stelmaszyńska, T., and Zgliczyński, J. M., 1978, N(2-Oxoacyl)amino acids and nitriles as final products of dipeptide chlorination mediated by the myeloperoxidase-H_2O_2-Cl$^-$ bactericidal system, *Eur. J. Biochem.* **92**:301.

Stjernholm, R. L., Allen, R. C., and Steele, R. H., 1973, Impaired chemiluminescence during phagocytosis of opsonized bacteria, *Infect. Immun.* **7**:313.

Strauss, R. R., Paul, B. B., Jacobs, A. A., and Sbarra, A. J., 1972a, *In vitro* bactericidal and associated metabolic activities of mouse spleen cells, *Infect. Immunity* **5**:114.

Strauss, R. R., Paul, B. B., Jacobs, A. A., and Sbarra, A. J., 1972b, Mouse splenic peroxidase and its role in bactericidal activity, *Infect. Immunity* **5**:120.

Taurog, A., Lothrop, M. L., and Estabrook, R. W., 1970, Improvements in the isolation procedure for thyroid peroxidase: Nature of the heme prosthetic group, *Arch. Biochem. Biophys.* **139**:212.

Thomas, J. A., Morris, D. R., and Hager, L. P., 1970, Chloroperoxidase, *J. Biol. Chem.* **245**:3135.

Weil, J., and Morris, J. C., 1949, Kinetic studies on the chloramines. I. The rates of formation of monochloramine, N-chlormethylamine and N-chlordimethylamine, *J. Am. Chem. Soc.* **71**:1664.

Widmann, J. J., Cotran, R. S., and Fahimi, H. D., 1972, Mononuclear phagocytes (Kupffer cells) and endothelial cells. Identification of two functional cell types in rat liver sinusoids by endogenous peroxidase activity, *J. Cell Biol.* **52**:159.

Wilson, D. L., and Manery, J. F., 1949, The permeability of rabbit leucocytes to sodium, potassium and chloride, *J. Cell. Comp. Physiol.* **34**:493.

Zatti, M., Rossi, R., and Patriarca, P., 1968, The H_2O_2-production by polymorphonuclear leucocytes during phagocytosis, *Experientia* **24**:669.

Zgliczyński, J. M., and Stelmaszyńska, T., 1975, Chlorinating ability of human phagocytosing leucocytes, *Eur. J. Biochem.* **56**:157.

Zgliczyński, J. M., Stelmaszyńska, T., Ostrowski, W., Naskalski, J., and Sznajd, J., 1968, Myeloperoxidase of human leukaemic leucocytes. Oxidation of amino acids in the presence of hydrogen peroxide, *Eur. J. Biochem.* **4**:540.

Zgliczyński, J. M., Stelmaszyńska, T., Domański, J., and Ostrowski, W., 1971, Chloramines as intermediates of oxidation reaction of amino acids by myeloperoxidase, *Biochim. Biophys. Acta* **235**:419.

Zgliczyński, J. M., Selvaraj, R. J., Paul, B. B., Stelmaszyńska, T., Poskitt, P. K. F., and Sbarra, A. J., 1977, Chlorination by the myeloperoxidase-H_2O_2-Cl$^-$ antimicrobial system at acid and neutral pH, *Proc. Soc. Exp. Biol. Med.* **154**:418.

11

Myeloperoxidase-Mediated Cytotoxic Systems

SEYMOUR J. KLEBANOFF

1. INTRODUCTION

Myeloperoxidase (MPO), H_2O_2, and a halide form a powerful antimicrobial system which is effective against a variety of microorganisms. This system is also toxic to certain mammalian cells, namely, spermatozoa, erythrocytes, leukocytes, platelets, and tumor cells. This review (see also Klebanoff, 1975a; Klebanoff and Clark, 1978) will consider the properties of the MPO-mediated antimicrobial system and its role in the intracellular and extracellular toxicity of the polymorphonuclear leukocyte (PMN).

2. BACKGROUND

Peroxidases do not have a direct toxic effect on microorganisms. However peroxidases catalyze the oxidation of a large number of hydrogen (or electron) donors by H_2O_2, and it is possible, at least theoretically, that an agent can be generated in this way which is toxic to adjacent cells. An early description of such a system was that of Kojima (1931) who reported that peroxidase (source not specified) catalyzed the peroxidatic conversion of phenols to the corresponding quinones with an increase in antimicrobial activity. Peroxidases were first implicated as a component of a biologically important antimicrobial system in milk and in saliva.

Abbreviations: cis-DBE, *cis*-dibenzoylethylene; CGD, chronic granulomatous disease; DABCo, 1,4-diazabicyclo[2.2.2]octane; D_2O, deuterium oxide; DPF, diphenylfuran; $HO_2^{\cdot}$, perhydroxy radical; LPO, lactoperoxidase; MPO, myeloperoxidase; $O_2^{\cdot-}$, superoxide anion; $^1O_2$, singlet oxygen; OH·, hydroxyl radical; PMN, polymorphonuclear leukocyte; SOD, superoxide dismutase; T_3, triiodothyronine; T_4, thyroxine.

SEYMOUR J. KLEBANOFF • Department of Medicine, University of Washington, Seattle, Washington 98195. Supported in part by United States Public Health Service grants AI07763, and HD02266.

2.1. PEROXIDASE-MEDIATED ANTIMICROBIAL SYSTEM OF MILK AND SALIVA

An antimicrobial system present in milk was reported by Hanssen (1924) to have the same heat sensitivity as the "oxydases and peroxydases" of milk. The similarity between this antimicrobial system ("lactenin") and the milk peroxidase, lactoperoxidase (LPO), was further emphasized by Wright and Tramer (1958), and, shortly thereafter, a number of investigators demonstrated the presence of antimicrobial activity in a highly purified preparation of LPO (Portmann and Auclair, 1959; Stadhouders and Veringa, 1962; Jago and Morrison, 1962). The milk antimicrobial system was found to require, in addition, a heat-stable dialyzable component which was identified by Reiter *et al.* (1964) to be thiocyanate ions. As these studies were being conducted, comparable studies were underway on the properties of an unusual antimicrobial system in saliva (Figure 1). This system was found by Zeldow (1959) to consist of a heat-stable dialyzable and a heat-labile nondialyzable component. The heat-stable dialyzable component was shown by Dogon *et al.* (1962) to be thiocyanate ions, and Klebanoff and Luebke (1965) reported that the heat-labile nondialyzable component was the salivary peroxidase. In initial studies, a highly purified preparation of LPO from bovine milk was employed (Klebanoff and Luebke, 1965), since the salivary peroxidase was reported to be immunologically and chemically indistinguishable from the milk peroxidase (Morrison *et al.*, 1965). In a subsequent study, the heat-labile component of the salivary antimicrobial system was replaced by a purified preparation of the peroxidase from human saliva (Slowey *et al.*, 1968).

In these early studies, H_2O_2 was not added to the reaction mixture. However, the organisms generally employed, *Streptococcus cremoris* in the milk system and *Lactobacillus acidophilus* in the salivary system, generate H_2O_2 (see Section 3.2.2), and the utilization of the H_2O_2 so formed as a component of the antimicrobial system was suggested by the inhibitory effect of catalase (Jago and Morrison, 1962; Klebanoff and Leubke, 1965). When H_2O_2 or a H_2O_2-generating system such as glucose and glucose oxidase was added, growth inhibition by LPO and thiocyanate ions was extended to non-H_2O_2-generating organisms (Klebanoff *et al.*, 1966).

3. MPO-MEDIATED ANTIMICROBIAL SYSTEM

In 1965 it was reported that LPO could be replaced by a purified preparation of MPO as the peroxidase component of the thiocyanate-dependent antilac-

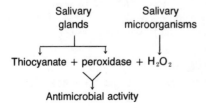

FIGURE 1. The salivary peroxidase–thiocyanate–H_2O_2 antimicrobial system. Peroxidase and thiocyanate ions secreted by the salivary glands combine with H_2O_2, possibly generated by salivary microorganisms, to form an antimicrobial system. From Klebanoff and Clark (1978), reproduced by permission of North-Holland Publishing Co.

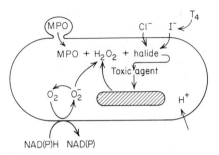

FIGURE 2. The myeloperoxidase-mediated antimicrobial system. Myeloperoxidase (MPO), released into the phagocytic vacuole during the degranulation process, reacts with H_2O_2, generated either by leukocytic or microbial metabolism, and a halide to form agents toxic to the ingested organism. From Klebanoff and Clark (1978), reproduced by permission of North-Holland Publishing Co.

tobacillus system (Klebanoff and Luebke, 1965), and, in the following year, the extension of the toxicity to non-H_2O_2-generating organisms by the addition of H_2O_2 was observed in the presence of MPO as well as LPO (Klebanoff et al., 1966). In 1967 it was reported that thiocyanate could be replaced in the MPO-mediated antimicrobial system by the halides, iodide, bromide, or chloride and by the thyroid hormones, thyroxine (T_4) or triiodothyronine (T_3) (Klebanoff, 1967a,b). The thyroid hormones were deiodinated by MPO and H_2O_2 (Klebanoff and Green, 1973) and presumably served as a source of inorganic iodide. The bactericidal activity of a guinea pig leukocyte extract and H_2O_2 was reported in the same year by McRipley and Sbarra (1967b). The requirement for a halide was confirmed in subsequent studies (Paul et al., 1970). LPO, in contrast to MPO, is effective with iodide and bromide, but not with chloride (Dogon et al., 1962; Zeldow, 1963; Klebanoff, 1968). A variety of bacterial species, both gram-positive and gram-negative, are susceptible to the MPO-mediated antimicrobial system (Klebanoff, 1967b; McRipley and Sbarra, 1967b; Klebanoff, 1968). The system also has potent fungicidal (Lehrer, 1969; Klebanoff, 1970a; Lehrer and Jan, 1970; Diamond et al., 1972; Howard, 1973), virucidal (Belding et al., 1970), and mycoplasmacidal (Jacobs et al., 1972) activity, and is toxic to certain mammalian cells (see Section 4). The MPO–H_2O_2–halide system also can inactivate certain soluble mediators, e.g., the chemotactic factors C5a and f-met-leu-phe (Clark and Klebanoff, 1979b) and $\alpha 1$ proteinase inhibitor (Matheson et al., 1979). The components of the MPO-mediated antimicrobial system are formed or released by the neutrophil at a time appropriate to the microbicidal act (Figure 2).

3.1. MPO

MPO is present in the PMN in exceptionally high concentrations; levels of 1–2% (Agner, 1941) and 5% (Schultz and Kaminker, 1962; Rohrer et al., 1966) of the wet weight of the cell have been reported. It is located entirely in the azurophil (primary) granules of the resting cell (Bainton and Farquhar, 1968a,b; Baggiolini et al., 1969; Spicer and Hardin, 1969), and its release into the phagocytic vacuole during the degranulation process is readily apparent in electron micrographs stained for peroxidase (Baehner et al., 1969; Klebanoff, 1970b). Some electron micrographs suggest the adherence of MPO to the surface of the ingested organisms (Klebanoff, 1970b), and enhanced killing of MPO-coated bacteria, by the peroxidase system, has been reported (Selvaraj *et al.*, 1978). MPO is a basic protein [isoelectric point of canine MPO > 10 (Agner, 1941)], and like other cationic proteins may bind to the negatively-charged cell surface of the organism.

3.2. H_2O_2

The H_2O_2 required for the MPO-mediated antimicrobial system can be supplied by reagent H_2O_2 or by an H_2O_2-generating system. H_2O_2, in relatively high concentrations, is inhibitory to MPO (Agner, 1941, 1963) and, as a result, H_2O_2 maintained at low steady-state concentrations by a H_2O_2-generating system is preferable to H_2O_2 added *in toto* at the beginning of the incubation. The H_2O_2-generating systems may be nonenzymatic or enzymatic and a distinction can be made between those systems which generate H_2O_2 via the intermediate formation of the superoxide anion ($O_2^{\cdot -}$) and those which form H_2O_2 directly from oxygen by divalent reduction without an apparent $O_2^{\cdot -}$ intermediate.

The perhydroxy radical ($HO_2^{\cdot}$) and its ionized form, the superoxide anion, are formed by the univalent reduction of oxygen (for review see Fridovich, 1976). The pK_a for the reaction $HO_2^{\cdot} \rightleftharpoons O_2^{\cdot -}$ is 4.8 so that the radical exists almost entirely as $O_2^{\cdot -}$ at neutral pH. When two molecules interact, one is oxidized and the other reduced in a dismutation reaction as follows

$$HO_2^{\cdot} + O_2^{\cdot -} + H^+ \rightarrow O_2 + H_2O_2$$

Spontaneous dismutation occurs most readily at pH 4.8, where the protonated and nonprotonated forms are present in equal concentrations. It is less efficient at more acid pH where the protonated form predominates, and is particularly slow at neutral or alkaline pH where $O_2^{\cdot -}$ predominates. Indeed, at very high pH, where $O_2^{\cdot -}$ is the sole species, spontaneous dismutation may not occur. Dismutation of $O_2^{\cdot -}$ is accelerated by superoxide dismutase (SOD), particularly at neutral or alkaline pH where spontaneous dismutation is relatively slow. Some H_2O_2-generating enzyme systems, for example, xanthine oxidase, form H_2O_2, at least in part, by the univalent reduction of oxygen followed by dismutation, whereas other enzymes, for example, glucose oxidase, appear to form H_2O_2 directly by divalent reduction. It is possible for $O_2^{\cdot -}$ to be generated within a confined space in an enzyme molecule unavailable to the proteins required for its assay (ferricytochrome C, SOD), with spontaneous dismutation occurring before $O_2^{\cdot -}$ can diffuse out of that space (Fridovich, 1970).

Both the glucose oxidase and xanthine oxidase systems can be employed as the source of H_2O_2 for the MPO-mediated antimicrobial system. When the xanthine oxidase system is used, microbicidal activity is inhibited by agents, such as ferricytochrome C, which decrease the production of H_2O_2 by reaction with $O_2^{\cdot -}$. This inhibition of microbicidal activity is reversed by SOD (Klebanoff, 1974; Figure 3). H_2O_2 generation via the intermediate formation of $O_2^{\cdot -}$ provides an opportunity for control of the MPO-mediated antimicrobial system through

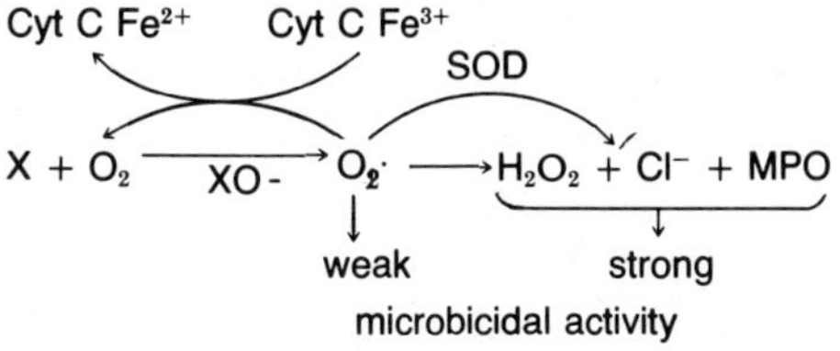

FIGURE 3.   Microbicidal activity of the xanthine (X)–xanthine oxidase (XO)–chloride–myeloperoxidase (MPO) system. The microbicidal activity of xanthine and xanthine oxidase is greatly increased by the addition of MPO and chloride. Ferricytochrome C inhibits the activity of the complete system and this inhibition is largely reversed by superoxide dismutase (SOD). From Klebanoff and Clark (1978); reproduced by permission of North-Holland Publishing Co.

$$\text{Cyt C Fe}^{2+} \quad \text{Cyt C Fe}^{3+}$$
$$\text{SOD}$$
$$X + O_2 \xrightarrow{\text{XO}} O_2^{\cdot} \longrightarrow H_2O_2 + Cl^- + MPO$$
$$\downarrow \qquad\qquad\qquad \downarrow$$
$$\text{weak} \qquad\qquad \text{strong}$$
$$\text{microbicidal activity}$$

the control of $O_2^{\cdot-}$ dismutation and thus H_2O_2 generation. In the PMN, H_2O_2 can be generated by either leukocytic or microbial metabolism.

3.2.1.   Leukocytic Metabolism

Phagocytosis by PMNs is associated with a burst of oxidative metabolism (Sbarra and Karnovsky, 1959) and much, if not all, of the extra oxygen consumed is converted to H_2O_2 (Iyer et al., 1961). H_2O_2 equivalent to 20–30% (Klebanoff and Hamon, 1972), 50–70% (Homan-Müller et al., 1975), 51% (Root and Metcalf, 1977), and nearly all (Zatti et al., 1968) of the oxygen consumed have been reported. Since the detection of H_2O_2 is affected by the competitive utilization of H_2O_2 by cellular systems, the addition of an agent such as azide or cyanide which inhibits utilization or degradation by catalase or MPO is required for maximum detection. However, even under these conditions, the amount of H_2O_2 detected should be considered a minimum value. H_2O_2 is generated in the PMN largely through a $O_2^{\cdot-}$ intermediate (Babior et al., 1973; Curnutte and Babior, 1975; Weening et al., 1975; Root and Metcalf, 1977).

It is not known with certainty where in the cell H_2O_2 is formed or the nature of the enzyme responsible. The prevailing view is that an NADH or an NADPH oxidase is involved; a consideration of the evidence which implicates one or other of these enzymes is beyond the scope of this review. Evidence for the localization of this enzyme in leukocyte granules (Rossi and Zatti, 1964; Patriarca et al., 1973; Hohn and Lehrer, 1975; DeChatelet et al., 1975; Auclair et al., 1976; Babior et al., 1976), in the cytosol (Cagan and Karnovsky, 1964; Baehner et al., 1970a), or on the cell membrane (Briggs et al., 1975a; Takanaka and O'Brien, 1975; Segal and Peters, 1977; Goldstein et al., 1977) has been presented; species differences may exist. Localization as an ectoenzyme on the cell surface, that is, an enzyme which on activation acts on a substrate in the extracellular fluid, is attractive since on invagination during phagocytosis, the enzyme would be in a position to release $O_2^{\cdot-}$, and by dismutation, H_2O_2, directly into the intravacuolar space (Figure 4). Further, the detection of the products of oxygen reduction in the extracellular fluid would favor a surface location. A portion of the H_2O_2 formed has been detected by a cytochemical technique in the phagosome, concentrated in one or two areas between the particle and the phagosome membrane (Briggs et al., 1975b).

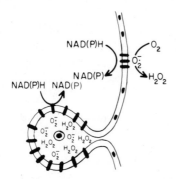

FIGURE 4.   Activation of a membrane NAD(P)H oxidase during phagocytosis. From Klebanoff and Clark (1978), reproduced by permission of North-Holland Publishing Co.

3.2.2.   Microbial Metabolism

Lactic acid bacteria are generally devoid of heme and, as a result, do not contain the cytochrome system. Terminal oxidations are catalyzed by flavoproteins, which convert oxygen to H_2O_2. These organisms also lack the heme protein, catalase, although nonheme peroxidases and catalases may be present (Whittenbury, 1964). As a result, H_2O_2 is produced in excess of the capacity of the organism to degrade it and can be detected in the medium. Among the organisms which can generate H_2O_2 in this way are the pneumococci, streptococci, and lactobacilli (McLeod and Gordon, 1922; Avery and Morgan, 1924; Annear and Dorman, 1952; Whittenbury, 1964).

The ability of certain microorganisms to generate the H_2O_2 required for their own destruction by the peroxidase system has already been alluded to in relation to the LPO-mediated antimicrobial system of saliva and milk. Similarly with the MPO system, H_2O_2 need not be added if a H_2O_2-generating organism is employed (Klebanoff, 1968). Further, in mixed cultures, H_2O_2 generated by one organism can be employed to kill a second non-H_2O_2-generating organism in the presence of MPO and a halide. This peroxidase-dependent antagonism can be demonstrated between two bacterial species (Klebanoff, 1970b; Klebanoff and Smith, 1970a; Hamon and Klebanoff, 1973), between bacteria and fungi (Hamon and Klebanoff, 1973), between bacteria and viruses (Klebanoff and Belding, 1974), and between bacteria and certain mammalian cells (Klebanoff and Smith, 1970b; Clark *et al.*, 1975). H_2O_2 of microbial origin may contribute significantly to the antimicrobial activity of the PMN, particularly when the leukocytic H_2O_2-generating systems are defective, as in chronic granulomatous disease (CGD) (see Section 6.1.2).

3.3.   HALIDE

The halide requirement for the MPO-mediated antimicrobial system can be met by chloride, iodide, bromide, or by the pseudohalide, thiocyanate. It is probable that in the intact cell, the halide requirement is met chiefly by chloride which is present in the cell in concentrations considerably greater than those required by the purified MPO system. The chloride concentration required varies with the MPO and H_2O_2 concentrations, the pH, and other conditions of the assay. Under one set of conditions, an effect of chloride was observed at a concentration of 5×10^{-6} M (0.005 meq/liter) with complete killing at 5×10^{-4} M (0.5 meq/liter) (Klebanoff, 1968). The intraleukocytic chloride concentration of horse and human leukocytes has been estimated to be 70 meq/liter (Endres and Herget, 1929) and 110.5 meq/liter (Baron and Ahmed, 1969), respectively. In the rabbit, the intraleukocytic chloride concentration varies with that of the extracellular fluid (Wilson and Manery, 1949); at an extracellular concentration equivalent to that of human serum (103 meq/liter), the intraleukocytic concentration is approximately 75 meq/liter. Chloride also may enter the vacuole in the layer of extracellular fluid which accompanies the particle.

Iodide is considerably more effective than chloride on a molar basis as the halide component of the MPO-mediated antimicrobial system. In the study described above (Klebanoff, 1968), an effect of iodide was observed at a concentration of 5×10^{-8} M (0.6 μg%) with complete killing at 5×10^{-6} M (60 μg%). The inorganic iodide concentration of serum is extremely low (< 1 μg%). Some investigators have reported a concentration of iodide by leukocytes (Segal and Sachs, 1964; Stolc, 1971), however, this has not been confirmed in other studies (Klebanoff and Hamon, 1972). Thus, the contribution of iodide to the total halide pool in the PMN may be small. Iodide can be replaced in the isolated MPO system by the thyroid hormones T_3 and T_4 (Klebanoff, 1967b), presumably due to their deiodination (or transiodination) by the MPO system (Klebanoff and Green, 1973). The thyroid hormones also are deiodinated by intact PMNs during phagocytosis (Klebanoff and Green, 1973; Woeber and Ingbar, 1973), and may thus serve as an additional source of iodide in the intact cell. In some systems (Klebanoff and Clark, 1975), the combined effect of chloride and either iodide, T_4, or T_3 is greater than additive. Although both bromide (Klebanoff, 1968) and thiocyanate ions (Klebanoff and Luebke, 1965) also can serve as cofactors in the isolated MPO system, their role in the intact cell is unknown. The effect of thiocyanate on the antimicrobial activity of the MPO system is complex; it can serve as the required cofactor (Klebanoff and Luebke, 1965; Klebanoff et al., 1966) and also inhibit microbicidal activity when iodide, bromide, or chloride is the halide employed (Klebanoff, 1967b, 1968), depending on the experimental conditions.

In summary, it is probable that the halides are present in the PMN in considerable excess. The chloride concentration is high, and the small amounts of iodide absorbed as such or released by deiodination of the thyroid hormones contribute to the halide pool. Other as yet unidentified cofactors also may be involved.

3.4. pH

The $MPO-H_2O_2$–chloride system can function over a wide pH range, with the pH optimum varying with the H_2O_2 and chloride concentrations (Zgliczyński *et al.,* 1977). Most studies have revealed an acid optimum at about pH 5.0 (Klebanoff, 1967b, 1968; McRipley and Sbarra, 1967b). The pH within the phagocytic vacuole also has been reported to be acid, with estimates ranging from 6.0–6.5 to 3.0 or below (Rous, 1925a,b; Sprick, 1956; Pavlov and Solev'ev, 1967; Mandell, 1970; Kakinuma, 1970; Jensen and Bainton, 1973). In one study in which the color change of indicator dyes bound to yeast particles was employed as a measure of pH, the intravacuolar pH fell following phagocytosis to a level of approximately 6.5 in 3.0 min and 4.0 in 7–15 min (Jensen and Bainton, 1973). Although the estimation of pH within intact cells is subject to errors introduced by the local microenvironment, the studies suggest that a fall in pH occurs within the vacuole which may be considerable.

3.5. INHIBITORS

A variety of agents can inhibit the MPO-mediated antimicrobial system, and some of these may do so *in situ.* Catalase would be expected to be inhibitory through its degradative action on H_2O_2, and indeed it is under most experimental conditions (Klebanoff, 1967b, 1968). Catalase, however, can substitute for MPO as the catalyst of the microbicidal system at pH 4.5 in the presence of iodide and low concentrations of H_2O_2 (Klebanoff, 1969). Catalase is present in the cytosol of guinea pig PMNs (Michell *et al.,* 1970), and has not been detected in the phagocytic vacuole (Stossel *et al.,* 1971). Human PMN catalase also has not been detected in the phagosome, although human catalase has been reported to be associated, at least in part, with granules (Breton-Gorius *et al.,* 1975; Nishimura *et al.,* 1976). The introduction of catalase into the phagocytic vacuole as a component of the ingested organism is considered in Section 6.1.2.

MPO also is inhibited by excess H_2O_2 (Agner, 1941, 1963; Evans and Rechcigl, 1967; DeChatelet *et al.,* 1971). MPO contains two heme groups per molecule and it has been proposed that when in excess, H_2O_2 forms complexes with both heme irons, the H_2O_2 is degraded to oxygen and water, and the enzyme is inactivated in the process (Agner, 1963). The concentration of H_2O_2 in the phagocytic vacuole is not known; however, the inactivation of MPO by H_2O_2 may offer a means of control of the MPO system when H_2O_2 is in excess. Estradiol and other phenolic estrogens prevent the decrease in microbicidal activity induced by high H_2O_2 concentrations (Klebanoff, 1979).

Other inhibitors of peroxidase-catalyzed reactions (Klebanoff, 1967b, 1968; McRipley and Sbarra, 1967b) include agents such as azide or cyanide which form complexes with the heme iron, reducing agents which can reduce the products of the peroxidase reaction back to its original form and thus appear to inhibit the reaction without actually doing so, agents which are oxidized by peroxidase and H_2O_2 and, thus, can compete for the available H_2O_2, and substances such as protein which compete with the target organism for the toxic products of the peroxidase system. Inhibitors of this sort may be introduced into the vacuole during the degranulation process, by diffusion from the cytosol or as a component of the ingested organism.

4. TOXICITY TO MAMMALIAN CELLS

The MPO–H_2O_2– halide system is toxic not only to microorganisms but also to a variety of mammalian cells (Edelson and Cohn, 1973; Klebanoff et al., 1976). These include spermatozoa, erythrocytes, leukocytes, platelets, and tumor cells.

4.1. SPERMATOZOA

The peroxidase–H_2O_2–halide system has spermicidal activity as measured by a loss of motility or by a decrease in pyruvate oxidation (Smith and Klebanoff, 1970; Klebanoff and Smith, 1970b). The peroxidase requirement could be met by either MPO, LPO, or a peroxidase present in uterine fluid from estrogen-primed rats, and the H_2O_2 requirement by either reagent H_2O_2, a peroxide-generating enzyme system, microbial metabolism (L. acidophilus), or spermatozoa in the presence of certain aromatic amino acids, and the halide requirement by either iodide or thiocyanate ions.

4.2. ERYTHROCYTES

Erythrocytes are hemolyzed by MPO when combined with a H_2O_2-generating system (glucose + glucose oxidase, xanthine + xanthine oxidase) and either chloride, iodide, thyroxine, or triiodothyronine (Klebanoff and Clark, 1975). Hemolysis by the iodide-dependent system is associated with the iodination of erythrocyte components (see Section 5.1). Hemolysis is inhibited by catalase and by the peroxidase inhibitors, azide or cyanide, and when the xanthine oxidase system is employed, is stimulated by SOD, presumably due to the accelerated formation of H_2O_2 from O_2^- (Klebanoff and Clark, 1975). MPO could be replaced by LPO in the iodide-dependent system (Edelson and Cohn, 1973; Klebanoff and Clark, 1975).

4.3. LEUKOCYTES

MPO, H_2O_2 (or a H_2O_2-generating system), and either chloride or iodide are cytotoxic to blood leukocytes (PMNs, mononuclear leukocytes) as measured by increased ^{51}Cr release and trypan blue dye exclusion (Clark and Klebanoff, 1977a). Cell damage is rapid, with maximal levels of ^{51}Cr release occurring in 30–60 min. Human mononuclear leukocytes are damaged by the corresponding LPO-dependent system (Edelson and Cohn, 1973).

4.4. PLATELETS

Exposure of platelets to the MPO–H_2O_2–halide system results in the release of the biological mediator 5-hydroxytryptamine (serotonin) (Clark and Klebanoff,

1979a). This release may be due to a cytotoxic effect on the platelets or to a triggering of the platelet release reaction. The concomitant release of small amounts of [^{14}C]adenine in double-label experiments suggests that at least a part of the serotonin release is due to a cytotoxic effect.

4.5. TUMOR CELLS

The peroxidase–H_2O_2–halide system has a cytotoxic effect on mammalian tumor cells (Edelson and Cohn, 1973; Philpott *et al.*, 1973; Clark *et al.*, 1975, 1976). In one study (Clark *et al.*, 1975), mouse ascites lymphoma cells designated LSTRA were exposed to the MPO–H_2O_2–halide system and the cytotoxic effect determined by ^{51}Cr release, trypan blue exclusion, inhibition of glucose C-1 oxidation, and loss of oncogenicity for mice. Either canine (Clark *et al.*, 1975) or human (Clark *et al.*, 1976) MPO could be employed. MPO could be replaced by LPO, H_2O_2 by H_2O_2-generating enzyme systems (glucose oxidase, xanthine oxidase), or bacteria (pneumococci, streptococci), and the halide requirement could be met by chloride, iodide, or the iodinated hormones. For example, when the oncogenicity assay was employed, 59 of 60 mice inoculated intraperitoneally with 100,000 tumor cells exposed to buffer alone developed tumors and died. In contrast, the same number of tumor cells exposed to the MPO–H_2O_2–chloride system did not cause tumors or death in any of the 36 mice injected. Deletion of any of the components of the system prevented damage to the tumor cells and nearly all animals died. Injection of serial dilutions of control cells exposed to buffer alone indicated that as few as 100 cells consistently caused tumors and death in approximately half of the recipient mice. Thus, the peroxidase system had to kill at least 99.9% of the cells to give the observed results. The toxicity of the peroxidase system could be made selective for tumor cells by using glucose oxidase conjugated to anti-tumor-cell antibody to generate H_2O_2 in the immediate vicinity of the target cell (Philpott *et al.*, 1973).

5. MECHANISM OF ACTION

MPO is a heme enzyme which catalyzes the oxidation of a large number of substances (electron or hydrogen donors) by H_2O_2 (peroxidatic reaction). Peroxidases, including MPO, can, in addition, catalyze the oxidation of certain electron donors by molecular oxygen (oxidatic reaction) and, more recently, peroxidases have been shown to react with the superoxide anion to form an oxyperoxidase (peroxidase compound III) in which, like oxyhemoglobin, oxygen is bound to the heme iron (for review of peroxidase enzyme–substrate complexes, see Yamazaki and Yokota, 1973). The halides are among the substances which can be oxidized by MPO and H_2O_2, and it is probable that a toxic agent or agents are formed which can attack the organism in a variety of ways, leading rapidly to cell death.

5.1. HALOGENATION

Peroxidase, H_2O_2, and iodide form a powerful iodinating system. Iodine is readily bound to free tyrosine or tyrosine residues of protein in covalent linkage to form monoiodo- and diiodotyrosine. Iodohistidine, sulfenyl iodides, and iodinated lipids can also be formed. Bacteria exposed to the $MPO-H_2O_2$-iodide system are iodinated as indicated by the conversion of iodide to a water-insoluble, trichloroacetic acid-precipitable form, and the iodine can be localized on the microorganism by radioautographic techniques (Klebanoff, 1967b). When the iodide to H_2O_2 ratio is low, iodination by the $MPO-H_2O_2$-iodide system is stimulated by chloride (Klebanoff, 1970b). Iodination of bacterial components also occurs when the thyroid hormones are employed as the cofactor rather than iodide (Klebanoff and Green, 1973), due either to deiodination with subsequent iodination or to transiodination without an apparent iodide intermediate. Bromination (Klebanoff, unpublished data) and chlorination (Zgliczyński and Stelmaszyńska, 1975; Stelmaszyńska and Zgliczyński, 1978; Thomas, 1979a) of bacterial components by the $MPO-H_2O_2$-halide system also have been observed (see Section 5.3 for chloramine formation).

Mammalian cells also are iodinated by MPO, H_2O_2, and iodide. When the initial iodide concentration and the total number of iodine atoms bound are low, little or no damage to the cell occurs, and the iodination reaction can be employed to label surface proteins. However, when the iodinating conditions are more severe and the total number of iodine atoms bound are correspondingly high, more extensive membrane damage with cell lysis can occur. This is particularly apparent with the erythrocyte as the target cell (Klebanoff and Clark, 1975). The cytotoxic effect of the $MPO-H_2O_2$-iodide system on mammalian tumor cells also is associated with the iodination of cellular components (Clark *et al.*, 1975).

Cell death may be due in part to the substitution of the bulky halogen for a hydrogen atom at crucial locations on the cell surface. It should be emphasized, however, that the association of halogenation and cell death does not necessarily imply a cause–effect relationship. Halogenation may be an innocent bystander to the toxic process or, more probably, one of a number of factors which contribute to cell death.

5.2. OXIDATION

The products of halide oxidation by the MPO system are powerful oxidizing agents capable of the oxidation of a variety of groups on the surface of target cells. In one study of the germicidal activity of iodine (Brandrick *et al.*, 1967), 80–90% of the iodine added to a bacterial suspension was reduced to iodide, suggesting a corresponding oxidation of chemical groups on the organisms, whereas the remainder was bound to the organisms, indicating iodination of cellular components. The oxidation of iodide by the peroxidase system and its subsequent reduction by bacterial components, with reutilization by the peroxidase system, has been proposed (Thomas and Aune, 1978a). This cyclic oxidation and reduction would allow iodide to function in catalytic concentrations. The oxidants formed from iodide may include free iodine (I_2), iodinium ion (I^+), and hypoiodous acid (HOI), and those formed from chloride may include chlorine (Cl_2), chloridium ion (Cl^+), and hypochlorous acid (HOCl). Similar agents are presumed to be formed from bromide, and recently the antimicrobial agent formed by the LPO–H_2O_2–thiocyanate system has been identified as the hypothiocyanite ion (Hoogendoorn *et al.*, 1977; Aune and Thomas, 1977). Other oxidants also have been proposed; one, singlet molecular oxygen ($^1O_2$), is considered in more detail in Section 5.4.

It is probable that the oxidant damage to the microorganism by the MPO system involves a number of different chemical groups on or in the organism. Sulfhydryl groups are particularly prone to oxidation and are essential to the action of a number of enzymes. Chlorine derivatives can inhibit a variety of sulfydryl enzymes at concentrations which are bactericidal, suggesting that this is one of the toxic mechanisms (Green and Stumpf, 1946; Knox *et al.*, 1948). Sulfhydryl oxidation (Thomas and Aune, 1977, 1978b), lipid peroxidation (Buege and Aust, 1976), and the oxidative cleavage of tryptophanyl peptides (Alexander, 1974) by peroxidase, H_2O_2, and iodide also have been reported. Chloramine formation and its consequences are considered in Section 5.3. Other oxidative reactions also should be anticipated.

5.3.   CHLORAMINE AND ALDEHYDE FORMATION

The MPO–H_2O_2–chloride system reacts with nitrogenous compounds to yield chloramines (Zgliczyński *et al.*, 1971, 1977). Hydrogen cyanide (HCN) and cyanogen chloride (ClCN) may be generated as by-products of this reaction (Zgliczyński and Stelmaszyńska, 1979). Chloramines may contribute to the toxicity in a number of ways. They may be directly toxic to the organisms, as has been proposed for taurine chloramine (Zgliczyński *et al.*, 1971). Taurine is present in PMNs in relatively high concentrations (Nour-Eldin and Wilkinson, 1955; Iyer, 1959), and its chloramine, unlike those of other amino acids, is stable (Zgliczyński *et al.*, 1971). However, taurine inhibits the MPO–H_2O_2–chloride antimicrobial system (Strauss *et al.*, 1971; Thomas, 1979b), suggesting that it is not required for toxicity. This may be due, in part, to the inactivation of MPO, associated with the catalysis of taurine monochloramine formation (Naskalski, 1977). Chloramines generally are unstable, hydrolyzing to release a reactive product ("available chlorine"), and, thus, can serve as storage forms of available chlorine, with prolongation of the toxicity of the MPO system. It has been proposed that the increase in the bactericidal activity of the MPO-H_2O_2-chloride system by NH_4^+ and certain guanidino compounds is due to the ability of the N-Cl derivatives to penetrate the hydrophobic cell membrane and oxidize intracellular components (Thomas, 1979b). Finally, the degradation of chloramines results in the formation of the corresponding aldehyde (Zgliczyński *et al.*, 1968, 1971), and Sbarra and co-workers have proposed that the aldehyde formed by the deamination and decarboxylation of amino acids may be toxic to adjacent cells (Strauss *et al.*, 1971; Sbarra *et al.*, 1977) or, if structural, may be harmful to the parent organism (Selvaraj *et al.*, 1974; Sbarra *et al.*, 1977).

5.4. SINGLET OXYGEN FORMATION

Electrons normally occur in pairs, stabilized by spins in the opposite direction. Ground state triplet oxygen is unusual in that it contains two single electrons with spins in the same direction (i.e., unpaired). Singlet oxygen is formed when an absorption of energy shifts one of these electrons to an orbital of higher energy with an inversion of spin (for reviews, see Kasha and Khan, 1970; Wilson and Hastings, 1970; Kearns, 1971; Foote, 1976). Two forms of $^1O_2$ are known, Δ and Σ. When excited to the Δ form, the newly paired electrons of opposite spin occupy the same orbital, whereas in Σ singlet oxygen the two electrons occupy different orbitals. Δ singlet oxygen has a lower energy above ground state, but a considerably longer lifetime than does Σ singlet oxygen. Singlet oxygen can dissipate its excess energy by the emission of light as it reverts to the triplet ground state or by increased chemical reactivity. That reactions induced by $^1O_2$ may be toxic to mammalian cells is suggested by the photodynamic action of dyes.

Photodynamic action refers to the damage to biological systems, including intact cells, induced by dye-sensitized photooxygenation reactions; one of the mechanisms proposed for this toxicity involves the intermediate formation of singlet oxygen. The sensitizing substance (dye) in the ground (singlet) state absorbs light and a valence electron is raised to an orbital of higher energy. The molecule is thus activated (excited singlet state). The excited singlet state of the dye converts to the longer-lived excited triplet state by the inversion of electron spin. The triplet sensitizer can initiate photosensitized reactions by one of two mechanisms designated Type I and Type II (Foote, 1976). In Type I reactions, the sensitizer reacts directly with the target, and in Type II reactions the interaction is initially with oxygen with the formation of $^1O_2$. The reaction of $^1O_2$ with sensitive groups on the surface of the cell is believed to contribute to the toxicity.

A well established mechanism for the generation of $^1O_2$ is the interaction of H_2O_2 and hypochlorite (Kasha and Khan, 1970; Wilson and Hastings, 1970; Kearns, 1971; Foote, 1976) as follows:

$$H_2O_2 + OCl^- \rightarrow {}^1O_2 + H_2O + Cl^-$$

Chloride is oxidized by MPO and H_2O_2 and the product formed appears to be hypochlorous acid (Agner, 1972; Harrison and Schultz, 1976). It might be expected therefore that the hypochlorite generated in this way can react with excess H_2O_2 to generate $^1O_2$. Is $^1O_2$ generated by the MPO system and if so, does it contribute to the toxicity? The evidence is as follows.

1. The MPO–H_2O_2–halide system emits light (Allen, 1975a,b; Rosen and Klebanoff, 1976). It should be emphasized that light emission, although compatible with, is not proof of $^1O_2$ formation. The emission of light indicates the formation of an electronically excited state but, without spectral analysis, does

not indicate the nature of the excited species. Spectral analysis of the light emitted by the MPO system reveals broad peak activity with a maximum near 570 nm rather than the characteristic spectrum of $^1O_2$ decay (Cheson et al., 1976; Andersen et al., 1977). Secondary excitations induced by $^1O_2$ are common. For example, reaction of $^1O_2$ with electron-rich olefins results in the formation of dioxetanes which are often unstable, decomposing to form an excited carbonyl product which decays with light emission (Lee and Wilson, 1973). Thus, the light emission by the MPO system appears, in large part, to be due either to secondary excitations initiated by $^1O_2$ or to other $^1O_2$–independent mechanisms.

2. The MPO- (Rosen and Klebanoff, 1977; Klebanoff et al., 1977) and LPO- (Piatt et al., 1977; Piatt and O'Brien, 1979) H_2O_2-halide systems convert diphenylfuran (DPF) to cis-dibenzoylethylene (cis-DBE). A number of chemical reactions can be initiated by $^1O_2$. Sometimes a unique product is formed, and this reaction can thus be employed as evidence for the presence of $^1O_2$. One such reaction is the conversion of DPF to cis-DBE (King et al., 1975). Conversion by the MPO system occurred most readily with bromide as the halide, with activity increasing sharply at concentrations above 0.1 mM. Activity was also high with chloride; however, concentrations greater than 10 mM were required. In contrast, with iodide, activity was low and limited to a narrow range of iodide concentrations (Rosen and Klebanoff, 1977). The conversion of DPF to cis-DBE by the MPO system is inhibited by $^1O_2$ scavengers such as β-carotene, bilirubin, histidine, and 1,4-diazabicyclo[2.2.2]octane (DABCO) and is stimulated by deuterium oxide (D_2O) (Rosen and Klebanoff, 1977). The substitution of D_2O for H_2O increases the lifetime of $^1O_2$ manyfold and generally stimulates $^1O_2$-dependent reactions (Merkel et al., 1972; Kajiwara and Kearns, 1973).

These findings are compatible with the generation of $^1O_2$ by the MPO system, particularly with bromide or chloride as the halide. It should be emphasized, however, that DPF conversion to cis-DBE (or the oxidation of other furans) is not specific for $^1O_2$. Indeed, recent evidence has suggested that the conversion of DFP to cis-DBE by the MPO–H_2O_2–halide system is due to the formation of oxidation products of chloride (or bromide) rather than $^1O_2$ (Held and Hurst, 1978; Harrison et al., 1978). Furthermore, inhibitor studies do not provide unequivocal proof of a $^1O_2$ mechanism since absolute specificity for that mechanism is not assured. This applies not only to the inhibition of DPF conversion, but also to the inhibition of the cytotoxic activity of the MPO-H_2O_2-halide system by $^1O_2$ quenchers (Klebanoff et al., 1976; Klebanoff, 1975b).

In summary, it is probable that MPO, H_2O_2, and a halide interact to form a powerful oxidant. With chloride, that oxidant is most probably hypochlorous acid, a powerful germicide which can attack the organism in a variety of ways: oxidation of sensitive surface groups, formation of chloramines, etc. The formation of $^1O_2$ by the chloride-dependent peroxidase system has yet to be unequivocally demonstrated. Very similar mechanisms appear to be operative with bromide as the halide; hypobromous acid may be the primary oxidant involved. With iodide the reactivity of the peroxidase system appears to be fundamentally different; oxidation and halogenation, however, remain the primary modes of attack.

5.5. NATURE OF THE TARGET CELL DYSFUNCTION

The nature of the functional defect resulting from the action of the MPO system on cells is not known. No evidence for an effect on nucleic acid or protein synthesis has yet been presented. The highly reactive nature of the toxic products of the MPO system would suggest that diffusion distances are short, thus focusing attention on the cell surface. One study has suggested that transport of metabolites across the cell membrane may be affected (Klebanoff and Clark, 1978). The uptake of a nonmetabolizable amino acid was inhibited by the MPO system, and, if uptake was allowed to occur, the subsequent addition of the MPO system resulted in a rapid discharge of the amino acid from the cell. Lysis of mammalian cells can occur (Klebanoff and Clark, 1975).

The nature of the cell surface influences susceptibility to the peroxidase system. Surface components may act as physical barriers between the toxic products of the MPO system and sensitive groups on the cell membrane. The cell membrane of some organisms may be more susceptible to oxidative damage than that of others, and the presence on the cell surface of inhibitors of the peroxidase system, for example, catalase, reducing groups, and singlet oxygen quenchers such as carotenoid pigments (Krinsky, 1974), may influence the interaction. For example, mutant strains of *S. typhimurium* with deficiencies in their cell wall lipopolysaccharides were found to be more sensitive to the $MPO-H_2O_2$–iodide system than was the parent strain (Tagesson and Stendahl, 1973). The degree of sensitivity increased with the lipopolysaccharide deficiency suggesting that the lipopolysaccharide core offered some resistance to the MPO system.

6. EVIDENCE FOR A ROLE *IN SITU*

6.1. MICROBICIDAL ACTIVITY

6.1.1. The Components of the MPO System Interact in the Phagocytic Vacuole

The presence of the components of the MPO-mediated antimicrobial system in the PMN has been considered in Section 3; the evidence that these components can interact in the phagocytic vacuole is considered here. As already indicated (Section 5.1), the microbicidal activity of the $MPO-H_2O_2$–iodide system is associated with the iodination of the microorganism (Klebanoff, 1967b). Iodination also occurs in the intact PMN following phagocytosis (Klebanoff, 1967b), and the fixed iodine has been localized, in part, in the phagocytic vacuole (Klebanoff, 1970b; Root and Stossel, 1974) on the surface of the ingested organism (Klebanoff, 1970b). Iodination is markedly decreased when leukocytes which lack H_2O_2 [CGD (Klebanoff and White, 1969)] or MPO [hereditary MPO deficiency (Pincus and Klebanoff, 1971; Klebanoff and Clark, 1977)] are employed. These findings suggest that MPO, H_2O_2, and a halide, in this instance iodide, can interact adjacent to the ingested organism. Further evidence for the interaction of the components of the MPO system in the PMN is the phagocytosis-induced chlorination reaction (Zgliczyński and Stelmaszyńska, 1975) and decarboxylation of amino acids (Strauss *et al.*, 1971), both of which are also catalyzed by MPO, H_2O_2, and chloride. A cytochemical technique has been employed for the detection of both peroxidase and H_2O_2 in the vacuolar space (Briggs *et al.*, 1975b).

6.1.2. H_2O_2 Is Required for Optimum Antimicrobial Activity

The generation of H_2O_2 by phagocytizing PMNs has been clearly demonstrated (Section 3.2.1) and there is considerable evidence to implicate it in the microbicidal activity of the cell.

In CGD, PMNs have a major microbicidal defect (Quie *et al.*, 1967) which is associated with the absence of the respiratory burst (Holmes *et al.*, 1967). The importance of H_2O_2 deficiency in the neutrophil dysfunction is emphasized by the reversal of the microbicidal defect on the introduction of a H_2O_2-generating system into the cell. Certain microorganisms, namely, pneumococci, streptococci, and lactobacilli, generate H_2O_2 (Section 3.2.2); these organisms are killed easily by CGD leukocytes (Kaplan *et al.*, 1968; Klebanoff and White, 1969; Mandell and Hook, 1969) and are rarely found in the lesions. Their susceptibility to the intracellular microbicidal systems is believed to be due to the replacement of a defective leukocytic H_2O_2-generating system with H_2O_2 of microbial origin (Klebanoff and White, 1969; Mandell and Hook, 1969). In support of this hypothesis is the finding that mutant strains of streptococci (Holmes and Good, 1972) and pneumococci (Pitt and Bernheimer, 1974) which form little or no H_2O_2 are killed less easily by CGD leukocytes than are the wild-type H_2O_2-generating strains. The metabolic (Baehner *et al.*, 1970b) and microbicidal (Johnston and Baehner, 1970) defect in CGD also can be reversed, at least in part, by the ingestion of latex particles coated with glucose oxidase and by extracellular H_2O_2 generated by glucose oxidase (Root, 1974). Glucose oxidase, in the presence of glucose, catalyzes the reduction of oxygen to H_2O_2 without an apparent superoxide intermediate (Massey *et al.*, 1969), suggesting that H_2O_2 rather than O_2^- mediates the observed effects.

These studies suggest that the introduction of H_2O_2, a H_2O_2-generating enzyme system, or H_2O_2-generating bacteria can reverse the defect in CGD and that this occurs in the apparent absence of O_2^- which presumably remains deficient in these cells. Since MPO is present in normal amounts in CGD leukocytes and is released normally into the phagocytic vacuole (Stossel *et al.*, 1972), the most likely explanation is the interaction of the added H_2O_2 with MPO to form a microbicidal system.

Catalase increases H_2O_2 degradation and thus inhibits the MPO–H_2O_2–halide system (see Section 3.5). When added to phagocytizing PMNs, catalase produces a small but significant inhibition of microbicidal activity (McRipley and Sbarra, 1967a; Klebanoff and Hamon, 1972; Johnston *et al.*, 1975), and this inhibition is increased when catalase is bound to latex particles (Johnson *et al.*, 1975). Catalase also can be introduced into the phagocytic vacuole as a component of the ingested organism and may influence the microbicidal activity of the cell. Thus, staphylococcal strains rich in catalase are more resistant to the microbicidal activity of PMNs *in vitro* and are more virulent to mice than are strains with a low catalase content (Mandell, 1975).

The microbicidal activity of PMNs also is inhibited by SOD bound to latex particles (Johnston *et al.*, 1975). This finding is difficult to reconcile with the considerable evidence in support of an important role for H_2O_2 in the microbicidal activity of the PMN. SOD catalyzes the dismutation of O_2^- and would be expected to increase the rate of formation of H_2O_2 and, thus, to favor H_2O_2-dependent antimicrobial systems. Microbicidal activity by intact PMNs is the end result of a complex series of events which include phagocytosis, degranulation, the metabolic burst, etc. It is not known whether the inhibition by SOD, under the experimental conditions employed by Johnston *et al.* (1975), is due solely to its effect on the oxygen products required for microbicidal activity.

6.1.3. MPO Is Required for Optimum Antimicrobial Activity

The evidence that MPO is required for optimum antimicrobial activity of PMNs comes from studies with peroxidase inhibitors and with PMNs which lack MPO.

6.1.3a. MPO Inhibitors. A variety of agents can inhibit peroxidase-catalyzed reactions (see Section 3.5); in no instance, however, has an inhibitor been shown to be absolutely specific for peroxidase. Azide, for example, while a potent inhibitor of peroxidase, also inhibits other heme protein enzymes and can quench $^1O_2$ (Hasty *et al.*, 1972), and cyanide is an efficient chelating agent. Nevertheless, inhibitor studies can be of value if their limitations are kept in mind and attempts made to evaluate their specificity in the system under study.

Azide inhibits the microbicidal activity of normal (Klebanoff, 1970a; Diamond *et al.*, 1972; Koch, 1974a,b) and CGD (Klebanoff, 1970a) PMNs at concentrations which have little or no effect on phagocytosis. With CGD leukocytes, an organism which is killed normally by these cells (e.g., *L. acidophilus*) must be employed. An inhibition is not observed when PMNs which lack MPO are used (Klebanoff, 1970a), suggesting that the effect on normal (or CGD) cells is due to an inhibition of MPO. Cyanide (Klebanoff, 1970a) and sulfonamides (Lehrer, 1971) also inhibit the microbicidal activity of normal but not MPO-deficient leukocytes. The microbicidal activity of normal PMNs is inhibited by the antithyroid agents, methimazole and propylthiouracil (Klebanoff and Hamon, 1972) and by the catecholamines, adrenalin, noradrenalin, and dihydroxyphenylalanine (Qualliotine *et al.*, 1972) *in vitro*; these agents also inhibit certain MPO-catalyzed reactions including iodination by PMNs. Ascorbic acid, however, inhibits iodination without affecting microbicidal activity (McCall *et al.*, 1971; Klebanoff and Hamon, 1972).

6.1.3b. MPO-Deficient Leukocytes. A number of patients with a genetic absence of MPO from neutrophils and monocytes have been found. Many of the patients with hereditary MPO deficiency have been in reasonably good health. Thus, the two patients described by Grignaschi *et al.* (1963) and one of the siblings described by Lehrer and Cline (1969) were not noted to have had infections; the patient of Undritz (1966) was reported by Schmid and Brune (1974) to have had recurrent candida vaginitis but to be otherwise in good health; one of the patients described by Breton-Gorius et al. (1975) was in good health except for a normocytic normochromic anemia, and three patients had severe acne vulgaris (Patriarca et al., 1975; Rosen and Klebanoff, 1976). The acne of the two patients seen by us, although severe enough to require therapy, was not unusual in its manifestations. Other patients with hereditary MPO deficiency, however, have had unusual infections. The patient studied extensively by Lehrer and Cline (1969) had disseminated candidiasis affecting bone, subcutaneous tissues, and probably lung. One of the patients described by Breton-Gorius et al. (1975) had a history of repeated infections, including staphylococcal osteomyelitis, numerous subcutaneous abscesses, two episodes of pneumonia, and pulmonary abscesses on two occasions. The patient of Moosmann and Bojanovsky (1975) had recurrent severe infections of the respiratory, gastrointestinal, and urinary tracts, and gall bladder and skin with both candida and bacterial pathogens, e.g., staphylococci. The patient described by Stendahl and Lindgren (1976) suffered from generalized pustular psoriasis for many years, and the patient of Cech et al. (1979a) had a C. albicans hepatic abscess. Thus, to summarize, although severe infections have been noted in some patients with MPO deficiency, the clinical picture is not as severe as in CGD, and a number of patients have been in good health.

The most compelling evidence in support of an important role for MPO in the microbicidal activity of the normal PMN is the marked microbicidal defect found in leukocytes from patients with hereditary MPO deficiency. The presence of systemic candidiasis in their patient prompted Lehrer and Cline (1969) to test MPO-deficient neutrophils for candidacidal activity and a striking defect was found. Normal leukocytes killed 30% of ingested C. albicans in 1 hr, whereas MPO-deficient leukocytes killed only 0.1% of the ingested candida during this period. Reduced fungicidal activity has been confirmed in a number of other studies (Klebanoff, 1970a; Lehrer, 1972, 1974; Schmid and Brune, 1974; Moosmann and Bojanovsky, 1975; Stendahl and Lindgren, 1976; Cech et al., 1979b). MPO-deficient neutrophils also kill many bacteria less efficiently than do normal leukocytes (Lehrer and Cline, 1969; Lehrer et al., 1969; Klebanoff, 1970a; Stendahl and Lindgren, 1976; Cech et al., 1979b), with the most pronounced abnormality seen during the early postphagocytic period. MPO deficiency is the only molecular lesion which has been detected in these cells; thus, the microbicidal defect is very strong evidence for MPO involvement in the microbicidal activity of normal PMNs, particularly during the early postphagocytic period. The bactericidal activity of MPO-deficient leukocytes, which is evident following the early defect Lehrer et al., 1969; Klebanoff and Hamon, 1972), indicates the presence in these cells of antimicrobial systems which do not require MPO. These systems develop slowly but are ultimately effective. This raises several questions.

1. What are the MPO-independent antimicrobial systems present in MPO-deficient leukocytes? It is probable that both oxygen-dependent and oxygen-

independent systems are operative. The evidence for oxygen-dependent antimicrobial systems is as follows.

i. The microbicidal defect in MPO-deficient leukocytes is not as severe as that of CGD leukocytes, suggesting the presence in leukocytes which lack MPO of antimicrobial systems which are dependent on the respiratory burst.

ii. In sharp contrast to CGD leukocytes (Holmes *et al.*, 1967), the respiratory burst of MPO-deficient leukocytes is not decreased (Lehrer and Cline, 1969); indeed, it appears to be greater than normal (see 3 below).

iii. The residual staphylocidal activity of MPO-deficient leukocytes is inhibited by hypoxia (Klebanoff and Hammon, 1972), indicating that the systems operative against this organism are dependent, in part, on oxygen.

iv. The products of oxygen reduction and excitation, namely, $O_2^{\cdot-}$, H_2O_2, hydroxyl radicals ($OH\cdot$), and $^1O_2$, have antimicrobial activity in the absence of MPO. The antimicrobial activity of these agents is considered elsewhere in this treatise and will not be considered in detail here. Briefly, a number of systems which generate $O_2^{\cdot-}$ and, by dismutation, H_2O_2, have antimicrobial activity (for review, see Klebanoff and Clark, 1978). In some instances the antimicrobial effect is inhibited by catalase but not SOD, implicating H_2O_2; in other instances, both catalase and SOD are inhibitory, and hydroxyl radicals have been proposed as the microbicidal agent. These radicals can be generated by a trace-metal-catalyzed interaction between $O_2^{\cdot-}$ and H_2O_2 (McCord and Day, 1978; Halliwell, 1978), as follows:

$$O_2^{\cdot-} + H_2O_2 \rightarrow O_2 + OH^- + OH\cdot$$

Finally, singlet oxygen may be formed by $O_2^{\cdot-}$-generating systems (Khan, 1970, 1977; Arneson, 1970; Kellogg and Fridovich, 1975; Kobayashi and Ando, 1979) and contribute to the toxicity.

The microbicidal activity of $O_2^{\cdot-}$-generating systems is increased considerably by MPO and a halide indicating that the toxicity of $O_2^{\cdot-}$ (or its products) is relatively weak as compared to the H_2O_2 formed from it when combined with the other components of the MPO system (Klebanoff, 1974; Figure 3). In an early study, xanthine and xanthine oxidase were employed as the source of $O_2^{\cdot-}$ and very little toxicity was observed in the absence of MPO and a halide (Klebanoff, 1974). More recently, acetaldehyde was employed rather than xanthine as the substrate for xanthine oxidase, and, under this condition, a bactericidal effect by the unsupplemented xanthine oxidase system was readily observed (Rosen and Klebanoff, 1979). The further addition of MPO and chloride, however, greatly increased the toxicity; indeed, 100 times as much acetaldehyde was required for comparable bactericidal activity in the absence of MPO and chloride than in their presence. The bactericidal activity of the acetaldehyde–xanthine oxidase system was inhibited by SOD, catalase, the $^1O_2$ quenchers azide, DABCO, and histidine, and to a lesser degree by the $OH\cdot$ scavengers mannitol and benzoate. This system also converts DPF to *cis*-DBE, which is compatible with the formation of $^1O_2$ (Rosen and Klebanoff, 1979).

These studies suggest that in the absence of MPO the products of oxygen reduction and excitation are produced in sufficient quantity to exert a toxic effect on ingested organisms. Oxygen-independent antimicrobial activity is also present in MPO-deficient leukocytes (Lehrer, 1972). This may result from the removal of the organisms from growth factors in the extracellular fluid, from the fall in intravacuolar pH, and from the secretion into the vacuolar space of antimicrobial agents such as lysozyme, lactoferrin, and cationic proteins (see Chapter 14 of this volume). The antimicrobial activity of MPO when supplemented with H_2O_2 and a halide is considerably greater than that of an unsupplemented acid extract of human PMNs (Klebanoff, 1975b), or the purified cationic proteins from these cells (Klebanoff and Clark, 1978).

2. Are the antimicrobial systems present in MPO-deficient leukocytes also present in normal PMNs? This question can be answered in the affirmative. Exposure of normal PMNs to an atmosphere of nitrogen does not abolish antimicrobial activity; indeed, some organisms are killed as rapidly under anaerobic as under aerobic conditions (Mandell, 1974). Clearly oxygen-independent antimicrobial systems are present in normal cells; a conclusion supported by the extraction of a number of antimicrobial agents, effective in the absence of oxygen. Similarly, treatment of normal neutrophils with the peroxidase inhibitor azide inhibits, but does not abolish, antimicrobial activity, indicating the presence of azide-insensitive (MPO-independent) antimicrobial systems in these cells (Klebanoff, 1970a).

The finding that an organism is killed when MPO is inhibited or absent does not necessarily indicate that MPO-independent antimicrobial systems are the lethal agents in the presence of enzymatically-active MPO, nor does the demonstration of antimicrobial activity under anaerobic conditions exclude the predominance of oxygen-dependent antimicrobial systems under aerobic conditions. These findings only indicate that when MPO or oxygen is absent, antimicrobial systems are present which can kill the organism. These may be back-up systems not normally needed.

3. Are MPO-independent antimicrobial systems equally active in normal and MPO-deficient PMNs? There is some evidence to suggest that this may not be the case. Azide inhibits the microbicidal activity of normal leukocytes without affecting MPO-deficient leukocytes (Klebanoff, 1970a). If, as this finding suggests, azide exerts its effect on normal cells by inhibiting MPO, then the microbicidal activity of azide-treated cells should be a measure of MPO-independent antimicrobial activity. When azide is added to normal PMNs, the antimicrobial activity falls to a level below that of similarly treated MPO-deficient cells (Klebanoff, 1970a). This suggests that the azide-insensitive antimicrobial systems of MPO-deficient PMNs are more highly developed than those of normal cells, i.e., that MPO-deficient leukocytes have adapted to the long-term absence of MPO with an increase in the activity of MPO-independent antimicrobial systems. In this regard, we have found the phagocytosis-induced increase in oxygen consumption (Klebanoff and Hamon, 1972), superoxide production (Rosen and Klebanoff, 1976), H_2O_2 generation (Klebanoff and Pincus, 1971), and glucose C-1 oxidation (Klebanoff and Pincus, 1971) to be greater in MPO-deficient than in normal PMNs. MPO appears to be required for the termination

of the respiratory burst (Jandl *et al.*, 1978); MPO-deficient patients therefore may be as well as they are because, in the absence of MPO, the respiratory burst is increased with an increased production of the toxic products of oxygen reduction and excitation.

6.2. EXTRACELLULAR CYTOTOXICITY

The striking morphologic and metabolic changes induced in the PMN by phagocytosis are geared to the concentration within the phagocytic vacuole of toxic agents which are directed against the ingested organism. However, leakage or secretion of these agents to the outside of the cell can occur with the potential for damage to adjacent cells. The toxic effect of the cell-free MPO–H_2O_2–halide system on mammalian cells is considered in Sections 4.1–4.5.

Intact PMNs are cytotoxic *in vitro* and, in some instances, this cytotoxicity is mediated by the MPO system. Thus, the mouse ascites lymphoma cells designated LSTRA were killed by PMNs when combined with a phagocytizable particle (opsonized zymosan) and a halide (chloride or iodide) (Clark and Klebanoff, 1975). Inhibition of cytotoxicity by the peroxidase inhibitors, azide and cyanide, and by catalase suggested the involvement of MPO and H_2O_2. This was supported by the absence of cytotoxicity when PMNs which lacked MPO or the respiratory burst (CGD) were employed. Cytotoxicity was restored to MPO-deficient leukocytes by the addition of purified MPO and to CGD leukocytes by the addition of H_2O_2 or a H_2O_2-generating system (glucose + glucose oxidase). A mechanism involving the phagocytosis-induced extracellular secretion of MPO and H_2O_2 and their interaction with a halide in the extracellular fluid to damage the target cell was proposed. Similarly, phagocytizing PMNs have been found to increase platelet serotonin release through the action of the MPO system (Clark and Klebanoff, 1977b), and the MPO-H_2O_2-halide system appears to be involved in the toxicity to LSTRA tumor cells of concanavalin-A-activated PMNs (Clark and Klebanoff, 1979c).

7. CONCLUSIONS

Among the antimicrobial systems of the PMN is one which consists of MPO, H_2O_2, and a halide. It is probable that the MPO-mediated antimicrobial system is involved in the killing of many, if not most, organisms in the normal PMN; it is, in my view, the predominant antimicrobial system during the early post-phagocytic period. Since ingested organisms can only be killed once, its action is generally sufficient. However, it is probable that most organisms are susceptible to more than one antimicrobial system in the phagocytic vacuole. Thus, an organism normally killed by the MPO system may be handled less efficiently, but adequately, by other systems, both oxygen-dependent and oxygen-independent, when MPO is absent, and severe infection is generally (but not always) averted. When all oxygen-dependent antimicrobial systems, both

MPO-dependent and MPO-independent, are absent, as in CGD, susceptibility to infection increases. Some organisms, however, can provide the needed product of oxygen metabolism under these conditions, and others are adequately controlled by oxygen-independent antimicrobial systems. This marked microbicidal reserve provides the PMN with an overkill capacity essential to our survival in a world populated with potentially harmful microorganisms.

REFERENCES

Agner, K., 1941, Verdoperoxidase. A ferment isolated from leukocytes, *Acta Physiol. Scand. (Suppl. 8)* **2**:1.

Agner, K., 1963, Studies on myeloperoxidase activity. I. Spectrophotometry of the MPO-H_2O_2 compound, *Acta Chem. Scand.* **17**:332.

Agner, K., 1972, Biological effects of hypochlorous acid formed by "MPO"-peroxidation in the presence of chloride ions, in: *Structure and Function of Oxidation-Reduction Enzymes* (A. Akeson and A. Ehrenberg, eds.), Vol. 18, pp. 329–335, Pergamon Press, New York.

Alexander, N.M., 1974, Oxidative cleavage of tryptophanyl peptide bonds during chemical- and peroxidase-catalyzed iodinations, *J. Biol. Chem.* **249**:1946.

Allen, R. C., 1975a, Halide dependence of the myeloperoxidase-mediated antimicrobial system of the polymorphonuclear leukocyte in the phenomenon of electronic excitation, *Biochem. Biophys. Res. Commun.* **63**:675.

Allen, R. C., 1975b, The role of pH in the chemiluminescent response of the myeloperoxidase-halide-HOOH antimicrobial system, *Biochem. Biophys. Res. Commun.* **63**:684.

Andersen, B. R., Brendzel, A. M., and Lint, T. F., 1977, Chemiluminescence spectra of human myeloperoxidase and polymorphonuclear leukocytes, *Infect. Immun.* **17**:62.

Annear, D. I., and Dorman, D. C., 1952, Hydrogen peroxide accumulation during growth of the pneumococcus, *Aust. J. Exp. Biol. Med. Sci.* **30**:191.

Arneson, R. M., 1970, Substrate-induced chemiluminescence of xanthine oxidase and aldehyde oxidase, *Arch. Biochem. Biophys.* **136**:352.

Auclair, C., Cramer, E., Hakim, J., and Bovin, P., 1976, Studies on the mechanism of NADPH oxidation by the granule fraction isolated from human resting polymorphonuclear blood cells, *Biochimie* **58**:1359.

Aune, T. M., and Thomas, E. L., 1977, Accumulation of hypothiocyanite ion during peroxidase-catalyzed oxidation of thiocyanate ion, *Eur. J. Biochem.* **80**:209.

Avery, O. T., and Morgan, H. J., 1924, The occurrence of peroxide in cultures of pneumococcus, *J. Exp. Med.* **39**:275.

Babior, B. M., Kipnes, R. S., and Curnutte, J. T., 1973, Biological defense mechanisms. The production by leukocytes of superoxide, a potential bactericidal agent, *J. Clin. Invest.* **52**:741.

Babior, B. M., Curnutte, J. T., and McMurrich, B. J., 1976, The particulate superoxide-forming system from human neutrophils. Properties of the system and further evidence supporting its participation in the respiratory burst, *J. Clin. Invest.* **58**:989.

Baehner, R. L., Karnovsky, M. J., and Karnovsky, M. L., 1969, Degranulation of leukocytes in chronic granulomatous disease, *J. Clin. Invest.* **48**:187.

Baehner, R. L., Gilman, N., and Karnovsky, M. L., 1970a, Respiration and glucose oxidation in human and guinea pig leukocytes: Comparative studies, *J. Clin. Invest.* **49**:692.

Baehner, R. L., Nathan, D. G., and Karnovsky, M. L., 1970b, Correction of metabolic deficiencies in the leukocytes of patients with chronic granulomatous disease, *J. Clin. Invest.* **49**:865.

Baggiolini, M., Hirsch, J. G., and de Duve, C., 1969, Resolution of granules from rabbit heterophil leukocytes into distinct populations by zonal sedimentation, *J. Cell Biol.* **40**:529.

Bainton, D. F., and Farquhar, M. G., 1968a, Differences in enzyme content of azurophil and specific granules of polymorphonuclear leukocytes. I. Histochemical staining of bone marrow smears, *J. Cell Biol.* **39**:286.

Bainton, D. F., and Farquhar, M. G., 1968b, Differences in enzyme content of azurophil and specific granules of polymorphonuclear leukocytes. II. Cytochemistry and electron microscopy of bone marrow cells, *J. Cell Biol.* **39**:299.

Baron, D. N., and Ahmed, S. A., 1969, Intracellular concentrations of water and of the principle electrolytes determined by the analysis of isolated human leucocytes, *Clin. Sci.* **37**:205.

Belding, M. E., Klebanoff, S. J., and Ray, C. G., 1970, Peroxidase-mediated virucidal systems, *Science* **167**:195.

Brandrick, A. M., Newton, J. M., Henderson, G., and Vickers, J. A., 1967, An investigation into the interaction between iodine and bacteria, *J. Appl. Bacteriol* **30**:484.

Breton-Gorius, J., Coquin, Y., and Guichard, J., 1975, Activités peroxydasiques de certaines granulations des neutrophiles dans deux cas de déficit congénital en myéloperoxydase, *C. R. Acad. Sci. (Paris)* **280**:1753.

Briggs, R. T., Drath, D. B., Karnovsky, M. L., and Karnovsky, M. J., 1975a, Localization of NADH oxidase on the surface of human polymorphonuclear leukocytes by a new cytochemical method, *J. Cell Biol.* **67**:566.

Briggs, R. T., Karnovsky, M. L., and Karnovsky, M. J., 1975b, Cytochemical demonstration of hydrogen peroxide in the polymorphonuclear leukocyte phagosomes, *J. Cell Biol.* **64**:254.

Buege, J. E., and Aust, S. D., 1976, Lactoperoxidase-catalyzed lipid peroxidation of microsomal and artificial membranes, *Biochim. Biophys. Acta* **444**:192.

Cagan, R. H., and Karnovsky, M. L., 1964, Enzymatic basis of the respiratory burst during phagocytosis, *Nature* **204**:255.

Cech, P., Stalder, H., Widmann, J. J., Rohner, A., and Miescher, P. A., 1979a, Leukocyte myeloperoxidase deficiency and diabetes mellitus associated with *Candida albicans* liver abscess, *Am. J. Med.* **66**:149.

Cech, P., Papathanassiou, A., Boreux, G., Roth, P., and Miescher, P. A., 1979b, Hereditary myeloperoxidase deficiency, *Blood* **53**:403.

Cheson, B. D., Christensen, R. L., Sperling, R., Kohler, B. E., and Babior, B. M., 1976, The origin of the chemiluminescence of phagocytosing granulocytes, *J. Clin. Invest.* **58**:789.

Clark, R. A., and Klebanoff, S. J., 1975, Neutrophil-mediated tumor cell cytotoxicity: Role of the peroxidase system, *J. Exp. Med.* **141**:1442.

Clark, R. A., and Klebanoff, S. J., 1977a, Myeloperoxidase-H_2O_2-halide system: Cytotoxic effect on human blood leukocytes, *Blood* **50**:65.

Clark, R. A., and Klebanoff, S. J., 1977b, Neutrophil-mediated release of serotonin from human platelets: Role of myeloperoxidase and H_2O_2, *Clin. Res.* **25**:474A.

Clark, R. A., and Klebanoff, S. J., 1979a, Myeloperoxidase-mediated platelet release reaction, *J. Clin. Invest.* **63**:177.

Clark, R. A., and Klebanoff, S. J., 1979b, Chemotactic factor inactivation by the myeloperoxidase-hydrogen peroxide-halide system: An inflammatory control mechanism, *J. Clin. Invest.* **64**:913.

Clark, R. A., and Klebanoff, S. J., 1979c, Role of the myeloperoxidase-H_2O_2-halide system in concanavalin A-induced tumor cell killing by human neutrophils, *J. Immunol.* **122**:2605.

Clark, R. A., Klebanoff, S. J., Einstein, A. B., and Fefer, A., 1975, Peroxidase-H_2O_2-halide system: Cytotoxic effect on mammalian tumor cells, *Blood* **45**:161.

Clark, R. A., Olsson, I., and Klebanoff, S. J., 1976, Cytotoxicity for tumor cells of cationic proteins from human neutrophil granules, *J. Cell Biol.* **70**:719.

Curnutte, J. T., and Babior, B. M., 1974, Biological defense mechanisms. The effect of bacteria and serum on superoxide production by granulocytes, *J. Clin. Invest.* **53**:1662.

DeChatelet, L. R., McCall, C. E., and Cooper, M. R., 1971, Direct measurement of iodine production by sonic extracts of polymorphonuclear leukocytes, *Clin. Chem.* **17**:392.

DeChatelet, L. R., McPhail, L. C., Mullikin, D., and McCall, C. E., 1975, An isotopic assay for NADPH oxidase activity and some characteristics of the enzyme from human polymorphonuclear leukocytes, *J. Clin. Invest.* **55**:714.

Diamond, R. D., Root, R. K., and Bennett, J. E., 1972, Factors influencing killing of *Cryptococcus neoformans* by human leukocytes *in vitro*, *J. Infect. Dis.* **125**:367.

Dogon, I. L., Kerr, A. C., and Amdur, B. H., 1962, Characterization of an antibacterial factor in human parotid secretions, active against *Lactobacillus casei*, *Arch. Oral Biol.* **7**:81.

Edelson, P. J., and Cohn, Z. A., 1973, Peroxidase-mediated mammalian cell cytotoxicity, *J. Exp. Med.* **138**:318.

Endres, G., and Herget, L., 1929, Mineralzusammensetzung der Blutplättchen und weissen Blutköperchen, *Z. Biol.* **88**:451.

Evans, W. H., and Rechcigl, M., Jr., 1967, Factors influencing myeloperoxidase and catalase activities in polymorphonuclear leukocytes, *Biochim. Biophys. Acta* **148**:243.

Foote, C. S., 1976, Photosensitized oxidation and singlet oxygen: Consequences in biological systems, in: *Free Radicals in Biology* (W. A. Pryor, ed.), Vol. 2, pp. 85–133, Academic Press, New York.

Fridovich, I., 1970, Quantitative aspects of the production of superoxide anion radical by milk xanthine oxidase, *J. Biol. Chem.* **245**:4053.

Fridovich, I., 1976, Oxygen radicals, hydrogen peroxide, and oxygen toxicity, in: *Free Radicals in Biology* (W. A. Pryor, ed.), Vol. 1, pp. 239–277, Academic Press, New York.

Goldstein, I. M., Cerqueira, M., Lind, S., and Kaplan, H. B., 1977, Evidence that the superoxide-generating system of human leukocytes is associated with the cell surface. *J. Clin. Invest.* **59**:249.

Green, D. E., and Stumpf, P. K., 1946, The mode of action of chlorine, *J. Am. Water Works Assoc.* **38**:1301.

Grignaschi, V. I., Sperperato, A. M., Etcheverry, M. J., and Macario, A. J. L., 1963, Un nuevo cuadro citoquimico: Negatividad espontanea de las reacciones de peroxidasas, oxidasas y lipido en la progenie neutrofilia y en los monocitos de dos hermanos, *Rev. Asoc. Med. Argent.* **77**:218.

Halliwell, B., 1978, Superoxide-dependent formation of hydroxyl radicals in the presence of iron chelates: Is it a mechanism for hydroxyl radical production in biochemical systems?, *FEBS Lett.* **92**:321.

Hamon, C. B., and Klebanoff, S. J., 1973, A peroxidase-mediated, *Streptococcus mitis*-dependent antimicrobial system in saliva, *J. Exp. Med.* **137**:438.

Hanssen, F. S., 1924, The bactericidal property of milk, *Br. J. Exp. Pathol.* **5**:271.

Harrison, J. E., and Schultz, J., 1976, Studies on the chlorinating activity of myeloperoxidase, *J. Biol. Chem.* **251**:1371.

Harrison, J. E., Watson, B. D., and Schultz, J., 1978, Myeloperoxidase and singlet oxygen: A reappraisal, *FEBS Lett.* **92**:327.

Hasty, N., Merkel, P. B., Radlick, P., and Kearns, D. R., 1972, Role of azide in singlet oxygen reactions: Reaction of azide with singlet oxygen, *Tetrahed. Lett.* **1**:49.

Held, A. M., and Hurst, J. K., 1978, Ambiguity associated with use of singlet oxygen trapping agents in myeloperoxidase-catalyzed oxidations, *Biochem. Biophys. Res. Commun.* **81**:878.

Hohn, D. C., and Lehrer, R. I., 1975, NADPH oxidase deficiency in X-linked chronic granulomatous disease, *J. Clin. Invest.* **55**:707.

Holmes, B., and Good, R. A., 1972, Laboratory models of chronic granulomatous disease, *J. Reticuloendothel. Soc.* **12**:216.

Holmes, B., Page, A. R., and Good, R. A., 1967, Studies of the metabolic activity of leukocytes from patients with a genetic abnormality of phagocytic function, *J. Clin. Invest.* **46**:1422.

Homan-Müller, J. W. T., Weening, R. S., and Roos, D., 1975, Production of hydrogen peroxide by phagocytizing human granulocytes, *J. Lab. Clin. Med.* **85**:198.

Hoogendoorn, H., Piessens, J. P., Scholtes, W., and Stoddard, L. A., 1977, Hypothiocyanite ion: The inhibitor formed by the system lactoperoxidase-thiocyanate-hydrogen peroxide. I. Identification of the inhibiting compound, *Caries Res.* **11**:77.

Howard, D. H., 1973, Fate of *Histoplasma capsulatum* in guinea pig polymorphonuclear leukocytes, *Infect. Immun.* **8**:412.

Iyer, G. Y. N., 1959, Free amino acids in leukocytes from normal and leukemic subjects, *J. Lab. Clin. Med.* **54**:229.

Iyer, G. Y. N., Islam, D. M. F., and Quastel, J. H., 1961, Biochemical aspects of phagocytosis, *Nature* **192**:535.

Jacobs, A. A., Low, I. E., Paul, B. B., Strauss, R. R., and Sbarra, A. J., 1972, Mycoplasmacidal activity of peroxidase-H_2O_2-halide systems, *Infect. Immun.* **5**:127.

Jago, G. R., and Morrison, M., 1962, Antistreptococcal activity of lactoperoxidase III, *Proc. Soc. Exp. Biol. Med.* **111**:585.

Jandl, R. C., André-Schwartz, J., Borges-Dubois, L., Kipnes, R. S., McMurrich, B. J., and Babior, B. M., 1978, Termination of the respiratory burst in human neutrophils, *J. Clin. Invest.* **61**:1176.

Jensen, M. S., and Bainton, D. F., 1973, Temporal changes in pH within the phagocytic vacuole of the polymorphonuclear neutrophilic leukocyte, *J. Cell Biol.* **56**:379.

Johnston, R. B., Jr., and Baehner, R. L., 1970, Improvement of leukocyte bactericidal activity in chronic granulomatous disease, *Blood* **35**:350.

Johnston, R. B., Jr., Keele, B. B., Jr., Misra, H. P., Lehmeyer, J. E., Webb, L. S., Baehner, R. L., and Rajagopalan, K. V., 1975, The role of superoxide anion generation in phagocytic bactericidal activity. Studies with normal and chronic granulomatous disease leukocytes, *J. Clin. Invest.* **55**:1357.

Kajiwara, T., and Kearns, D. R., 1973, Direct spectroscopic evidence for a deuterium solvent effect on the lifetime of singlet oxygen in water, *J. Am. Chem. Soc.* **95**:5886.

Kakinuma, K., 1970, Metabolic control and intracellular pH during phagocytosis of polymorphonuclear leucocytes, *J. Biochem.* **68**:177.

Kaplan, E. L., Laxdal, T., and Quie, P. G., 1968, Studies of polymorphonuclear leukocytes from patients with chronic granulomatous disease of childhood: Bactericidal capacity for streptococci, *Pediatrics* **41**:591.

Kasha, M., and Khan, A. U., 1970, The physics, chemistry and biology of singlet molecular oxygen, *Ann. N.Y. Acad. Sci.* **171**:5.

Kearns, D. R., 1971, Physical and chemical properties of singlet molecular oxygen, *Chem. Rev.* **71**:395.

Kellogg, E. W., III, and Fridovich, I., 1975, Superoxide, hydrogen peroxide and singlet oxygen in lipid peroxidation by a xanthine oxidase system. *J. Biol. Chem.* **250**:8812.

Khan, A. U., 1970, Singlet molecular oxygen from superoxide anion and sensitized fluorescence of organic molecules, *Science* **168**:476.

Khan, A. U., 1977, Theory of electron transfer generation and quenching of singlet oxygen [$^1\Sigma g^+$ and $^1\Delta g$] by superoxide anion. The role of water in the dismutation of O_2^-, *J. Am. Chem. Soc.* **99**:370.

King, M. M., Lai, E. K., and McCay, P. B., 1975, Singlet oxygen production associated with enzyme-catalyzed lipid peroxidation in liver microsomes, *J. Biol. Chem.* **250**:6496.

Klebanoff, S. J., 1967a, A peroxidase-mediated antimicrobial system in leukocytes, *J. Clin. Invest.* **46**:1078.

Klebanoff, S. J., 1967b, Iodination of bacteria: A bactericidal mechanism, *J. Exp. Med.* **126**:1063.

Klebanoff, S. J., 1968, Myeloperoxidase-halide-hydrogen peroxide antimicrobial system, *J. Bacteriol.* **95**:2131.

Klebanoff, S. J., 1969, Antimicrobial activity of catalase at acid pH, *Proc. Soc. Exp. Biol. Med.* **132**:571.

Klebanoff, S. J., 1970a, Myeloperoxidase: Contribution to the microbicidal activity of intact leukocytes, *Science* **169**:1095.

Klebanoff, S. J., 1970b, Myeloperoxidase-mediated antimicrobial systems and their role in leukocyte function, in: *Biochemistry of the Phagocytic Process: Localization and the Role of Myeloperoxidase and the Mechanism of the Halogenation Reaction* (J. Schultz, ed.), pp. 89–110, North-Holland, Amsterdam.

Klebanoff, S. J., 1974, Role of the superoxide anion in the myeloperoxidase-mediated antimicrobial system, *J. Biol. Chem.* **249**:3724.

Klebanoff, S. J., 1975a, Antimicrobial mechanisms in neutrophilic polymorphonuclear leukocytes, *Semin. Hematol.* **12**:117.

Klebanoff, S. J., 1975b, Antimicrobial systems of the polymorphonuclear leukocyte, in: *The Phagocytic Cell in Host Resistance* (J. A. Bellanti and D. H. Dayton, eds.), pp. 45–59, Raven Press, New York.

Klebanoff, S. J., 1979, Effect of estrogens on the myeloperoxidase-mediated antimicrobial system, *Infect. Immun.* **25**:153.

Klebanoff, S. J., and Belding, M. E., 1974, Virucidal activity of H_2O_2-generating bacteria: Requirement for peroxidase and a halide, *J. Infect. Dis.* **129**:345.

Klebanoff, S. J., and Clark, R. A., 1975, Hemolysis and iodination of erythrocyte components by a myeloperoxidase-mediated system, *Blood* **45**:699.

Klebanoff, S. J., and Clark, R. A., 1977, Iodination by human polymorphonuclear leukocytes: A re-evaluation, *J. Lab. Clin. Med.* **89**:675.

Klebanoff, S. J., and Clark, R. A., 1978, *The Neutrophil: Function and Clinical Disorders*, North-Holland, Amsterdam.

Klebanoff, S. J., and Green, W. L., 1973, Degradation of thyroid hormones by phagocytosing human leukocytes, *J. Clin. Invest.* **52**:60.

Klebanoff, S. J., and Hamon, C. B., 1972, Role of myeloperoxidase-mediated antimicrobial systems in intact leukocytes, *J. Reticuloendothel. Soc.* **12**:170.

Klebanoff, S. J., and Luebke, R. G., 1965, The antilactobacillus system of saliva. Role of salivary peroxidase, *Proc. Soc. Exp. Biol. Med.* **118**:483.

Klebanoff, S. J., and Pincus, S. H., 1971, Hydrogen peroxide utilization in myeloperoxidase-deficient leukocytes: A possible microbicidal control mechanism, *J. Clin. Invest.* **50**:2226.

Klebanoff, S. J., and Smith, D. C., 1970a, Peroxidase-mediated antimicrobial activity of rat uterine fluid, *Gynec. Invest.* **1**:21.

Klebanoff, S. J., and Smith, D. C., 1970b, The source of H_2O_2 for the uterine fluid-mediated sperm-inhibitory system, *Biol. Reprod.* **3**:236.

Klebanoff, S. J., and White, L. R., 1969, Iodination defect in the leukocytes of a patient with chronic granulomatous disease of childhood, *N. Engl. J. Med.* **280**:460.

Klebanoff, S. J., Clem, W. H., and Luebke, R. G., 1966, The peroxidase-thiocyanate-hydrogen peroxide antimicrobial system, *Biochim. Biophys. Acta* **117**:63.

Klebanoff, S. J., Clark, R. A., and Rosen, H., 1976, Myeloperoxidase-mediated cytotoxicity, in: *Cancer Enzymology* (J. Schultz and F. Ahmad, eds.), pp. 267–288, Academic Press, New York.

Klebanoff, S. J., Rosen, H., and Clark, R. A., 1977, Formation of singlet oxygen by the myeloperoxidase-mediated antimicrobial system, in: *Movement, Metabolism and Bactericidal Mechanisms of Phagocytes* (F. Rossi, P. L. Patriarca, and D. Romeo, eds.), pp. 295–305, Piccin, Padua.

Knox, W. E., Stumpf, P. K., Green, D. E., and Auerbach, V. H., 1948, The inhibition of sulfhydryl enzymes as the basis of the bactericidal action of chlorine, *J. Bacteriol* **55**:451.

Kobayashi, S., and Ando, W., 1979, Co-oxidation of 1,3-diphenylisobenzofuran by the Haber-Weiss reaction: Is singlet oxygen concerned in this oxidation?, *Biochem. Biophys. Res. Commun.* **88**:676.

Koch, C., 1974a, Neutrophil granulocyte function *in vitro*. Evaluation of fluid-phase leukocyte-bacteria reaction system, *Acta Pathol. Microbiol. Scand.* **82**:127.

Koch, C., 1974b, Effect of sodium azide upon normal and pathological granulocyte function, *Acta Pathol. Microbiol. Scand.* **82**:136.

Kojima, S., 1931, Studies on peroxidase. II. The effect of peroxidase on the bactericidal action of phenols, *J. Biochem.* **14**:95.

Krinsky, N. I., 1974, Singlet excited oxygen as a mediator of the antibacterial action of leukocytes, *Science* **186**:363.

Lee, D. C.-S., and Wilson, T., 1973, Oxygen in chemiluminescence. A competitive pathway of dioxetane decomposition catalyzed by electron donors, in: *Chemiluminescence and Bioluminescence* (M. J. Cormier, D. M. Hercules, and J. Lee, eds.), pp. 265–283, Plenum Press, New York.

Lehrer, R. I., 1969, Antifungal effects of peroxidase systems, *J. Bacteriol.* **99**:361.

Lehrer, R. I., 1970, Measurement of candidacidal activity of specific leukocyte types in mixed cell populations. I. Normal, myeloperoxidase-deficient, and chronic granulomatous disease neutrophils, *Infect. Immun.* **2**:42.

Lehrer, R. I., 1971, Inhibition by sulfonamides of the candidacidal activity of human neutrophils, *J. Clin. Invest.* **50**:2498.

Lehrer, R. I., 1972, Functional aspects of a second mechanism of candidacidal activity by human neutrophils, *J. Clin. Invest.* **51**:2566.

Lehrer, R. I., and Cline, M. J., 1969, Leukocyte myeloperoxidase deficiency and disseminated candidiasis: The role of myeloperoxidase in resistance to *Candida* infection, *J. Clin. Invest.* **48**:1478.

Lehrer, R. I., and Jan, R. G., 1970, Interaction of *Aspergillus fumigatus* spores with human leukocytes and serum, *Infect. Immun.* **1**:345.

Lehrer, R. I., Hanifin, J., and Cline, M. J., 1969, Defective bactericidal activity in myeloperoxidase deficient human neutrophils, *Nature* **223**:78.

Mandell, G. L., 1970, Intraphagosomal pH of human polymorphonuclear neutrophils, *Proc. Soc. Exp. Biol. Med.* **134**:447.

Mandell, G. L., 1974, Bactericidal activity of aerobic and anaerobic polymorphonuclear neutrophils, *Infect. Immun.* **9**:337.

Mandell, G. L., 1975, Catalase, superoxide dismutase, and virulence of *Staphylococcus aureus*. In vitro and *in vivo* studies with emphasis on staphylococcal-leukocyte interaction, *J. Clin. Invest.* **55**:561.

Mandell, G. L., and Hook, E. W., 1969, Leukocyte bactericidal activity in chronic granulomatous disease: Correlation of bacterial hydrogen peroxide production and susceptibility to intracellular killing, *J. Bacteriol* **100**:531.

Massey, V., Strickland, S., Mayhew, S. G., Howell, L. G., Engel, P. C., Matthews, R. G., Schuman, M., and Sullivan, P. A., 1969, The production of superoxide anion radicals in the reaction of reduced flavins and flavoproteins with molecular oxygen, *Biochem. Biophys. Res. Commun.* **36**:891.

Matheson, N. R., Wong, P. S., and Travis, J., 1979, Enzymatic inactivation of human alpha-1-proteinase inhibitor by neutrophil myeloperoxidase, *Biochem. Biophys. Res. Commun.* **88**:402.

McCall, C. E., DeChatelet, L. R., Cooper, M. R., and Ashburn, P., 1971, The effects of ascorbic acid on bactericidal mechanisms of neutrophils, *J. Infect. Dis.* **124**:194.

McCord, J. M., and Day, E. D. Jr., 1978, Superoxide-dependent production of hydroxyl radical catalyzed by iron-EDTA complex, *FEBS Lett.* **86**:139.

McLeod, J. M., and Gordon, J., 1922, Production of hydrogen peroxide by bacteria, *Biochem. J.* **16**:499.

McRipley, R. J., and Sbarra, A. J., 1967a, Role of the phagocyte in host-parasite interactions. XI. Relationship between stimulated oxidative metabolism and hydrogen peroxide formation, and intracellular killing, *J. Bacteriol.* **94**:1417.

McRipley, R. J., and Sbarra, A. J., 1967b, Role of the phagocyte in host-parasite interactions. XII. Hydrogen peroxide-myeloperoxidase bactericidal system in the phagocyte, *J. Bacteriol.* **94**:1425.

Merkel, P. B., Nilsson, R., and Kearns, D. R., 1972, Deuterium effects on singlet oxygen lifetimes in solutions. A new test of singlet oxygen reactions, *J. Am. Chem. Soc.* **94**:1030.

Michell, R. H., Karnovsky, M. J., and Karnovsky, M. L., 1970, The distribution of granule-associated enzymes in guinea pig polymorphonuclear leucocytes, *Biochem. J.* **116**:207.

Moosmann, K., and Bojanovsky, A., 1975, Rezidivierende Candidosis bei Myeloperoxydasemangel, *Mschr. Kinderheilk.* **123**:408.

Morrison, M., Allen, P. Z., Bright, J., and Jayasinghe, W., 1965, Lactoperoxidase. V. Identification and isolation of lactoperoxidase from salivary gland. *Arch. Biochem. Biophys.* **111**:126.

Naskalski, J. W., 1977, Myeloperoxidase inactivation in the course of catalysis of chlorination of taurine, *Biochim. Biophys. Acta* **485**:291.

Nishimura, E. T., Whest, G. M., and Yang, H-Y., 1976, Ultrastructural localization of peroxidatic catalase in human peripheral blood leukocytes, *Lab. Invest.* **34**:60.

Nour-Eldin, F., and Wilkinson, J. F., 1955, Amino-acid content of white blood cells in human leukaemias, *Br. J. Haematol.* **1**:358.

Patriarca, P., Cramer, R., Dri, P., Fant, L., Basford, R. E., and Rossi, F., 1973, NADPH oxidizing activity in rabbit polymorphonuclear leukocytes: Localization in azurophil granules, *Biochem. Biophys. Res. Commun.* **316**:830.

Patriarca, P., Cramer, R., Tedesco, F., and Kakinuma, K., 1975, Studies on the mechanism of metabolic stimulation in polymorphonuclear leukocytes during phagocytosis. II. Presence of the $NADPH_2$ oxidizing activity in a myeloperoxidase-deficient subject, *Biochim. Biophys. Acta* **385**:387.

Paul, B. B., Jacobs, A. A., Strauss, R. R., and Sbarra, A. J., 1970, Role of the phagocyte in host-parasite interactions. XXIV. Aldehyde generation by the myeloperoxidase-H_2O_2-chloride antimicrobial system: A possible *in vivo* mechanism of action, *Infect. Immun.* **2**:414.

Pavlov, E. P., and Solov'ev, V. N., 1967, Changes in the hydrogen ion concentration of cytoplasm during the phagocytosis of microbes stained with indicator dyes, *Byull. Eksp, Biol. Med.* **63**:78.

Philpott, G. W., Bower, R. J., and Parker, C. W., 1973, Selective iodination and cytotoxicity of tumor cells with an antibody-enzyme conjugate, *Surgery* **74**:51.

Piatt, J. F., and O'Brien, P. J., 1979, Singlet oxygen formation by a peroxidase, H_2O_2, and a halide system, *Eur, J. Biochem.* **93**:323.

Piatt, J. F., Cheema, A. S., and O'Brien, P. J., 1977, Peroxidase catalyzed singlet oxygen formation from hydrogen peroxide, *FEBS Lett.* **74**:251.

Pincus, S. H., and Klebanoff, S. J., 1971, Quantitative leukocyte iodination, *N. Engl. J. Med.* **284**:744.

Pitt, J., and Bernheimer, H. P., 1974, Role of peroxide in phagocytic killing of pneumococci, *Infect. Immun.* **9**:48.

Portmann, A., and Auclair, J. E., 1959, Relation entre la lacténine L_2 et la lactoperoxydase, *Lait* **39**:147.

Qualliotine, D., DeChatelet, L. R., McCall, C. E., and Cooper, M. R., 1972, Effect of catecholamines on the bactericidal activity of polymorphonuclear leukocytes, *Infect. Immun.* **6**:211.

Quie, P. G., White, J. G., Holmes, B., and Good, R. A., 1967, *In vitro* bactericidal capacity of human polymorphonuclear leukocytes: Diminished activity in chronic granulomatous disease in childhood, *J. Clin. Invest.* **46**:668.

Reiter, B., Pickering, A., and Oram, J. D., 1964, An inhibitory system—lactoperoxidase/thiocyanate/peroxide—in raw milk, in: *Proceedings of the Fourth International Symposium on Food Microbiology*, pp. 297–305, Göteborg, Sweden.

Rohrer, G. F., von Wartburg, J. P., and Aebi, H., 1966, Myeloperoxidase aus menschlichen Leukozyten. I. Isolierung and Charakterisierung des Enzymes, *Biochem. Z.* **344**:478.

Root, R. K., 1974, Correction of the function of chronic granulomatous disease (CGD) granulocytes (PMN) with extracellular H_2O_2, *Clin. Res.* **22**:452A.

Root, R. K., and Metcalf, J. A., 1977, H_2O_2 release from human granulocytes during phagocytosis. Relationships to superoxide anion formation and cellular catabolism of H_2O_2: Studies with normal and cytochalasin B-treated cells, *J. Clin. Invest.* **60**:1266.

Root, R. K., and Stossel, T. P., 1974, Myeloperoxidase-mediated iodination by granulocytes. Intracellular site of operation and some regulating factors, *J. Clin. Invest.* **53**:1207.

Rosen, H., and Klebanoff, S. J., 1976, Chemiluminescence and superoxide production by myeloperoxidase-deficient leukocytes, *J. Clin. Invest.* **58**:50.

Rosen, H., and Klebanoff, S. J., 1977, Formation of singlet oxygen by the myeloperoxidase-mediated antimicrobial system, *J. Biol. Chem.* **252**:4803.

Rosen, H., and Klebanoff S. J., 1979, Bactericidal activity of a superoxide anion-generating system. A model for the polymorphonuclear leukocyte, *J. Exp. Med.* **149**:27.

Rossi, F., and Zatti, M., 1964, Changes in the metabolic pattern of polymorphonuclear leucocytes during phagocytosis, *Br. J. Exp. Pathol.* **45**:548.

Rous, P., 1925a, The relative reaction within living mammalian tissues. I. General features of vital staining with litmus, *J. Exp. Med.* **41**:379.

Rous, P., 1925b, The relative reaction within living mammalian tissues. II. On the mobilization of acid material within cells, and the reaction as influenced by the cell state, *J. Exp. Med.* **41**:399.

Sbarra, A. J., and Karnovsky, M. L., 1959, The biochemical basis of phagocytosis. I. Metabolic changes during the ingestion of particles by polymorphonuclear leukocytes, *J. Biol. Chem.* **234**:1355.

Sbarra, A. J., Selvaraj, R. J., Paul, B. B., Mitchell, G. W., Jr., and Louis, F., 1977, Some newer insights of the peroxidase mediated antimicrobial system, in: *Movement, Metabolism and Bactericidal Mechanisms of Phagocytes* (F. Rossi, P. L. Patriarca, and D. Romeo, eds.), pp. 295–304, Piccin, Padua.

Schmid, L., and Brune, K., 1974, Assessment of phagocytic and antimicrobial activity of human granulocytes, *Infect. Immun.* **10**:1120.

Schultz, J., and Kaminker, K., 1962, Myeloperoxidase of the leucocyte of normal human blood. I. Content and localization, *Arch. Biochem. Biophys.* **96**:465.

Segal, A. W., and Peters, T. J., 1977, Analytical subcellular fractionation of human granulocytes with special reference to the localization of enzymes involved in microbicidal mechanisms, *Clin. Sci. Mol. Med.* **52**:429.

Selvaraj, R. J., Paul, B. B., Strauss, R. R., Jacobs, A. A., and Sbarra, A. J., 1974, Oxidative peptide cleavage and decarboxylation by the $MPO-H_2O_2-Cl^-$ antimicrobial system, *Infect. Immun.* **9**:255.

Selvaraj, R. J., Zgliczyński, J. M., Paul, B. B. and Sbarra, A. J., 1978, Enhanced killing of myeloperoxidase-coated bacteria in the myeloperoxidase-H_2O_2-Cl^- system, *J. Infect. Dis.* **137**:481.

Siegel, E., and Sachs, B. A., 1964, *In vitro* leukocyte uptake of ^{131}I labeled iodide, thyroxine and triiodothyronine, and its relation to thyroid function, *J. Clin. Endocrinol.* **24**:313.

Slowey, R. R., Eidelman, S., and Klebanoff, S. J., 1968, Antibacterial activity of the purified peroxidase from human parotid saliva, *J. Bacteriol.* **96**:575.

Smith, D. C., and Klebanoff, S. J., 1970, A uterine fluid-mediated sperm-inhibitory system, *Biol. Reprod.* **3**:229.

Spicer, S. S., and Hardin, J. H., 1969, Ultrastructure, cytochemistry, and function of neutrophil leukocyte granules. A review, *Lab. Invest.* **20**:488.

Sprick, M. G., 1956, Phagocytosis of *M. tuberculosis* and *M. smegmatis* stained with indicator dyes, *Am. Rev. Tuberc.* **74**:552.

Stadhouders, J., and Veringa, H. A., 1962, Some experiments related to the inhibitory action of milk peroxidase on lactic acid streptococci, *Neth. Milk Dairy J.* **16**:96.

Stelmaszyńska, T., and Zgliczyński, J. M., 1978, N-(2-Oxoacyl) amino acids and nitriles as final products of dipeptide chlorination mediated by the myeloperoxidase/H_2O_2/Cl^- system, *Eur. J. Biochem.* **92**:301.

Stendahl, O., and Lindgren, S., 1976, Function of granulocytes with deficient myeloperoxidase-mediated iodination in a patient with generalized pustular psoriasis, *Scand. J. Haematol.* **16**:144.

Stolc, V., 1971, Stimulation of iodoproteins and thyroxine formation in human leukocytes by phagocytosis, *Biochem. Biophys. Res. Commun.* **45**:159.

Stossel, T. P., Pollard, T. D., Mason, R. J., and Vaughan, M., 1971, Isolation and properties of phagocytic vesicles from polymorphonuclear leukocytes, *J. Clin. Invest.* **50**:1745.

Stossel, T. P., Root, R. K., and Vaughan, M., 1972, Phagocytosis in chronic granulomatous disease and the Chediak-Higashi syndrome, *N. Engl. J. Med.* **286**:120.

Strauss, R. R., Paul, B. B., Jacobs, A. A., and Sbarra, A. J., 1971, Role of the phagocyte in host-parasite interactions. XXVII. Myeloperoxidase-H_2O_2-Cl-mediated aldehyde formation and its relationship to antimicrobial activity, *Infect. Immun.* **3**:595.

Tagesson, C., and Stendahl, O., 1973, Influence of the cell surface lipopolysaccharide structure of *Salmonella typhimurium* on resistance to intracellular bactericidal systems, *Acta Pathol. Microbiol. Scand.* **81**:473.

Takanaka, K., and O'Brien, P. J., 1975, Mechanisms of H_2O_2 formation of leukocytes. Evidence for a plasma membrane location, *Arch. Biochem. Biophys.* **169**:428.

Thomas, E. L., 1979a, Myeloperoxidase, hydrogen peroxide, chloride antimicrobial system: Nitrogen-chlorine derivatives of bacterial components in bactericidal action against *Escherichia coli*, *Infect. Immun.* **23**:522.

Thomas, E. L., 1979b, Myeloperoxidase-hydrogen peroxide-chloride antimicrobial system: Effect of exogenous amines on antibacterial action against *Escherichia coli*, *Infect. Immun.* **25**:110.

Thomas, E. L., and Aune, T. M., 1977, Peroxidase-catalyzed oxidation of protein sulfhydryls mediated by iodine, *Biochemistry* **16**:3581.

Thomas, E. L., and Aune, T. M., 1978a, Cofactor role of iodide in peroxidase antimicrobial action against *Escherichia coli*, *Antimicrob. Agents Chemother.* **13**:1000.

Thomas, E. L., and Aune, T. M., 1978b, Oxidation of *Escherichia coli* sulfhydryl components by the peroxidase-hydrogen peroxide-iodide antimicrobial system, *Antimicrob. Agents Chemother.* **13**:1006.

Undritz, E., 1966, Die Alius-Grignaschi-Anamolie: Der erblich-konstitutionelle Peroxydasedefekt der Neutrophilen und Monozyten, *Blut* **14**:129.

Weening, R. S., Wever, R., and Roos, D., 1975, Quantitative aspects of the production of superoxide radicals by phagocytizing human leukocytes, *J. Lab. Clin. Med.* **85**:245.

Whittenbury, R., 1964, Hydrogen peroxide formation and catalase activity in lactic acid bacteria, *J. Gen. Microbiol.* **35**:13.

Wilson, D. L., and Manery, J. F., 1949, The permeability of rabbit leucocytes to sodium, potassium and chloride, *J. Cell. Comp. Physiol.* **34**:493.

Wilson, T., and Hastings, J. W., 1970, Chemical and biological aspects of singlet excited molecular oxygen, *Photophysiology* **5**:49.

Woeber, K. A., and Ingbar, S. H., 1973, Metabolism of L-thyroxine by phagocytosing human leukocytes, *J. Clin. Invest.* **52**:1796.

Wright, R. C., and Tramer, J., 1958, Factors influencing the activity of cheese starters. The role of milk peroxidase, *J. Dairy Res.* **25**:104.

Yamazaki, I., and Yokota, K-N., 1973, Oxidation states of peroxidase, *Mol. Cell. Biochem.* **2**:39.

Zatti, M., Rossi, F., and Patriarca, P., 1968, The H_2O_2-production by polymorphonuclear leucocytes during phagocytosis, *Experientia* **24**:669.

Zeldow, B. J., 1959, Studies on the antibacterial activity of human saliva. I. A bactericidin for lactobacilli. *J. Dent. Res.* **38**:798.

Zeldow, B. J., 1963, Studies on the antibacterial action of human saliva. III. Cofactor requirements of a *Lactobacillus* bactericidin, *J. Immunol.* **90**:12.

Zgliczyński, J. M., and Stelmaszyńska, T., 1975, Chlorinating ability of human phagocytosing leucocytes, *Eur. J. Biochem.* **56**:157.

Zgliczyński, J. M., and Stelmaszyńska, T., 1979, Hydrogen cyanide and cyanogen chloride formation by the myeloperoxidase-H_2O_2-Cl^- system, *Biochim. Biophys. Acta* **567**:309.

Zgliczyński, J. M., Stelmaszyńska, T., Ostrowski, W., Naskalski, J., and Sznajd, J., 1968, Myeloperoxidase of human leukemic leucocytes. Oxidation of amino acids in the presence of hydrogen peroxide, *Eur. J. Biochem.* **4**:540.

Zgliczyński, J. M., Stelmaszyńska, T., Domański, J., and Ostrowski, W., 1971, Chloramines as intermediates of oxidative reaction of amino acids by myeloperoxidase, *Biochim. Biophys. Acta* **235**:419.

Zgliczyński, J. M., Selvaraj, R. J., Paul, B. B., Stelmaszyńska, T., Poskitt, P. K. F., and Sbarra, A. J., 1977, Chlorination by the myeloperoxidase-H_2O_2-Cl^- antimicrobial system at acid and neutral pH, *Proc. Soc. Exp. Biol. Med.* **154**:418.

12

Free-Radical Production by Reticuloendothelial Cells

ROBERT C. ALLEN

> One of the principle objects of theoretical research in any department of knowledge is to find the point of view from which the subject appears in its greatest simplicity.
>
> —J. W. Gibbs, in a letter to the American Academy of Arts and Sciences (1881)

1. INTRODUCTION

Questions concerning metabolism imply interest in the pattern of change or dynamics of a system. Energy is required for change, and the pattern of change is determined by how energy is spent. The microbicidal function of the polymorphonuclear leukocyte (PMN) is a dynamic process, and will be considered from the viewpoint of greatest simplicity, energetics. Special emphasis will be placed on transducing mechanisms whereby energy is converted and directed so as to effect microbicidal action.

    The first transduction to be considered is the conversion of carbohydrate bond energy to the potential of reducing equivalents. These reducing equivalents are used in univalent redox reactions resulting in the generation of free radicals. Consequently, a general explanation of free radical chemistry will be given. Special emphasis will be directed to electronic manipulations resulting in the generation of elecronically excited singlet molecular oxygen. Such reactions are best understood through the concept of disproportionation, and, therefore, this concept will be fully explained. The participation of various oxygen species in microbicidal action will be considered, and the mechanism whereby electronically excited carbonyl chromophores are generated will be discussed. The final

ROBERT C. ALLEN • U.S. Army Institute of Surgical Research and Clinical Investigation Service, Brooke Army Medical Center, Fort Sam Houston, Texas 78234. The opinions or assertions contained herein are the private views of the author and are not to be construed as reflecting the views of the Department of the Army or the Department of Defense.

transduction involves the relaxation of electronically excited molecules to their ground state by photon emission. The phenomenon of chemiluminescence is the consequence of this transduction. An elementary understanding of certain basic principles of quantum mechanics is necessary for full appreciation of the association between chemiluminescence and oxidative microbicidal action. These principles will be presented in a simplified manner that does not assume a prior familiarity with this field.

The following is not a complete review of the literature. Instead, it will attempt a unified approach to the subject of microbicidal energetics. The relationship of chemiluminescence to microbicidal action will be fully considered. The author's major objective is to broaden the outlook of the reader, and hopefully, to stimulate further questioning and investigation of the fundamentals of oxidative microbicidal action.

2. CARBOHYDRATE BOND ENERGY AND REDUCING EQUIVALENTS

In 1933 Baldridge and Gerard reported that phagocytosis by PMN resulted in an increased rate of O_2 consumption over a relatively short period of time. Twenty-three years later Stahelin *et al.* (1956) reported a curious observation regarding the pathway of glucose metabolism in PMN. Phagocytosis resulted in an increase in glucose oxidation by the dehydrogenases of the hexose monophosphate (HMP) shunt (Stahelin *et al.*, 1956; Sbarra and Karnovsky, 1959; Rossi and Zatti, 1964, 1966; Morton *et al.* 1969). These postphagocytic phenomena, increased O_2 consumption and increased HMP shunt activity, were associated with a third observation, the postphagocytic generation of hydrogen peroxide (Iyer, *et al.*, 1961; Zatti *et al.*, 1968; Paul and Sbarra, 1968).

Several mechanisms have been proposed to explain these phenomena (Evans and Karnovsky, 1962; Karnovsky *et al.*, 1971); however, the most satisfactory mechanism is that put forward by Rossi *et al.* (1972). These investigators reported the presence of both NADH oxidase and NADPH oxidase in the granular (20,000g) fraction of PMN. Both cyanide-insensitive oxidases are activated by phagocytosis, as reflected by decrease in the Michaelis constant (K_m). However, NADPH oxidase is by far the more active enzyme with a K_m tenfold less than NADH oxidase. Oxidase activation is measurable within 30 sec after addition of opsonized bacteria to PMN.

Let us now consider the preceding observations from an energetic viewpoint. In essence, reducing equivalents, in the form of NADPH, are mobilizied by oxidation of carbohydrate via the dehydrogenases of the HMP shunt.

$$\text{Glucose 6-phosphate} + \text{NADP}^+ \xrightarrow{\text{Dehydrogenase}} \text{6-Phosphogluconate} + \text{NADPH} + \text{H}^+ \quad (1)$$

$$\text{6-Phosphogluconate} + \text{NADP}^+ \xrightarrow{\text{Dehydrogenase}} \text{Ribulose 5-phosphate} + CO_2 + \text{NADPH} + \text{H}^+ \quad (2)$$

Net reaction:

$$\text{Glucose 6-phosphate} \xrightarrow{\hspace{3cm}} \text{Ribulose 5-phosphate} + CO_2$$
$$+ 2NADP^+ \hspace{4cm} + 2NADPH + 2H^+ \hspace{1cm} (3)$$

For each CO_2 liberated, four reducing equivalents are mobilized. This transduction also serves to give direction to the utilization of energy in that NADPH, the carrier of reducing equivalents, is the substrate for the oxidase. This enzyme effects the next step in the sequence of reactions resulting in microbicidal action.

3. IMPORTANCE OF NADPH OXIDASE ACTIVATION

The kinetic properties of the dehydrogenases of the HMP shunt are not changed with phagocytosis (Stjernholm, 1968). The increase in postphagocytic HMP shunt activity results from the change in ratio of oxidized to reduced nucleotide ($NADP^+$/NADPH). Activated NADPH oxidase effects this change by spending the reducing potential of NADPH, and in so doing, generates the rate limiting $NADP^+$. Patriarca et al. (1971a) have reported a change in the $NADP^+$/NADPH ratio from 1.10/9.57 (values in nanomoles per 10^8 PMN) in the resting state to 2.27/7.38 in the postphagocytic state.

Alteration of the plasma membrane of the PMN is required for activation of NADPH oxidase (Selvaraj and Sbarra, 1966; Zatti and Rossi, 1967; Rossi et al., 1971; Patriarca et al., 1971b). It is reasonable to assume that this oxidase resides in the plasma membrane. This assumption is consistent with the conclusion drawn by Salin and McCord (1974), based on their studies of superoxide ($O_2^{\cdot -}$) generation of PMN.

Oxidases employing NADH and NADPH as their source of reducing equivalent are typically flavoproteins, and although NADPH oxidase of PMN is poorly characterized with respect to prosthetic group, the assumption that this enzyme is a flavoprotein is consistent with its substrate requirement and its proposed mechanism of action in the generation of $O_2^{\cdot -}$ (Allen et al., 1973, 1974; Johnston et al., 1975; Babior et al., 1975, 1976).

Consider the standard reduction potential (E_0') of the redox couples involved in the NADPH oxidase reactions. The E_0' is actually the midpoint potential and is related to the observed reduction potential (E_h) in a manner analogous to the Henderson–Hasselbalch relationship of the pK of an acid to its observed pH. The E_0' is equal to -0.32 volt for the NADPH/$NADP^+$ couple, and is in the range of -0.28 to -0.01 for flavoprotein H_2/flavoprotein couples. The change in E_0' ($\Delta E_0'$) is therefore in the range of 0.04–0.31 volt (v). The $\Delta E_0'$ is related to the change in standard free energy ($\Delta G_0'$) by the equation, $\Delta G_0' = -nF\Delta E_0'$, where n is the number of electrons transferred and F is the caloric equivalent of a faraday, approximately 23 kcal (Clark, 1960; Mahler and Cordes, 1971). Therefore, the oxidase reaction would be thermodynamically allowed with a possible $\Delta G_0'$ in the range of -1.8 to -14.3 kcal/mol.

Mechanistically, flavoproteins have a relatively unique redox character.

They are stable in the semiquinone or univalently ($1e^-$ equivalent) reduced state. As a semiquinone, the flavoprotein is a stable free radical. Many flavoprotein enzymes have been reported to produce the superoxide radical ($O_2^{\cdot-}$) as a reaction product (Knowles *et al.*, 1969; Massey *et al.*, 1969). An appreciation of the semiquinone concept is important to the mechanistic understanding of how flavoproteins catalyze the univalent reduction of O_2 to $O_2^{\cdot-}$.

The reduction of the flavoprotein by NADPH appears to involve "the direct transfer of a substrate hydrogen plus two electrons in a rapid step" (Mahler and Cordes, 1971). If the oxidase contains two flavins per enzyme (E-2flav), reduction by NADPH plus H^+ might generate an enzyme containing two flavin semiquinones (2H-E-2flav$^{\cdot}$), and the univalent reduction of two O_2 would yield two $O_2^{\cdot-}$ plus two H^+.

$$\text{E-2flav} + \text{NADPH} + H^+ \longrightarrow 2H\text{-E-2flav}^{\cdot} + \text{NADP}^+ \tag{4}$$

$$2H\text{-E-2flav}^{\cdot} + O_2 \longrightarrow H\text{-E-(flav)-flav}^{\cdot} + O_2^{\cdot-} + H^+ \tag{5}$$

$$H\text{-E-(flav)-flav}^{\cdot} + O_2 \longrightarrow \text{E-2flav} + O_2^{\cdot-} + H^+ \tag{6}$$

Net reaction:

$$\text{NADPH} + H^+ + 2O_2 \longrightarrow \text{NADP}^+ + 2H^+ + 2O_2^{\cdot-} \tag{7}$$

The $O_2^{\cdot-}$ may be thought of as an intermediate redox state between O_2 and H_2O_2. This is analogous to the relationship of the semiquinone to the fully oxidized and reduced state of the flavin. The $O_2^{\cdot-}$ is also analogous to semiquinone in that it is a relatively stable free radical.

An additional function for NADPH oxidase may be the acidification of the phagocytic vacuole or extracellular space by a proton pump mechanism (Allen, 1979). The stoichiometry of the net NADPH oxidase reaction describes the generation of $2H^+$ as well as $2O_2^{\cdot-}$. Accordingly, the electrons necessary for O_2 reduction, and the H^+ necessary for acidification of the vacuole, originate from the metabolism of glucose via the hexose monophosphate shunt. As such, activation of the oxidase provides a dynamic means for removing metabolically generated protons from the cytoplasm by linking this process with the extracytoplasmic generation of $O_2^{\cdot-}$.

One possibility is that the perhydroxyl (hydrodioxyl) radical, $HO_2^{\cdot}$, is the intermediate or membrane-associated product of oxidase activity. Dissociation of this acid is governed by a pK_a of 4.8 (Behar *et al.*, 1970). Discharge of $HO_2^{\cdot}$ into an environment of approximate neutrality would insure dissociation to yield H^+ with $O_2^{\cdot-}$ as the conjugate base. Note that electrical neutrality is maintained. The uncharged $HO_2^{\cdot}$ would provide a means for membrane transit, and its subsequent dissociation would provide a mechanism for ion trapping of $O_2^{\cdot-}$.

4. FREE RADICALS

The word *radical* is derived from the Latin word *radix* meaning root. In the terminology of organic chemistry it has come to mean that portion of a molecule that is replaceable by a single atom, or that portion of a molecule that remains unchanged during chemical reaction. It is usually represented by the letter R.

For example, a common general abbreviation for alcohols is R-OH, where R represents the radical (e.g., R: CH_3, C_2H_5, C_3H_7) and OH represents the reactive or functional group. A free radical in the original sense meant a radical free of its functional group (Pryor, 1976). Consider the hypothetical example R-OH; if bond cleavage were to occur by a homolytic mechanism, the resulting products would be R· and OH·. There is a departure from the original terminology in that both R· and OH· are considered to be free radicals.

In the modern terminology, a *free radical* is defined as an atom or molecule having at least one unpaired electron. It may be neutral, positive, or negative in charge, and its reactivity may vary from relatively stable, such as the flavoprotein semiquinone, to extremely reactive, such as the hydroxyl radical (OH·). In the following discussion, the term *radical* will be used to imply free radical unless stated otherwise.

The most feasible method for radical generation in biological chemistry is through univalent redox reactions. Two such reactions, the univalent reduction of an oxidized flavin to its semiquinone and the univalent reduction of O_2 to O_2^-, have already been considered. The O_2 that we breath is actually the most ubiquitous of all biologically important radicals. In its ground state O_2 has two unpaired electrons each filling a different orbital. Consequently O_2 has paramagnetic and diradical properties.

As will be subsequently discussed, the electronic excitation of a molecule usually results from promotion of an electron from its ground state orbital to an orbital of higher energy. This may effect unpairing of electrons, and consequently, electronically excited molecules usually have radical character.

5. FUNDAMENTAL PRINCIPLES OF QUANTUM MECHANICS

The nature of the material to be subsequently discussed requires that a brief presentation of some basic quantum principles be given. The following section is not meant to be a complete discussion of this subject, but rather is intended to introduce the reader to the terminology and fundamental principles necessary for the future description of electronic excitation phenomena and chemiluminescence. More complete treatments of this subject are available elsewhere (Gray, 1965; Phillips, 1965; Turro, 1965; Orchin and Jaffé, 1967, 1971).

At the turn of the century Planck proposed that radiation occurs as discrete quantities of energy, which he termed *quanta*. The energy, E, of the individual quantum could be expressed by the equation, $E = h\nu$, where ν represents the frequency, and h is a proportionality constant. Five years later, Einstein proposed the theory of photoelectric interaction, which states that the energy of light is absorbed as concentrated packets called *photons*. This energy is highly concentrated and can be transferred completely to a single electron. Thus, light appears to exhibit both wave and particle behavior. In a similar manner, electrons can also be described by their wave and particulate behavior. As an electron moves about the nucleus of an atom, it does so in a three-dimensional manner. The amplitude of this movement or *wavefunction* (ϕ), can thus be described using a three dimensional coordinate system, such as the wavefunction, $\phi_{x,y,z}$. These functions describe the *orbitals* of electrons.

A particular wavefunction is specified by its quantum numbers of which there are four. Figure 1 can be used as a tangible guide in describing these quantum numbers. This figure shows the contribution of the orbitals of two atomic oxygens to the molecular bonding of O_2. Each atomic oxygen has five orbital possibilities. These orbitals are schematically represented by the five circles lying to the right or left of, and separated by thin vertical lines from, the ten molecular orbital possibilities of O_2. The following description of quantum numbers will focus attention on the atomic orbitals of oxygen.

The principal quantum number, n, serves to grossly describe the energy level of the orbital. Its value is usually expressed in arabic numerals; the energy content of the level is increased as the value of the number increases. Reference to the orbitals of atomic oxygen demonstrates that the principle quantum numbers 1 and 2 are possible.

The shape of the orbital is determined by the value of the angular momentum or azimuthal quantum number, l. It may have an integer value from 0 to n-1. Where l equals 0, 1, or 2, the orbitals are described by the letters s, p, or d respectively. Thus, for the oxygen atom where n equals 1, and l equals 0,

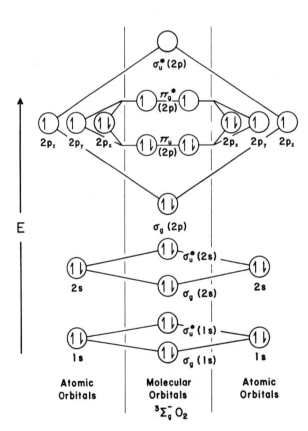

FIGURE 1.   Unhybridized molecular orbital-energy diagram of ground state, triplet O_2 ($^3\Sigma_g^- O_2$). The interaction of the five orbitals of each atomic oxygen, illustrated by the circles to the left and right of the thin vertical lines, results in generation of the ten orbitals of O_2. The antibonding orbitals are indicated by an asterisk. Electrons are represented by the small arrows within the circles. The large arrow to the left of the diagram represents increase in energy.

this describes the 1s orbital. Where n equals 2, and l equals 0 or 1, these values describe the 2s and 2p orbitals respectively.

The magnetic quantum number, m, reflects the different vectoral orientations of the angular momentum and defines the orbital with reference to a given direction. Its integer values are $+l$, 0, and $-l$. Therefore, for atomic oxygen where l equals 1, m may be $+1$, 0, or -1. These values define the $2p_x$, $2p_y$, and $2p_z$ orbital possibilities.

The electron spin quantum number, s, deals with the contribution of spin direction of the electron to the angular momentum. Its value may be either $+1/2$, or $-1/2$. The value of the spin quantum number is described in Figure 1 by the direction of the arrows that schematically represent the electrons.

It would be advantageous at this point to discuss three rules regarding the tendency of electrons to fill orbitals. When these rules are obeyed, the atom or molecule will be in its ground state or lowest energy electronic configuration. The Aufbau principle states that as electrons are added to orbitals, the lowest energy orbitals will be filled first. In the oxygen atom, the lower energy 1s and 2s orbitals are filled, but the 2p orbital remains unfilled. The Pauli exclusion principle states that a maximum of two electrons can occupy an orbital, and then only if the electron spins are opposed. In reducing O_2 to $O_2^{\cdot-}$, the electron placed in the $\pi_g^*,2p$ orbital must have a spin number opposite to the original occupying electron. Hund's rule states that given a set of degenerate orbitals, one electron will occupy each orbital before two electrons can occupy a given orbital of the degenerate set. The term degenerate implies that each orbital of the set has an equal energy level, such as the $2p_x$, $2p_y$, and $2p_z$ orbitals of atomic oxygen. The rule further states that the electrons of singly occupied orbitals will have parallel spins. Obedience to this rule by ground-state molecular oxygen results in two unpaired electrons each filling a different $\pi_g^*,2p$ orbital, and, thus, results in the diradical and paramagnetic properties of O_2.

The atomic orbitals, represented by phi (ϕ), of different atoms may interact through the binuclear sharing of electrons of appropriate energy levels. The resulting molecular orbital, represented by psi (ψ), is therefore a combination of the atomic orbitals, and will have characteristics possessed by the contributing atomic orbitals. In generating a chemical bond, orbitals are not destroyed. As Figure 1 demonstrates, each atomic oxygen contributes five orbitals, and, therefore, the total orbitals of molecular oxygen are ten. The linear combination of atomic orbitals is obtained by the simple addition or subtraction of the combining atomic orbitals. Addition of the atomic orbitals results in the generation of a *bonding* molecular orbital, ψ_b. Subtraction results in the generation of an *antibonding* molecular orbital, ψ_a.

When a molecular orbital is cylindrically symmetrical about its binding axis, it is termed sigma, σ. Likewise, a molecular orbital formed by the overlapping of p orbitals above and below the plane of the atoms is termed a pi, π, orbital. Although generated from the combination of two $2p_z$ atomic orbitals, the resulting molecular orbital of O_2 is a $\sigma_g,2p$ because this molecular orbital is symmetric with respect to rotation around the internuclear axis and, thus, has symmetry properties compatible with an s atomic orbital.

Bonding molecular orbitals, ψ_b, are written as σ or π. An asterisk is used to denote the antibonding character of an orbital. The subscripts g and u represent the german words gerade (even) and ungerade (uneven), respectively, and denote the symmetric and antisymmetric behavior of the orbital with respect to the symmetry operation of inversion. For a full presentation of the importance of symmetry, the reader is referred to Orchin and Jaffé (1971).

For any given molecule, the sum of bonding orbitals will equal the sum of antibonding orbitals. However, this does not imply that all orbitals are filled. Perusal of the ten orbitals of O_2 in Figure 1 demonstrates that there are five bonding and five antibonding orbitals. The bonding orbitals are completely filled, but this is not the case for the antibonding orbitals. The filling of antibonding orbitals tends to cancel the effect of bonding orbitals, and, therefore, the attraction between atoms. In reference to Figure 1, note that the bonding effect of σ_g, $1s$ is canceled by the antibonding effect of σ_u^*, $1s$; likewise, σ_g, $2s$ is canceled by σ_u^*, $2s$. The bonding of σ_g, $2p$ is unopposed by the empty σ_u^*, $2p$ orbital and the four bonding electrons of π_u, $2p$ are opposed by the two antibonding electrons of π_g^*, $2p$. Therefore, the net bonding of O_2 is the sum of six bonding electrons minus two antibonding electrons for a total of four bonding electrons or two bonds. As can be seen in Figure 1, antibonding orbitals are of higher energy than their corresponding bonding orbitals, and electrons will fill the bonding orbitals first.

An example of antibonding effect is obtained by comparison of the molecular orbital diagrams of O_2 and O_2^- in Figures 1 and 2. In the reduction of O_2 to O_2^- one electron is added to the π_g^*, $2p$ level of O_2, resulting in an increase in the antibonding property. Using bond distance as an index of strength, reduction to O_2^- results in a change in bond distance from 1.21 Å for O_2 to 1.26 Å for O_2^-.

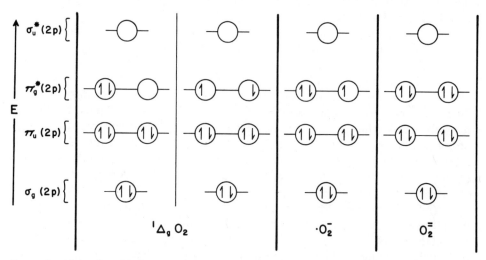

FIGURE 2.   Molecular orbital-energy diagrams of electronically excited, singlet O_2 ($^1\Delta_g O_2$), superoxide anion (O_2^-), and deprotonated hydrogen peroxide (O_2^{2-}). The σ, $1s$ and σ, $2s$ orbitals have been deleted from this diagram. The two possible orbital configurations for $^1\Delta_g O_2$ are given in accordance with the "real orbital" calculations of Ogryzlo (1970).

Further reduction by addition of a second electron to the $\pi_g^*,2p$ level results in a complete cancellation of π bonding effect. This is the molecular orbital picture of $O_2{}^{2-}$, the deprotonated form of H_2O_2. This molecular species is bonded only through $\sigma_g,2p$ and has a bond distance of 1.49 Å (Orchin and Jaffé, 1971).

6. CONCEPT OF DISPROPORTIONATION AND THE GENERATION OF SINGLET MOLECULAR OXYGEN

Disproportionation is defined as the transformation of a substance into two or more dissimilar substances by simultaneous oxidation and reduction. Such reactions provide a mechanism for liberating the relatively large quantities of energy required for electronic excitation of O_2. An appreciation of the concept of disproportionation will provide an approach to the understanding of microbicidal energetics.

The disproportionation type of reaction readily lends itself to pictorial description using redox-energy diagrams (Phillips and Williams, 1965; Koppenol, 1976). The acid disproportionation of $O_2{}^-$ is described by Figure 3. In this reaction the change in standard reduction potential, $\Delta E_0'$, resulting from the one electron reduction of $O_2{}^-$ to H_2O_2 is reflected by the difference in position of these species with respect to the abscissa and the left-sided ordinate. The $\Delta E_0'$ is related to the change in standard free energy, $\Delta G_0'$, by the previously described function, $\Delta G_0' = -nF\Delta E_0'$. The n describes the change in formal charge per oxygen atom, not per oxygen molecule. The electron required for the reduction of $O_2{}^-$ to H_2O_2 is provided through the oxidation of a second $O_2{}^-$ to O_2. The free energy liberated by the reduction component of the disproportionation is schematically described by the thin vector. The energy difference between the ground triplet state, $^3\Sigma_g^- O_2$, and the lowest electronically excited singlet state, $^1\Delta_g O_2$, of oxygen is described by the relative positions of these species with respect to the right-sided ordinate. Again, the energy is per atom oxygen, not per molecule oxygen.

Note that the energy required for the generation of $^1\Delta_g O_2$ lies below the thin vector describing the total free energy available. Therefore, the generation of $^1\Delta_g O_2$ by this reaction is thermodynamically allowed.

$$2O_2{}^- + 2H^+ \rightarrow H_2O_2 + {}^1\Delta_g O_2, \Delta G_0' = -6.4 \text{ kcal/mol} \tag{8}$$

The chemical and biological generation of $^1\Delta_g O_2$ by this mechanism has been previously considered (Stauff, *et al.*, 1963; Khan, 1970; Allen *et al.*, 1973, 1974; Mayeda and Bard, 1974; Koppenol, 1976).

The importance of acidity for O_2^- disproportionation can be appreciated through a consideration of the proton dependence of the reaction rates. In aprotic environments the rate constant, k, is less than 0.5 $M^{-1}sec^{-1}$; at pH 7.0, the k is 4.5×10^5 M^{-1} sec^{-1}; at pH 4.8, the pK_a of HO_2^-, the k is maximal at 8.5×10^7 M^{-1} sec^{-1}; and at pH 2.0, the k decreases to 1.0×10^6 M^{-1} sec^{-1} (Behar *et al.*, 1970; Bielski and Allen, 1977). The reaction may, therefore, be more correctly written:

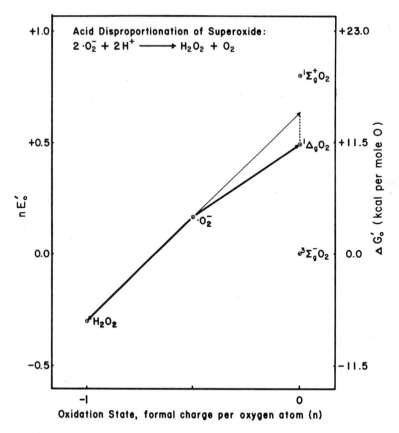

FIGURE 3.   Redox-energy diagram of the acid disproportionation of $O_2^{\cdot -}$ at pH 7.0, $T25°C$, and P_{O_2} at one atmosphere. The values for E_0 are those compiled by Koppenol (1976).

$$O_2^{\cdot -} + HO_2^{\cdot} \rightarrow HO_2^{-} + O_2 \; (^1\Delta_g O_2) \tag{9}$$

The hydrodioxyl radical, $HO_2^{\cdot}$, is the conjugate acid of $O_2^{\cdot -}$.

A consideration of the electronic interaction involved in acid disproportionation of O_2^{-} is a valuable aid in mechanistically explaining the generation of $^1\Delta_g O_2$. Review of Figure 2 shows that $O_2^{\cdot -}$ has one filled and one half-filled $\pi_g^*, 2p$ orbital. Reaction of $O_2^{\cdot -}$ with $HO_2^{\cdot}$ involves overlap between the half-filled orbitals of both species resulting in electron transfer (Koppenol and Butler, 1977). Therefore, the reduced product, H_2O_2, or its conjugate base, HO_2^{-}, will have two completely filled $\pi_g^*, 2p$ orbitals. The electron donating species will have one filled and one vacant $\pi_g^*, 2p$ orbital; this is the electronic structure of $^1\Delta_g O_2$.

Superoxide dismutase (SOD) provides an enzymatic mechanism for the disproportionation or dismutation of $O_2^{\cdot -}$ (McCord and Fridovich, 1969). The proposed mechanism of catalysis involves the cyclic reduction and oxidation of the copper of SOD (Fee and DiCorleto, 1973).

$$E\text{-}Cu^{2+} + O_2^{\cdot-} \longrightarrow H\text{-}E\text{-}Cu^+ + O_2, \Delta G_0 = -17.7 \text{ kcal/mol} \quad (10)$$
$$H\text{-}E\text{-}Cu^+ + O_2^{\cdot-} + H^+ \longrightarrow E\text{-}Cu^{++} + H_2O_2, \Delta G_0 = -11.9 \text{ kcal/mol} \quad (11)$$

Net reaction:

$$2O_2^{\cdot-} + 2H^+ \longrightarrow H_2O_2 + O_2 \; (^3\Sigma_g^- O_2) \quad (12)$$

The rate constant, k, for reaction (12) is 1.9×10^9 M^{-1} sec^{-1} within the pH range 5.0 to 9.5 (Fridovich, 1976). Note that the SOD reaction occurs sequentially with piecemeal liberation of free energy (Koppenol, 1976). The partitioning of energy results in insufficient energy for the generation of $^1\Delta_g O_2$ at any given reaction, and thus ground state O_2 will be the product (Mayeda and Bard, 1974). The synthesis of SOD by a bacterium might serve to protect the organism from the action of $^1\Delta_g O_2$ through competitive depletion of $O_2^{\cdot-}$. Weser *et al.* (1975) suggested that SOD may also catalyze the direct relaxation of $^1\Delta_g O_2$ to $^3\Sigma_g^- O_2$. In essence SOD would serve to absorb the electronic excitation energy of $^1\Delta_g O_2$. This hypothesis is attractive when one considers the relative heat stability of SOD. The enzyme is also devoid of tryptophan (Weser *et al.*, 1972; Steinman *et al.*, 1974); this amino acid is highly susceptible to $^1\Delta_g O_2$ attack (Nilsson *et al.*, 1972).

Hydrogen peroxide, the reduced product of $O_2^{\cdot-}$ disproportionation, is the reactant of the second disproportionation reaction. The redox-energy relationship for H_2O_2 disproportionation is described schematically in Figure 4. This disproportionation has a greater associated exergonicity. In fact, generation of the second electronically excited singlet state of O_2, $^1\Sigma_g^+ O_2$, is thermodynamically allowed. However, the half-life of $^1\Sigma_g^+ O_2$ is relatively short, and if generated, this species would tend to relax to $^1\Delta_g O_2$ before reacting (Foote, 1976).

Of possible biological relevance is the proposal by Kellogg and Fridovich (1975) that $^1\Delta_g O_2$ is generated as a product of the Haber–Weiss reaction. This radical chain mechanism, as originally proposed by Haber and Weiss (1934), consists of two sequential reactions, describing the OH·-mediated decomposition of H_2O_2.

$$OH\cdot + H_2O_2 \longrightarrow H_2O + O_2^{\cdot-} + H^+ \quad (13)$$
$$H^+ + O_2^{\cdot-} + H_2O_2 \longrightarrow OH\cdot + H_2O + O_2 \quad (14)$$

Net reaction:

$$2H_2O_2 \xrightarrow{\text{OH·}} 2H_2O + O_2 \; (^1\Delta_g O_2) \quad (15)$$

In essence, these reactions describe the OH·-catalyzed disproportionation of H_2O_2. The generation of $^1\Delta_g O_2$ is thermodynamically allowed. However, McClune and Fee (1976) and Halliwell (1976) were unable to demonstrate the occurrence of reaction (14) over the pH range 6.0–10.6. Smith and Kulig (1976) have presented evidence that $^1\Delta_g O_2$ is a product of the base catalyzed disproportionation of H_2O_2.

In 1970, Arneson proposed that the radical components, OH· and $O_2^{\cdot-}$, of reactions (13) and (14) could react to generate $^1\Delta_g O_2$.

$$O_2^{\cdot-} + OH\cdot + H^+ \rightarrow H_2O + O_2 \; (^1\Delta_g O_2) \quad (16)$$

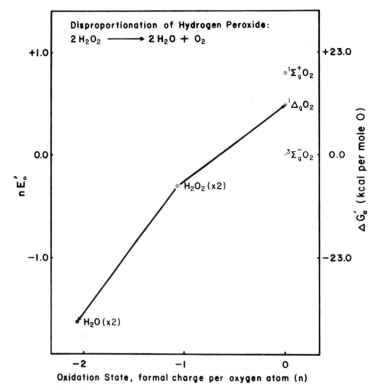

FIGURE 4. Redox-energy diagram of the disproportionation of H_2O_2 at pH 7.0, $T25^\circ C$, and P_{O_2} at one atmosphere. The values for E_0 are those compiled by Koppenol (1976).

This reaction is fast with a k of 1×10^{10} M^{-1} sec^{-1} (Sehested et al., 1968). The importance of orbital overlap in the mechanistic generation of $^1\Delta_g O_2$ by reaction (16) has been considered by Koppenol and Butler (1977).

Myeloperoxidase (MPO) constitutes approximately 5% of the dry weight of PMN (Schultz and Kaminker, 1962). The antimicrobial activity of this enzyme in the presence of H_2O_2 and an oxidizable cofactor is well documented (Klebanoff et al., 1966; Klebanoff, 1968, 1970; McRipley and Sbarra, 1967). It has been proposed that the MPO system is microbicidal through its generation of $^1\Delta_g O_2$ (Allen et al., 1974; Allen, 1975a,b). The proposed mechanism for the generation of OCl^- and $^1\Delta_g O_2$ is described by the following reactions.

$$H_2O_2 + Cl^- \xrightarrow[\text{Acid}]{\text{MPO}} H_2O + OCl^- \tag{17}$$

$$OCl^- + H_2O_2 \longrightarrow H_2O + Cl^- + {}^1\Delta_g O_2 \tag{18}$$

$$\text{Net reaction: } 2H_2O_2 \longrightarrow 2H_2O + {}^1\Delta_g O_2 \tag{19}$$

Note that the MPO reaction is, in essence, a H_2O_2 disproportionation. Oxidations catalyzed by MPO are associated with photon emission or chemiluminescence (CL). This CL is proposed to result from the relaxation of the electronically excited products of MPO-mediated oxidations. The effector oxidant is most probably $^1\Delta_g O_2$; however, oxidation may involve OCl^- or Cl_2 as a direct oxidant

(Allen, 1975a,b). Recently, Rosen and Klebanoff (1977), using a chemical detection system, presented further evidence that $^1\Delta_g O_2$ is generated by the MPO system.

Catalase also catalyzes the disproportionation, or dismutation of H_2O_2, but under normal circumstances the oxygen liberated is $^3\Sigma_g^- O_2$, and not $^1\Delta_g O_2$ (Porter and Ingraham, 1974). Therefore, in a functional sense, catalase removes H_2O_2 in a manner analogous to the removal of $O_2\cdot^-$ by SOD and, thus, can be thought of as a H_2O_2 dismutase. Catalase is also a relatively stable enzyme capable of absorbing the rather large free energy liberated in the reaction without denaturation. The importance of catalase in microbial protection against the oxidative microbicidal mechanisms of the PMN has been demonstrated by Mandell (1975).

Throughout the previous discussion, $^1\Delta_g O_2$ has been used to describe the electrophilic-O_2 product of the disproportionation reactions being considered. In the physical-chemical literature, this notation is usually assigned to free, gas-phase O_2 of singlet multiplicity. As such, $^1\Delta_g O_2$ may not be the most appropriate notation for O_2 of singlet multiplicity in solution. For example, dimolecular emission from two $^1\Delta_g O_2$ is highly improbable in solution.

In solution, the more probable condition is that O_2 of singlet multiplicity ($^1O_2$) is stabilized through interaction with an appropriate ground state carrier (^1M) to yield a complex, ^1M-O_2. Such complexes would yield differing degrees of $^1O_2$-type electrophilic reactivity, as governed by the electronic structure and electronegativity of ^1M. In the MPO-catalyzed reaction, O_2 might be carried as hydrogen chloroperoxide ($HClO_2$) or its conjugate base, chloroperoxide anion (ClO_2^-).

Therefore, for the sake of clarity, it should be understood that $^1\Delta_g O_2$ is used to imply an O_2 or O_2-complex with singlet multiplicity and electrophilic reactivity. The $^1\Delta_g O_2$ notation is used here in the more general sense, and does not necessarily imply free, singlet molecular oxygen in the gas phase.

7. WHAT ARE ELECTRONIC STATES?

The energy of a molecule can be defined relative to three energy levels or states: electronic, vibrational, and rotational. The relationship between these energy levels is visually depicted in the Jablonski energy diagram presented in Figure 5. The height of the ordinate reflects the increase in energy or frequency. The various horizontal lines reflect the possible energy states of the molecule. It should be pointed out that in the diagram only those energy levels necessary to illustrate energy crossing and transfer have been included, and that the space separating electronic levels is populated by vibrational levels, and in turn, these spaces by rotational levels. Transition from one energy level to another can be quantified; that is, the energy difference is discrete.

The thermal energy equivalent of the frequency can be obtained by the equation:

$$E(\text{kcal/mol}) = 28.6 \times v(\text{cm}^{-1}) \times 10^{-4} = 28.6 \times 10^3 / \lambda(\text{nm}) \qquad (20)$$

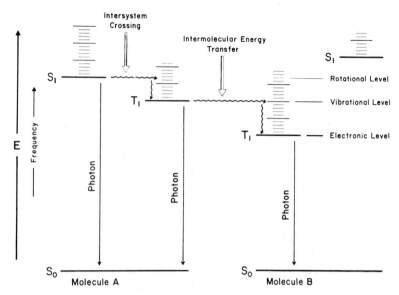

FIGURE 5.   Jablonski diagram illustrating the relationship of electronic, vibrational, and rotational energy levels. The diagram demonstrates intersystem crossing from the lowest excited singlet state (S_1) to the lowest excited triplet state (T_1) in molecule A. Intermolecular energy transfer from the lowest excited triplet state (T_1) of molecule A to generate the lowest excited triplet state (T_1) of molecule B is also demonstrated. The various possibilities for relaxation to ground state with photon emission are illustrated. The frequency of the emitted photon is dependent upon the energy difference between the excited and ground states.

where λ is the wavelength expressed in nanometers. The greatest difference in energy is associated with change in electronic level. These energies are within the ultraviolet and visible range of the spectrum, 200–780 nm. The thermal equivalent is approximately 143–38 kcal/mol, respectively. The vibrational levels of a molecule also express quantum differences in energy. These energies are in the infrared region of the spectrum. This is the range of energy classically considered in biochemical energetics. Of even lower energies are the various rotational levels that subdivide the vibrational differences.

Molecules possess relatively few electronic levels. In Figure 5, S_0 denotes the singlet ground state or electronic state of lowest energy. Virtually all biological molecules have singlet ground states. The exceptions to this rule are certain free radicals and molecules containing transition metals. Recall that diradical, triplet ground state O_2 is such an exception. If a photon of appropriate energy is absorbed by an electron, it will be promoted to a higher energy or electronically excited state. In the figure, S_1 represents the lowest singlet excited state of a molecule.

The multiplicity of a molecule reflects its total spin quantum number (J) for any given electronic state. Recall that the spin quantum number reflects the spin direction of an electron, and is given a value of either plus or minus ½. Multiplicity can be established by the equation $J = |2S| + 1$, where S represents the sum of electron spins. When all spins cancell, $S = O$ and $J = 1$, and the molecule is said to be of singlet multiplicity. The $\pi_g^*,2p$ structure of $^1\Delta_g O_2$ defines its singlet multiplicity, represented by the superscript one. The upper case delta and subscript g are symmetry notations (Orchin and Jaffe, 1971). If the spins are unopposed as in $^3\Sigma_g^- O_2$, the value of $S = \pm 1$ and $J = 3$, and the multiplicity is said to be triplet. The excited triplet state of a molecule (i.e., T_1) is always of lower energy than its corresponding excited singlet state (i.e., S_1). Hund's rule dictates that the lowest energy state results when electrons in singly occupied orbitals have parallel spins. As previously stated, the triplet multiplicity of $^3\Sigma_g^- O_2$ is in accordance with Hund's rule.

A molecule may relax from its excited to its ground electronic state by photon emission. Usually this transition is from the lowest excited singlet or triplet state to the ground singlet state. The $S_1 \rightarrow S_0$ transition is responsible for the phenomenon of fluorescence. This is a relatively fast transition reflecting the unstable half-life of S_1, $10^{-9} - 10^{-8}$ sec.

The transition from S_1 to T_1 ($S_1 \rightarrow T_1$) can occur by the process of intersystem crossing (ISC). This is a low probability transition, and is, thus, said to be "forbidden." When it occurs, the resulting T_1, or lowest triplet electronic excitation state, is metastable. Its relatively long half-life, many orders of magnitude greater than that of S_1, results from the "forbidden" nature of its relaxation from T_1 to S_0. Transitions of the $S_1 \rightarrow T_1 \rightarrow S_0$ type are responsible for the phenomenon of delayed emission or phosphorescence. The energy difference between the S_1 and T_1 states is responsible for the observation that for a given molecule, the fluorescence ($S_1 \rightarrow S_0$) will be of greater frequency than its phosphorescence ($T_1 \rightarrow S_0$).

Nonradiative, intermolecular energy transfer is of probable importance in the phenomenon of PMN-associated chemiluminescence. Such energy transfer between molecules may be through the interaction of electric dipole fields, and can occur over relatively great intermolecular distances, 50–100 Å.

$$S_1 \text{ Mol.}_A + S_0 \text{ Mol.}_B \longrightarrow S_0 \text{ Mol.}_A + S_1 \text{ Mol.}_B \tag{21}$$

A second possibility is triplet nonradiative transfer by the mechanism:

$$T_1 \text{ Mol.}_A + S_0 \text{ Mol.}_B \longrightarrow S_0 \text{ Mol.}_A + T_1 \text{ Mol.}_B \tag{22}$$

This type of transfer, depicted in Figure 5, requires closer contact between the interacting molecules such that overlap of orbitals is possible. More complete treatments of this subject are available (Turro, 1965; Calvert and Pitts, 1966; Förster, 1959).

In 1900, Raab reported that living cells could be killed by exposure to certain dyes in the presence of light. Five years later, O_2 was shown to be a necessary component for killing by this dye–light system (Jodlbauer and von Tappeiner, 1905). Light-sensitized dye–O_2 killing and substrate oxidation phenomena have been described collectively as photodynamic action (Blum, 1941).

In 1939, Kautsky proposed that electronically excited O_2 was the mediator of photodynamic oxidations by the mechanisms:

$$S_0 \text{ Dye} + \text{Photon} \longrightarrow S_1 \text{ Dye} \xrightarrow{\text{ISC}} T_1 \text{ Dye} \qquad (23)$$

$$T_1 \text{ Dye} + T_0O_2 \ (^3\Sigma_g^- O_2) \longrightarrow S_0 \text{ Dye} + S_1O_2 \ (^1\Delta_g O_2) \qquad (24)$$

This proposal was not well received at the time. However, twenty-five years later, Foote and Wexler (1964), using the reaction of H_2O_2 and OCl^- to chemically generate $^1\Delta_g O_2$ (Khan and Kasha, 1963), demonstrated that the products of oxidation were identical to those obtained using most photodynamic systems. In the same year, Cory and Taylor (1964), using radiofrequency discharge to generate $^1\Delta_g O_2$, arrived at the same conclusion.

8.   REACTIVITY OF O_2, $O_2^{\cdot -}$, H_2O_2, AND OH·

In considering the reactivity of the various oxygens associated with PMN microbicidal activity, a review of Figures 1 and 2 will be useful. In the ground electronic state both of the $\pi_g^*,2p$ orbitals of O_2 are partially filled. As a consequence, O_2 may readily participate in radical reactions; that is, it can readily react with other radicals also possessing partially filled orbitals. However, most molecules of biological importance are not radicals. Biological molecules commonly have singlet ground states, and, thus, have either fully filled or unfilled orbitals. This condition impairs orbital overlap with the partially filled orbitals of O_2, and, consequently, oxidations mediated by ground state O_2 require a high activation energy, i.e., burning. Appreciation of this concept leads to an understanding of why the thermodynamically favorable oxidation of organic molecules does not spontaneously occur. It is the diradical nature of O_2 that affords biological systems protection from disintegrative oxidation.

In a homologous manner, $O_2^{\cdot -}$, having one fully filled and one partially filled $\pi_g^*,2p$ orbital, is also limited in its capacity for orbital overlap with organic molecules. However, its radical electronic structure does not prevent $O_2^{\cdot -}$ from reacting with the partially filled orbitals of other radicals. The reactivity of $O_2^{\cdot -}$ with itself and with OH· has been previously considered. It has been proposed that superoxide is "a highly reactive substance" and "a likely candidate as a bactericidal agent in leukocytes" (Babior *et al.*, 1973). It should be appreciated that "super" does not necessarily reflect the reactivity of the radical, and it should also be recalled that $O_2^{\cdot -}$ generation is detected and quantified by cytochrome c and nitroblue tetrazolium reduction. These molecules are not oxidized. This is not intended to underrate the importance of $O_2^{\cdot -}$. Rather, it is intended to focus attention on $O_2^{\cdot -}$ as a key intermediate in the generation of microbicidal species.

The $\pi_g^*,2p$ orbitals of H_2O_2 are fully filled, and consequently this molecule does not have radical character. In the absence of a proper catalyst, H_2O_2 is relatively unreactive. This is demonstrated through a consideration of H_2O_2-producing bacteria. Many microorganisms lacking the ability to synthesize cytochromes are incapable of reducing O_2 to H_2O, and consequently H_2O_2 accumulates as the terminal product of O_2 reduction. These H_2O_2-producing bacteria, such as streptococci and lactobacilli, are capable of surviving relatively large concentrations of H_2O_2. For a review of this subject, see Dolin (1961).

Figure 6 demonstrates the CL responses obtained from addition of MPO to five live broth cultures of aerobically grown bacteria. Three of the organisms were cytochrome-negative streptococci. These streptococci yielded rather large CL responses on addition of MPO without exogenous H_2O_2. It should be pointed out that the high initial bursts of CL probably reflect the accumulation of bacterial-generated H_2O_2 in the media, and although large CL responses were obtained, microbicidal activity was relatively poor. In this instance, CL probably reflects the $^1\Delta_g O_2$-mediated oxidation of substrates present in the tryptose broth, resulting in competitive inhibition of microbicidal oxidation. Addition of MPO to viable organisms suspended in normal saline did result in potent microbicidal action. CL was again observed but was of a lower magnitude. This CL did correlate with streptococcal killing.

No significant CL or killing was observed after addition of MPO without exogenous H_2O_2 to the two catalase-positive bacteria, *Escherichia coli* and *Staphylococcus aureus*. This observation held true for both broth and saline suspensions of the organisms. These findings are consistent with other investigations of MPO microbicidal action (Klebanoff, 1968).

The hydroxyl radical, OH·, has also been considered as a possible effector of PMN microbicidal action (Johnston *et al.*, 1975). These investigators proposed that OH· might be generated in the PMN by the reaction of O_2^- with H_2O_2. However, the introduction of OH· scavengers, such as mannitol, ethanol, and

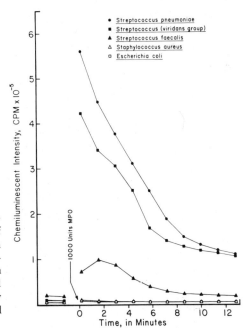

FIGURE 6. Temporal traces of CL following addition of 0.5 ml of MPO, containing 1000 units of activity, to 4.5-ml cultures of variable bacteria in tryptose broth. Each bacterial culture was adjusted to a concentration having an absorption of 0.25 at 525 nm ($A_{525} = 0.25$) when blanked against the broth. Measurement of CL was by the method previously described (Howes and Steele, 1971; Allen, 1973, 1977).

benzoate to the PMN resulted in only weak inhibition of microbicidal action. Klebanoff (1974) had reported similar findings in his investigation of microbicidal action by the xanthine–xanthine oxidase system. Furthermore, glucose is as effective as either mannitol or ethanol in inhibiting the CL response from phagocytically stimulated PMN (Allen and Salin, unpublished observation).

In 1970, Beauchamp and Fridovich reported that addition of methional to the xanthine–xanthine oxidase system resulted in its oxidation to ethylene. They proposed that $O_2{}^-$ and H_2O_2 reacted to produce $OH\cdot$, and that this radical reacted with methional to produce ethylene. However, in 1975 Kellogg and Fridovich reported that peroxidation of linolenate by the xanthine–xanthine oxidase system was not inhibited by $OH\cdot$ scavengers. This oxidation was inhibited by scavengers of $^1\Delta_g O_2$.

Recently the oxidation of methional has been applied to the study of PMN microbicidal oxidations (Weiss *et al.*, 1977; Tauber and Babior, 1977). Introduction of methional to phagocytically activated PMN resulted in the generation of ethylene. This was taken as evidence for the participation of $OH\cdot$ in microbicidal action. However, there is little evidence that this reaction is specific for $OH\cdot$, and ethylene liberation may be the result of a variety of reactions. Recently, Klebanoff and Rosen (1978) presented evidence that methional oxidation by PMN is largely dependent upon myeloperoxidase activity, and that $^1\Delta_g O_2$ is involved in the reaction. Clarification of the role of $OH\cdot$ as either a direct or indirect effector of PMN microbicidal action will require further investigation. The previous sentence should be extended to include all of the oxygen species.

9. SINGLET MOLECULAR OXYGEN, ELECTRONICALLY EXCITED CARBONYL CHROMOPHORES, AND CHEMILUMINESCENCE

In its electronically excited singlet state, O_2 is a good electrophilic reactant capable of attacking a broad spectrum of biologically important molecules. The reactivity of $^1\Delta_g O_2$ with lipids, nucleic acids, and amino acids has been demonstrated (Rawls and van Santen, 1970; Anderson and Krinsky, 1973; Hallett *et al.*, 1970; Nilsson *et al.*, 1972). The rate constants for $^1\Delta_g O_2$-mediated oxidation of histidine, tryptophan, and methionine are in the order of 10^7 M^{-1} sec^{-1} (Nilsson *et al.*, 1972).

The electrophilic behavior of $^1\Delta_g O_2$ is a consequence of its vacant $\pi_g^*, 2p$ orbital. This vacancy allows overlap with regions of high electron density on target molecules. For example, consider the reaction of $^1\Delta_g O_2$ with the π orbital of a hypothetical target molecule, $R_2C=CR_2$. If reaction proceeds through a dioxetane intermediate, one of the resulting product carbonyls, $R_2C=O$, will be electronically excited (Fenical *et al.*, 1969; Wilson and Schaap, 1971; Turro and Lechtken, 1973; Turro and Devaquet, 1975).

What is the thermodynamic feasibility of such a reaction? Cleavage of a $C=C$ and a $O=O$ bond required +146 and +119 kcal/mol of energy respectively, for a total of +265 kcal/mol. The formation of two product carbonyls is exothermic liberating 2×-176 kcal/mol (Pauling, 1939; Roberts and Caserio,

1964). Thus, the net energy yield for the reaction is 265 + (−352), or −87 kcal/mol. The −22 kcal/mol energy of $^1\Delta_g O_2$ can be thought of as providing the activation energy necessary for reaction. If this energy is added, the total energy yield becomes −109 kcal/mol. These energies, −87 and −109 kcal/mol, are well within the range necessary for electronic excitation of carbonyl chromophores, and correspond to wavelengths of approximately 330 and 260 nm, respectively.

The molecular orbital relationship between oxygen and carbon in the carbonyl chromophore is presented in Figure 7. The diagram depicts the energy differences between the carbonyl in its n,π^* elecronically excited singlet and its ground state configuration. The n,π^* notation implies quantum transition of an electron from the n or nonbonding orbital of the oxygen atom to the vacant π^* orbital of the carbonyl. The term nonbonding is used to indicate that the electrons of this $2p$ orbital of oxygen do not participate in bonding with carbon to form the carbonyl. Change in spin multiplicity through intersystem crossing could also result in generation of the corresponding triplet state of the n,π^* carbonyl.

The reactivity and physical properties of molecules are a reflection of their electronic configuration. Electronically excited molecules may differ from their respective ground states in ionizability, redox potential, bond distances and angles, dipole moment, reactivity, radical character, etc. (Orchin and Jaffé, 1971).

The phenomenon of PMN chemiluminescence is proposed to reflect the oxidation of the microbial substrates resulting in the generation of electronically excited carbonyl chromophores. If the excited product of oxidation relaxes to ground state by photon emission, chemiluminescence will result. The particular emission from a given carbonyl chromophore will be influenced by the complex-

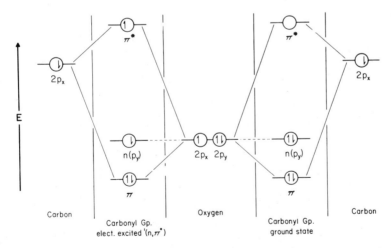

FIGURE 7. Molecular orbital-energy diagram of the carbonyl chromophore in its n,π^* excited singlet state, shown on the left, and in its ground singlet state, shown on the right. The arrow with E represents increase in energy.

ity of the molecule of which it is a functional group. The absorption and emission characteristics of a chromophore are also influenced by the external physical environment. Any given molecular excitation will be associated with a given emission spectrum. However, since microbicidal oxidations involve a multitude of substrates, a multitude of spectra will result. The net spectrum observed will reflect the contributions of these spectra. The numerous possibilities for intersystem crossing and excitation energy transfer also insure that the emission spectrum in microbicidal action is broad-banded (Cheson *et al.*, 1976; Steele *et al.*, 1976; Andersen *et al.*, 1977). Emission phenomena are further complicated by the fact that each different microbe and even the same microbe under different growth conditions, will possess different molecular constitutions. Each microbe should, therefore, present a different net chemiluminescent spectrum. Microbes are also structurally complex, and therefore, different substrates may be oxidized at different times.

10. CERTAIN CHEMILUMINESCENT OBSERVATIONS

Chemiluminescence (CL) is the consequence of PMN microbicidal oxidations, and as such, it reflects the participation of all energy transductions involved in effecting this process (Allen *et al.*, 1972). Figure 8 provides a schematized description of microbicidal oxidation. Ultimately, the CL response of PMN is related to the metabolism of glucose. A study correlating integral value of CL with integral value of HMP shunt activity is presented in Figure 9. This correlation reflects the necessity for transducing the energy of the carbohydrate bond into the reducing currency of NADPH. The quantity of NADPH generated should and does correlate with the intensity of photon emission (Allen, 1973; Allen and Lint, 1979).

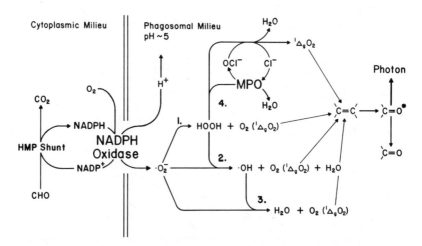

FIGURE 8. Postulated energy transductions involved in microbicidal action. Four mechanisms for the generation of $^1\Delta_g O_2$ are shown. The C=C represents a hypothetical target molecule on the phagocytized microbe.

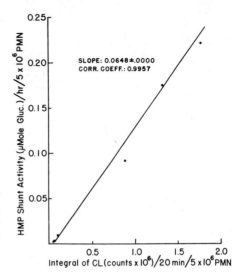

FIGURE 9. Correlation between CL and HMP shunt activity. Both CL and HMP shunt oxidation of glucose were measured from 5×10^6 PMN. The variation in responses reflects the postvenipuncture age of the PMN preparations (Allen *et al.*, 1972; Allen, 1973; Allen and Lint, 1979).

Although O_2 is not necessary for the mechanics of phagocytic engulfment (Sbarra and Karnovsky, 1959), it is necessary for microbicidal activity (Selvaraj and Sbarra, 1966). Likewise, O_2 is necessary for CL. In the experiment, the results of which are presented in Figure 10, opsonized bacteria were added to a PMN suspension, and the airspace above the suspension was gently flushed with N_2 in an attempt to decrease the P_{O_2} of the system. The vial was then sealed airtight, and continuously monitored for CL. After 30 min the CL fell to baseline. Shaking the sealed vial at that time resulted in a small spike of CL. The vial was shaken 10 min and 20 min later, and the system was considered to be O_2-free

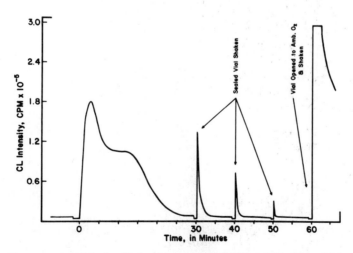

FIGURE 10. The necessity of O_2 for CL from activated PMN. The vial contained 1.5×10^7 PMN in 7.0 ml of phosphate buffered saline with 80 mg/dl glucose. The vial was gently flushed with N_2 for 15 sec, and the cap was tightly sealed after addition of opsonized *Propionibacterium shermanii* (Allen, 1973).

when shaking produced no CL. At that point, the vial was opened to room air, resealed, and counted. This maneuver resulted in a very large burst of CL (Allen, 1973). This is consistent with the role of O_2 as described in Figure 8. If O_2 is depleted, the phagocytically activated PMN will become metabolically poised in a reduced state, and O_2 will be rate limiting for oxidative microbicidal action and its associated CL. Presentation of O_2 to this system, therefore, would be expected to result in a very large CL response.

Chronic granulomatous disease (CGD) provides a clinical demonstration of the importance of NADPH oxidase in realizing the potential of NADPH (McPhail *et al.*, 1977; Allen *et al.*, 1977). Dysfunction in the activation of this oxidase in CGD-PMN results in impairment of all associated microbicidal transductions as described in Figure 8. Inability to oxidize NADPH to $NADP^+$ results in a stabilization of the $NADP^+$/NADPH ratio, and as a consequence, HMP shunt activity is inhibited. The generation of O_2^- (Curnutte *et al.*, 1974) and H_2O_2 (Holmes *et al.*, 1967) is also inhibited. Therefore, no reactants are available for disproportionations and other reactions resulting in the generation of oxidants such as OCl^-, $OH\cdot$, and $^1\Delta_gO_2$. Consequently, there is an absence of oxidative microbicidal activity and CL (Stjernholm *et al.*, 1973; Allen *et al.*, 1977).

Unlike staphylococci and coliforms, streptococcal organisms are effectively killed by CGD-PMN (Kaplan *et al.*, 1968, Mandell and Hook, 1969). As previously discussed, the lack of a cytochrome system requires that streptococci use flavoproteins in their terminal reduction of O_2. It is possible that O_2^- is generated at this step. These organisms produce SOD which would protect against its accumulation (McCord *et al.*, 1971). The end products of streptococcal metabolism are acid and H_2O_2. Phagocytosis of live streptococci by CGD-PMN

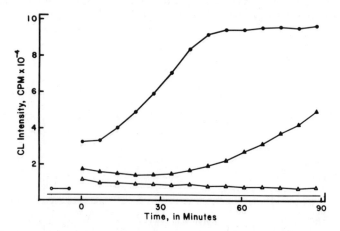

FIGURE 11.   The CL response from CGD-PMN following phagocytosis of viable streptococci. Each vial contained 5×10^6 PMN from either CGD patients or normal donors. The cells were suspended in 4.0 ml phosphate buffered saline containing 100 mg/dl glucose. Phagocytosis was initiated by addition of 1.0 ml of a suspension contining 0.5 ml *Streptococcus viridans* in tryptose broth (A_{525} = 0.25) and 0.5 ml of autologous serum. The circles and triangles represent the responses from normal and CGD-PMN, respectively. The open triangles represent the CL response from CGD-PMN following phagocytosis of heat-killed (autoclaved) streptococci.

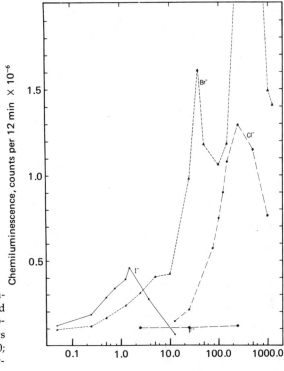

FIGURE 12. Semilog plot of halide concentration against integral of CL obtained from the MPO–halide–H_2O_2 antimicrobial system (Allen, 1975a). 250 units MPO in 2 ml acetate buffer, pH 5.0; H_2O_2 (50 μmol) added to initiate reaction.

results in microbicidal activity, and is associated with CL as is demonstrated in Figure 11. This CL reflects the functioning of the MPO microbicidal system as previously described in Figure 8. The PMN of CGD patients contain normal concentrations of MPO (Klebanoff and White, 1969), and the disproportionation of streptococcal-generated H_2O_2 by MPO can result in the generation of O_2Cl^-,

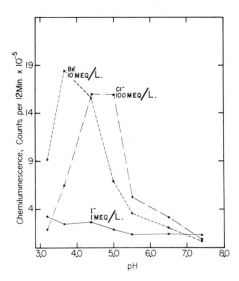

FIGURE 13. Plot of pH against integral of CL obtained from the MPO–halide–H_2O_2 antimicrobial system using either Cl^-, Br^-, or I^- as the halide cofactor (Allen, 1975b). 500 units MPO in 2 ml acetate buffer; H_2O_2 (50 μmol) added to initiate reaction.

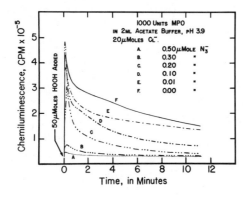

FIGURE 14.   The inhibitory effect of various concentrations of NaN$_3$ on the CL response from the MPO–Cl$^-$–H$_2$O$_2$ antimicrobial system (Allen, 1973). 1000 units MPO in 2 ml acetate buffer, pH 3.9; 20 μmol Cl$^-$. (A) 0.50 μmol N$_3^-$; (B) 0.30 μmol N$_3^-$; (C) 0.20 μmol N$_3^-$; (D) 0.10 μmol N$_3^-$; (F) no N$_3^-$.

OCl$^-$, or $^1\Delta_g O_2$. Thus, the products of streptococcal metabolism provide a secondary pathway to the utilization of the MPO system. The phagocytosis of heat-killed streptococci does not result in CL.

Extracted MPO is highly microbicidal if the requirement of acid pH, H$_2$O$_2$, and oxidizable halide are met. This microbicidal action is measurable as CL (Allen, 1975a,b; Andersen *et al.*, 1977). Figures 12 and 13 demonstrate the relationships of MPO-associated CL to the type and quantity of halide present, and to the pH, respectively. Inhibitors of MPO microbicidal action also effectively inhibit CL. The inhibition of MPO associated CL by varying concentrations of azide, a potent inhibitor of MPO microbicidal activity, is shown in Figure 14 (Allen, 1973). The CL response is also substrate dependent (Nelson *et al.*, 1977), and can be increased by addition of substrate susceptible to $^1\Delta_g O_2$ attack. This observation is consistent with the concept of excited carbonyl chromophores. Recently, Rosen and Klebanoff (1977) presented further evidence that $^1\Delta_g O_2$ is the mediator of MPO microbicidal oxidation. They were able to isolate the $^1\Delta_g O_2$-specific product, *cis*-dibenzoylethylene, from the oxidation of 2,5-diphenylfuran by MPO. Their data on halide and pH requirements for oxidation are comparable to the CL data presented in Figures 12 and 13.

Study of PMN from patients with hereditary MPO deficiency is also consistent with the mechanism proposed in Figure 8. These PMN are efficient producers of O$_2^{\cdot-}$, but lack MPO. Their microbicidal activity is characterized by a lag period, but if given adequate time these PMN will kill bacteria. This microbicidal action is associated with CL; but the CL is of a decreased magnitude especially during the initial postphagocytic period (Rosen and Klebanoff, 1976). The CL response from MPO-deficient PMN is consistent with the role of O$_2^{\cdot-}$ disproportionation in the generation of $^1\Delta_g O_2$.

In 1974, Krinsky reported a completely different approach to the question of $^1\Delta_g O_2$ involvement in PMN microbicidal activity. He observed that a bacterial mutant lacking carotenoid pigment was more rapidly killed by PMN than its carotene containing parent organism. Carotenoids are thought to quench or inactivate $^1\Delta_g O_2$ through an energy transfer mechanism.

$$^1\Delta_g O_2 + {}^1Car \rightarrow {}^3\Sigma_g^- O_2 + {}^3Car \tag{25}$$

Krinsky concluded that the presence of carotene might serve the parent bacterium by protecting it from the action of $^1\Delta_g O_2$ generated by PMN.

11. OTHER APPLICATIONS OF THE CHEMILUMINESCENCE APPROACH

The CL approach to the study of oxidative phenomena has been extended to other phagocytes. Monocytes and macrophages are also characterized by post-phagocytic stimulation of metabolism in a manner homologous to that of PMN (Gee *et al.*, 1970; Rossi *et al.*, 1972). Both monocytes and macrophages have been reported to produce a postphagocytic CL, but of a lower intensity than that from PMN (Sagone *et al.*, 1976; Nelson *et al.*, 1976; Beall *et al.*, 1977; Hatch *et al.*, 1978). Studies of the oxidative activity of these phagocytes have been facilitated through the use of sensitive oxidizable substrates such as luminol. Oxidation of molecules of this type results in generation of electronically excited carbonyl chromophores in high yield with a correspondingly high yield of CL, and as such, can be used as extremely sensitive probes (Allen and Loose, 1976).

Eosinophils also have been reported to yield a CL upon activation (Klebanoff *et al.*, 1977; DeChatelet *et al.*, 1977). These granulocytes contain peroxidase and have HMP shunt activity greater than neutrophil PMN. More recently, Cl has been observed from basophils and mast cells upon activation with anti-IgE and compound 48/80 (Henderson and Kaliner, 1978).

Of probably greater future importance is the extension of the CL approach to the study of the humoral component of the humoral–phagocyte axis in host defense (Hemming *et al.*, 1976; Allen, 1977; Stevens and Young, 1977). This approach implies the necessity for opsonization of the microbe if phagocytosis is to occur. Therefore, if the parameters of PMN and bacterial concentration are held constant, the differences in the CL responses should and do reflect the different opsonic capacities of the sera tested. The use of CL affords a functional, sensitive, accurate, and rapid method for the quantification of opsonic activity, and offers the added advantage of providing kinetic information. This approach may result in important information regarding the relative participation of immunoglobulins and the classical and alternative complement pathways in opsonization.

ACKNOWLEDGMENTS.   I would like to express gratitude to Dr. Richard H. Steele for initially stimulating my interest in excited state chemistry. I would also like to express my appreciation to Dr. Theodore R. McNitt, Dr. Dennis L. Stevens, Mrs. Mildred C. Bratten, Ms. Geralyn L. Strong, and my wife, Joan, for their assistance in the preparation of this manuscript. I thank the U.S. Army for its present support of this research.

REFERENCES

Allen, R. C., 1973, Studies on the generation of electronic excitation states in human polymorphonuclear leukocytes and their participation in microbicidal activity, Dissertation, Tulane University, New Orleans, pp. 74–291, Xerox University Microfilms, Ann Arbor, Mich.

Allen, R. C., 1975a, Halide dependence of the myeloperoxidase-mediated antimicrobial system of the polymorphonuclear leukocyte in the phenomenon of electronic excitation, *Biochem. Biophys. Res. Commun.* **63**:675.

Allen, R. C., 1975b, The role of pH in the chemiluminescent response of the myeloperoxidase-halide-HOOH antimicrobial system, *Biochem. Biophys. Res. Commun.* **63**:684.

Allen, R. C., 1977, Evaluation of serum opsonic capacity by quantitating the initial chemiluminescent response from phagocytizing polymorphonuclear leukocytes, *Infect. Immun.* **15**:828.

Allen, R. C., 1979, Reduced, radical, and excited state oxygen in leukocyte microbicidal activity, in: *Frontiers in Biology*, Vol. 48 (J. T. Dingle, P. J. Jacques, and I. H. Shaw), pp. 197–233, North Holland, Amsterdam.

Allen, R. C., and Lint, T. F., 1979, Correlation of chemiluminescence to microbicidal metabolic response from polymorphonuclear leukocytes: A study of *in vitro* aging, in: *Analytical Applications of Bioluminescence and Chemiluminescence* (E. Schram and P. Stanley, eds.), pp. 589–600, State Printing and Publishing, Westlake Village, Calif.

Allen, R. C., and Loose, L. D., 1976, Phagocytic activation of a luminol-dependent chemiluminescence in rabbit alevolar and peritoneal macrophages, *Biochem. Biophys. Res. Commun.* **69**:245.

Allen, R. C., Stjernholm, R. L., and Steele, R. H., 1972, Evidence for the generation of (an) electronic excitation state(s) in human polymorphonuclear leukocytes and its participation in bactericidal activity, *Biochem. Biophys. Res. Commun.* **47**:679.

Allen, R. C. Stjernholm, R. L., Benerito, R. R., and Steele, R. H., 1973, Functionality of electronic excitation states in human microbicidal activity, in: *Chemiluminescence and Bioluminescence* (M. J. Cormier, D. M. Hercules, and J. Lee, eds.), p. 498, Plenum Press, New York.

Allen, R. C., Yevich, S. J., Orth, R. W., and Steel, R. H., 1974, The superoxide anion and singlet molecular oxygen: Their role in the microbicidal activity of the polymorphonuclear leukocyte, *Biochem. Biophys. Res. Commun.* **60**:909.

Allen, R. C., Stjernholm, R. L., Reed, M. A., Harper, T. B., III, Gupta, S., Steele, R. H., and Waring, W. W., 1977, Correlation of metabolic and chemiluminescent responses of granulocytes from three female siblings with chronic granulomatous disease, *J. Infect. Dis.* **136**:510.

Andersen, B. R., Brendzel, A. M., and Lint, T. F., 1977, Chemiluminescence spectra of human myeloperoxidase and polymorphonuclear leukocytes, *Infect. Immun.* **17**:62.

Anderson, S. M., and Krinsky, N. I., 1973, Protective action of carotenoid pigments against photodynamic damage to liposomes, *Photochem. Photobiol.* **18**:403.

Arneson, R. M., 1970, Substrate-induced chemiluminescence of xanthine oxidase and aldehyde oxidase, *Arch. Biochem Biophys.* **136**:352.

Babior, B. M., Kipnes, R. S., and Curnutte, J. T., 1973, Biological defense mechanisms: The production by leukocytes of superoxide, a potential bactericidal agent, *J. Clin. Invest.* **52**:741.

Babior, B. M., Curnutte, J. T., and Kipnes, R. S., 1975, Pyridine nucleotide-dependent superoxide production by a cell-free system from human granulocytes, *J. Clin. Invest.* **56**:1035.

Babior, B. M., Curnutte, J. T., and McMurrich, B. J., 1976, The particulate superoxide-forming system from human neutrophils: Properties of the system and further evidence supporting its participation in the respiratory burst, *J. Clin. Invest.* **58**:989.

Baldridge, C. W., and Gerard, R. W., 1933, The extra respiration of phagocytosis, *Am. J. Physiol.* **103**:235.

Beall, G. D., Repine, J. E., Hoidal, J. R., and Rasp, F. L., 1977, Chemiluminescence by human alveolar macrophages: Stimulation with heat-killed bacteria or phorbol myristate acetate, *Infect. Immun.* **17**:117.

Beauchamp, C., and Fridovich, I., 1970, A mechanism for production of ethylene from methional. The generation of the hydroxyl radical by xanthine oxidase, *J. Biol. Chem.* **245**:4641.

Behar, D., Czapski, G., Rabani, J., Dorfman, L. M., and Schwarz, H. A., 1970, The acid dissociation constant and decay kinetics of the perhydroxyl radical, *J. Phys. Chem.* **74**:3209.

Bielski, B. H. J., and Allen, A. O., 1977, Mechanism of the disproportionation of superoxide radicals, *J. Phys. Chem.* **81**:1048.

Blum, H. R., 1941, *Photodynamic Action and Diseases Caused by Light,* Reinhold, New York.

Calvert, J. G., and Pitts, J. N., 1966, *Photochemistry,* Wiley, New York.

Cheson, B. D., Christensen, R. L., Sperling, R., Kohler, B. E., and Babior, B. M., 1976, The origin of the chemiluminescence of phagocytosing granulocytes, *J. Clin. Invest.* **58**:789.

Clark, W. M., 1960, Oxidation-reduction potentials of organic systems, Williams and Wilkins, Baltimore.

Corey, E. J., and Taylor, W. C., 1964, A study of the peroxidation of organic compounds by externally generated singlet oxygen molecules, *J. Am. Chem. Soc.* **86**:3881.

Curnutte, J. T., Whitten, D. M., and Babior, B. M., 1974, Defective superoxide production by granulocytes from patients with chronic granulomatous disease, *N. Engl. J. Med.* **290**:593.

DeChatelet, L. R., Shirley, P. S., McPhail, L. C., Huntley, C. C., Muss, H. B., and Bass, D. A., 1977, Oxidative metabolism of the human eosinophil, *Blood* **50**:525.

Dolin, M. I., 1961, Cytochrome-independent electron transport enzymes of bacteria, in: *The Bacteria,* Vol. 2, *Metabolism* (I. C. Gunsalus and R. Y. Stanier, eds.), pp. 425–460, Academic Press, New York.

Evans, H. W., and Karnovsky, M. L., 1962, The biochemical basis of phagocytosis, *Biochemistry* **1**:159.

Fee, J. A., and DiCorleto, 1973, Observations on the oxidative-reduction properties of bovine erythrocyte superoxide dismutase, *Biochemistry* **12**:4893.

Fenical, W., Kearns, D. R., and Radlick, P., 1969, The mechanism of the addition of $^1\Delta_g$ excited oxygen to olefins. Evidence for a 1,2-dioxetane intermediate, *J. Am. Chem. Soc.* **91**:3396.

Foote, C. S., 1976, Photosensitized oxidation and singlet oxygen: Consequences in biological systems, in: *Free Radicals in Biology* (W. A. Pryor, ed.), Vol. 2, pp. 85–133, Academic Press, New York.

Foote, C. S., and Wexler, S., 1964, Olefin oxidation with excited singlet molecular oxygen, *J. Am. Chem. Soc.* **86**:3879.

Förster, T., 1959. Transfer mechanisms of electronic excitation, *Disc. Faraday Soc.* **27**:7.

Fridovich, I., 1976, Oxygen radicals, hydrogen peroxide, and oxygen toxicity, in: *Free Radicals in Biology* (W. A. Pryor, ed.), Vol. 1, pp. 239–277, Academic Press, New York.

Gee, J. B. L., Vassallo, C. L., Bell, P., Kaskin, J., Basford, R. E., and Field, J. B., 1970, Catalase-dependent peroxidative metabolism in alveolar macrophage during phagocytosis, *J. Clin. Invest.* **49**:1280.

Gray, H. B., 1965, *Electrons and Chemical Bonding,* Benjamin, New York.

Haber, F., and Weiss, J., 1934, The catalytic decomposition of hydrogen peroxide by iron salts, *Proc. Roy. Soc. Ser. A* **147**:332.

Hallett, F. R., Hallett, B. P., and Snipes, W., 1970, Reactions between singlet oxygen and the constituents of nucleic acids, *Biophys. J.* **10**:305.

Halliwell, B., 1976, An attempt to demonstrate a reaction between superoxide and hydrogen peroxide, *FEBS Lett.* **72**:8.

Hatch, G. E., Gardner, D. E., and Menzel, D. B., 1978, Chemiluminescence of phagocytic cells caused by N-formylmethionyl peptides, *J. Exp. Med.* **147**:182.

Hemming, V. G., Hall, R. T., Rhodes, P. G., Shigeoka, A. O., and Hill, H. R., 1976, Assessment of group B streptococcal opsonins in human and rabbit serum by neutrophil chemiluminescence, *J. Clin. Invest.* **58**:1379.

Henderson, W. R., and Kaliner, M., 1978, Immunologic and nonimmunoligic generation of superoxide from mast cells and basophils, *J. Clin. Invest.* **61**:187.

Holmes, B., Page, A. R., and Good, R. A., 1967, Studies of the metabolic activity of leukocytes from patients with a genetic abnormality of phagocytic function, *J. Clin. Invest.* **46**:1422.

Howes, R. M., and Steele, R. H., 1971, Microsomal chemiluminescence induced by NADPH and its relation to lipid peroxidation, *Res. Commun. Chem. Pathol. Pharmacol.* **2**:619.

Iyer, G. Y. N., Islam, D. M. F., and Quastel, J. H., 1961, Biochemical aspects of phagocytosis, *Nature* **192**:535.

Jodlbauer, A., and von Tappeiner, H., 1905, Die Beteiligung des Sauerstoffs bei der Wirkung fluorescierender Stoffe, *Deut. Arch. Klin. Med.* **82**:520.

Johnston, R. B., Keele, B. B., Misra, H. P., Lehmeyer, J. E., Webb, L. S., Baehner, R. L., and Rajagopalan, K. V., 1975, The role of superoxide anion generation in phagocytic bactericidal activity: Studies with normal and chronic granulomatous disease leukocytes, *J. Clin. Invest.* **55**:1357.

Kaplan, E. L., Laxdal, T., and Quie, P. Q., 1968, Studies of polymorphonuclear leukocytes from patients with chronic granulomatous disease of childhood: Bactericidal capacity for streptococci, *Pediatrics* **41**:591.

Karnovsky, M. L., Simmons, S., Karnovsky, M. J., Noseworthy, J., and Glass, E. A., 1971, Comparative studies on the metabolic basis of bactericidal activity in leukocytes, in: *Phagocytic Mechanisms in Health and Disease* (R. C. Williams and H. H. Fudenberg, eds.), pp. 67–82, International Medical Book, New York.

Kautsky, H., 1939, Quenching of luminescence by oxygen, *Trans. Faraday Soc.* **35**:216.

Kellogg, E. W., Fridovich, I., 1975, Superoxide, hydrogen peroxide, and singlet oxygen in lipid peroxidation by a xanthine oxidase system, *J. Biol. Chem.* **250**:8812.

Khan, A. U., 1970, Singlet molecular oxygen from superoxide anion and sensitized fluorescence of organic molecules, *Science* **168**:476.

Khan, A. U., and Kasha, M., 1963, Red chemiluminescence of molecular oxygen in aqueous solution, *J. Chem. Phys.* **39**:2105.

Klebanoff, S. J., 1968, Myeloperoxidase-halide-HOOH antibacterial system, *J. Bacteriol.* **95**:2131.

Klebanoff, S. J., 1970, Myeloperoxidase: Contribution to the microbicidal activity of intact leukocytes, *Science* **169**:1095.

Klebanoff, S. J., 1974, Role of the superoxide anion in the myeloperoxidase-mediated antimicrobial system, *J. Biol. Chem.* **249**:3724.

Klebanoff, S. J., and Rosen, H., 1978, Ethylene formation by polymorphonuclear leukocytes. Role of myeloperoxidase, *J. Exp. Med.* **148**:490.

Klebanoff, S. J., and White, L. R., 1969, Iodinating defect in the leukocytes of a patient with chronic granulomatous disease of childhood. *N. Engl. J. Med.* **280**:460.

Klebanoff, S. J., Clem, W. H., and Luebke, R. G., 1966, The peroxidase-thiocyanate hydrogen peroxide antimicrobial system, *Biochim. Biophys. Acta* **117**:63.

Klebanoff, S. J., Durack, D. T., Rosen, H., and Clark, R. A., 1977, Functional studies on human peritoneal eosinophils, *Infect. Immun.* **17**:167.

Knowles, P. J., Gibson, J. F., Pick, F. M., and Bray, R. C., 1969, Electron-spin-resonance evidence for enzymic reduction of oxygen to a free radical, the superoxide ion, *Biochem. J.* **111**:53.

Koppenol, W. H., 1976, Reactions involving singlet oxygen and the superoxide anion, *Nature* **262**:420.

Koppenol, W. H., and Butler, J., 1977, Mechanism of reactions involving singlet oxygen and the superoxide anion, *FEBS Lett.* **83**:1.

Krinsky, N. I., 1974, Singlet excited oxygen as a mediator of the antibacterial action of leukocytes, *Science* **186**:363.

McClune, G. J., and Fee, J. A., 1976, Stopped flow spectrophotometric observation of superoxide dismutation in aqueous solution, *FEBS Lett.* **67**:294.

McCord, J. M., and Fridovich, I., 1969, Superoxide dismutase. An enzymatic function for erythrocuprein, *J. Biol. Chem.* **244**:6049.

McCord, J. M., Keele, B. B., Jr., and Fridovich, I., 1971, An enzyme-based theory of obligate anaerobiosis: The physiological function of superoxide dismutase, *Proc. Natl. Acad. Sci. USA* **68**:1024.

McPhail, L. C., DeChatelet, L. R., Shirley, P. S., Wilfert, C., Johnston, R. B., Jr., and McCall, C. E., 1977, Deficiency of NADPH oxidase activity in chronic granulomatous disease, *J. Pediatr.* **90**:213.

McRipley, R. J., and Sbarra, A. J., 1967, Role of the phagocyte in host-parasite interaction. XII. Hydrogen peroxide-myeloperoxidase bactericidal system in the phagocyte, *J. Bacteriol.* **94**:1425.

Mahler, H. R., and Cordes, E. H., 1971, *Biological Chemistry,* Harper and Row, New York.

Mandell, G. L., 1975, Catalase, superoxide dismutase, and virulence of *Staphylococcus aureus: In vitro* and *in vivo* studies with emphasis on staphylococcal leukocyte interaction, *J. Clin. Invest.* **55**:561.

Mandell, G. L., and Hook, E. W., 1969, Leukocyte bactericidal activity in chronic granulomatous disease: Correlation of bacterial hydrogen peroxide production and susceptibility to intracellular killing, *J. Bacteriol.* **100**:531.

Massey, V., Strickland, S., Mayhew, S. G., Howell, L. G., Engel, P. C., Matthew, R. G., Schuman, M., and Sullivan, P. A., 1969, The production of superoxide radicals in the reaction of reduced flavins and flavoproteins with molecular oxygen, *Biochem. Biophys. Res. Commun.* **36**:891.

Mayeda, E. A., and Bard, A. J., 1974, Singlet oxygen. The suppression of its production in dismutation of superoxide ion by superoxide dismutase, *J. Am. Chem. Soc.* **96**:4023.

Morton, D. J., Moran, J. F., and Stjernholm, R. L., 1969, Carbohydrate metabolism in leucocytes. XI. Stimulation of eosinophils and neutrophils, *J. Reticuloendothel. Soc.* **6**:525.

Nelson, R. D., Mills, E. L., Simmons, R. L., and Quie, P. G., 1976, Chemiluminescence response of phagocytizing human monocytes, *Infect. Immun.* **14**:129.

Nelson, R. D., Herron, M. J., Schmidtke, J. R., and Simmons, R. L., 1977, Chemiluminescence response of human leukocytes: Influence of medium components on light production, *Infect. Immun.* **17**:513.

Nilsson, R., Merkel, P. B., and Kearns, D. R., 1972, Unambiguous evidence for the participation of singlet oxygen in photodynamic oxidation of amino acids, *Photochem. Photobiol.* **16**:117.

Ogryzlo, E. A., 1970, Physical properties of singlet oxygen, *Photophysiology* **5**:35.

Orchin, M., and Jaffé, H. H., 1967, *The Importance of Antibonding Orbitals*, Houghton Mifflin, Boston.

Orchin, M., and Jaffé, H. H., 1971, *Symmetry, Orbitals, and Spectra*, Wiley-Interscience, New York.

Patriarca, P., Cramer, R., Moncalvo, S., Rossi, F., and Romeo, D., 1971a, Enzymatic basis of metabolic stimulation in leucocytes during phagocytosis: The role of activated NADPH oxidase, *Arch. Biochem. Biophys.* **145**:255.

Patriarca, P., Cramer, R., Marussi, M., Moncalvo, S., and Rossi, F., 1971b, Phospholipid splitting and metabolic stimulation in polymorphonuclear leucocytes, *J. Reticuloendothel. Soc.* **10**:251.

Paul, B. B., and Sbarra, A. J., 1968, The role of the phagocyte in host-parasite interactions. XIII. The direct quantitative estimation of H_2O_2 in phagocytizing cells, *Biochim. Biophys. Acta.* **156**:168.

Pauling, L., 1939, *The Nature of the Chemical Bond*, Cornell University Press, Ithaca.

Phillips, C. S. G., and Williams, R. J. P., 1965, *Inorganic Chemistry*, Vol. 1, *Principles and Non-Metals*, Oxford University Press, New York and Oxford.

Porter, D. J. T., and Ingraham, L. L., 1974, Concerning the formation of singlet O_2 during the decomposition of H_2O_2 by catalase, *Biochim. Biphys. Acta* **334**:97.

Pryor, W. A., 1976, The role of free radical reactions in biological systems, in: *Free Radicals in Biology* (W. A. Pryor, ed.), Vol. 1, pp. 1–49, Academic Press, New York.

Raab, O., 1900, Ueber die Wirkung fluoreszirender Stoffe auf Infusorien, *Z. Biol.* **39**:524.

Rawls, H. R., and van Santen, P. J., 1970, Possible role for singlet oxygen in the initiation of fatty acid autoxidation, *J. Am. Oil Chem. Soc.* **47**:121.

Roberts, J. D., and Caserio, M. C., 1964, *Basic Principles of Organic Chemistry*, Benjamin, New York.

Rosen, H., and Klebanoff, S. J., 1976, Chemiluminescence and superoxide production by myeloperoxidase-deficient leukocytes, *J. Clin. Invest.* **58**:50.

Rosen, H., and Klebanoff, S. J., 1977, Formation of singlet oxygen by the myeloperoxidase-mediated antimicrobial system, *J. Biol. Chem.* **252**:4803.

Rossi, F., and Zatti, M., 1964, Changes in the metabolic pattern of polymorphonuclear leukocytes during phagocytosis, *Br. J. Exp. Pathol.* **45**:548.

Rossi, F., and Zatti, M., 1966, Effect of phagocytosis on the carbohydrate metabolism of polymorphonuclear leucocytes, *Biochem. Biophys. Acta* **121**:110.

Rossi, F., Patriarca, P., and Cramer, R., 1971, Effect of specific antibodies on the metabolism of guinea pig leukocytes, *J. Reticuloendothel. Soc.* **9**:67.

Rossi, F., Romeo, D., and Patriarca, P., 1972, Mechanism of phagocytosis-associated oxidative metabolism in polymorphonuclear leukocytes and macrophages, *J. Reticuloendothel. Soc.* **12**:127.

Sagone, A. L., King, G. W., and Metz, E. N., 1976, A comparison of the metabolic response to phagocytosis in human granulocytes and monocytes, *J. Clin. Invest.* **57**:1352.

Salin, M. L., and McCord, J. M., 1974, Superoxide dismutase in polymorphonuclear leukocytes, *J. Clin. Invest.* **54**:1005.

Sbarra, A. J., and Karnovsky, M. L., 1959, The biochemical basis of phagocytosis. I. Metabolic changes during the ingestion of particles by polymorphonuclear leukocytes, *J. Biol. Chem.* **234**:1355.

Schultz, J., and Kaminker, K., 1962, Myeloperoxidase of the leukocyte of normal human blood. I. Content and localization, *Arch. Biochem. Biophys.* **96**:465.

Sehested, K., Rasmussen, O. L., and Fricke, H., 1968, Rate constants of OH with HO_2, O_2^-, $H_2O_2^+$ from hydrogen peroxide formation in pulse-irradiated oxygenated water, *J. Phys. Chem.* **72**:626.

Selvaraj, R. J., and Sbarra, A. J., 1966, Relationship of glycolytic and oxidative metabolism to particle entry and destruction in phagocytosing cells, *Nature (London)* **211**:1272.

Smith, L. L., and Kulig, M. J., 1976, Singlet molecular oxygen from hydrogen peroxide disproportionation, *J. Am. Chem. Soc.* **98**:1027.

Stahelin, H., Suter, E., and Karnovsky, M. L., 1956, Studies on the interaction between phagocytes and tubercle bacilli. I. Observation on the metabolism of guinea pig leukocytes and the influence of phagocytosis, *J. Exp. Med.* **104**:121.

Stauff, J., Schmidkeenz, H., and Hartmann, G., 1963, Weak chemiluminescence of oxidation reactions, *Nature* **198**:281.

Steele, R. H., Allen, R. C., and Reed, M. A., 1976, Estimation of the spectra of chemiluminescence from polymorphonuclear leukocytes, *J. Reticuloendothel. Soc.* **20**:25a.

Steinman, H. M., Naik, V. R., Abernathy, J. L., and Hill, R. L., 1974, Bovine erythrocyte superoxide dismutase complete amino acid sequence, *J. Biol. Chem.* **249**:7326.

Stevens, P., and Young, L. S., 1977, Quantitative granulocyte chemiluminescence in the rapid detection of impaired opsonization of *Escherichia coli*, *Infect. Immun.* **16**:796.

Stjernholm, R. L., 1968, The Metabolism of Human Leucocytes, *Congr. Int. Soc. Hematol. 12th Plenary Session New York*, **5**, 175.

Stjernholm, R. L., Allen, R. C., Steele, R. H., Waring, W. W., and Harris, J. A., 1973, Impaired chemiluminescence during phagocytosis of opsonized bacteria, *Infect. Immun.* **7**:313.

Tauber, A. I., and Babior, B. M., 1977, Evidence for hydroxyl radical production by human neutrophils, *J. Clin. Invest.* **60**:374.

Turro, N. J., 1965, *Molecular Photochemistry*, Benjamin, New York.

Turro, N. J., and Devaquet, A., 1975, Chemiexcitation mechanisms. The role of symmetry and spin-orbit coupling in diradicals, *J. Am. Chem. Soc.* **97**:3859.

Turro, N. J., and Lechtken, P., 1973, Thermal generation of organic molecules in electronically excited states. Evidence for a spin forbidden, diabatic pericyclic reaction, *J. Am. Chem. Soc.* **95**:264.

Weiss, S. J., King, G. W., and LoBuglio, A. F., 1977, Evidence for hydroxyl radical generation by human monocytes, *J. Clin. Invest.* **60**:370.

Weser, U., Barth, G., Djerassi, C., Hartman, H. J., Krauss, P., Voelcken, G., Voelter, W., and Voetsch, W., 1972, A study on purified apo-erythrocuprein, *Biochim. Biophys. Acta.* **278**:28.

Weser, U., Paschen, W., and Younes, M., 1975, Singlet oxygen and superoxide dismutase, *Biochem. Biophys. Res. Comm.* **66**:769.

Wilson, T., and Schaap, A. P., 1971, The chemiluminescence from *cis*-diethoxy-1,2-dioxetane. An unexpected effect of oxygen, *J. Am. Chem. Soc.* **93**:4126.

Zatti, M., and Rossi, F., 1967, Relationship between glycolysis and respiration in surfactant treated leukocytes, *Biochim. Biophys. Acta.* **148**:553.

Zatti, M., Rossi, F., and Patriarca, P., 1968, The H_2O_2-production by polymorphonuclear leukocytes during phagocytosis, *Experientia* **24**:669.

13

The Role of Oxygen Radicals in Microbial Killing by Phagocytes

BERNARD M. BABIOR

1. INTRODUCTION

Recognition of the possibility that phagocytes might employ oxygen radicals as microbicidal agents dates from the discovery by McCord and Fridovich (1969) of the superoxide dismutases, a nearly ubiquitous group of enzymes, which catalyze the destruction of the oxygen radical O_2^- (superoxide) according to the reaction:

$$2O_2^- + 2H^+ \rightarrow O_2 + H_2O_2$$

The presence of one or more of these enzymes in virtually every living organism implies an indispensable role for superoxide dismutase in living systems, and from the nature of the dismutase-catalyzed reaction it may be inferred that this role is to protect such systems against fatal damage by O_2^- and its chemical descendants. The initial evidence supporting this assertion—namely, the reported absence of superoxide dismutase from obligate anaerobes (McCord et al., 1971)—has now been found to be incorrect, since newer and more sensitive methods for measuring the activity of the enzyme have shown it to be present, albeit in low concentrations, in *Clostridia* and other species which are killed by oxygen (Lumsden and Hall, 1975; Hewitt and Morris, 1975). However, the existence of a dismutaseless mutant of *Escherichia coli* which cannot survive in air (McCord et al., 1973), and the induction of superoxide dismutase in organisms (mammals as well as bacteria) stressed by oxygen or O_2^- (Hassan and Fridovich, 1977; Stevens and Autor, 1977), provides firm support for the notion that all organisms must protect themselves against damage by oxygen radicals, and that they do so at least in part by means of superoxide dismutase.

A corollary of the idea that organisms are obliged to protect themselves against oxygen radicals is that organisms when necessary may use oxygen radi-

---

BERNARD M. BABIOR • Department of Medicine and Blood Research Laboratory, Tufts-New England Medical Center, Boston, Massachusetts 02111.

cals for the destruction of pathogens. A role for oxygen in defense against pathogens has been recognized since the work of Baldridge and Gerard (1933), who showed that oxygen uptake by phagocytes rises sharply when these cells are exposed to ingestible particles. That this oxygen consumption reflects the generation of oxygen-derived antimicrobial agents, as opposed to nonspecific stimulation of the metabolic processes of the phagocyte, was shown by Sbarra and Karnovsky (1959), who demonstrated that this oxygen was not used to provide energy for the cell, and by Iyer *et al.* (1961), who found that at least some of it was converted to H_2O_2, and surmised correctly that this H_2O_2 was employed by the cell to kill bacteria. The immediate source of this H_2O_2 has proved to be O_2^-, which has been shown by several groups of workers (Babior *et al.*, 1973; Drath and Karnovsky, 1975; Weening *et al.*, 1975; Johnston *et al.*, 1975; Goldstein *et al.*, 1975) to be produced in impressive amounts by suitably stimulated phagocytes. Thus, oxygen radicals, known to be lethal to living systems, are turned to good use by phagocytic cells in their activities against pathogenic microorganisms.

2. KILLING BY OXYGEN RADICALS IN MODEL SYSTEMS

2.1. HIGH ENERGY RADIATION

The destruction of living systems by high-energy radiation is so well-known as to be almost a cliché. Rapidly proliferating cells such as myeloid precursors and gastrointestinal epithelium are particularly susceptible to damage by X rays and γ rays. This damage is manifested by the characteristic clinical findings seen in patients and animals exposed to such radiation. These include pancytopenia and gastroenterological symptoms such as nausea, vomiting, and diarrhea. In very large doses, high energy radiation kills by general toxicity affecting non-proliferating as well as proliferating cells, clinically manifested as an acutely fatal central nervous system syndrome (Prasad, 1974).

The passage of high energy radiation through tissues may cause damage through a direct interaction between the incident radiation and biological macromolecules (Altman, 1973). An example of such an interaction is the production of thymine dimers in DNA irradiated with ultraviolet light of suitable wavelengths (Beukers and Berends, 1961). However, a large proportion of the damage inflicted by high energy radiation is thought to be mediated by oxygen radicals. These originate from the radiolysis of water by the incident high energy photons, a process which generates hydrogen atoms and hydroxyl radicals (Dorfman and Adams, 1973).

$$H_2O \rightarrow H\cdot + O_2 \rightarrow HO_2^- \rightarrow H^+ + O_2^-$$

Dismutation of O_2^- gives rise to H_2O_2 and (possibly) singlet oxygen (Kellogg and Fridovich, 1975), a highly reactive electronically excited form of oxygen:

$$2O_2^- + 2H^+ \rightarrow H_2O_2 + {}^1O_2$$

As a result of these processes, highly reactive oxidizing species are formed which are capable of reacting with tissue constituents.

Evidence that oxygen radicals account for at least part of the damage caused by high energy radiation has been provided by a number of studies. Michelson and his associates showed that both mammalian cells [myeloblasts in culture (Michelson and Buckingham, 1974)] and a bacterial species [*Photobacterium leiognathi* (Lavelle *et al.*, 1973)] were protected against radiation-induced killing by superoxide dismutase. The radiation-mediated peroxidation of crude soybean phospholipids was found to be greatly attenuated in the presence of superoxide dismutase (Petkau and Chelack, 1976). The inactivation of the DNA of bacteriophage øX 174 by γ irradiation was antagonized to a substantial extent by superoxide dismutase as well as by catalase (van Hemmen and Meuling, 1975), indicating not only that oxidizing species were in part responsible for the inactivation of the macromolecule, but also that O_2^- and H_2O_2 were both among the substances mediating the inactivation reaction. These findings clearly implicate oxygen radicals as components of the mechanisms by which cells are killed by high energy radiation, and suggest that the lethal effects of these radicals result from their ability to damage DNA.

2.2. CHEMICAL O_2^--GENERATING SYSTEMS

Oxygen radicals have also been implicated in the action of a number of toxic agents of medical or scientific interest. Following are examples of such agents, together with the mechanism by which they are thought to generate the harmful radicals through which their effects are exerted.

2.2.1. Xanthine Oxidase

Xanthine oxidase has long been known to generate O_2^- during the course of its action (Fridovich and Handler, 1962). Dating even farther back is the knowledge of its antimicrobial activity, which was demonstrated forty years ago by Green and Pauli (1943). It is only with the advent of superoxide dismutase, however, that it has been possible to connect these two facts and show that oxygen radicals are important in the killing of bacteria by this system.

The first demonstration of a role for O_2^- in the bactericidal activity of xanthine oxidase was that of Fridovich and associates. These workers found that *Lactobacillus plantarum,* an organism devoid of superoxide dismutase, was highly susceptible to killing by xanthine oxidase, but that killing was prevented by the addition to that system of exogenous superoxide dismutase (Gregory and Fridovich, 1974). Similar observations were made with *E. coli* grown in limiting iron, conditions under which this organism fails to manufacture an iron-containing superoxide dismutase with which it is normally supplied (Gregory *et al.*, 1973). O_2^- is thus implicated in bacterial killing in these two experiments, in both of which microorganisms deficient in superoxide dismutase were employed.

The story is different with microorganisms containing a normal complement of superoxide dismutase (Gregory *et al.*, 1973; Babior *et al.*, 1975). With these organisms, killing is abolished by catalase, an enzyme which has no effect on the killing of *L. plantarum* or iron-starved *E. coli*. This observation identified H_2O_2 as a constituent of the antimicrobial activity of the xanthine oxidase system. With iron-replete *E. coli*, this was the only oxygen-derived agent necessary for bacterial killing, since exogenous superoxide dismutase afforded no protection. *Staphylococcus epidermidis*, however, was protected by superoxide dismutase as well as by catalase, indicating that the destruction of this organism required both O_2^- and H_2O_2. This evidence suggested that the true bactericidal agent in this case was the product of a reaction between O_2^- and H_2O_2. In other experiments, it had been shown that a potent oxidizing species was produced from O_2^- and H_2O_2 in the xanthine oxidase reaction (Beauchamp and Fridovich, 1970). This oxidizing species was postulated to be $OH\cdot$, produced by the so-called Haber–Weiss reaction:

$$O_2^- + H_2O_2 \rightarrow OH\cdot + OH^- + O_2$$

On this basis, it was proposed that $OH\cdot$ was the agent responsible for the destruction of *S. epidermidis*.

Whether $OH\cdot$ is actually produced by the reaction between O_2^- and H_2O_2 has been the subject of considerable controversy, owing to the exceedingly low rate of the Haber–Weiss reaction *in the absence of catalyst* (Halliwell, 1976). The presence in biological systems of transition metal ions [e.g., Fe (II)] which could serve as catalysts of this reaction softens these objections to a certain extent, but the occurrence of the Haber–Weiss reaction in biological systems has to date been neither proven nor disproven. Be that as it may, the fact remains that *S. epidermidis* incubated with the xanthine oxidase system is killed by the combination of O_2^- and $H2O_2$, compounds known to react together to generate a powerful oxidant which is likely to be the actual microbicidal agent. Whether or not this oxidant is $OH\cdot$ remains to be seen.

2.2.2. Antibiotics

Antibiotics can be broadly defined as compounds produced by one organism which have lethal effects on other living systems. Most of these exert their effects either through the alteration of the transport properties of biological membranes [e.g., valinomycin, nigericidin (Pressman, 1976), atractyloside (Klingenberg, 1970)] or by the inhibition of an enzyme whose activity is indispensable for life [e.g., amygdalin (Sayre and Kaymakcalan, 1964), diphtheria toxin (Gill *et al.*, 1969), α-amanitin (Kedinger *et al.*, 1970)]. Of interest in the present context, however, is a group of antibiotics which appear to kill by virtue of oxygen radical generation.

The prototype of this group of agents is streptonigrin. This compound, an antibiotic produced by the fungus *Streptomyces flocculus*, is an aminoquinone whose structure, shown in Figure 1, was determined by Rao *et al.* (1963). Studies on the bactericidal activity of this agent, using *E. coli* as the test organism,

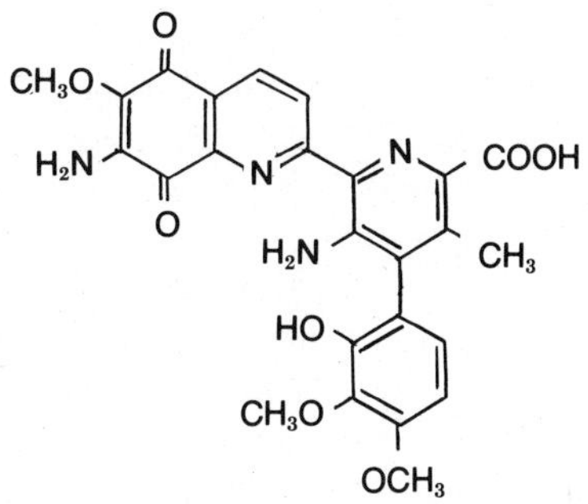

FIGURE 1.   Streptonigrin.

disclosed that killing was much slower under nitrogen than in air, and that organisms whose levels of superoxide dismutase had been augmented by growth under 100% oxygen were very resistant to the action of this antibiotic (Gregory and Fridovich, 1973). These observations strongly suggest that the lethal action of streptonigrin involves oxygen radicals. It had been known from earlier work that catalytic quantities of streptonigrin greatly accelerated the oxidation of reduced pyridine nucleotides by DT diaphorase, a mitochondrial flavoprotein (Hochstein *et al.*, 1965), and that the streptonigrin semiquinone radical could be observed in living organisms incubated with this agent in the absence of oxygen (White and Dearman, 1965). From these observations, it appears that in biological systems, streptonigrin undergoes a cyclic one-electron oxidation–reduction cycle whose net result is the formation of $O_2^{\cdot -}$ at the expense of reduced pyridine nucleotides:

$$2 \text{ Streptonigrin}_{ox} + \text{NAD(P)H} \rightarrow 2 \text{ Streptonigrin}_{semi}^{\cdot} + \text{NAD(P)}$$
$$\text{Streptonigrin}_{semi}^{\cdot} + O_2 \rightarrow \text{Streptonigrin}_{ox} + O_2^{\cdot -}$$

The $O_2^{\cdot -}$ produced in this sequence would then undergo secondary reactions leading to the formation of other oxidizing agents of even greater reactivity, including (among others) H_2O_2 and possibly $OH\cdot$ and $^1O_2$.

The oxidants generated by streptonigrin appear to kill by the destruction of DNA, a mode of action resembling that of high energy radiation. Thus, streptonigrin incubated with DNA in the presence of a reducing agent (which serves as the ultimate source of electrons for oxygen radical production) introduces single-strand breaks in the macromolecule by a mechanism which is blocked by superoxide dismutase or catalase (White and White, 1966; Mizuno and Gilboe, 1970; Cone *et al.*, 1976). Synthetic aminoquinones chemically related to streptonigrin can also carry out this process (Lown and Sim, 1976). The killing of *E. coli* by streptonigrin is accompanied by the conversion of a large fraction of the bacterial DNA to mononucleotides (White and White, 1968). This is thought to result from the hydrolysis of oxygen-radical-damaged DNA strands by the DNA-repairing systems of the cell.

Similar though less complete evidence has been presented to suggest that other antibiotics may also kill by means of oxygen radicals. Mitomycin C, an agent best known for its ability to cross-link DNA, can also introduce single-strand breaks into DNA by a mechanism which requires the presence of a reducing agent and is inhibited by catalase and the radical scavenger isopropanol (but not by superoxide dismutase) (Lown *et al.*, 1976). On the basis of these observations, hydroxyl radical (OH·) generated by the cyclic oxidation and reduction of mitomycin C has been proposed as the proximate DNA-cleaving species. Daunomycin and adriamycin, two closely related anthrocycline antibiotics currently employed in the chemotherapy of cancer, have also been shown to cleave DNA in the presence of a reducing agent by a superoxide-dismutase- and catalase-sensitive reaction (Lown *et al.*, 1977), a set of findings which implicate oxygen radicals in their action. These two agents have also been shown to catalyze NADPH oxidation, O_2^- production, and lipid peroxidation by a microsomal system containing cytochrome P-450 (Goodman and Hochstein, 1977). These results are best explained by postulating their participation as one-electron carriers between NADPH and oxygen. Bleomycin, another DNA-cleaving antibiotic used in the treatment of cancer, has also been postulated to act through oxygen radical generation. The evidence for this proposal is that the cleavage of DNA by bleomycin requires Fe^{2+} (Fe^{3+} will not work) and oxygen (Sausville *et al.*, 1976); it is highly likely that large quantities of oxygen radicals will be produced by this system, and the damage inflicted on DNA by such radicals has been amply documented. There are other examples of antibiotics and other chemicals whose harmful biological effects have been shown to relate to their ability to generate oxygen radicals in living systems. A detailed discussion of this topic, however, is beyond the scope of this review.

3. OXYGEN RADICALS AND PHAGOCYTE FUNCTION

The foregoing indicates that oxygen radicals have the capacity to destroy living systems. The lethal effects of this class of compounds have been exploited by phagocytes, which use oxygen radicals and related oxidizing agents to kill ingested microorganisms.

3.1. PRODUCTION OF OXYGEN RADICALS BY PHAGOCYTES: THE RESPIRATORY BURST

On contract with microorganisms, phagocytes undergo a profound change in oxygen metabolism, the function of which is to provide the oxygen radicals employed by these cells for the destruction of the bacteria. This metabolic change, which is termed the *respiratory burst* (Klebanoff, 1975; Cheson *et al.*, 1977; Babior, 1978), is characterized by large increases in oxygen uptake, O_2^- and H_2O_2 production, and hexosemonophosphate shunt activity in the stimulated cells. These changes are thought to be due to the activation of an enzyme, dormant in resting cells, which catalyzes the one-electron reduction of oxygen to O_2^- using a reduced pyridine nucleotide (most likely, NADPH) as electron donor (Patriarca *et al.*, 1971; Babior *et al.*, 1976; Babior and Kipnes, 1977):

$$NADPH + 2O_2 \rightarrow NADP^+ + H^+ + 2O_2^-$$

H_2O_2 production results from the dismutation of O_2^- (Root and Metcalf, 1977), a reaction which can occur spontaneously or under catalysis by superoxide dismutase:

$$2O_2^- + 2H^+ \rightarrow O_2 + H_2O_2$$

The increase in hexosemonophosphate shunt activity, generally measured as an increase in $^{14}CO_2$ production from [1-^{14}C]glucose, reflects the increase in $NADP^+$ formation in stimulated neutrophils. Part of this augmented $NADP^+$ production results from the activity of the O_2^--forming enzyme (see above), and part from the action of a glutathione-requiring system which serves to detoxify any H_2O_2 which may leak back into the cytoplasm of the activated cell (Reed, 1969). In the reactions of the hexosemonophosphate shunt, this newly formed $NADP^+$ is reconverted to NADPH by two sequential reactions whose net result is the oxidation of glucose-6-phosphate to CO_2 and a 5-carbon sugar phosphate:

$$Glucose\text{-}6\text{-}phosphate + NADP^+ \rightarrow 6\text{-}Phosphogluconic\ acid + NADPH$$
$$6\text{-}Phosphogluconic\ acid + NADP^+ \rightarrow Ribulose\text{-}5\text{-}phosphate + CO_2 + NADPH$$

Thus, the key reaction in the respiratory burst is the one-electron reduction of oxygen to O_2^-, a reaction which explains the increase in oxygen uptake and O_2^- production by stimulated phagocytes. H_2O_2 formation and hexosemonophosphate shunt activation are secondary consequences of the O_2^--forming reaction.

The evidence that oxygen radicals are important for the destruction of bacteria by phagocytes was obtained largely through studies of patients whose cells cannot mount a respiratory burst. There are two inherited conditions in which respiratory burst activity of phagocytes is absent: chronic granulomatous disease (CGD) (Berendes *et al.*, 1957; Quie *et al.*, 1967; Baehner and Nathan, 1967; Mandell and Hook, 1969 a,b; Baehner *et al.*, 1972; Gray *et al.*, 1973; Curnutte *et al.*, 1974) and severe glucose-6-phosphate dehydrogenase deficiency (Baehner *et al.*, 1972; Gray *et al.*, 1973). In both of these conditions, neutrophils fail to show the normal increase in oxygen uptake, O_2^- and H_2O_2 production, and hexosemonophosphate shunt activity on exposure to stimuli which activate the respiratory burst in normal cells (Quie *et al.*, 1967; Baehner and Nathan, 1967; Mandell and Hook, 1969a,b; Baehner *et al.*, 1972; Gray *et al.*, 1973; Curnutte *et al.*, 1974). The affected neutrophils show a characteristic bactericidal defect, in that they are unable to kill *Staphylococcus aureus* and Enterobacteriaceae. Both of these organisms dispose of endogenously manufactured H_2O_2 by means of catalase. They are, however, able to act efficiently against pneumococci and streptococci, catalase-negative bacteria which dispose of H_2O_2 by excreting it into the medium, and thereby provide the defective neutrophils with the oxidizing agents which they themselves cannot generate (Mandell and Hook, 1969a,b; Johnston and Baehner, 1970). This *in vitro* bactericidal defect is reflected in the clinical behavior of patients with these conditions. These patients are very susceptible to bacterial infections by organisms which their neutrophils cannot kill, but have no trouble with catalase-negative bacteria. It is clear from these findings that respiratory burst activity with its attendant oxygen radical production is necessary for the normal bactericidal activity of the phagocyte.

The enzyme responsible for the oxygen-consuming reaction of the respiratory burst has been the subject of controversy for many years. This controversy has centered around which of two candidate pyridine nucleotide oxidases is the burst oxidase. One of the candidate oxidases is found as an activity in the particulate fraction of neutrophil homogenates which seems to prefer NADPH as the reducing agent (generally referred to as the "NADPH oxidase"), while the other is a soluble enzyme specific for NADH (referred to as the "NADH oxidase") (Cheson et al., 1977). The discovery that O_2^- was the first product of the respiratory burst and hence had to be the direct product of the burst oxidase, a finding which was made relatively recently in terms of the duration of the controversy, and the observation that a requirement for Mn^{2+} originally assigned to the particulate oxidase was an experimental artifact (Curnutte et al., 1976; Patriarca et al., 1975b), has led, in the writer's view, to a resolution of the controversy. The preponderance of evidence supports the view that the burst oxidase is in fact the NADPH oxidase. This activity, which can be demonstrated in particles isolated from normal neutrophils which have been stimulated with suitable activating agents, but is missing from particles isolated from resting neutrophils or from stimulated neutrophils obtained from patients with CGD (Babior et al., 1976; Babior and Kipnes, 1977; Hohn and Lehrer, 1975; Curnutte et al., 1975), appears to be a plasma-membrane-associated flavoprotein which catalyzes the reaction:

$$2O_2 + NAD(P)H \rightarrow 2O_2^- + NADP^+ + H^+$$

Either pyridine nucleotide will serve as electron donor, but the enzyme seems to favor NADPH, as indicated by the finding that the K_m for NADPH (20 μM at pH 6.0) is one-fiftieth that for NADH (Babior et al., 1976). Further evidence that NADPH is the physiological electron donor is the defect in respiratory burst activity seen in glucose-6-phosphate dehydrogenase deficiency. In this condition there is a marked decrease in the intracellular NADPH concentration, but a much smaller fall in the intracellular concentration of NADH, a species provided mainly by a pathway (glycolysis) in which glucose-6-phosphate dehydrogenase does not participate (Baehner et al., 1972). Glucose-6-phosphate dehydrogenase deficiency thus impairs the respiratory burst through a deficiency of substrate, while the respiratory burst defect in CGD is due to a deficiency of the oxidase itself.

3.2. HOW PHAGOCYTES USE RESPIRATORY BURST PRODUCTS TO KILL BACTERIA

It was the discovery that H_2O_2 is produced by stimulated neutrophils that led to the postulate that the function of the respiratory burst was to provide the phagocyte with microbicidal agents. The antibacterial action of H_2O_2 had been known for some time, and it was logical to postulate that this compound served in this capacity in phagocytes. The presence of large quantities of ascorbic acid in phagocytes suggested an extension of this notion, namely, that the relatively weak microbicidal activity of H_2O_2 could be potentiated in the phagocyte by the combination of ascorbic acid and a suitable metal ion (particularly Cu^{2+}) a phenomenon which had been demonstrated *in vitro* (Drath and Karnovsky, 1974). However, the most powerful microbicidal system in which H_2O_2 participates, and the one whose role in the microbicidal action of phagocytes has been most convincingly demonstrated, is that involving myeloperoxidase (Klebanoff, 1967a,b).

Myeloperoxidase is a heme enzyme of molecular weight 150,000 (Ehrenberg and Agner, 1958) which is present in the azurophilic granules of neutrophils (Bainton and Farquhar, 1968). It is there in very large amounts, comprising nearly 5% of the dry weight of human neutrophils. During phagocytosis, the myeloperoxidase-containing granules fuse with the primary phagosome, delivering the enzyme into the intraphagosomal space around the ingested microorganism. H_2O_2 is also being delivered into the phagosome in large quantities, produced by the action of the burst oxidase, an enzyme known to be a component of the plasma membrane, and, consequently, of the wall of the phagosome, which consists in part of invaginated plasma membrane. In the presence of a suitable halide or pseudohalide ion, the combination of myeloperoxidase and H_2O_2 form an exceedingly powerful microbicidal system, much more powerful than H_2O_2 by itself or in combination with ascorbate plus Cu^{2+} (Klebanoff, 1967a,b; McRipley and Sbarra, 1967; Klebanoff, 1968). Any of several halide or halide-like ions will serve in this reaction, including Cl^-, Br^-, I^-, or SCN^-; the physiological halide is probably Cl^-, since it is present in large quantities within the phagocyte.

Several mechanisms have been proposed to explain the killing of microorganisms by the peroxide–halide–myeloperoxidase system, but none have been proven beyond doubt. The action of this system on ingested bacteria has been shown to result in halogenation of the microorganisms [demonstrated with both I^- (Pincus and Klebanoff, 1971) and Cl^- (Zgliczynski *et al.*, 1968)], conversion of their constituent amino acids to aldehydes by oxidative decarboxylation (Strauss *et al.*, 1971), and peroxidation of their unsaturated fatty acids (Shohet *et al.*, 1974; Stossel *et al.*, 1974). The extent to which these processes contribute to the death of the organism remains to be determined, though it seems exceedingly unlikely that effects as disruptive as, for example, the iodination of major portions of the cell membrane would be without far-reaching deleterious effects.

Although the peroxide–halide–myeloperoxidase system is securely established as an important oxygen-dependent bactericidal system of phagocytes, studies on patients with familial myeloperoxidase deficiency have demonstrated the existence of oxygen-dependent bactericidal systems which do not require myeloperoxidase for their operation. The evidence for this is that patients with myeloperoxidase deficiency, in contrast to those with congenital respiratory burst defects (CGD and glucose-6-phosphate dehydrogenase deficiency), tend to have few problems with bacterial infections (Lehrer and Cline, 1969; Patriarca *et al.*, 1975a; Stendahl and Lindgren, 1976). Moreover, their neutrophils, though showing delayed killing when tested with *S. aureus,* will eventually destroy all the ingested microorganisms (Klebanoff, 1970), while survival of ingested *S. aureus* is characteristic of CGD neutrophils. Absence of the respiratory burst thus leads to a much more serious defect in bacterial killing than does the absence of myeloperoxidase, indicating that there must be bactericidal mechanisms which do not need myeloperoxidase but which require oxidizing agents produced in the respiratory burst.

Four oxidizing agents have been considered in connection with these myeloperoxidase-independent but oxygen-requiring bacterial killing mechanisms of phagocytes: O_2^-, H_2O_2, singlet oxygen, and $OH\cdot$ (the hydroxyl radical). H_2O_2 had been discussed above. O_2^- was proposed some time ago as a candidate bactericidal agent (Babior *et al.*, 1973), but the results of several studies of bacterial killing by O_2^--generating systems *in vitro* have cast serious doubt on this idea (Gregory and Fridovich, 1974; Gregory *et al.*, 1973; Babior *et al.*, 1975). Singlet oxygen and $OH\cdot$, however, both remain under consideration, and there is a considerable amount of evidence regarding the possible participation of each of these species in bacterial killing by phagocytes.

Singlet oxygen was the first of these two to be studied experimentally. This is a highly reactive, electronically excited form of oxygen which could, in principle, be generated by several reactions involving intermediates produced by stimulated phagocytes, including the reaction between H_2O_2 and the hypochlorite produced in the myeloperoxidase reaction (Selinger, 1960)

$$H_2O_2 + OCl^- \rightarrow {}^1O_2 + H_2O + Cl^-$$

the spontaneous dismutation of O_2^- (Kahn, 1970)

$$2O_2^- + 2H^+ \rightarrow {}^1O_2 + H_2O_2$$

and the Haber–Weiss reaction (Haber and Weiss, 1934), a postulated reaction between O_2^- and H_2O_2 whose existence is under considerable dispute (Halliwell, 1976)

$$O_2^- + H_2O_2 \rightarrow {}^1O_2 + OH\cdot + OH^-$$

The first observations interpreted in terms of singlet oxygen production by phagocytes was the finding that light was emitted by phagocytizing neutrophils (Allen *et al.*, 1972). This light was postulated to represent the photons emitted when the electronically excited singlet oxygen relaxed to the ground state (atmospheric oxygen). Subsequent work, however, showed that this light emission merely represented an oxidation of the phagocytized particle by the products of the respiratory burst, and provided no information as to the identity of the oxidizing agent (Cheson *et al.*, 1976). The finding that a carotenoid-deficient mutant of *Sarcina lutea* was much more susceptible to killing by neutrophils than the wild type was also interpreted as evidence for singlet oxygen production, on the grounds that carotenoids are efficient singlet oxygen scavengers, and their absence would be expected to sensitize the microorganism toward killing by this agent (Krinsky, 1974). This interpretation, however, requires that carotenoids react specifically with singlet oxygen, and not at all with any of the other oxidants that might be formed during the respiratory burst; such specificity has not as yet been demonstrated. A similar lack of specificity confuses the interpretation of other studies on singlet oxygen production by phagocytes and by systems derived from phagocytes, most of which are based on the use of singlet oxygen scavengers or singlet oxygen trapping reagents of less than absolute specificity to demonstrate that singlet oxygen is produced in the reactions under study (Krinsky, 1974). In conclusion, it is probably fair to state that the evidence for the involvement of singlet oxygen in bacterial killing by phagocytes is somewhat ambiguous.

In contrast to singlet oxygen, whose production by phagocytes is open to considerable doubt, the generation of $OH\cdot$ (or at least of a species more reactive than either O_2^- or H_2O_2, whose chemical properties resemble those of $OH\cdot$) has been demonstrated in several studies. The evidence that this species is produced by phagocytes is based on the finding that $OH\cdot$ will release ethylene from methional (3-thiomethylpropionaldehyde) and 2-keto-4-thiomethylbutyric acid (KMB). When either of these agents is incubated with stimulated phagocytes [both neutrophils (Tauber and Babior, 1977; Weiss *et al.*, 1978) and monocytes (Weiss *et al.*, 1977) have been used], ethylene is released into the overlying atmosphere. Ethylene release is inhibited by $OH\cdot$ scavengers, and does not occur in the absence of a stimulus, or in the presence of a stimulus if the cells used are from a patient with CGD. These results are consistent with the idea that $OH\cdot$ is produced by phagocytes undergoing the respiratory burst. Whether $OH\cdot$ itself is responsible for the ethylene production observed in these systems is open to some question, because many other reactive radical species, including alkyl radicals $(R\cdot)$, alkoxyl radicals $(RO\cdot)$ and alkyl hydroperoxyl radicals $(ROO\cdot)$ also release ethylene from methional (Pryor and Tang, 1978). Regardless of the actual identity of the methional-oxidizing radical, however, the available data suggest that it is likely to have ultimately been derived from $OH\cdot$ produced by activated phagocytes.

In other O_2^--generating systems in which evidence for $OH\cdot$ production has been obtained, the Haber–Weiss reaction was proposed as its source. To determine whether this reaction could account for $OH\cdot$ production by phagocytes, the effect of superoxide dismutase and catalase on ethylene production was examined. The Haber–Weiss reaction predicts that $OH\cdot$ production should be abated by either of these enzymes. This was, in fact, what was observed when KMB was used as the detection reagent (Weiss *et al.*, 1978). With methional, however, catalase had little effect on ethylene production, though superoxide dismutase was as effective with methional as with the keto acid (Tauber and Babior, 1977; Weiss *et al.*, 1977). These observations could be explained if it is assumed that the real $OH\cdot$-forming reaction employs as reactants O_2^- and an alkyl or acyl hydroperoxide, rather than O_2^- and H_2O_2 (Pryor, 1978). Methional, as an aldehyde, would be contaminated with a certain amount of acyl hydroperoxide, formed by the spontaneous reaction of the aldehyde with atmospheric oxygen, so an additional source of peroxide would not be required and $OH\cdot$ production would be sensitive to catalase. With KMB, however, H_2O_2

would be necessary to produce the hydroperoxide used for OH· formation, so that with this detecting system catalase as well as superoxide dismutase would abolish OH· formation.

What is the evidence for the involvement of OH·, or radicals derived from OH·, in bacterial killing by phagocytes? Extensive observations with chemical models have shown that bacteria are killed by systems which generate O_2^-, and that killing by these systems is inhibited by superoxide dismutase and catalase. These findings suggest that reactive radicals which are produced by O_2^--generating systems are able to act as microbicidal agents. Similar observations made with phagocytes have implicated such radicals in killing by these cells. Experiments with neutrophils showed that the killing of three different pathogenic microorganisms—*S. aureus*, *E. coli* and *Streptococcus viridans*—was almost abolished in the presence of superoxide dismutase or catalase, provided latex particles were present in the reaction mixture to aid the uptake of the enzyme into the phagocytic vesicle (Johnston *et al.*, 1975). From these results, it seems likely that OH· or similarly reactive species participate along with the peroxide–halide–myeloperoxidase system in oxygen-dependent microbial killing by phagocytes.

REFERENCES

Allen, R. C., Stjernholm, R. L., and Steele, R. H., 1972, Evidence for the generation of an electronic excitation state(s) in human polymorphonuclear leukocytes and its participation in bactericidal activity, *Biochem. Biophys. Res. Comm.* **47**:679.

Altman, K. I., 1973, Radiation chemistry, in: *Medical Radiation Biology* (G. V. Dalrymple, M. E. Gaulden, G. M. Kollamorgen, and H. H. Vogel, eds.), p. 15, Saunders, Philadelphia.

Babior, B. M., 1978, Oxygen-dependent microbial killing by phagocytes, *N. Engl. J. Med.* **298**:659, 721.

Babior, B. M., and Kipnes, R. S., 1977, Superoxide-forming enzyme from human neutrophils: Evidence for a flavin requirement, *Blood* **50**:517.

Babior, B. M., Kipnes, R. S., and Curnutte, J. T., 1973, Biological defense mechanisms. The production by leukocytes of superoxide, a potential bactericidal agent, *J. Clin. Invest.* **52**:741.

Babior, B. M., Curnutte, J. T., and Kipnes, R. S., 1975, Biological defense mechanisms. Evidence for the participation of superoxide in bacterial killing by xanthine oxidase, *J. Lab. Clin. Med.* **85**:235.

Babior, B. M., Curnutte, J. T., and McMurrich, B. J., 1976, The particulate superoxide-forming system from human neutrophils. Properties of the system and further evidence supporting its participation in the respiratory burst, *J. Clin. Invest.* **58**:989.

Baehner, R. L., and Nathan, D. G., 1967, Leukocyte oxidase: Defective activity in chronic granulomatous disease, *Science* **155**:835.

Baehner, R. L., Johnston, R. B., Jr., and Nathan, D. G., 1972, Comparative study of the metabolic and bactericidal characteristics of severely glucose-6-phosphate dehydrogenase deficient polymorphonuclear leukocytes and leukocytes from children with chronic granulomatous disease, *J. Reticuloendothel. Soc.* **12**:150.

Bainton, D. F., and Farquhar, M. G., 1968, Differences in enzyme content of azurophil and specific granules of polymorphonuclear leukocytes. II. Cytochemistry and electron microscopy of bone marrow cells, *J. Cell Biol.* **39**:299.

Baldridge, C. W., and Gerard, R. W., 1933. The extra respiration of phagocytosis, *Am. J. Physiol.* **103**:235.

Beauchamp, C., and Fridovich, I., 1970, A mechanism for the production of ethylene from methional. The generation of hydroxyl radical by xanthine oxidase, *J. Biol. Chem.* **245**:4641.

Berendes, H., Bridges, R. A., and Good, R. A., 1957, Fatal granulomatosis of childhood. Clinical study of a new syndrome, *Minn. Med.* **40**:309.

Beukers, R., and Berends, W., 1961, The effects of UV-irradiation on nucleic acids and their components, *Biochem. Biophys. Acta* **49**:181.

Cheson, B. D., Christensen, R. L., Sperling, R., Kohler, B. E., and Babior, B. M., 1976, The origin of the chemiluminescense of phagocytosing granulocytes, *J. Clin. Invest.* **58**:789.

Cheson, B. D., Curnutte, J. T., and Babior, B. M., 1977, The oxidative killing mechanisms of the neutrophils, *Prog. Clin. Immunol.* **3**:1.

Cone, R., Hasan, S. K., Lown, J. W., and Morgan, A. R., 1976, The mechanism of the degradation of DNA by streptonigrin, *Can. J. Biochem.* **54**:219.

Curnutte, J. T., J. T., Whitten, D. M., and Babior, B. M., 1974, Defective superoxide production by granulocytes from patients with chronic granulomatous disease. *N. Engl. J. Med.* **290**:593.

Curnutte, J. T. Kipnes, R. S., and Babior, B. M., 1975, Defect in pyridine nucleotide dependent superoxide production by a particulate fraction from the granulocytes of patients with chronic granulomatous disease, *N. Engl. J. Med.* **293**:628.

Curnutte, J. T., Karnovsky, M. L., and Babior, B. M., 1976, Manganese-dependent NADPH oxidation by granulocyte particles. The role of superoxide and the nonphysiological nature of the manganese requirement, *J. Clin. Invest.* **57**:1059.

Dorfman, L. M., and Adams, G. E., 1973, Reactivity of the hydroxyl radical in aqueous solutions, NSRDS-NBS No. 46, U.S. Dept. of Comm. Natl. Bur. of Stand., pp. 4–6.

Drath, D. B., and Karnovsky, M. L., 1974, Bactericidal activity of metal-mediated peroxide-ascorbate systems, *Infect. Immun.* **10**:1077.

Drath, D. B., and Karnovsky, M. L., 1975, Superoxide production by phagocytic leukocytes, *J. Exp. Med.* **141**:257.

Ehrenberg, A., and Agner, K., 1958, The molecular weight of myeloperoxidase, *Acta Chem. Scand.* **12**:95.

Fridovich, I., and Handler, P., 1962, Xanthine oxidase. V. Differential inhibition of the reduction of various electron acceptors, *J. Biol. Chem.* **237**:916.

Gill, D. M., Pappenheimer, A. M., Brown, R., and Kurnick, J. T., 1969, Studies on the mode of action of diphtheria toxin. VII. Toxin-stimulated hydrolysis of nicotinamide adenine dinucleotide in mammalian cell extracts, *J. Exp. Med.* **129**:1.

Goldstein, I. M., Roos, D., Kaplan, H. B., and Weissman, G., 1975, Complement and immunoglobulins stimulate superoxide production by human neutrophils independently of phagocytosis, *J. Clin. Invest.* **56**:1155.

Goodman, J., and Hochstein, P., 1977, Generation of free radicals and lipid peroxidation by redox cycling of adriamycin and daunomycin, *Biochem. Biophys. Res. Comm.* **77**:797.

Gray, G. R., Stomatoyannopoulos, G., Naiman, S. C., Kilman, N. R., Klebanoff, S. J., Austin, T., Yoshida, A., and Robinson, G. C. F., 1973, Neutrophil dysfunction, chronic granulomatous disease, and nonspherocytic hemolytic anemia caused by complete deficiency of glucose-6-phosphate dehydrogenase, *Lancet* **2**:530.

Green, D. E., and Pauli, R., 1943, The antibacterial action of the xanthine oxidase system, *Proc. Soc. Exp. Biol. Med.* **54**:148.

Gregory, E. M., and Fridovich, I., 1973, Oxygen toxicity and superoxide dismutase, *J. Bacteriol.* **114**:1193.

Gregory, E. M., and Fridovich, I., 1974, Oxygen metabolism in *Lactobacillus plantarum*, *J. Bacteriol.* **117**:166.

Gregory, E. M., Yost, F. J., Jr., and Fridovich, I., 1973, Superoxide dismutases of *Escherichia coli*: Intracellular localization and functions, *J. Bacteriol.* **115**:987.

Haber, F., and Weiss, J., 1934, The catalytic decomposition of hydrogen peroxide by iron salts, *Proc. R. Soc. Edinburgh Sect. A* **147**:332.

Halliwell, B., 1976, An attempt to demonstrate a reaction between superoxide and hydrogen peroxide, *FEBS Lett.* **72**:8.

Hassan, H. M., and Fridovich, I., 1977, Enzymatic defenses against the toxicity of oxygen and of streptonigrin in *Excherichia coli*, *J. Bacteriol.* **129**:1574.

352   BERNARD M. BABIOR

Hewitt, J., and Morris, J. G., 1975, Superoxide dismutase in some obligately anaerobic bacteria, *FEBS Lett.* **50**:315.

Hochstein, P., Laszlo, J., and Miller, D., 1965, A unique, dicoumarol-sensitive, nonphosphorylating oxidation of DPNH and TPNH catalyzed by streptonigrin, *Biochem. Biophys. Res. Comm.* **19**:289.

Hohn, D. C., and Lehrer, R. I., 1975, NADPH oxidase deficiency in X-linked chronic granulomatous disease, *J. Clin. Invest.* **55**:707.

Iyer, G. Y. N., Islam, M. F., and Quastel, J. H., 1961, Biochemical aspects of phagocytosis, *Nature* **192**:535.

Johnston, R. B., Jr., and Baehner, R. L., 1970, Improvement of leukocyte bactericidal activity in chronic granulomatous disease, *Blood* **35**:350.

Johnston, R. B., Jr., Keele, B. B., Jr., Misra, H. P., Lehmeyer, J. E., Webb, L. S., Baehner, R. L., and Rajagopalan, K. V., 1975, The role of superoxide anion generation in phagocytic bactericidal activity. Studies with normal and chronic granulomatous disease leukocytes, *J. Clin. Invest.* **55**:1357.

Kedinger, C., Gniazdowski, M., Mandel, J. L., Gissinger, F., and Chambon, P., 1970, α-Amanitin: A specific inhibitor of one of two DNA-dependent RNA polymerase activities from calf thymus, *Biochem. Biophys. Res. Comm.* **38**:165.

Kellogg, E. W., and Fridovich, I., 1975, Superoxide, hydrogen peroxide, and singlet oxygen in lipid preoxidation by a xanthine oxidase system, *J. Biol. Chem.* **250**:8812.

Kahn, A. U. 1970. Singlet molecular oxygen from superoxide anion and sensitized fluorescence of organic molecules, *Science* **168**:476.

Klebanoff, S. J., 1967a, A peroxidase-mediated antimicrobial system in leukocytes, *J. Clin. Invest.* **46**:1078.

Klebanoff, S. J., 1967b, Iodination of bacteria: A bactericidal mechanism, *J. Exp. Med.* **126**:1063.

Klebanoff, S. J., 1968, Myeloperoxidase-halide-hydrogen peroxide antibacterial system, *J. Bacteriol.* **95**:2131.

Klebanoff, S. J., 1970, Myeloperoxidase: Contribution to the microbicidal activity of intact leukocytes, *Science* **169**:1095.

Klebanoff, S. J., 1975, Antimicrobial mechanisms in neutrophilic polymorphonuclear leukocytes, *Semin. Hematol.* **12**:117.

Klingenberg, M., 1970, Mitochondria metabolite transport, *FEBS Lett.* **6**:145.

Krinsky, N. I., 1974, Singlet excited oxygen as a mediator of the antibacterial action of leukocytes, *Science* **186**:363.

Lavelle, F., Michelson, A. M., and Dimitrijevic, L., 1973, Biologocial protection by superoxide dismutase, *Biochem. Biophys. Res. Comm.* **55**:350.

Lehrer, R. I., and Cline, M. J., 1969, Leukocyte myeloperoxidase deficiency and disseminated candidiasis: The role of myeloperoxidase in resistance to *Candida* infection, *J. Clin. Invest.* **48**:1478.

Lown, J. W., and Sim, S. K., 1976, Studies related to antitumor antibiotics. Part VIII. Cleavage of DNA by streptonigrin analogues and the relationship to antineoplastic activity, *Can. J. Biochem.* **54**:446.

Lown, J. W., Begleiter, A., Johnson, D., and Morgan, A. R., 1976, Studies related to antitumor antibiotics. Part V. Reactions of mitomycin C with DNA examined by ethidium fluorescence assay, *Can. J. Biochem.* **54**:110.

Lown, J. W., Sim, S. K., Majumdar, K. C., and Chang, R. Y., 1977, Strand scission of DNA by bound adriamycin and daunorubicin in the presence of reducing agents, *Biochem. Biophys. Res. Comm.* **76**:705.

Lumsden, J., and Hall, D. O., 1975, Superoxide dismutase in photosynthetic organisms provides an evolutionary hypothesis, *Nature* **257**:670.

Mandell, G. L., and Hook, E. W., 1969a, Leukocyte bactericidal activity in chronic granulomatous disease: Correlation of bacterial hydrogen peroxide production and susceptibility to intracellular killing, *J. Bacteriol.* **100**:531.

Mandell, G. L., and Hook, E. W., 1969b, Leukocyte function in chronic granulomatous disease of childhood. Studies on a seventeen year old boy, *Am. J. Med.* **47**:473.

McCord, J. M., and Fridovich, I., 1969, Superoxide dismutase, an enzymatic function for erythrocuprein (hemocuprein), *J. Biol. Chem.* **244**:6049.

McCord, J. M., Keele, B. B., Jr., and Fridovich, I., 1971, An enzyme-based theory of obligate anaerobiosis: The physiological function of superoxide dismutase, *Proc. Natl. Acad. Sci. USA* **68**:1024.

McCord, J. M., Beauchamp, C. O., Goscin, S., Misra, H. P., and Fridovich, I. Superoxide and superoxide dismutase, 1973, in: *Oxidases and Related Redox Systems*, Vol. 1 (T. E. King, H. S. Mason, and M. Morrison, eds.), pp. 51–76. University Park Press, Baltimore.

McRipley, R. J., and Sbarra, A. J., 1967, Role of the phagocyte in host-parasite interactions. XII. Hydrogen peroxide-myeloperoxidase bactericidal system in the phagocyte, *J. Bacteriol.* **94**:1425.

Michelson, A. M., and Buckingham, M. E., 1974, Effects of superoxide radicals on myoblast growth and differentiation, *Biochem. Biophys. Res. Comm.* **58**:1079.

Mizuno, N. S., and Gilboe, D. P., 1970, Bindings of streptonigrin to DNA, *Biochim. Biophys. Acta* **224**:319.

Patriarca, P., Cramer, R., Moncalvo, S., Rossi, F., and Romeo, D., 1971, Enzymatic basis of metabolic stimulation in leukocytes during phagocytosis: The role of activated NADPH oxidase, *Arch. Biochem. Biophys.* **145**:255.

Patriarca, P., Cramer, R., Tedesco, F., and Kakinuma, K., 1975a, Studies on the mechanism of metabolic stimulation in polymorphonuclear leukocytes during phagocytosis. II. Presence of the $NADPH_2$ oxidizing activity in a myeloperoxidase-deficient subject, *Biochim. Biophys. Acta* **385**:387.

Patriarca, P., Dri, P., Kakinuma, K., Tedesco, F., and Rossi, F., 1975b, Studies on the mechanism of metabolic stimulation in polymorphonuclear leukocytes during phagocytosis. I. Evidence for superoxide anion involvement in the oxidation of $NADPH_2$, *Biochim. Biophys. Acta* **385**:380.

Petkau, A., and Chelack, W. S., 1976, Radioprotective effect of superoxide dismutase on model phospholipid membranes, *Biochim. Biophys. Acta* **433**:445.

Pincus, S. H., and Klebanoff, S. J., 1971, Quantitative leukocyte iodination, *N. Engl. J. Med.* **284**:744.

Prasad, K. N., 1974, *Human Radiation Biology*, Harper and Row, New York, pp. 149–190.

Pressman, B. C., 1976, Biological applications of ionophores, *Annu. Rev. Biochem.* **45**:501.

Pryor, W. A., 1978, The formation of free radicals and the consequences of their reactions *in vivo*, *Photochem. Photobiol.* **28**:787.

Pryor, W. A., and Tang, R. H., 1978, Ethylene formation from methional, *Biochem. Biophys. Res. Comm.* **81**:498.

Quie, P. G., White, J. G., Holmes, B., and Good, R. A., 1967, *In vitro* bactericidal capacity of human polymorphonuclear leukocytes: Diminished activity in chronic granulomatous disease of childhood, *J. Clin. Invest.* **46**:668.

Rao, K. V., Biemann, K., and Woodward, R. B., 1963, The structure of streptonigrin, *J. Am. Chem. Soc.* **85**:2532.

Reed, P. W., 1969, Glutathione and the hexosemonophosphate shunt in phagocytizing and hydrogen peroxide-treated rat leukocytes, *J. Biol. Chem.* **244**:2459.

Root, R. K., and Metcalf, J. A., 1977, H_2O_2 release from human granulocytes during phagocytosis. Relationship to superoxide anion formation and cellular catabolism of H_2O_2. Studies with normal and cytochalasin B treated cells, *J. Clin. Invest.* **60**:1266.

Sausville, E. A., Peisach, J., and Horwitz, S. B., 1976, A role for ferrous ion and oxygen in the degradation of DNA by bleomycin, *Biochem. Biophys. Res. Comm.* **73**:814.

Sayre, J. W., and Kaymakcalan, S., 1964, Cyanide poisoning from apricot seeds among children in Central Turkey, *N. Engl. J. Med.* **270**:1113.

Sbarra, A. J., and Karnovsky, M. L., 1959, The biochemical basis of phagocytosis. I. Metabolic changes during the ingestion of particles by polymorphonuclear leukocytes, *J. Biol. Chem.* **234**:1355.

Selinger, H. H., 1960, A photoelectric method for the measurement of spectra of light sources of rapidly varying intensities, *Anal. Biochem.* **1**:60.

Shohet, S. B., Pitt, J., Baehner, R. L., and Poplack, D. G., 1974, Lipid peroxidation in the killing of phagocytized pneumococci, *Infect Immun.* **19**:1321.

Stendahl, O., and Lindgren, S., 1976, Function of granulocytes with deficient myeloperoxidase-mediated iodination in a patient with generalized pustular psoriasis, *Scand. J. Haematol.* **16**:144.

Stevens, J. B., and Autor, A. P., 1977, Induction of superoxide dismutase by oxygen in neonatal rat lung, *J. Biol. Chem.* **232**:3509.

Stossel, T. P., Mason, R. J., and Smith, A. L., 1974, Lipid peroxidation by human blood phagocytes, *J. Clin. Invest.* **54**:638.

Strauss, R. R., Paul, B. B., Jacobs, A. A., and Sbarra, A. J., 1971, Role of the phagocyte in host-parasite interactions. XXVII. Myeloperoxidase-H_2O_2-Cl^--mediated aldehyde formation and its relationship to antimicrobial activity, *Infect. Immun.* **3**:595.

Tauber, A. I., and Babior, B. M., 1977, Evidence for hydroxyl radical production by human neutrophils, *J. Clin. Invest.* **60**:374.

van Hemmen, J. J., and Meuling, W. J. A., 1975, Inactivation of biologically active DNA by γ-ray-induced superoxide radicals and their dismutation products, singlet molecular oxygen and hydrogen peroxide, *Biochim. Biophys. Acta* **402**:133.

Weening, R. S., Wever, R., and Roos, D., 1975, Quantitative aspects of the production of superoxide radicals by phagocytizing human granulocytes, *J. Lab. Clin. Med.* **85**:245.

Weiss, S. J., King, G. W., and LoBuglio, A. F., 1977, Evidence for hydroxyl radical generation by human monocytes, *J. Clin. Invest.* **60**:370.

Weiss, S. J., Rustagi, P. K., and LoBuglio, A. F., 1978, Human granulocyte generation of hydroxyl radical, *J. Exp. Med.* **147**:316.

White, H. L., and White, J. R., 1966, Interaction of streptonigrin with DNA *in vitro*, *Biochim. Biophys. Acta* **132**:648.

White, H. L., and White, J. R., 1968, Lethal action and metabolic effects of streptonigrin on *Escherichia coli*, *Mol. Pharmacol.* **4**:549.

White, J. R., and Dearman, H. H., 1965, Generation of free radicals from phenazine methosulfate, streptonigrin and rubiflavin in bacterial suspensions, *Proc. Natl. Acad. Sci. USA* **54**:887.

Zgliczyński, J. M., Stelmaszynska, T., Ostrowski, W., Naskalski, J., and Sznajd, J., 1968, Myeloperoxidase of human leukaemic leukocytes. Oxidation of amino acids in the presence of hydrogen peroxide, *Eur. J. Biochem.* **4**:540.

14

Oxygen-Independent Antimicrobial Systems in Polymorphonuclear Leukocytes

JOHN K. SPITZNAGEL

1. INTRODUCTION

It has long been recognized that neutrophil polymorphonuclear granulocytes (PMN) exercise antimicrobial action only at very short range, and that they must attach to and ingest their targets in order to kill them with maximum efficiency. Metchnikoff (1905) first recognized this. Kanthack and Hardy (1894) extended Metchnikoff's ideas to show the importance of degranulation of cytoplasmic granules. They showed that bacilli stopped growing only when they came in contact with and degranulated guinea pig granulocytes. Bacilli in the same fluid untouched by granulocytes continued to grow logarithmically. Remarkably, Kanthack's work was forgotten. But Hirsch (1962) and Zucker-Franklin (Zucker-Franklin and Hirsch, 1964) rediscovered and extended these concepts, emphasizing the importance of the phagolysosomes formed about microbes as they are endocytized by heterophil (or neutrophil) polymorphonuclear granulocytes. Their studies focused attention on the importance of substances carried by cytoplasmic granules and deposited in phagolysosomes of PMN. It is now generally agreed that there are two principal granule classes, the specific granules and the azurophil granules (Bainton and Farquhar, 1968; Ullyot *et al.*, 1973). The granules are now known to comprise several antibacterial substances including cationic antibacterial protein (CAP), lysozyme (LYZ), lactoferrin (LF), and myeloperoxidase (MPO). The distribution of several of these substances in the granules is shown in Table 1. Lysozyme and lactoferrin are found in the specific granules. Lactoferrin is confined to the specifics. Lysozyme, cathepsin

JOHN K. SPITZNAGEL • Department of Microbiology, Emory University, Atlanta, Georgia 30322. This is a revised and updated review based on one presented at a symposium sponsored by the American Red Cross and published in *The Granulocyte: Function and Clinical Utilization* (T. J. Greenwalt and G. A. Jamieson, eds.), 1977, pp. 103–131.

TABLE 1.   SUBCELLULAR LOCALIZATION AND MOLECULAR OXYGEN REQUIREMENTS FOR PUTATIVE ANTIMICROBIAL SUBSTANCES OF HUMAN NEUTROPHIL GRANULOCYTES

Specific (granule type)	Light azurophil (buoyant density = 1.206 g/ml)	Heavy azurophil (buoyant density = 1.22 g/ml)	Heavy granule (buoyant density = 1.25 g/ml)
Lactoferrin[b,c] Lysozyme[d,e,f]	Lysozyme[c,d,e] Cationic proteins Cathepsin[g,h] Elastase[c,g,h]	O_2 independent[a] Lysozyme[c,e] Cationic proteins Cathepsin[h] Elastase[h] O_2 dependent	Partially identified
	MPO[i]	MPO[i]	NADPH[j] oxidase[f]

[a] Molecular oxygen requirement for antimicrobial action.
[b–h] References: [b] Leffell and Spitznagel (1972); [c] Spitznagel *et al.* (1974); [d] Bretz and Baggiolini (1974); [e] West *et al.* (1974); [f] Iverson *et al.* (1978); [g] DeWald *et al.* (1975); [h] Ohlsson *et al.* (1977).
[i] MPO, myeloperoidase.
[j] NADPH, reduced nicotinamide adenine dinucleotide phosphate.

G, and elastase are in the light azurophils. Cathepsin G is principally in the light azurophils. Lysozyme and the proteases are also found in the heavy azurophils. All these substances will be considered as having O_2-independent antibacterial action. Myeloperoxidase (MPO) is confined to the azurophils—especially the light azurophils—and is a principal oxygen-dependent antimicrobial substance. Recently, reduced nucleotide oxidase has been localized in a granule that is slightly heavier than the heavy azurophils. Since this enzyme may be responsible for generation of superoxide anion, it appears a candidate for a position as an oxygen-dependent antimicrobial factor. Several substances such as CAP and LYZ have primary antimicrobial capacity. MPO on the other hand requires H_2O_2, a product of the reduction of molecular oxygen. Thus, there are both oxygen-independent and oxygen-dependent mechanisms (Klebanoff, 1975). In addition to phagocytosis and intracellular events, events initiated outside the phagocyte prior to ingestion are undoubtedly important. In body fluids, protein solutes such as complement, properdin, antibody, and combinations of these act on microbial cells, both preparing their surfaces for subsequent attachment by phagocytic components and rendering some microbial species more vulnerable to antimicrobial mechanisms later to be encountered within the phagocyte. In this chapter, I consider various antimicrobial systems that act intracellularly and independently of molecular oxygen. Selected evidence for their role in defense against infection in animals and humans will be briefly discussed.

2.   ROLE OF HYDROGEN ION IN INTRALEUKOCYTIC KILLING

It has long been known that the color of litmus particles phagocytized by PMN changes, and suggests the intraphagosomal hydrogen ion concentration shifts to high levels (Rous, 1925). This has been repeatedly confirmed. An especially elegant study by Jensen (Jensen and Bainton, 1973) suggested that within 10 min the intravacuolar pH may be 4.0–4.5. It has generally been inferred that this change is necessary to activate the lysosomal hydrolases entering the phagolysosome. This inference may be true for heterophil PMN of mouse, guinea pig, and rabbit which have been used in many studies. In the cytoplasmic granules of rabbit, for example, most of the enzymes, the proteinases included, tend to have an acid pH optimum (Cohn and Hirsch, 1960; Zeya and Spitznagel, 1971). In human PMN there are principally neutral proteinases in the granules (Folds *et al.*, 1972; Spitznagel *et al.*, 1974). It is unsurprising to find that human PMN, according to Mandell, lower their intravacuolar pH, but only to 6.0–6.5 (Mandell, 1970). This pH would be better suited for the pH optima of several of the granule enzymes that enter human phagolysosomes (Leffell and Spitznagel, 1974), but it would be easily tolerated by most nonpathogenic bacteria so that hydrogen ion as such would be less likely to function as an antimicrobial agent. Jacques and Bainton (1978) in a recent study have extended Mandell's findings and have shown that pH drop may indeed occur, but in only 10% of phagocytizing cells; moreover, timing is important so that within 10 min after phagocytosis is initiated very few cells have low pH vacuoles, but within an hour approximately 25% of cells will have pH values between 4.5–5.0. The reason results differ between different investigators are probably complex. The complexities are related to the very nature of the mechanisms that maintain or modulate intralysosomal or intravacuolar pH. Especially important are permeability of lysosomal membranes to protons and other cations. Mechanisms that may contribute to formation and maintenance of a proton gradient are important (see review by Goldman, 1976). The results with indicator dyes in estimation of hydrogen-ion concentration are inherently imprecise. The solutes and surrounding structures within the phagosome possess ion exchange properties, and the conditions under which the cells are maintained may influence the results. All these and other factors may influence the color of indicator dyes. Indicator dyes themselves probably influence pH. For all of these reasons more precise measurements await improved methods.

2.1. ORGANIC ACIDS

Organic acids are toxic for various bacteria (Dubos, 1953). These substances, for example, lactic acid, are more toxic at hydrogen ion concentrations that suppress dissociation of the protons from their carboxylic groups. The bacterial membrane is most permeable to the undissociated acids. This toxic action is enhanced under anaerobic conditions. Our own experiments (unpublished) show lactic acid is a rather weak antimicrobial agent under circumstances that resemble putative *in vivo* conditions.

2.2. ORGANIC ACIDS AND ANTIMICROBIAL ACTION OF PMN

The increased glycolysis during phagocytosis is accompanied by lactic acid production (Karnovsky, 1962). It seems likely that some of this acid reaches the phagocytic vacuole (Mandell, 1970). It is also likely that lactic acid would perform antimicrobial functions if the hydrogen ion concentration rose sufficiently to suppress dissociation of the proton from its carboxylic group. As noted above, organic acids are antimicrobial when undissociated. The pK_a of lactic acid is about 4.75. Therefore, carboxyl groups would only be undissociated at pH values substantially less than the 6.0–6.5 found by Mandell. The values published by Jacques and Bainton (1978), however, suggest that, with passage of time, lactic acid could contribute in an increasing percentage of phagocytizing cells to intravacuolar antimicrobial activities.

3. LYSOZYME

This enzyme, long known for its ability to lyse certain bacteria, was first systematically studied and its distribution in tissues and body fluids described by Sir Alexander Fleming (Fleming, 1922). A small (mol. wt. 14,000) cationic protein with one peptide chain, it has come to be the most completely understood of all enzymes. Its natural substrate is the bacterial cell wall. Lysozyme is an endoacetylmuramidase specific for the β-(1,4)-N-acetylmuramyl-N-acetyl-glucosamine linkages in the peptidoglycans of bacterial cell walls (Strominger and Tipper, 1974). It is noteworthy that not all bacteria appear to provide equally accessible substrate. This may be because in most bacteria the N-acetylmuramic acid residues are substituted by tetrapeptides involved in cross-linking the hexosamine chains. This substitution is absent in *Micrococcus lysodeikticus* and certain other bacteria that lysozyme lyses readily. In their cell walls some of the N-acetylmuramic acid residues are naked. It is thought this makes the polysaccharide fit more easily into the enzyme cleft. It should be noted that cell walls where the N-acetylmuramic acid is substituted are cleaved by lysozyme, but only after the action of another enzyme. In some instances, for example, with group A streptococci, removal of C polysaccharide or N-acetylation coupled with de-O-acylation enhance lysozyme action (Gallis *et al.*, 1976).

Other characteristic microbial features may interfere with lysozyme action. In *Escherichia coli*, for example, the outer membrane or some portion of it must be removed. Only then can lysozyme attack the cell wall. Normal serum can damage *E. coli* cell walls. The action involves complement and either naturally-occurring antibody or properdin (Wilson and Spitznagel, 1968; Muschel and Jackson, 1963). This action renders the peptidoglycan accessible to lysozyme which then cleaves it. Similar effects can be produced with *E. coli* and Triton X-100, ethylene-diaminetetraacetic acid, and lysozyme (Schnaitman, 1971).

3.1. ROLE OF LYSOZYME IN THE ANTIMICROBIAL ACTIONS OF PMN

Human PMN contain about 16 μg of lysozyme per 10^7 cells. The elegant studies of Brumfitt have shown that wild types of *M. lysodeikticus* which are sensitive to lysozyme are quickly lysed in PMN. Selected for resistance to lysozyme, this organism becomes resistant to lysis in PMN (Brumfitt and Glynn,

1961). Bacteria such as *Streptococcus pyogenes*, however, have exceedingly resistant cell walls which tend to survive phagocytosis (Glick *et al.*, 1972). We have found recently, through analysis of reducing groups in streptococcal cell walls (groups A and B), that lysozyme probably cleaves many appropriate bonds *in situ*, although it is unable to solubilize appreciable amounts of the cell wall constitutents, a process we have tentatively named cryptic lysis. The effect is much greater against group B than against group A cell walls (Spitznagel and Babcock, in preparation). The work of Wilson shows that during opsonization the complement attacks the outer membrane of Enterobacteriaceae paving the way for lysozyme to attack peptidoglycan. That this probably occurs *in vivo* is reflected by the rounding up and outer membrane damage seen in endocytized *E. coli* (Wang-Iverson *et al.*, 1978).

It may be concluded that lysozyme plays an important role in killing and degradation of susceptible bacteria. However, some of its effects may be rather subtle and dependent on preparation with other factors. The range of bacteria that can be so affected and auxiliary mediators such as antibody complement that may extend this range deserve further investigation.

4. LACTOFERRIN

This protein has a molecular weight of 77,000 and although it has no heme groups, each molecule avidly binds two molecules of iron. It readily forms complexes with other proteins and polysaccharides. Because apo-lactoferrin is a constituent of two well-known biological sources of antimicrobial activity, milk and PMN, its capacity to kill bacteria has been tested. High concentrations, about 1 mg/ml, inhibit growth of *Staphylococcus aurens, S. epidermidis*, and *Pseudomonas aeruginosa*. Its antimicrobial action may depend on its ability to chelate iron (Masson, 1970; Gladstone and Walton, 1970). Gladstone has suggested that apo-lactoferrin may protect the antibacterial action of PMN from the inhibitory action of iron (Gladstone, 1973). Recently, Arnold *et al.* (1977) have shown that *Streptococcus mutans* and *Vibrio cholerae*, but not *E. coli*, are killed by $\leq 2 \ \mu M$ apo-lactoferrin. Lesser concentrations caused marked inhibition of cultures. They concluded that the bactericidal effect was contingent on the iron-chelating capacities of lactoferrin.

4.1. ROLE OF LACTOFERRIN IN ANTIMICROBIAL ACTION OF POLYMORPHONUCLEAR LEUKOCYTES

The antibacterial action of lactoferrin is weak for some species and under some circumstances. It seems likely that it plays a supportive rather than a primary role. Suggestive evidence that it is important comes from a patient studied by us and our colleagues at Bowman Gray School of Medicine. A 40-year old man suffered repeated severe infections and myeloproliferative disease. His PMN completely lacked any detectable lactoferrin and had only 50% of the normal amount of lysozyme. There was complete absence of the specific granules from his PMN. His PMN had reduced killing efficiency for *Proteus morganii* and enterococci. They killed staphylococci normally. Phagocytosis was unimpaired; the oxidative concomitants of phagocytosis were normal as were the azurophil granules and their constituent MPO and β-glucuronidase. The azurophils degranulated normally (Spitznagel *et al.*, 1972). The complete absence of specific granules may have been associated with absence of some crucial component which we are not yet able to detect, measure, or identify. The situation suggested, nevertheless, that absence of lactoferrin could have been the problem since it is a major constituent of specific granules. We have observed one other patient similar in all these features but who died before studies of intraleukocytic killing could be completed.

5. GRANULAR CATIONIC PROTEINS

Antimicrobial cationic proteins, so designated because they are bactericidal and have cathodal electrophoretic mobilities greater than lysozyme at pH4, have been found in the cytoplasmic granules of guinea pigs (Zeya and Spitznagel, 1963; Spitznagel and Zeya, 1964), rabbits (Zeya and Spitznagel, 1966, 1968, 1971), chickens (Brune and Spitznagel, 1974), and humans (Welsh and Spitznagel, 1971; Olsson and Venge, 1972; Odeberg and Olsson, 1975; Weiss *et al.*, 1978). Cationic proteins have also been described that possess antifungal capacities against *Candida* species and against *Cryptococcus* (Lehrer *et al.*, 1975; Drazin and Lehrer, 1977; Lehrer and Ladra, 1977). In this discussion we are only concerned with granule proteins. That is to say, with proteins that have been extracted from granules that have first been separated from the polymorphornuclear leukocytes and then treated to yield a granule protein. It is necessary to use proteins thus prepared, because the cationic proteins of the nucleus, the histones, are well-known to possess antimicrobial properties. In those instances where cationic proteins are prepared, for example, from frozen and thawed preparations of polymorphonuclear leukocytes the chance of contamination with histone is so great that antimicrobial activities of the preparations cannot be considered in terms of the granular or lysosomal proteins of the cytoplasm. It is also important during the isolation of the granules that care be taken to do as little damage to the nuclei of the PMN as possible. We have found that this can be accomplished by striving only to break 50–80% of the PMN in preparation for isolating the granules; while this is somewhat wasteful of the cytoplasmic material, it does reduce the nuclear breakage substantially.

All of the cationic proteins that can be isolated from PMN granules so collected appear to possess some antimicrobial activity, but some are far more active weight for weight than are others. They also have dissimilar antimicrobial spectra (Zeya and Spitznagel, 1968). It should be clearly understood that from one species of animal to another these proteins will differ in molecular weight, amino acid composition, and enzymic properties. The best characterized examples are those that have been taken from leukemic human PMN and from normal rabbit PMN. The human cationic proteins range in molecular weight from 10,000 to 25,000 (Odeberg and Olsson, 1975) except for one recently reported by Weiss *et al.* (1978). This protein is said to have a molecular weight of 75,000. The amino acid analysis of a mixture of these proteins from human PMN showed 14 mol% of arginine, 7.7 mol% of lysine and 7.7 mol% of histidine (Olsson and Venge, 1972). The more recently reported cationic protein from human PMN has a reported amino acid composition of 3% arginine, 7.9% lysine, and 3.6% histidine. In contrast to the human proteins, the three most cationic proteins from rabbit PMN ranged in molecular weight from 4,000 to 8,000. They had much more arginine (over 30 mol% for the most cationic one) than has been found in the human proteins. Lysine and histidine were also much less than the published figures for human proteins (Zeya and Spitznagel, 1968). Clearly the cationic proteins of human and rabbit PMN differ considerably in molecular weight. It appears that the reason for the cationicity of the rabbit proteins is their high content of basic amino acids. One can only guess that perhaps the human cationic proteins are cationic because they have a high percentage of amide groups. The content of amide groups in the human proteins has not to our knowledge been determined as yet.

The relative amounts of these highly cationic proteins are less in human PMN than they are in rabbit PMN. The antibacterial activity of the human cationic proteins varies from 0.5 to 5 μg/ml, depending heavily on the organism tested. *Streptococcus faecalis*, *S. aureus*, and *E. coli* all are sensitive. *P. aeruginosa* seems resistant to 50 μg/ml of the human cationic proteins. Optimum activity of these proteins which have chymotryptic activity and have been called cathepsin G (Starkey and Barrett, 1975) is about pH 7.0 in 0.15 M sodium chloride (Odeberg and Olsson, 1975). The various cationic proteins from rabbit PMN show considerable selectivity inhibiting different species of bacteria. The most sensitive organisms, the enterococci were killed by about 0.3 μg/ml. Other organisms such as *Proteus morganii* were killed by 10 times that concentration of the specific protein (Zeya and Spitznagel, 1968). Certain of the human antimicrobial proteins show sharp selectivity for the gram-negative rods, such as the Enterobacteriaceae. Other proteins kill the gram-positive cocci (Weiss *et al.*, 1978; Modrzakowski *et al.*, 1979). The antimicrobial action of the rabbit cationic proteins was originally shown, with *E. coli* as a test organism, to depend on damage to the membranes of the cells. Nucleic acids and metabolically incorporated ^{32}P were released in increased amounts by *E. coli* suspended in toxic amounts of the cationic proteins. Injured bacteria showed immediate reduced respiratory activity, and all respiratory activity stopped at about 150 min after exposure of the bacteria to the protein. Ability to replicate, however, was lost before respiration stopped and fell to less than 14 in 30–60 min. Anionic macromolecules inhibited killing presumably through coulombic reactions. The rabbit proteins had large amounts of hydrophobic amino acids such as leucine in addition to the cationic amino acids. Thus they may have damaged membranes not just because of their strongly cationic character but also because of hydrophobic interactions with membrane lipids. Their action may in some degree have resembled that of quaternary ammonium detergents. The cationic proteins from rabbit PMN granules were selective in their antimicrobial action. Taken together, however, they showed a wide spectrum of activity. Odeberg and Olsson (1975) studied the mechanisms of microbicidal action of human granulocyte cationic proteins and found that the chymotryptic-like or cathepsin G cationic proteins of human leukemic granulocytes inhibited incorporation of radioactive precursors into protein, RNA, and DNA in both *S. aureus* and *E. coli.* The cationic protein also inhibited $^{86}Rb^+$ influx, but did not increase the leakage of intracellular rubidium. They interpreted this to indicate inhibition of energy-dependent membrane transport without a corresponding breakdown in the semipermeable character of the membrane. Magnesium and calcium ions displayed protective effects against the microbicidal activity indicating operation of charge interactions between cationic protein and bactericidal surface.

In a recent series of experiments, Rest *et al.* (1978) and Modrzakowski *et al.* (1979) have studied the effects of the surface lipopolysaccharide (LPS) of *E. coli* and *Salmonella typhimurium* on the antimicrobial actions of the human granular proteins of the polymorphonuclear leukocytes. Their results can be contrasted with those of Friedberg and Shilo (1970) and of Tagesson and Stendahl (1973). As a result of studies comparing the sensitivities of smooth strains of *S. typhimurium* and *E. coli* with the sensitivities of their deep rough mutants, it was found that smooth and Ra chemotypes were relatively resistant to human granule proteins. As the bacteria lost more and more of their LPS carbohydrate they became gradually more sensitive. Thus, Rc and Re chemotypes were most sensitive of all. Moreover the inhibitory action of granule extracts on the smooth and Ra chemotypes tended to be bacteriostatic. The action on the Rc, Rd, and Re chemotypes was bactericidal. Friedberg and Shilo (1970) and Tagesson and Stendahl (1973) used guinea pig and rabbit granules, respectively, and saw abrupt change to increased sensitivity accompany any loss of core polysaccharide. Rest *et al.* (1978) found that the antimicrobial action of the granule protein involved two steps. One step, the binding of the antimicrobial substance(s) to the bacteria occurred rapidly at 4°C. The second step of actual killing only proceeded at 37°C, and was intermediate at 22°C between that at 4°C and that at 37°C. Thus, two steps—one probably independent of enzyme activity and another probably dependent on enzyme action, were identified. It remains to be discovered whether granule or bacterial enzymes, or perhaps both, are involved in the second step. The greatest antimicrobial activity appeared to reside in the azurophil or primary granules of the human PMN and was bactericidal. The activity in the specific granules was relatively weak, and only expressed itself against the deepest rough mutants and was bacteriostatic. Further fractionation of the active antimicrobial proteins by means of gel filtration and preparative electrophoresis in polyacrylamide gels revealed that there were in fact several fractions which were antimicrobial. Some of them were active against Enterobacteriaceae, some had different activities against the various outer membrane mutants of Enterobacteriaceae, and others were reactive against gram-positive organisms. These antimicrobial activities appeared independent of oxygen. The activities have been found in granules from several leukemic patients and also in the granules from numerous normal individuals' polymorphonuclear leukocytes. The components responsible for the antimicrobial activity was shown to be absorbed at 4°C to the bacterial cells. Extensive washing did not dislodge this substance and when the bacteria were returned to the incubation temperature the cells were killed. Within a very short time they were unable to reproduce (Rest *et al.,* 1977, 1978; Modrzakowski *et al.,* 1979). In studies on a 65,000 molecular-weight cationic protein isolated from a patient with chronic myelogenous leukemia, Weiss *et al.* (1978) found evidence for membrane activity during the antimicrobial action of the protein which appeared to render Enterobacteriaceae permeable and sensitive to actinomycin D.

It seems reasonable at this time to conclude that the cationic proteins, whether these be from rabbit or from human sources, exert at least part of their action on the outer membranes of the Enterobacericeae and perhaps also on the membrane of the staphylococcus. The nature of the interaction of the proteins with the lipopolysaccharide of the outer membrane of the Enterobacericeae is still unclear, but it is evident that the polysaccharides of the LPS are of importance in determining the susceptibility of the Enterobacteriaceae to these proteins. Thus the deep rough mutants that have the greatest loss of sugar chains are the most exquisitely sensitive to the antimicrobial action of these substances. On the other hand, the parent organisms that possess their full panoply of polysaccharide in the LPS are the most resistant. It seems entirely likely that the earlier ideas that centered on the coulombic attraction between the cationic proteins and the outer membrane constituents represent an oversimplification. Not only is it likely that the process of antimicrobial action is more complex, but it probably differs somewhat from one antimicrobial substance to another.

5.1. CATIONIC PROTEINS AND ANTIMICROBIAL ACTIONS OF PMN

The antibacterial properties of pus have long been recognized (Skarnes and Watson, 1957; Spitznagel and Zeya, 1964). Skarnes extended the observations of early workers when he extracted, from rabbit PMN autolyzed in lactic acid, a substance toxic for gram-positive bacteria. He called it leukin. Leukin was characterized as a histone on the basis of amino acid analysis. The preparative procedure would have produced a complex mixture and included substances from both nucleus and cytoplasm (Skarnes and Watson, 1956). At the same time Hirsch prepared an antimicrobial extract from the cytoplasm of rabbit polymorphs. He called this material phagocytin, showed that it was not a histone, and at first concluded that it might be a globulin (Hirsch, 1956). He later stated in a second paper that it was unclear whether or not phagocytin was a protein (Hirsch, 1960). Cohn and Hirsch then showed that phagocytin is confined to the cytoplasmic granules of rabbit PMN (Cohn and Hirsch, 1960).

Spitznagel, who became aware of Hirsch's earlier work with antimicrobial action of histones, which are cationic nuclear proteins (Hirsch, 1958), and of Skarnes' work with leukin (Skarnes and Watson, 1956), adapted histochemical techniques to study cytochemical and antibacterial effects of cationic proteins (Spitznagel, 1961a,b). Spitznagel and Chi used these techniques and showed that bacteria become coated *in vivo* with cationic, arginine-rich protein in guinea pig and rabbit experimental skin abcesses. Moreover, and most important, the cationic proteins were shown to come from the cytoplasmic granules of the polymorphonuclear leukocytes and not from the nucleus (Spitznagel and Chi, 1963). Zeya and Spitznagel (1963, 1966, 1968) then isolated these substances from the cytoplasmic granules and characterized them as described above. They called them arginine-rich, antibacterial cationic proteins because that best described their biochemical and biological features. To this date no enzymic activity has been ascribed to these cationic proteins from animal PMN. For a recent study, see Dewald et al. (1975). Electrophoretic analysis of phagocytin (kindly provided by Dr. J. G. Hirsch), and comparison of this extract with the cationic proteins in our laboratory (unpublished) showed phagocytin was a complex mixture that included proteins that co-electrophoresed with the granule cationic proteins.

The question naturally arose at that time whether human PMN possessed such cationic proteins. This work referred to above (Spitznagel and Zeya, 1964; Welsh and Spitznagel, 1971) makes it clear that there are cationic proteins in human PMN, but subsequent work shows that they have chymotryptic activity and are substantially different from those of rabbits (Olsson and Venge, 1972). The other cationic protein recently described by Weiss et al. (1978) likewise differs substantially from the highly antimicrobial cationic proteins of rabbit PMN granular proteins. Despite these differences it is clear that some analogy of action does exist between the proteins.

There is no direct evidence for the cationic proteins of human polymorphs playing a direct role in killing bacteria. There are findings which suggest that they or some unidentified but oxygen-independent mechanisms do play a direct role. The myeloperoxidase-deficient PMN kills bacteria and yeast almost as well as PMN that are normal with respect to MPO, but it takes somewhat longer to do it (Lehrer and Cline, 1969; Lehrer et al., 1969; Klebanoff, 1970). Moreover, MPO-deficient PMN yield a candidacidal cationic protein with esterase activity (Lehrer et al., 1975). Mandell (1974) working with PMN, showed that even under strictly anaerobic conditions human PMN can regularly kill certain bacteria. Finally, even chronic granulomatous disease PMN retain some ability to suppress or to kill S. aureus and Paracolon hafnia (Quie et al., 1967).

So far no instance of congenitally impaired intraphagocytic killing has been associated with reduction in amount or activity of cationic proteins. The principal biological role of these proteins is unknown.

In animals, especially in rabbits and in chickens, there is perhaps more evidence for the antibacterial role of cationic proteins. Not only has light microscopy shown that cationic proteins react with bacteria as the bacteria die in the phagosome (Spitznagel and Chi, 1963) but MacRae has shown elegantly that cationic proteins from the matrix of primary granules degranulate into the phagolysosome and bind to the bacterial cell walls (MacRae and Spitznagel, 1975). Moreover, chicken heterophil PMN have cationic proteins, including two kinds of lysozyme, but they totally lack peroxidase. Moreover, they failed to show hydrogen peroxide production although their hexose monophosphate shunt is quite active, and when they phagocytize their cyanide-resistant oxygen consumption increases (Penniall and Spitznagel, 1975). This suggests killing in chicken PMN may depend heavily on oxygen-independent mechanisms.

Walton (1978) has shown that cationic proteins are important for staphylocidal action by rabbit PMN and that ferrous iron and hematin which inhibit antimicrobial action of cationic proteins, also inhibit killing of staphylococci by rabbit PMN. Azide and catalase, which block MPO and hydrogen peroxide, were without influence on the staphylocidal activities of rabbit PMN. This finding showed that neither myeloperoxidase or hydrogen peroxide were involved in this particular killing mechanism. A considerable proportion of the killing took place outside the PMN. Gladstone suggested that the cationic proteins emerged from the cell, killing bacteria extracellularly. The most recent paper by Walton (1978) elaborates upon these themes and suggests that the antimicrobial action of rabbit granular cationic proteins against a selected strain of *S. aureus* are due to the cationicity of the protein and depend upon the active oxidative phosphorylation due to the staphylococci.

The weight of evidence suggests that cationic proteins of different kinds play a role in the antimicrobial activities of PMN of various species. They may be most important in chickens and more important in rabbits than in man, but nonetheless, they are probably important even in man to some extent. As with all the antimicrobial mechanisms discussed, rigorous proof of their involvement *in vivo* is still wanting.

6. NEUTRAL PROTEINASES

Human PMN carry neutral proteinases in their primary granules (Folds *et al.*, 1972; Spitznagel *et al.*, 1974; Dewald *et al.*, 1975; Ohlsson *et al.*, 1977). It has been shown by Janoff that the cell walls of certain bacteria may be lysed by some of these neutral proteinases, which he has characterized as esterases with elastolytic capacity (Janoff and Blondin, 1973). The range of antimicrobial actions for these esterases deserves further investigation. If they do not participate in intraleukocytic killing of bacteria, it seems highly likely that they do engage in the degradation of the bacterial peptides.

7. GENERAL COMMENTS

Substances with primary antimicrobial capacity are carried in the cytoplasmic granules of PMN of every species studied including man. Accumulating evidence suggests that during phagocytosis the granules of the polymorphonuclear leukocyte cytoplasm rapidly transport these antimicrobial substances to the phagolysosomes, fuse their membranes with them, and empty their contents into these vacuoles. The antimicrobial action of many of these proteins is direct and apparently has no need for the complex mechanisms required by the oxygen-dependent systems. On the other hand, the oxygen-independent systems are only active in concentrations that are substantially greater than those required for example for myeloperoxidase by the oxygen dependent systems. This requirement for higher concentration is seen for example with the action of cationic proteins, especially the action which is independent of enzymatic activity in the cationic molecules. The concentration requirement may, however, be easily met in phagolysosomes where volumes are small and water relatively scarce. Goldman in her review suggests that the water content of lysosomes is about 1 μg/ml of protein (Goldman, 1976). Since the toxicity of these substances is primary, the requirement for high concentrations may well serve a protective function. When these substances escape from the PMN, the dilutional effect of the body fluids could mitigate any adverse action they might have on tissues. The role of these substances in the economy of the polymorph is poorly understood. They probably provide a reserve or supplementary antimicrobial system which can support killing in situations where partial pressures of oxygen are very low and in PMN that lack MPO or normal oxidative metabolism. Their function may be crucial to the cell but this has yet to be demonstrated. No instance of genetically determined deficiency of cationic proteins has been described which might indicate that their absence is incompatible with life.

8.   SUMMARY

Antimicrobial actions of polymorphonuclear leukocytes depend on a panoply of substances carried in their cell membranes and cytoplasmic granules. These substances mediate killing in several systems. Some operate independently of molecular oxygen while others have an obligatory requirement for it. The different systems may act in primary or in reserve capacities, backing each other up in times of stress or failure. Thus, a deficiency in one system would not necessarily leave the PMN completely incapable of antimicrobial action, although impairment might be severe, as in chronic granulomatous disease. In general a poorly functioning PMN is better than no PMN at all.

REFERENCES

Arnold, R. R., Cole, M. F., and McGhee, J. R., 1977, A bactericidal effect for human lactoferrin, *Science* 197:263–265.

Bainton, D. F., and Farquhar, M. G., 1968, Differences in enzyme content of azurophil and specific granules of PMN leukocytes, *J. Cell Biol.* 39:229–317.

Brumfitt, W., and Glynn, A. A., 1961, Intracellular killing of *Micrococcus lysodeikticus* by macrophages and polymorphonuclear leucocytes: A comparative study, *Br. J. Exp. Pathol.* 42:408–423.

Brune, K., and Spitznagel, J. K., 1974, Perioxidaseless chicken leukocytes: Isolation and characterization of antibacterial granules, *J. Infect. Dis.* 127:84–94.

Cohn, Z. A., and Hirsch, J. G., 1960, The isolation and properties of the specific cytoplasmic granules of rabbit polymorphonuclear lcucocytes, *J. Exp. Med.* 112:983–994.

Dewald, B., Rindler-Ludwig, R., Bretz, U., and Baggiolini, M., 1975, Subcellular localization and heterogenicity of neutral proteases in neutrophilic polymorphonuclear leukocytes, *J. Exp. Med.* 141:709–723.

Drazin, R. E., and Lehrer, R. I., 1977, Fungicidal properties of a chymotrypsin-like cationic protein from human neutrophils: Adsorption to *Candida parapsilosis*, *Infect. Immun.* 17:382–388.

Dubos, R. J., 1953, Effect of ketone bodies and other metabolites on the survival and multiplication of staphylococci and tubercle bacilli, *J. Exp. Med.* 98:145–147.

Fleming, A., 1922, On a remarkable bacteriolytic element found in tissues and secretions, *Proc. R. Soc. London* 93:306.

Folds, J. D., Welsh, I. R. H., and Spitznagel, J. K., 1972, Neutral proteases confined to one class of lysosomes of human polymorphonuclear leukocytes, *Proc. Soc. Exp. Biol. Med.* 139:461–463.

Friedberg, D., and Shilo, M., 1970, Interaction of gram-negative bacteria with the lysosomal fraction of polymorphonuclear leukocytes, *Infect. Immun.* 1:305–510.

Gallis, H. A., Miller, S. E., and Wheat, R. W., 1976, Degradation of ¹⁴C-labeled streptococcal cell walls by egg white lysozyme and lysosomal enzymes, *Infect. Immun.* 13:1459–1466.

Gladstone, G. P., 1973, The effect of iron and haematin on the killing of staphylococci by rabbit polymorphonuclear leucocytes, *Contrib. Microbiol. Immunol.* 1:222.

Gladstone, G. P., and Walton, E., 1970, Effect of iron on the bactericidal proteins from rabbit polymorphonuclear leucocytes, *Nature* 227:849–851.

Gladstone, G. P., and Walton, E., 1971, The effect of iron and haematin on the killing of staphylococci by rabbit polymorphs, *Br. J. Exp. Pathol.* 52:452.

Goldman, R., 1976, Ion distribution and membrane permeability in lysosomal suspensions, in: *Lysosomes in Biology and Pathology*, Vol. 5 (J. T. Dingle, ed.), pp. 309–336, North-Holland, Amsterdam.

Hirsch, J. G., 1956, Phagocytin: A bactericidal substance from polymorphonuclear leucocytes, *J. Exp. Med.* 103:589–592.

Hirsch, J. G., 1958, Bactericidal action of histone, *J. Exp. Med.* 108:925.

Hirsch, J. G., 1960, Further studies on preparation and properties of phagocytin, *J. Exp. Med.* 111:323.

Jacques, Y. V., and Bainton, D. F., Changes in pH within the phagocytic vacuoles of human neutrophils and monocytes, 1978, *Lab. Invest.* 39:179.

Janoff, A., and Blondin, J., 1973, The effect of human granulocyte elastase on bacterial suspensions, *Lab. Invest.* 29:454–457.

Jensen, M. S., and Bainton, D. F., 1973, Temporal changes in pH within the phagocytic vacuole of the polymorphonuclear neutrophilic leukocyte, *J. Cell Biol.* 56:379.

Karnovsky, M. L., 1962, Metabolic basis of phagocytic activity, *Physiol. Rev.* 42:143–168.

Klebanoff, S. J., 1970, Myeloperoxidase contribution to the microbicidal activity of intact leukocytes, *Science* 169:1096.

Klebanoff, S. J., 1975, Antimicrobial mechanisms in neutrophilic polymorphonuclear leukocytes, *Semin. Hematol.* 12:117.

Leffell, M. S., and Spitznagel, J. K., 1974, Intracellular and extracellular degranulation of human polymorphonuclear azurophil and specific granules induced by immune complexes, *Infect. Immun.* 10:1241–1249.

Lehrer, R. I., and Cline, M. J., 1969, Leukocyte myeloperoxidase deficiency and disseminated candidiasis: The role of myeloperoxidase in resistance to *Candida* infection, *J. Clin. Invest.* 48:1478–1488.

Lehrer, R. I., Hanifin, J., and Cline, M. J., 1969, Defective bactericidal activity in myeloperoxidase-deficient human neutrophils, *Nature* 223:78–79.

Lehrer, R. I., Ladra, K. M., and Hake, R. B., 1975, Nonoxidative fungicidal mechanisms of mammalian granulocytes: Demonstration of components with candidacidal activity in human, rabbit, and guinea pig leukocytes, *Infect. Immun.* 11:1226–1234.

MacRae, E. K., and Spitznagel, J. K., 1975, Ultrastructural localization of cationic proteins in cytoplasmic granules of chicken and rabbit polymorphonuclear leukocytes, *J. Cell Sci.* 17:79–83.

Mandell, G. L., 1970, Intraphagosomal pH of human polymorphonuclear neutrophils, *Proc. Soc. Exp. Biol. Med.* 134:447–451.

Mandell, G. L., 1974, Bactericidal activity of aerobic and anaerobic polymorphonuclear neutrophils, *Infect. Immun.* 9:337–341.

Masson, P., 1970, *La Lactoferrin: Protéine les Secretions Externes et des Leukocytes Neutrophiles*, Librairie Malonie, Paris.

Methnikoff, E., 1905, *Immunity in Infective Diseases*, Cambridge University Press, London.

Modrzakowski, M. C., Cooney, M. H., Martin, L. E., and Spitznagel, J. K., 1979, Bactericidal activity of fractionated granule contents from human polymorphonuclear leukocytes, *Infect. Immun.* 23:587.

Muschel, J. H., and Jackson, J. E., 1963, Activity of the antibody complement system and lysozyme against rough gram negative organisms, *Proc. Soc. Exp. Biol. Med.* 113:881–883.

Odeberg, H., and Olsson, E., 1975, Antibacterial activity of cationic proteins from human granulocytes, *J. Clin. Invest.* 56:1118–1124.

Olsson, I., and Venge, P., 1972, Cationic proteins of human granulocytes, *Scand. J. Haematol.* 9:204–214.

Penniall, R., and Spitznagel, J. K., 1975, Chicken neutrophils: Oxidative metabolism in phagocytic cells devoid of myeloperoxidase, *Proc. Natl. Acad. Sci. USA* 72:5012–5015.

Quie, P. G., White, J. G., Holmes, B., and Good, R. A., 1967, *In vitro* bactericidal capacity of human polymorphonuclear leukocytes: Diminished activity in chronic granulomatous disease of childhood, *J. Clin. Invest.* 46:668–679.

Rest, R. F., Cooney, M. H., and Spitznagel, J. K., Bactericidal activity of specific and azurophil granules from human neutrophils: Studies with outer-membrane mutants of *Salmonella typhimurium* LT-2, 1978, *Infec. Immun.* 19:131.

Rous, P., 1925, The relative reaction within living mammalian tissues, *J. Exp. Med.* 41:399–405.

Schnaitman, C. A., 1971, Effect of ethylenediaminetetra-acetic acid, triton X-100, and lysozyme on the morphology and chemical composition of isolated cell walls of *Escherichia coli, J. Bacteriol.* 108:553–556.

Skarnes, R. C., and Watson, D. W., 1956, Characterization of leukin: An antibacterial factor from leucocytes active against gram-positive pathogens. *J. Exp. Med.* 104:829–845.

Skarnes, R. C., and Watson, D. W., 1957, Antimicrobial factors of normal tissues and fluid, *Bacteriol. Rev.* 21:273.

Spitznagel, J. K., 1961a, The effects of mammalian and other cationic polypeptides on the cytochemicaal character of bacterial cells, *J. Exp. Med.* 114:1063–1078.

Spitznagel, J. K., 1961b, Antibacterial effects associated with changes in bacterial cytology produced by cationic polypeptides, *J. Exp. Med.* 114:1079.

Spitznagel, J. K., and Chi, H. Y., 1963, Cationic proteins and antibacterial properties of infected tissues and leukocytes, *Am. J. Pathol.* 43:697.

Spitznagel, J. K., Cooper, M. R., McCall, C. E., DeChatelet, L. R., and Welsh, I. R. H., 1972, Character of azurophil and specific granules purified from human polymorphonuclear leukocytes, *J. Clin. Invest.* 93:(Abstr.).

Spitznagel, J. K., Dalldorf, F. G., Leffell, M. S., Folds, J. D., Welsh, I. R. H., Cooney, M. H., and Martin, L. E., 1974, Character of azurophil and specific granules purified from human polymorphonuclear leukocytes, *Lab. Invest.* 30:774–785.

Strominger, J. L., and Tipper, D. J., 1974, Structural of bacterial cell walls: The lysozyme substrate, in: *Lysozyme* (E. F. Osserman, R. E. Canfield, and S. Beychock, eds.), pp. 169–184, Academic Press, New York.

Tagesson, C., and Stendahl, O., 1973, Phagocytosis of *Salmonella typhimurium* 395 MR10 by rabbit polymorphonuclear leucocytes: Formation of phagolysosomes demonstrated by zonal separation, *Acta Pathol. Microbiol. Scand. Sect. B* 81:481–486.

Walton, E., 1978, The preparation, properties and action on *Staphylococcus aureus* of purified fractions from the cationic proteins of rabbit polymorphonuclear leucocytes, *Br. J. Exo. Pathol.* 59:416–431.

Wang-Iverson, P., Pryzwansky, K. B., Spitznagel, J. K., and Cooney, 1978, Bactericidal capacity of phorbol myristate acetate-treated human polymorphonuclear leukocytes, *Infect. Immun.* 22:945.

Weiss, J., Elsbach, P. E., Olsson, I., and Odeberg, H., 1978, Purification and characterization of a potent bactericidal and membrane active protein from the granules of human polymorphonuclear leukocytes, *J. Biol. Chem.* 253:2664–2672.

Welsh, I. R. H., and Spitznagel, J. K., 1971, Distribution of lysosomal enzymes, cationic proteins and bactericidal substances in subcellular fractions of human polymorphonuclear leukocytes, *Infect. Immun.* 4:97–102.

Wilson, L. A., and Spitznagel, J. K., 1968, Molecular and structural damage to *Escherichia coli* produced by antibody, complement, and lysozyme systems, *J. Bacteriol.* 96:1339–1348.

Zeya, H. I., and Spitznagel, J. K., 1963, Antibacterial and enzymic basic proteins bearing granules of polymorphonuclear leukocytes, *Science* 142:1085–1087.

Zeya, H. I., and Spitznagel, J. K., 1966, Antimicrobial specificity of leukocyte lysosomal cationic proteins, *Science* 154:1049–1051.

Zeya, H. I., and Spitznagel, J. K., 1968, Arginine-rich proteins of polymorphonuclear leukocyte lysosomes, *J. Exp. Med.* 127:927–941.

Zeya, H. I., and Spitznagel, J. K., 1971, Characterization of cationic protein-bearing granules of polymorphonuclear leukocytes, *Lab. Invest.* 24:229–236.

Zucker-Franklin, D., and Hirsch, J. G., 1964, Electron microscope studies on the degranulation of rabbit peritoneal leukocytes during phagocytosis, *J. Exp. Med.* 120:569–572.

15

Antimicrobial Functions of Phagocytes and Microbial Countermeasures

PETER DENSEN
and GERALD L. MANDELL

1. INTRODUCTION

Phagocytic cells are an essential part of a system that functions to protect the host against microbial invasion. Neutrophils (and on occasion, eosinophils) are the first cells of the immune system to respond to a challenge such as microbial penetration of the skin or mucous membranes. The fact that of many thousands of species of microbes only a few are pathogenic suggests that there is something unique about these organisms. These microbes possess virulence factors, alternatively called impedins or aggressins, which may not only enable them to establish a site of infection but also to maintain that infection in the face of host defenses.

Whether or not phagocytic cells will emerge victorious after their encounter with a microbe depends on a number of factors including the inoculum size, the ability of the phagocyte to ingest organisms at a rate faster than they can multiply, and the ability of the phagocyte to kill ingested or attached organisms.

The purpose of this chapter is to discuss the interaction between microbes and phagocytes. The study of microbe–phagocyte interaction is particularly rewarding since it not only affords information about mechanisms of disease, but also yields insights into the normal and abnormal function of cells. Much of what follows pertains to neutrophils, but some aspects of the interaction of the monocyte-macrophage system and eosinophils with various microbes will be discussed. The discussion will draw heavily from the work of many investigators as well as past and present investigations in our own laboratory. The format will

PETER DENSEN and GERALD L. MANDELL • Department of Medicine, University of Virginia School of Medicine, Charlottesville, Virginia 22901.

be to briefly review phagocyte development, morphology, and physiology and then to show how various organisms have evolved specific factors which may inhibit one or more of the phagocytic responses.

2. CELLULAR DEVELOPMENT, MORPHOLOGY, AND PHYSIOLOGY

2.1. NEUTROPHILS

2.1.1. Neutrophil Development

Neutrophil maturation in the bone marrow occurs in two one-week-long phases: a mitotic phase followed by a nonmitotic phase (Cartwright *et al.*, 1964). During the mitotic phase the cells progress from myeloblasts to promyelocytes, and acquire primary or azurophil granules (Bainton and Farquhar, 1966). These granules contain myeloperoxidase, lysozyme, and various hydrolases (Spitznagel *et al.*, 1974). During the nonmitotic phase promyelocytes develop into mature neutrophils and acquire secondary or specific granules. These granules contain lactoferrin, lysozyme, and several cationic proteins (Spitznagel *et al.*, 1974) and functionally resemble secretory granules. In the process of acquiring a full complement of granules, the neutrophil becomes an end-stage cell with few mitochondria, little ribosomal material, and a retracted, segmented nucleus. However, it is remarkably well-suited for its role as a bactericidal phagocyte. Its cytoplasm is rich with glycogen as an energy source; a membrane that was rigid in precurser cells is now highly flexible (Lichtman and Weed, 1972), and microtubules and microfilaments important for cell motility and degranulation (Hoffstein *et al.*, 1977, Moore *et al.*, 1976) are present in abundance.

2.1.2. Delivery to the Site of Infection

Mature neutrophils enter the circulation from the marrow and, after circulating tor about 8 hr (Craddock, 1972), leave the circulation by adhering to the endothelium and entering the tissues by diapedesis. Adherence appears to be the result of the generation of a plasma adherence factor. During infection an additional change in the surface characteristics of the neutrophils augments adherence (Lentnek *et al.*, 1976). Once the neutrophils enter the tissue they survive for only 1–2 days. It is unlikely that neutrophils reenter the circulation. Outside the circulation the neutrophils randomly patrol the vast tissue spaces in search of microbial invaders. The recognition of microbial invasion by neutophils occurs as a result of the release of microbial substances which attract the neutrophils directly (chemotaxins) or indirectly (chemotaxigens) by first activating the complement cascade with the subsequent generation of the chemotactic fragments C3a and C5a. It is of interest that neutrophil granules contain substances capable of modulating the inflammatory response (Ward and Hill, 1970; Wright and Gallin, 1975, 1977).

2.1.3. Opsonization

Although neutrophils are capable of ingesting some microbes in the absence of serum factors, most virulent bacterial species must be opsonized before the neutrophil is capable of ingesting them. However, phagocytosis of virulent bacteria may occur in the absence of opsonins if the interaction occurs on a rough surface where the neutrophil can trap the organism. Wood termed this phenomenon surface phagocytosis (Wood *et al.*, 1946).

The substances in serum which opsonize microbes are the C3b fragment of complement and specific immunoglobulin. Optimal phagocytosis usually requires the participation of both of these substances. However, different bacterial species as well as mutants within the same species may vary in their requirements for optimal phagocytosis (Peterson *et al.*, 1976a).

The physicochemical characteristics of bacteria which determine opsonic requirements for optimal phagocytosis appear to be related to the nature of the surface of the bacteria relative to that of the neutrophil. Bacterial opsonization by complement or immunoglobulin or both, changes the affinity of the bacterium for water such that the bacterium which was previously more hydrophilic is now more hydrophobic relative to the neutrophil. Organisms which are more hydrophobic than neutrophils to begin with probably do not require opsonization for ingestion (van Oss and Gillman, 1972).

In addition to altering the surface characteristics of bacteria, complement and immunoglobulin act as ligands, connecting the bacterium to the neutrophil by adherence of the C3b and Fc regions of these molecules to their "receptors" on the neutrophil membrane. Since neutrophils have Fc "receptors" for IgG only, opsonization of bacteria by other immunoglobulins is not effective in promoting bacterial ingestion by neutrophils (Stossel, 1974).

2.1.4. Phagocytosis

The mechanisms whereby microbes are enclosed within a membrane-bound phagosome by the neutrophil is probably analogous to those described by Griffin *et al.* (1976) in their elegant experiments on macrophage phagocytosis. These investigators demonstrated that ingestion was not an all-or-none phenomenon elicited by the opsonic attachment of the particle to the macrophage membrane. Rather, ingestion was the result of the sequential interaction between opsonic ligands, distributed homogeneously over the surface of the particle, and their "receptors" on the macrophage membrane. This sequential interaction resulted in the circumferential flow of the macrophage membrane around the particle and its subsequent enclosure within the phagosome (Griffin *et al.*, 1976).

Implicit in the act of phagocytosis is the concept of the fluidity of the phagocyte membrane. Membrane movement is presumably due to contraction of microfilaments against a microskeleton of microtubules. It has been postulated that the interaction of the opsonic ligands with their "receptors" activates the aggregation of the microfilaments in the underlying cytoplasm, resulting in contraction and membrane movement (Griffin *et al.*, 1976).

2.1.5. The Metabolic Burst

In addition to initiating ingestion, opsonic attachment of microbes to the neutrophil markedly stimulates cellular metabolism. The resulting "metabolic burst" is characterized by a marked increase in oxygen consumption by the neutrophil and the production of active metabolites of oxygen including superoxide, singlet oxygen, the hydroxyl radical, and hydrogen peroxide. There is concomitant increased utilization of glucose by the hexose monophosphate shunt and increased lipid turnover. By virtue of the localized nature of opsonic attachment, stimulation is limited to the membrane involved in the formation of the phagosome. This segmental stimulation, as observed by the localized reduction of nitroblue tetrazolium (Nathan *et al.*, 1969), is to be contrasted with the diffuse stimulation which occurs in the presence of soluble substances such as endotoxin (Ochs and Igo, 1973).

2.1.6. Degranulation

Opsonic attachment is also a sufficient stimulus for the release of granule contents, both into the phagosome and extracellularly (Goldstein *et al.*, 1975b, 1976, Henson and Oades, 1975, Roos *et al.*, 1976). The release of the different granule types appear to be independent phenomena (Leffell and Spitznagel, 1975; Wright *et al.*, 1977). The contents of the secondary granules are released earlier, and up to 90% of the secondary granules released appear extracellularly, consistent with the suggestion that secondary granules function as secretory granules. In contrast, greater than 50% of the primary granule contents released appears in the phagosome (Leffell and Spitznagel, 1975).

Degranulation is associated with a rise in cyclic GMP levels and microtubule formation. A rise in cyclic AMP appears to inhibit degranulation and the assembly of microtubules (Goldstein *et al.*, 1975a; Ignarro, 1973; Ignarro and Cech, 1976; Weissmann *et al.*, 1975; Zurier *et al.*, 1975).

2.1.7. Neutrophil Bactericidal Response

Microbial killing by neutrophils is favored by the incorporation of the organism within a confined space, the phagosome, which permits concentration of the bactericidal substances without inducing neutrophil damage. Neutrophil bactericidal mechanisms have been classified as nonoxidative and oxidative. The nonoxidative mechanisms involve the action of lysozyme, lysozyme and complement, acid (pH 4.5–5.0), unsaturated lactoferrin, and cationic proteins (Klebanoff, 1975). The importance of these mechanisms is evidenced by the fact that killing of certain bacteria by neutrophils does occur in anaerobic environments (Mandell, 1974).

Oxidative bactericidal mechanisms revolve around the generation of active metabolites of oxygen (superoxide, singlet oxygen, hydroxyl radical, and hydrogen peroxide) in the phagosome. These metabolites by themselves are capable of killing some bacteria. The killing potential of this system is markedly enhanced by the discharge of myeloperoxidase into phagosomes from the primary granule (Klebanoff, 1968, 1975). The interaction of myeloperoxidase and hydrogen peroxide in the presence of a halide ion results in the incorporation of the halide ion onto the bacterial surface. This is associated with the death of the bacterium. Other oxidative reactions with other oxygen intermediates may also contribute to the bactericidal activity of neutrophils (Klebanoff, 1968, 1975).

2.2. MONOCYTES-MACROPHAGES

The mononuclear phagocyte system is comprised of marrow and circulating monocytes and tissue macrophages. Monocytes rapidly develop from committed stem cells in the bone marrow in 2–3 days. These cells are released into the circulation and after approximately 8 hr leave the circulation to enter the soft tissues, where they become macrophages and survive for months (Cline *et al.*, 1978).

Unlike the neutrophil which leaves the marrow as a mature end-stage phagocyte, the monocyte undergoes the majority of its functional development outside the bone marrow. The composition and functional properites of the mature monocyte-macrophage vary from animal to animal, and within the same animal from organ to organ. Furthermore, tissue-cultivated macrophages may change properties under different conditions of growth. Activated macrophages are generally larger, more active metabolically, contain more granules, and more efficiently dispose of ingested microbes than their nonactivated counterparts (Cline *et al.*, 1978). Activation may occur as a result of lymphocyte products or perhaps after an encounter with a specific microorganism. The expression of macrophage activation may be nonspecific, so that there is enhanced microbicidal activity for organisms other than the one specifically responsible for activation (Mackaness, 1964). Nonspecific activation occurs in peritoneal macrophages induced by an intraperitoneal injection of casein hydrolysate. The mature macrophage has "receptors" for both C3b and the Fc portion of IgG. Unlike the neutrophil, mononuclear phagocytes are long-lived and may reenter the circulation, thereby acting as vehicles for the dissemination of ingested but viable microorganisms.

Macrophages also play a role in antitumor activity and in the disposal of cellular debris (Cline *et al.*, 1978). Compared to neutrophils, macrophages, even if activated, ingest and kill conventional microorganisms inefficiently. However, it is against the obligate and facultative intracellular organisms, which neutrophils either do not ingest or are incapable of killing, that macrophages appear to function as the primary line of host defense. Relatively little is known concerning the microbicidal mechanisms of macrophages although these cells possess the chemical machinery for both oxidative and nonoxidative killing mechanisms (Cline *et al.*, 1978).

2.3. EOSINOPHILS

Although the association of metazoan parasitic infections with eosinophilia has long been appreciated, direct evidence for the eosinophil in the killing of these parasites has been obtained only recently. It, seems appropriate to discuss briefly the role of these cells in antimicrobial activity. Despite the presence of a segmented nucleus and numerous cytoplasmic granules, which would appear to link the eosinophil with the neutrophil, the fact that eosinophilia is a prominent feature of congenital agranulocytosis suggests that these cells arise from different precursor cells. After approximately a week of maturation in the marrow, the eosinophil briefly enters the circulation and finally comes to rest primarily as a tissue-based cell lying in close proximity to body surfaces. Like neutrophils, eosinophils are not believed to reenter the circulation (Beeson and Bass, 1977).

The cytoplasm of eosinophils contains at least two distinct granule types; a large, more numerous granule, with an electron-dense cystalloid core, and a smaller, homogeneous granule. The crystalloid core is composed of an arginine-rich "major basic protein" comprising the bulk of the granule protein. Although the protein can bind to acidic molecules such as DNA and heparin its function remains unknown (Gleich, 1977). The crystalloid core is surrounded by a less dense, regularly arranged matrix which contains eosinophil myeloperoxidase (Gleich, 1977). The smaller granules are rich in aryl sulfatase and acid phosphatase.

Eosinophil maturation is accompanied by the development of "receptors" for complement and the Fc portion of IgG_1 and IgG_3, the number and activity of which may increase with age or stimulation (Beeson and Bass, 1977). Resting and stimulated oxidative metabolism is higher in eosinophils than in neutrophils, but the significance of this observation remains conjectural (Beeson and Bass, 1977).

Although eosinophils can be induced to phagocytize bacteria *in vitro*, phagocytic function is sluggish. The eosinophil probably plays only a small role in protection of the host from bacteria.

Evidence that the eosinophil performs an anti-helmenthic function stems from the *in vitro* work of Butterworth *et al.* (1975), who demonstrated that chromium release and death of labeled schistosomules was dependent upon the presence of eosinophils and immune serum. Mahmoud *et al.* (1975) demonstrated that the passive transfer of immunity to schistosomules in mice was dependent upon the presence of eosinophils in the recipient. Removal of these eosinophils, using antieosinophil serum, ablated their acquired immunity. These investigators have also implicated the eosinophil in protection against *Trichinella spiralis* (Grove *et al.*, 1977).

The mechanism by which eosinophils kill these multicellular helminths is an example of antibody-mediated cell-dependent cytotoxicity. Killing of the schistosomules is presumed to be the result of exocytosis of eosinophil granule contents onto the surface of the worm while it is tightly apposed to the eosinophil (Densen *et al.*, 1978a; McClaren *et al.*, 1977). Eosinophil exocytosis has also been demonstrated on the surface of other helminthic parasites (McClaren *et al.,* 1977).

3. MICROBIAL MECHANISMS OF RESISTANCE TO THE MICROBICIDAL ACTIVITY OF PHAGOCYTES

The overwhelming majority of microbes are nonpathogenic for healthy human beings. A relative few cause the bulk of infectious diseases in man. Pathogenic bacteria possess virulence factors which enable them to overcome host defense mechanisms and cause disease. These factors and their role in the microbial interaction with phagocytic cells have been the subject of several recent reviews (Densen and Mandell, 1978b; Goren, 1977) and the importance of the microbial surface in modulating these interactions has received emphasis (Smith, 1977). Organisms may behave as obligate extracellular, faculative intracellular, or obligate intracellular pathogens (Suter, 1956). Obligate extracellular pathogens tend to produce acute infections and once the immune response of the host facilitates their ingestion they are usually rapidly killed and the disease controlled. Neutrophils are the primary phagocytes responsible for the defense of the host against these microbial invaders. In contrast, organisms possessing the capability of surviving within phagocytes tend to cause a more chronic illness. The monocyte-macrophage phagocytes are the primary cells responsible for the host defense against these organisms. These intracellular pathogens may or may not be phagocytized by neutrophils. Resistance of these organisms to intracellular killing by neutrophils provides for their eventual escape with the demise of the short-lived neutrophil and subsequent encounter with monocytes arriving at the site of microbial invasion. The multicellular helminths are obviously extracellular pathogens by virtue of their size but tend to produce chronic infections.

Microbial virulence factors may aid the microbe by allowing it to avoid recognition, inhibit chemotaxis, interfere with attachment, inhibit ingestion, depress the metabolic burst, alter degranulation, mediate their own entry into cells, or resist bacterial substances (Table 1).

3.1. AVOIDING RECOGNITION

Recognition of microbes by phagocytes is dependent on the presentation of microbial antigens foreign to the host. An organism having surface antigens identical to the host would not be recognized as foreign. The adult schistosome is such an organism. Goldring *et al.* (1976) have demonstrated that schistosomules grown in the presence of red cells of a specific blood type may be agglutinated by antisera directed against the specific blood group. Schistosomules were able to acquire host red cell determinants A, B, H, and Lewis B+ but not Rh, M, N, S, or Duffy. This suggested that parasite antigens might be

TABLE 1. MICROBIAL VIRULENCE FACTORS AFFECTING PHAGOCYTES

   I.   Avoiding recognition.
       Schistosomula
  II.   Inhibition of chemotaxis.
       Salmonella typhi, Neisseria meningitides, serratia, *Pseudomonas aeruginosa, Mycobacterium tuberculosis, Staphylococcus aureus, Escherichia coli*
  III.   Inhibition of attachment.
       Streptococcus pneumoniae, Streptococcus pyogenes, Neisseria gonorrhoeae, Neisseria meningitidis, Klebsiella pneumoniae, Staphylococcus aureus, Hemophilus influenzae, Escherichia coli, Yersinia pestis, Bacillus anthracis, Campylobacter fetus, Cryptococcus neoformans, Psuedomonas aeruginosa
  IV.   Attachment with inhibition of ingestion.
       Neisseria gonorrhoeae, mycoplasma, influenza virus
   V.   Depression of the metabolic burst.
       Salmonella typhi
  VI.   Leukotoxicity.
       Streptococcus pneumoniae, Streptococcus pyogenes, Staphylococcus aureus, Pseudomonas aeruginosa, Entamoeba histolytica
  VII.   Inhibition of degranulation.
       Mycobacterium tuberculosis, Toxoplasma gondii, Mycobacterium microti
 VIII.   Entry into cells.
       Chlamydiae, *Toxoplasma gondii*
  IX.   Resistance to bactericidal activity.
       Staphylococcus aureus, Escherichia coli, Salmonella typhimurium, Salmonella minnesota, Mycobacterium leprae, Mycobacterium lepraemurium, Mycobacterium tuberculosis
   X.   Escape from the phagosome.
       Mycobacterium bovis, reovirus, vaccinia virus

---

masked by the acquisition of host red cell antigens, and the resulting failure of recognition might explain why eosinophils do not appear to be important for immunity to the adult worm. The mechanism by which the worm assumes host red blood cell antigens is unknown, but current evidence favors a direct acquisition from the host as opposed to microbial synthesis of host antigens (Goldring *et al.,* 1976).

3.2. INHIBITION OF CHEMOTAXIS

Although bacterial products frequently act as chemotaxins or chemotaxigens, it has long been appreciated that leukocytes are not readily attracted to some bacterial strains. Delay in the appearance of leukocytes at the site of microbial invasion may affect the outcome of the infection by permitting bacterial multiplication, thus effectively increasing the inoculum size. The mechanism by which bacteria inhibit chemotaxis is not known in most cases. There may be failure to activate complement, alteration of chemotactic bacterial products, or a direct inhibitory effect on the leukocyte.

Early investigators utilized a relatively crude system, the migration of leukocytes from a clot or buffy coat, to measure chemotaxis in the presence or absence of bacteria. *Salmonella typhi, Neisseria meningitidis,* serratia, and *Pseudomonas aeruginosa* impeded neutrophil migration (Martin and Chandhuri, 1952). *Pseudomonas aeruginosa* may owe its antichemotactic effect to the elaboration of an elastase (see below) which destroys complement (Schultz and Miller, 1974).

Studies evaluating leukocyte chemotaxis in the presence of virulent and avirulent strains of the same bacteria demonstrated notable differences. Thus virulent but not avirulent brucella inhibited leukocyte migration from clots. Interestingly, no inhibition of leukocyte migration was noted if the leukocytes were obtained from animals that had been infected with brucella for at least 1 hr (Elberg and Schreider, 1953).

Virulent and avirulent strains of *Mycobacterium tuberculosis* differ in their growth pattern *in vitro;* virulent but not avirulent strains grow in parallel cords. A surface lipid called cord factor is responsible for this characteristic growth pattern and may also be responsible for the inhibition of chemotaxis observed with the virulent strains *in vitro* (Algower and Bloch, 1949, Bloch, 1950).

A mucopeptide from the cell walls of both virulent and avirulent strains of *Staphylococcus aureus* was antichemotactic for mouse leukocytes when tested in the Boyden chamber (Weksler and Hill, 1968). Virulent strains could be differentiated from avirulent strains by the inhibition of the early edema formation induced by the mucopeptide from the virulent but not the avirulent strains (Hill, 1968). Inhibition of early edema formation occurred as a result of an inhibitory effect on the kinin system (Easmon *et al.,* 1973). Thus virulent *Staphylococcus aureus* inhibited both the early inflammatory and cellular response to staphylococcal invasion in the mouse.

More recently, Bergman and his co-investigators working in our laboratory, studied the interaction of human neutrophils with *Escherichia coli* (Bergman *et al.,* 1978). Enterotoxigenic *E. coli* produce a noninflammatory diarrhea by stimulation of adenylate cyclase, with a resultant increase in intracellular cyclic adenosine 3'-5'-monophosphate (cAMP). Since cAMP was known to inhibit several aspects of neutrophil function, the effect of several toxigenic and nontoxigenic *E. coli* strains on several aspects of neutrophil function was examined. Phagocytic, metabolic, and bactericidal function of neutrophils was equivalent in response to presentation of either toxigenic or nontoxigenic *E. coli.* However, when neutrophil chemotaxis was measured using the agarose well method, filtrates from the enterotoxigenic *E. coli* were less chemotactic for neutrophils than filtrates from the nonenterotoxigenic strains. The inhibition of neutrophil migration by enterotoxigenic *E. coli* appeared to be due to the direct effect of the toxin on the neutrophils. This was likely since elimination of the inhibition of migration was noted after destruction of the toxin by heating or inactivation by the addition of specific antitoxin. The mediator seemed to be cyclic AMP since there was a significant increase in the concentration of intracellular cAMP in the neutrophils in response to the enterotoxin. Enteroxtoxin effect could be mimicked with dibutyryl cyclic AMP or enhanced with phosphodiesterase inhibitors.

3.3. INHIBITION OF ATTACHMENT

3.3.1. Factors Effective only in the Absence of Specific Opsonic Antibody

The presence of capsular antigens or K antigens endows bacteria with antiphagocytic properties because they prevent attachment to the neutrophil. However, in the presence of specific capsular antibody, neutrophils readily engulf and kill these organisms. Bacterial surface antigens which inhibit both attachment and ingestion are almost always polysaccharides. Only the streptococcal M determinant, the K-88 *E. coli* antigen, and the polyglutamic acid capsule of *Bacillus anthracis* are known to be proteins. The polysaccharide capsule is largely water, which renders the bacterium hydrophilic relative to the neutrophil in accord with the physicochemical explanation of resistance to phagocytosis advanced by van Oss *et al.* (1972). Why the hydrophilic state is antiphagocytic has not been explained adequately.

Minor chemical alterations in surface structure may produce major differences in the ability of an organism to resist attachment and ingestion by neutrophils. Medearis *et al.* (1968) studied two O antigen mutants of the *E. coli* strain 0111:B-4, the first differing only in the absence of colitose and the second by the absence of glucose, *N*-acetyl glucosamine, and colitose from the O antigen of the parent *E. coli*. The first mutant was 100 and the second mutant 1000 times less virulent for mice than the parent strain. Loss of bacterial virulence correlated with the relative ease with which these mutants could be phagocytized by neutrophils. Differences in the chemical composition of capsules may account for the differences in susceptibility to phagocytosis and variations in virulence of the capsular types of *Streptococcus pneumoniae*. However, the resistance to attachment and phagocytosis, by mutants of a given capsular type, correlated directly with the amount of capsular material produced (MacLeod and Krauss, 1950).

The list of microbes possessing surface structures which functionally inhibit attachment to and ingestion by neutrophils continues to increase and includes *Streptococcus pneumoniae, Streptococcus pyogenes* (both the streptococcal M protein and hyaluronic acid capsule are antiphagocytic (Foley and Wood, 1959, Wood *et al.*, 1946), *Neisseria gonorrhoeae* (Hendley *et al.*, 1977; James and Swanson, 1977; Richardson and Sadoff, 1977), *Neisseria meningitidis* (Honston and Rankin, 1907, Roberts, 1967, 1970), *Klebsiella pneumoniae* (Smith and Wood, 1947), the Smith strain of *Staphylococcus aureus* (Melly *et al.*, 1974; Yoshida *et al.*, 1974, *Haemophilus influenzae* group B (Suter, 1956), *Escherichia coli* (Howard and Glynn, 1971), *Yersinia pestis* (Williams *et al.*, 1972), *Bacillus anthracis* (Keppie *et al.*, 1963), *Campylobacter fetus* (McCoy *et al.*, 1975), and *Cryptococcus neoformans* (Bulmer and Sons, 1968; Diamond *et al.*, 1972; Kozel and Mastroianni, 1976).

The K antigens of *E. coli* are unique among bacterial surface antigens in that they are antiphagocytic, are associated with tissue tropism by the organism (Robbins *et al.* 1974), and are responsible for the resistance of the organism to complement. *E. coli* missing the K antigen can be agglutinated by specific O (somatic) antisera. The presence of the K antigen renders the organisms nonagglutinable by O antisera, suggesting that the K antigen masks the O determinant. Although complement is activated, and the lytic portion of complement deposited on *E. coli*, complement deposition in those strains possessing K antigens occurs distant to the bacterial membrane or in an ineffective lattice work so that bacterial lysis does not occur (Glynn and Howard, 1970; Lentnek *et al.*, 1976; Rottini *et al.*, 1975).

Cryptococcus neoformans (Bulmer and Tacker, 1975; Farhi *et al.*, 1970) and *Yersinia pestis* are distinguished by the acquisition of antiphagocytic structures only after they have infected man. This fact influences the outcome of their encounter with phagocytes. When either avirulent or virulent strains of *Yersinia pestis* are injected into nonimmune guinea pigs, they are both readily phagocytized. After an hour phagocytosis of the virulent but not avirulent strains ceases and extracellular multiplication occurs. If these virulent organisms are harvested and injected into a second guinea pig no phagocytosis is observed (Burrus, 1955). Subsequently, it has been demonstrated (Cavanaugh and Randall, 1959) that virulent *Yersinia pestis* exists in a nonencapsulated state in the flea. The organisms transferred to the victim of the flea bite are rapidly phagocytized and killed by neutrophils. However, a few are ingested by monocytes which do not kill the organism. Intracellular multiplication occurs leading to death of the monocyte. Encapsulated organisms are released which then resist ingestion by both neutrophils and monocytes.

Pseudomonas aeruginosa is capable of elaborating a protease ("elastase") which inactivates C1, C3, C5, C8, and C9 in the fluid phase and C1 and C3 in the cell bound phase, thereby inactivating both the classical and alternate complement pathways. Shultz and Miller (1974) examined neutrophil chemotaxis and phagocytosis in the presence of pseudomonas elastase. Both these parameters were diminished when compared to similar experiments performed in the absence of elastase.

3.3.2. Factors Operating Even in the Presence of Specific Opsonic Antibody

The presence of specific antibody to bacterial surface antigens is instrumental in negating the antiphagocytic factors. In order for bacteria to continue to evade phagocytosis, specific opsonins and/or complement must either be avoided or inactivated. *Staphylococcus aureus* has evolved two such virulence factors, coagulase and protein A. Coagulase-positive staphylococci are rapidly phagocytized *in vitro* in the presence of serum. However, if the incubation is carried out in plasma, coagulase-positive staphylococci convert fibrinogen to fibrin and become enmeshed in a fibrin net which appears to interfere with efforts by the neutrophil to ingest the bacteria (Foster, 1962).

Protein A is a versatile substance found on the surface of, and secreted by, many strains of staphylococci (Dossett *et al.*, 1969, Peterson *et al.*, 1977). Protein A binds specifically to the Fc portion of IgG, a feature which may account in part for the appearance of "natural" antibodies to staphylococci (Forsgren and Sjoquiest, 1966). Therefore, protein A may compete with the neutrophil Fc receptor for IgG, and once bound to IgG may sterically block the interaction of IgG with its Fc receptor (Dossett *et al.*, 1969). In addition, protein A activates the classical complement pathway (Peterson *et al.*, 1976b).

Peterson *et al.* (1977) studied the interaction of human neutrophils with 10 different strains of staphylococci. They noted that the strains with the greatest amount of protein A were the least well phagocytized in the presence of normal serum or purified IgG, but the best phagocytized in IgG-deficient serum. The addition of free protein A to all three sera inhibited staphylococcal phagocytosis. They concluded that in the absence of IgG, complement was activated by bacterial protein A and deposited on the surface of the staphylococci, effectively opsonizing them. If free protein A were present in the sera, complement was activated but not deposited on the bacterial surface, with resultant decreased opsonization and phagocytosis as complement was depleted. Presumably steric hindrance inhibited phagocytosis of the protein-A-rich strains of staphylococci in the presence of both IgG and complement.

3.4.   INHIBITION OF INGESTION

Some microbial organisms mediate their own attachment to the surface of phagocytic cells and once attached, resist ingestion in the absence of specific antibody. *Neisseria gonorrhoeae* (Dilworth *et al.*, 1975, Swanson *et al.*, 1974, 1975, Thongthai and Sawyer, 1973, Watt, 1970) and mycoplasma (Jones and Hirsch, 1971, Simberkoff and Elsbach, 1971, Zucker-Franklin *et al.*, 1966) are examples of such organisms. The attachment of these organisms to cells is self-mediated, and, in the case of *Neisseria gonorrhoeae*, is mediated at least in part through pili which act as lectins, attaching to a specific chemical moiety on the phagocytic membrane. Virulent but not avirulent gonococci possess pili and attach to, but resist ingestion and killing by neutrophils, as we and others have shown (Dilworth *et al.*, 1975; Swanson *et al.*, 1974, 1975; Watt, 1970) (Figure 1). Avirulent gonococci are ingested and killed. It is possible that the mechanisms by which gonococci attach to cells and resist ingestion are independently mediated, since, in addition to pili, gonococci have recently been shown to have a capsule (Hendley *et al.*, 1977; James and Swanson, 1977; Richardson and Sadoff, 1977).

Buchanan has recently furnished evidence suggesting that, in the presence of specific antipilus antibody, mouse peritoneal macrophages can ingest piliated gonococci (Buchanan *et al.*, 1978a). Brooks *et al.* (1976) using convalescent high-titered immune serum noted killing of gonococci by neutrophils. However, we (Densen *et al.*, 1978b) and others (Gibbs, 1974), using phase contrast microscopy have been unable to observe ingestion of piliated gonococci in the presence of high-titer convalescent serum from patients with disseminated gonococcal infection. We did not, however, measure killing, so the possibility exists that the attached but uningested gonococci were killed by neutrophils in the presence of specific antibody.

A novel explanation for the ability of attached piliated gonococci to avoid ingestion has been suggested by Senff *et al.* (1977). These investigators spin-labeled the membrane of neutrophils and measured electon spin resonance spectra before and after incubation of the neutrophils with virulent (piliated) and avirulent (nonpiliated gonococci. The virulent but not the avirulent gonococci increased the rigidity of the neutrophil membrane. Since membrane fluidity is essential for ingestion, these results suggest that gonococcal-mediated resistance to ingestion may be mediated by a decrease in membrane fluidity caused by the interaction of the gonococcal pili with the neutrophil membrane. To see if neutophil phagocytic function is impaired by contact with type I gonocci we examined the effect of gonococcal attachment to neutrophils on the subsequent ingestion of *Candida albicans*. Using phase contrast microscopy we observed qualitatively normal phagocytosis by neutrophils with attached type I gonococci (Densen *et al.*, 1978b).

Simberkoff and Elsbach (1971) addressed similar concerns when they evaluated killing of *E. coli* by neutrophils coated with mycoplasma. These authors detected reduced killing of *E. coli* by neutrophils with attached mycoplasma compared to the killing of *E. coli* by normal neutrophils. However, it is unclear whether the reduced killing of *E. coli* was due to reduced ingestion. The antiphagocytic effect of an increase in membrane rigidity, if any, would appear to be localized to the area of attachment and not to a generalized alteration of the neutrophil membrane.

We have also been interested in the explanation for the failure of neutrophils to kill attached but uningested gonococci. Neutrophil oxidative metabolism is stimulated equally by the attachment of virulent gonococci and by the ingestion of avirulent gonococci. However, myeloperoxidase-mediated iodination of attached organisms was markedly reduced compared to ingested organisms (Densen and Mandell, 1978a). This finding suggested that attached gonococci survived on the surface of oxidatively active neutrophils because release of myeloperoxidase from the primary granule did not occur. Senff and co-workers have recently confirmed this hypothesis (Senff and Sawyer, 1977) but in addition demonstrated that in contrast to the primary granule, secondary granule release did occur. These results are consistent with the work of Rest *et al.* (1978) and Klebanoff (1968) who demonstrated the importance of the contents of the primary granule in the generation of both nonoxidative and oxidative bactericidal activity.

Whether or not the failure of the neutrophil to release the primary granule contents is due to a failure to stimulate primary granule release or an inhibition of release is unknown. The recent findings of Buchanan *et al.* (1978b) may have some bearing on this question. He reported that the receptor on the neutrophil membrane for the gonococcal pilus was the ganglioside GM-1.

Ganglioside molecules by virtue of their amphipathic nature are thought to be important in the transmission of signals across mammalian membranes, and interaction of the gonococcal pilus with such a molecule could provide a signal important in the control of degranulation. In addition, the GM-1 ganglioside is known to be the receptor for choleratoxin, whose action at this receptor stimulates adenylate cyclase and an increase in intracellular cAMP, which in neutrophils is associated with inhibition of degranulation.

Neutrophil response to the attachment of mycoplasma (Simberkoff and

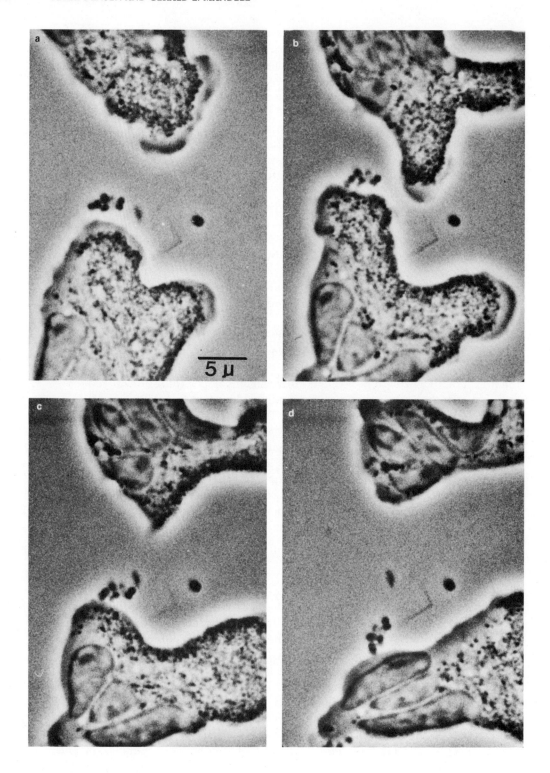

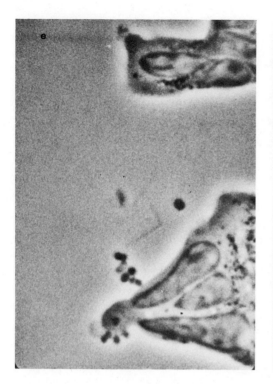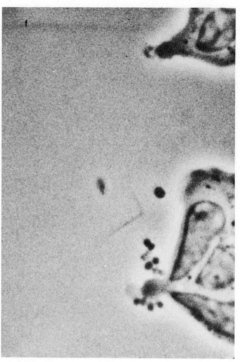

FIGURE 1.   (a–f) Neutrophil interaction with virulent gonococci. This phase contrast photomicrograph sequence portrays the attachment of the gonococci to the neutrophil membrane without ingestion by the neutrophil. The gonococci "slide" along the membrane to the uropod where they join other gonococci previously encountered by the neutrophil. In this location the gonococci are not killed despite activating the neutrophil metabolically. Time, 90 sec. ×3200. Reduced 13% for reproduction.

Elsbach, 1971) appears analogous to that described above for piliated virulent gonococci. Neutrophil oxidative metabolism was stimulated by adherent mycoplasma but the attached, uningested organisms remained viable. In the presence of specific antisera, ingestion and killing occurred rapidly. Degranulation was not measured.

Sawyer (1969) investigated the effect of influenza virus attachment to neutrophils on the ingestion of virus, neutrophil metabolism, and subsequent ingestion of bacteria. As observed with the gonococcus and mycoplasma, attachment of the virus to neutrophils appeared to be mediated by the virus and not by humoral substances or by the neutrophil. Attached virus was either rapidly eluted by action of viral neuraminidase or ingested. Elution of live virus occurred from rat and mouse neutrophils and macrophages, but ingestion with subsequent destruction of the virus occurred only in guinea pig macrophages. Leukocyte glycolysis and subsequent ingestion and killing of bacteria were depressed follwing viral attachment and elution from neutrophils. However, depression of glycolysis by the virus correlated poorly with the depression of bacterial phagocytosis.

3.5. DEPRESSION OF THE METABOLIC BURST

The burst of metabolic oxidative activity following ingestion is essential for normal neutrophil bactericidal activity. Miller *et al.* (1972) reported that neutrophil oxygen consumption following ingestion of the virulent Quailes strain of *Salmonella typhi* was diminished when compared to the increase in neutrophil oxygen consumption following ingestion of an avirulent strain of *S. typhi*. Kossack, working in our laboratory has confirmed and extended these initial observations (Kossack *et al.*, 1978). He demonstrated that human neutrophils ingest equivalent numbers of avirulent and virulent *S. typhi* at equivalent rates of ingestion. Neutrophils ingesting these strains of *S. typhi* remained viable during the course of the experiment. However, neutrophil oxygen consumption and chemiluminescence were depressed when the virulent but not the avirulent strains were ingested. A normal increase in neutrophil oxidative metabolism was observed when neutrophils containing either the virulent or avirulent strains were stimulated a second time using zymosan. This result suggests that the depression by virulent *S. typhi* of the oxidative metabolic burst is a local effect, presumably exerted on the membrane enclosing the organism within the phagosome. We were surprised to find that despite the reduced oxidative response, neutrophils ingesting virulent *S. typhi* kill these organisms as well as ingested avirulent strains. Since *S. typhi* is a faculative intracellular pathogen, depression of phagocyte oxidative metabolism may be more relevant to macrophage function and may enhance its survival within monocytes or macrophages.

3.6. ALTERATION OF DEGRANULATION

3.6.1. Augmentation of Degranulation

Leucotoxins may be cell bound or soluble and have been found in association with *Streptococcus pneumoniae* (Oram, 1934), *Streptococcus pyogens* (Gay and Oram, 1933), *Staphyloccus aureus* (Gladstone and van Heyningen, 1957), *Pseudomonas aeruginosa* (Scharmann *et al.*, 1976), and *Entamoeba histolytica* (Artigas, 1968). Although massive degranulation and neutrophil death are the end results of these cytotoxins, this should not be taken as evidence that the mechanisms of toxicity are the same.

Woodin (1968) has extensively studied the action of staphylococcal leucocidin and evidence to date suggests that the action of the pseudomonas leucocidin is similar (Scharmann *et al.*, 1976). This action involves the attachment of leucocidin to the neutrophil membrane probably by binding to a specific phospholipid. Binding is associated with depolarization of the neutrophil membrane with extracellular loss of potassium, an influx of sodium, and a rounding of the neutrophil. Depolarization is followed by massive exocytosis of granule contents. This phenomenon is dependent on the presence of calcium in the medium. Degranulation was inhibited in the absence of calcium, but the morphological changes of swelling and rounding of the neutrophil were unchanged. Thus, cell death may be related to ion and water fluxes induced by the action of the toxin on the neutrophil membrane (Scharmann *et al.*, 1976).

Interaction of streptococcal toxins and neutrophils has been extensively studied by Wilson, Ginsberg, Ofek, (Hirsch *et al.*, 1963, Ofek *et al.*, 1970, Wilson, 1957). *Streptococcus pyogenes* has at least two cytotoxins, streptolysin O and streptolysin S. The action of these two toxins differs, and they are both dissimilar to staphylococcal leucocidin. Streptolysin O appears to act by binding to membrane cholesterol and "punching holes" in the neutrophil external membrane which results in cell lysis (Bernheimer, 1954).

Sullivan *et al.* (1978) working in our laboratory investigated the effect of streptolysin S on neutrophils. Streptolysin S is a cell-bound toxin which requires contact between the target and the toxin for its action. Streptolysin S appears to induce cytotoxicity in a manner similar to leucocidin. Shortly after contact the neutrophils stop moving and begin to round up and swell. Granules disappear and the cell dies (Figure 2). Exocytosis of granule contents occurs in the presence of calcium. The cytotoxicity of streptolysin S may be diminished by raising the external potassium concentration, lowering the sodium concentration, and omitting calcium from the extracellular milieu. These changes minimize ionic fluxes induced by the toxin and inhibit degranulation.

Bernheimer and Schwartz (1964) studied the ability of many different bacterial toxins to lyse both liver and neutrophil lysosomes. The α toxin of staphylococcal and *Clostridia perfringens* as well as both streptolysin O and S were capable of lysing granules from both tissues; however, staphylococcal leucocidin had no effect on either lysosome. These authors noted that toxins which had hemolytic activity lysed both types of lysosomes, but toxins unable to lyse lysosomes had no hemolytic activity. This fact appeared to support previous work indicating that lysosomal and erythrocytic membranes were similar (Bernheiner and Schwartz, 1964).

Similar ionic fluxes, cellular swelling, and cell death has been noted as a result of the action of antibody and complement on Krebs ascites tumor cells (Green *et al.*, 1959). Cellular swelling could be prevented by raising the osmotic pressure of the extracellular environment. These authors were unable to demonstrate membrane disruption to account for the leakage of cellular contents. They proposed that antibody and complement created "functional holes" in the membrane that permitted an influx of water due to the higher intracellular osmotic pressure.

3.6.2. Inhibition of Degranulation

If facultative and obligate intracellular pathogens are to survive in host phagocytic cells, the bactericidal action of granule substances must be prevented. The short-lived nature of the neutrophil makes assessment of the role of these cells in the pathogenesis of disease caused by intracellular pathogens difficult.

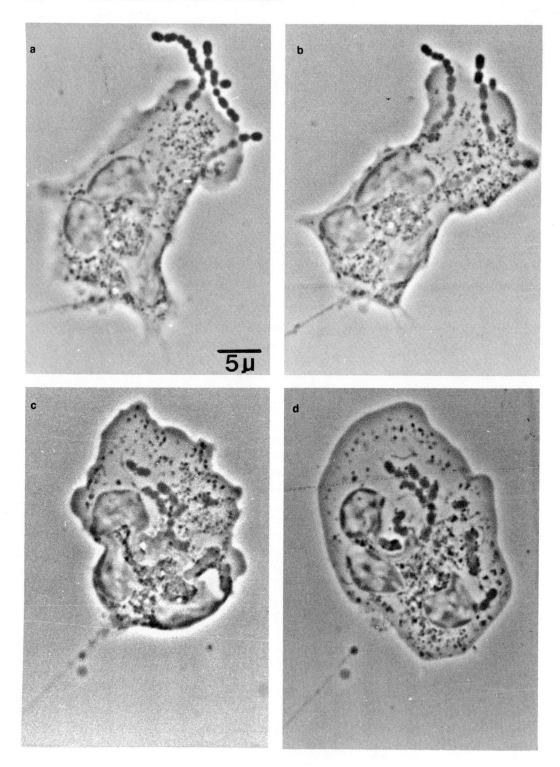

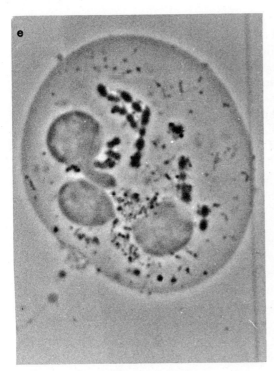
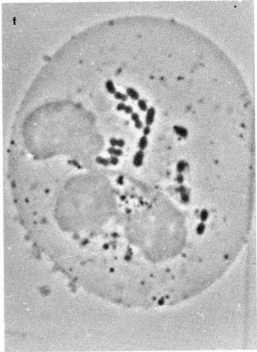

FIGURE 2. (a–f) Neutrophil interaction with leukotoxic *Streptococcus pyogenes*. Phase contrast photomicrography reveals the rapid ingestion of the streptococci by the neutrophil. The neutrophil undergoes progressive swelling and rounding with massive degranulation, cytoplasmic clearing, and nuclear swelling. The neutrophil and streptococci are both killed during this interaction. Time, 180 sec. ×2500. Reduced 6% for reproduction.

Armstrong and D'Arcy Hart (1971) investigated the intracellular survival of *Mycobacterium tuberculosis* in cultivated mouse peritoneal macrophages. When the macrophages were cultured in media containing ferritin, there was a selective concentration of ferritin in macrophage lysosomes. When these macrophages ingested *M. tuberculosis*, failure of ferritin to enter the phagosome was noted by electron microscopic techniques. Granule fusion with the phagosomes did occur if the organism was killed before ingestion. These authors suggested that intracellular survival correlated with inhibition of degranulation and that this inhibition was dependent on the viability of the tubercle bacillus. Subsequently, they have demonstrated that degranulation does occur when *M. tuberculosis* is phagocytized in the presence but not in the absence of immune serum. However, the organisms survived intracellularly despite degranulation (Armstrong and D'Arcy Hart, 1975).

Similar results have been reported by Jones and Hirsch (1974) and Jones *et al.* (1975), investigating the interaction of *Toxoplasma gondii* with mouse peritoneal macrophages. Living toxoplasma, ingested by macrophages, containing Thorotrast-labeled lysosomes, inhibited granule fusion with the phagosome. If the organisms were killed or first incubated with heat-inactivated immune serum, granule fusion occurred rapidly. In contrast to *M. tuberculosis*, macrophage degranulation resulted in the death of the toxoplasma.

The binding of the Fc portion of IgG to its receptor on the phagocyte membrane is an effective trigger for degranulation (Goldstein *et al.*, 1976). This suggests the possibility that the failure of macrophage degranulation, in the absence of specific antibody to *M. tuberculosis* and *T. gondii*, may be a consequence of the absence of the proper ligand between host cell and parasite rather than a specific inhibition of degranulation. However, a specific degranulation-inhibiting sulfatide has been isolated from the cell wall of *M. tuberculosis*. The substance is a trehalose glycolipid that is markedly anionic. Yeast particles coated with a small amount of this substance and ingested by macrophages, inhibit granule release into the macrophage phagosome (Goren *et al.*, 1976). The importance of polyanions in modifying lysosomal release has received recent emphasis (Goren, 1977).

Lowrie and co-workers measured the intracellular levels of cAMP in macrophages ingesting *Mycobacterium microti*. They found elevated levels of cAMP in macrophages ingesting live but not heat killed *M. microti* and suggested that this organism might inhibit macrophage degranulation by altering intracellular cyclic nucleotide levels (Lowrie *et al.*, 1975).

3.7. ENTRY OF PATHOGENS INTO CELLS

Another feature unique to many obligate intracellular pathogens is the ability to stimulate their own incorporation into the cells. Incorporation may be the result of fusion of membranes, as occurs with some viruses, by injection into the cytoplasm or by inducing endocytosis. Endocytosis does not appear to require any humoral ligands. Both chlamydiae (Friis, 1972) and *Toxoplasma gondii* (Arkowa, 1977) have been found to have specialized structures in close apposition to the membrane at one end of the cell which facilitates their attachment and penetration into cells.

Jones *et al.* (1975) examined *Toxoplasma gondii* entry into cells. In the absence of specific antibody, *Toxoplasma gondii* entered HeLa cells, fibroblasts, and macrophages. In the presence of specific antibody, *Toxoplasma gondii* was phagocytized by macrophages, but was no longer capable of entry into HeLa cells and fibroblasts. This interesting observation suggests that the process of toxoplasma entry into cells in the nonimmune situation is mediated by the parasite and not the host cell. In addition, it prompts speculation that the presence of antibody may in some way block parasite-mediated endocytosis while facilitating phagocytosis by professional phagocytes.

3.8. RESISTANCE TO BACTERICIDAL ACTIVITY

3.8.1. Resistance to Oxidative Attack: Detoxification

Rogers and Tompsett (1952) observed that a small but significant number of coagulase-positive but not coagulase-negative staphylococci survived in neutrophils following ingestion. Mandell (1975) noted that staphylococcal strains rich in catalase survived better than low-catalase strains after being ingested by neutrophils. In addition, high-catalase staphylococci were not iodinated as well as low-catalase staphylococci once they were ingested. These findings were explained by inactivation of hydrogen peroxide by staphylococcal catalase within the confines of the phagosome. The addition of exogenous catalase to neutrophils phagocytizing low-catalase strains increased their intracellular survival. In addition, when the low-catalase strains were injected into mice along with exogenous catalase, their virulence for the mice was enhanced. These experiments suggest that catalase is a staphylococcal virulence factor, and that it may act in part by protecting ingested staphylococci from neutrophil oxidative bactericidal mechanisms.

3.8.2. Resistance to Lysosomal Attack

Recent work by Rest *et al.* (1977) has demonstrated that minor changes in the carbohydrate structure of lipopolysaccharides may greatly alter the susceptibility of bacteria to lysosomal contents. Acetate extracts of granules from human neutrophils were tested for bactericidal activity against smooth and rough mutants of *E. coli, S. typhimurium,* and *S. minnesota.* The rough strains were all much more sensitive to the lysosomal extracts than the smooth strains. Furthermore, rough strains were killed efficiently at pH 7–8, compared to smooth strains which were killed only in the pH 5 range.

Mycobacterium leprae and *Mycobacterium lepraemurium* are intracellular pathogens whose intraphagocytic survival may depend upon a coating with an electron-transparent material as visualized by electron microscopy. These organisms do not inhibit macrophage degranulation and are presumably protected from the bactericidal action of granule contents by the material in this electron-transparent zone. The nature of this material is unknown (Draper and Rees, 1970).

As noted above, *Mycobacterium tuberculosis* is capable of intraphagosomal survival even when macrophage degranulation occurs. Indeed, it has been suggested that the lysosomal contents emptied into phagosomes containing *M. tuberculosis* may be beneficial to the survival of these organisms (Brown *et al.* 1969).

3.9. ESCAPE FROM THE PHAGOSOME

Organsims lying free in the cytoplasm of cells apparently are not recognized as foreign by the host cell and are free to utilize cellular machinery and nutrients for their own needs. An electron microscopy study of the fate of *Mycobacterium bovis* in rabbit macrophages suggested that 25% of *M. bovis* ingested by normal alveolar macrophages lay free in the cytoplasm of the cell. However, when *M. bovis* was ingested by activated macrophages the organisms were retained and digested within the phagosome. It was suggested that phagosomal disruption might be a mechanism of mycobacterial survival within cells (Tenner-Racz *et al.,* 1977).

The active participation of lysosomal contents of fibroblasts in the digestion of the outer coat of both reovirus and vaccinia has been suggested (Dales, 1963, Silverstein *et al.* 1972). Following hydrolysis of the outer coat of the virus, the phagosomal membrane is disrupted and the viral particle released into the cytoplasm. Presumably a similar event could occur in macrophages ingesting these viruses.

4. CONCLUSIONS

Phagocytic cells have evolved a multifaceted mechanism for killing microbial invaders. However, microorganisms have evolved equally effective means of avoiding or neutralizing phagocyte microbicidal function, or, in some instances, utilizing to their own advantage the same mechanisms designed to destroy them. Study of microbial defenses against phagocytes provides a powerful tool for gaining insight into the normal function of the cells that are vital for host defense.

REFERENCES

Allgower, M., and Bloch, H., 1949, The effect of tubercle bacilli on the migration of phagocytes *in vitro*, *Am. Rev. Tuber.* **59**:562.

Arkowa, M., Komata, Y., Asai, T., and Midorikawa, O., 1977, Transmission and scanning electron microscopy of host cell entry by *Toxoplasma gondii*, *Am. J. Pathol.* **87**:285.

Armstrong, J. A., and D'Arcy Hart, P., 1971, Response of cultured macrophages to *Mycobacterium tuberculosis* with observations on fusion of lysosomes with phagosomes, *J. Exp. Med.* **134**:713.

Armstrong, J. A., and D'Arcy Hart, P., 1975, Phagosome-lysosome interactions in cultured macrophages infected with virulent tubercle bacilli, *J. Exp. Med.* **142**:1.

Artigas, J., 1968, Accion de *E. histolytica* sobre leucocites, *Bol. Chil. Parasitol.* **21**:114.

Bainton, D. F., and Farquhar, M. Q., 1966, Origin of granules in polymorphonuclear leukocytes. Two types derived from opposite faces of the golgi complex in developing granulocytes, *J. Cell Biol.* **28**:277.

Beeson, P. B., and Bass, D. A., 1977, *The Eosinophil*, Saunders, Philadelphia.

Bergman, M. J., Guerrant, R. L., Murad, F., Richardson, S. H., Weaver, D., and Mandell, G. L., 1978, The interaction of polymorphonuclear neutrophils with *Escherichia coli*: Effect of enterotoxin on phagocytosis, killing, chemotaxis and cyclic AMP, *J. Clin. Invest.* **61**:227.

Bernheimer, A. W., 1954, Streptolysins and their inhibitors, in: *Streptococcal Infection* (M. McCarty, ed.), pp. 19–38, Columbia University Press, New York.

Bernheimer, A. W., and Schwartz, L. L., 1964, Lysosomal disruption by bacterial toxins, *J. Bacteriol.* **87**:1100.

Bloch, H., 1950, Studies on the virulence of tubercule bacilli: Isolation and biological properties of a constituent of virulent organisms, *J. Exp. Med.* **91**:197.

Brooks, G. F., Israel, K. S., and Petersen, B. H., 1976, Bactericidal and opsonic activity against *Neisseria gonorrhoeae* in sera from patients with disseminated gonococcal infection, *J. Infect. Dis.* **134**:450.

Brown, C. A., Draper, P., and D'Arcy Hart, P., 1969, Mycobacteria and lysosomes: A paradox, *Nature* **221**:658.

Buchanan, T., Chen, Kirk C. S., Jones, R. S., Hildebrandt, J. F., Pearce, W. A., Hermodson, M. A., Newland, J. C., and Luchtel, D. L., 1978a, Pili and principal outer membrane protein of *Neisseria gonorrhoeae*: Immunochemical, structural, and pathogenic aspects, in: *Immunobiology of Neisseria gonorrhoeae*, pp. 145–154, American Society of Microbiology, Washington, D.C.

Buchanan, T., Pearce, W. A., and Chen, K. C. S., 1978b, Attachment of *Neisseria gonorrhoeae* pili to human cells, and investigations of the chemical nature of the receptor for gonococcal pili, in: *Immunobiology of Neisseria gonorrhoeae*, pp. 242–249, American Society of Microbiology, Washington, D.C.

Bulmer, G. S., and Sans, M. D., 1968, *Cryptococcus neoformans*. III. Inhibition of phagocytosis, *J. Bacteriol.* **95**:5.

Bulmer, G. S., and Tacker, J. R., 1975, Phagocytosis of *Cryptococcus neoformans* by alveolar macrophages, *Infect. Immun.* **11**:73.

Burrus, T. W., 1955, The basis of virulence for mice of *P. pestis*, in: *Mechanisms of Microbiological Pathogenicity*, Fifth Symposium of the Society of General Microbiology, pp. 152–175, Cambridge University Press, London.

Butterworth, A. E., Sturrock, R. F., Houba, V., Mahmoud, A. A. F., Sher, A., Rees, P. H., 1975, Eosinophils as mediators of antibody-dependent damage to schistosomula, *Nature* **256**:727.

Cartwright, G. E., Athens, J. W., and Wintrobe, M. M., 1964, The kinetics of granulopoiesis in normal man, *Blood* **24**:780.

Cavanaugh, D. C., and Randall, R., 1959, The role of multiplication of *Pasteurella pestis* in mononuclear phagocytes in the pathogenesis of flea borne plague, *J. Immunol.* **83**:348.

Cline, M. J., Lehrer, R. J., Territo, M. C., and Golde, D. W., 1978, Monocytes and macrophages: Functions and diseases, *Ann. Int. Med.* **88**:78.

Craddock, C. G., 1972, Production, distribution and fate of granulocytes, in: *Hematology* (W. J. Williams, E. Beritter, and A. J. Ersler, ed.), pp. 607–618, McGraw-Hill, New York.

Dales, S., 1963, The uptake and development of vaccinia virus in Strain L cells followed with labeled viral deoxyribonucleic acid, *J. Cell Biol.* **18**:51.

Densen, P., and Mandell, G. L., 1978a, Gonococcal interactions with polymorphonuclear neutrophils: Importance of the phagosome for bactericidal activity, *J. Clin. Invest.* **62**:1161.

Densen, P., and Mandell, G. L., 1978b, Phagocytic cells versus microbes, *Infect. Dis. Rev.* **5**:43.

Densen, P., Mahmoud, A. A. F., Sullivan, J., Warren, K. S., and Mandell, G. L., 1978a, Demonstration of eosinophil degranulation on the surface of opsonized schistosomules by phase-contrast cinemicrography, *Infect. Immun.* **22**:282.

Densen, P., Rein, M. F., Sullivan, J. A., and Mandell, G. L., 1978b, Morphologic observations of neutrophil-gonococcus interaction, in: *Immunology of Neisseria gonorrhoeae*, pp. 213–220, American Society of Microbiology Publications Office, Washington, D.C.

Diamond, R. D., Root, R. K., and Bennett, J. E., 1972, Factors influencing killing of *Cryptococcus neoformans* by human leukocytes in vitro, *J. Infect Dis.* **125**:367.

Dilworth, J. A., Hendley, J. O., and Mandell, G. L., 1975, Attachment and ingestion of gonococci by human neutrophils, *Infect. Immun.* **11**:512.

Dossett, J. H., Kronvall, G., Williams, R. C., Jr., and Quie, P. G., 1969, Antiphagocytic effects of staphylococcal protein A, *J. Immunol.* **103**:1405.

Draper, P., and Rees, R. J. W., 1970, Electron transparent zone of mycobacteria may be a defense mechanism, *Nature* **228**:860.

Easmon, C. S. F., Hamilton, J., and Glynn, A. A., 1973, Mode of action of a staphylococcal anti-inflammatory factor, *Br. J. Exp. Pathol.* **54**:638.

Elberg, S. S., and Schneider, P., 1953, Directed leukocyte migration in response to infection and other stimuli, *J. Infect. Dis.* **83**:36.

Farhi, F., Bulmer, G. S., and Tacker, J. R., 1970, *Cryptococcus neoformans*. IV. The not-so-encapsulated yeast, *Infect. Immun.* **1**:526.

Foley, M. J., and Wood, W. B., Jr., 1959, Studies on the pathogenicity of group A streptococcus. II. The antiphagocytic effects of the M protein and the capsular gel, *J. Exp. Med.* **110**:617.

Forsgren, A., and Sjoquiest, J., 1966, Protein A from *Staphylococcus aureus*. I. Pseudoimmune reaction with human gamma globulin, *J. Immunol.* **97**:822.

Foster, W. D., 1962, The role of coagulase in the pathogenicity of *Staphyloccus aureus*, *J. Pathol. Biol.* **83**:287.

Friis, R. R., 1972, Inactivation of L cells and *Chlamydia psittaci*: Entry of the parasite and host responses to its development, *J. Bacteriol.* **110**:706.

Gay, F. P., and Oram, F., 1933, *Streptococcus* leucocidin, *J. Immunol.* **25**:501.

Gibbs, D. L., 1974, The interaction between human polymorphonuclear leukocytes and *Neisseria gonorrhoeae* cultivated in the chick embryo, Ph.D. Thesis, Graduate School of Cornell University, New York.

Gladstone, C. P., and van Heyningen, W. E., 1957, Staphylococcal leucocidin, *Br. J. Exp. Pathol.* **38**:124.

Gleich, F. J., 1977, The eosinophil: New aspects of structure and function, *J. Allergy Clin. Immunol.* **60**:73.

Glynn, A. A., and Howard, C. J., 1970, The sensitivity of *Escherichia coli* related to their K antigens, *Immunology* **18**:331.

Goldring, O. L., Clegg, J. A., Smithers, S. R., and Terry, R. J., 1976, Acquisition of human blood group antigens by *Schistosoma mansoni, Clin. Exp. Immunol.* **26**:181.

Goldstein, I. M., Hoffstein, S. T., and Weissmann, G., 1975a, Mechanisms of lysosomal enzyme release from human polymorphonuclear leukocytes, *J. Cell Biol.* **66**:647.

Goldstein, I. M., Roos, D., Kaplan, H. B., and Weissmann, 1975b, Complement and immunoglobulin stimulate superoxide production by human leukocytes independently of phagocytosis, *J. Clin. Invest.* **46**:1155.

Goldstein, I. M., Kaplan, H. B., Radin, A., and Frosch, M., 1976, Independent effects of IgG and complement upon human polymorphonuclear leukocyte function, *J. Immunol.* **117**:1282.

Goren, M. B., 1977, Phagocyte-lysosomes: Interactions with infectious agents, phagosomes, and experimental perturbations in function, *Annu. Rev. Microbiol.* **31**:507.

Goren, M. B., D'Arcy Hart, P., Young, M. R., and Armstrong, J. A., 1976, Prevention of phagosome-lysosome fusion in cultured macrophages by sulfatides of *Mycobacterium tuberculosis, Proc. Natl. Acad. Sci. USA* **73**:2510.

Green, H., Barrow, P., and Goldberg, B., 1959, Effect of an antibody and complement on permeability control in ascites tumor cells and erythrocytes, *J. Exp. Med.* **110**:699.

Griffin, F. M., Griffin, J. A., and Silverstein, S. C., 1976, Studies on the mechanism of phagocytosis. II. The interaction of macrophages with anti-immunoglobulin IgG-coated bone marrow-derived lymphocytes, *J. Exp. Med.* **144**:788.

Grove, D. I., Mahoud, A. A. F., and Warren, K. S., 1977, Eosinophils and resistance to *Trichinella spiralis, J. Exp. Med.* **145**:755.

Hendley, J. O., Powell, K. R., and Rodewald, R., Holzgrefe, H. H., and Lyles, R., 1977, Demonstration of a capsule on *Neisseria gonorrhoeae, N. Engl. J. Med.* **296**:608.

Henson, P. M., and Oades, Z. G., 1975, Stimulation of human neutrophils by soluble and insoluble immunoglobulin aggregates. Secretion of granule constituents and increased oxidation of glucose, *J. Clin. Invest.* **56**:1053.

Hill, M. J., 1968, A staphylococcal aggression, *J. Med. Microbiol.* **1**:33.

Hirsch, J. G., Bernheimer, A. W., and Weissmann, G., 1963, Motion picture study of the toxic action of streptolysins on leukocytes, *J. Exp. Med.* **118**:223.

Hoffstein, S., Goldstein, I. M., and Weissmann, G., 1977, Role of microtubule assembly in lysosomal enzyme secretion from human polymorphonuclear leukocytes, *J. Cell Biol.* **73**:242.

Houston, T., and Rankin, J. C., 1907, The virulence for mice of strains of *Escherichia coli* related to the effect of K antigens on their resistance to phagocytosis and killing by complement, *Immunology* **20**:767.

Howard, C. J., and Glynn, A. A., 1971, The virulence for mice of strains of *Escherichia coli* related to the effect of K antigens on their resistance to phagocytosis and killing by complement, *Immunology* **20**:767.

Ignarro, L. J., 1973, Neutral protease release from human leukocytes regulated by neurohormones and cyclic nucleotides, *Nature New Biol.* **245**:151.

Ignarro, L. J., and Cech, S. Y., 1976, Bidirectional regulation of lysosomal enzyme secretion and phagocytosis in human neutrophils by guanosine 3′,5′-monophosphate and adenosine 3′,5′-monophosphate, *Proc. Soc. Exp. Biol. Med.* **151**:448.

James, J. F., and Swanson, J., 1977, The capsule of the gonococcus, *J. Exp. Med.* **145**:1082.

Jones, T. C., and Hirsch, J. G., 1971, The interaction of *Mycoplasma pulmonis* with mouse peritoneal macrophages and L cells, *J. Exp. Med.* **133**:231.

Jones, T. C., and Hirsch, J. G., 1974, The interaction between *Toxoplasma gondii* and mammalian cells. II. The absence of lysosomal fusion with phagocytic vacuoles containing living parasites, *J. Exp. Med.* **136**:1173.

Jones, T. C., Len, L., and Hirsch, J. G., 1975, Assessment *in vitro* of immunity against *Toxoplasma gondii*, *J. Exp. Med.* **141**:466.

Keppie, J., Harris-Smith, P. W., and Smith, H., 1963, The chemical basis of the virulence of *Bacillus anthracis*, *Br. J. Exp. Pathol.* **44**:446.

Klebanoff, S. J., 1968, Myeloperoxidase-halide-hydrogen peroxide antibacterial system, *J. Bacteriol.* **95**:2131.

Klebanoff, S. J., 1975, Antimicrobial mechanisms in neutrophilic polymorphonuclear leukocytes, *Semin. Hematol.* **12**:117.

Kossack, R. E., Schadelin, J., Guerrant, R. L., Densen, P., and Mandell, G. L., 1978, Diminished neutrophil oxidative metabolism following phagocytosis of virulent *Salmonella typhi*, *Clin. Res.* **26**:28A.

Kozel, T. R., and Mastroianni, R. P., 1976, Inhibition of phagocytosis by cryptococcal polysaccharide: Dissociation of the attachment and ingestion phases of phagocytosis, *Infect. Immun.* **14**:62.

Leffell, M. S., and Spitznagel, J. K., 1975, Fate of human lactoferrin and myeloperoxidase in phagocytizing human neutrophils: Effects of immunoglobulin G subclasses and immune complexes coated on latex beads, *Infect. Immun.* **12**:812.

Lentnek, A. L., Schreiber, A. D., and MacGregor, R. R., 1976, The induction of augmented granulocyte adherence by inflammation, *J. Clin. Invest.* **57**:1098.

Lichtman, M. A., and Weed, R. I., 1972, Alteration of the cell periphery during granulocyte maturation: Relationship to cell function, *Blood* **39**:301.

Lowrie, D. B., Jackett, B. S., and Ratliffe, N. A., 1975, *Mycobacterium microti* may protect itself from intracellular destruction by releasing cyclic AMP into phagosomes, *Nature* **254**:600.

Mackaness, G. B., 1964, The immunological basis of acquired cellular resistance, *J. Exp. Med.* **120**:105.

Macleod, C. M., and Krauss, M. R., 1950, Relation of virulence of pneumococcal strains for mice to the quantity of capsular polysaccharide formed *in vitro*, *J. Exp. Med.* **92**:1.

Mahmoud, A. A. F., Warren, K. S., and Peters, P. A., 1975, A role for the eosinophil in acquired resistance to *Schistosoma mansoni* infection as determined by antieosinophil serum, *J. Exp. Med.* **142**:805.

Mandell, G. L., 1974, Bactericidal activity of aerobic and anaerobic polymorphonuclear neutrophils, *Infect. Immun.* **9**:337.

Mandell, G. L., 1975, Catalase, superoxide dismutase, and virulence of *Staphylococcus aureus*, *J. Clin. Invest.* **55**:561.

Martin, S. P., and Chaudhuri, S. N., 1952, Effect of bacteria and their products on migration of leukocytes, *Proc. Soc. Exp. Biol. Med.* **81**:286.

McCoy, E. R., Doyle, D., Burda, K., Corbeil, L. B., and Winter, A. J., 1975, Superficial antigens of *Campylobacter* (*Vibrio*) *fetus*: Characterization of an antiphagocytic component, *Infect. Immun.* **11**:517.

McLaren, D. J., MacKenzie, C. D., and Ramalho-Pinto, F. J., 1977, Ultrastructural observations on the *in vitro* interaction between rat eosinophils and some parasitic helminths (*Schistosoma mansoni*, *Trichinella spiralis*, and *Nippostrongylus brasiliensis*), *Clin. Exp. Immunol.* **30**:105.

Medearis, D. N., Camitta, B. M., and Heath, E. C., 1968, Cell wall composition and virulence in *Escherichia coli*, *J. Exp. Med.* **128**:399.

Melly, M. A., Suter, L. J., Liau, D., Hash, J. H., 1974, Biological properties of the encapsulated *Staphylococcus aureus* M, *Infect. Immun.* **10**:389.

Miller, R. M., Gartus, J., and Hornick, R. B., 1972, Lack of enhanced oxygen consumption by polymorphonuclear leukocytes on phagocytosis of virulent *Salmonella typhi*, *Science* **175**:1010.

Moore, P. L., Bank, H. L., Brissie, N. T., and Spicer, S. S., 1976, Association of microfilament bundles with lysosomes in polymorphonuclear leukocytes, *J. Cell Biol.* **71**:659.

Nathan, D. G., Baehner, R. L., and Weaver, D. K., 1969, Failure of nitroblue tetrazolium reduction in the phagocytic vacuoles of leukocytes in chronic granulomatous disease, *J. Clin. Invest.* **48**:1895.

Ochs, H. D., and Igo, R. P., 1973, The NBT slide test: A simple screening method for detecting chronic granulomatous disease and female carriers, *J. Pediatr.* **83**:77.

Ofek, J., Bergner-Rabinowitz, S., and Ginsberg, I., 1970. Oxygen-stable hemolysins of group A streptococci. The relation of the leukotoxic factor to streptolysin S, *J. Infect. Dis.* **122**:517.

Oram, F., 1934, *Pneumococcus* leucocidin, *J. Immunol.* **26**:233.

Peterson, P. K., Verhoef, J., Kim, Y., and Quie, P. G., 1976a, Heterogeneity of opsonic requirements for phagocytosis of different staphylococcal strains, *Abstr. 115B, 16th Interscience Conference on Antimicrobial Agents and Chemotherapy*, American Society for Microbiology Publications Office, Washington, D.C.

Peterson, P. K., Verhoef, J., Sabbath, L. D., and Quie, P. G., 1976b, Extracellular and bacterial factors influencing staphylococcal phagocytosis and killing by human polymorphonuclear leukocytes, *Infect. Immun.* **14**:496.

Peterson, P. K., Verhoef, J., Sabbath, L. D., and Quie, P. G., 1977, Effect of protein A on staphylococcal opsonization, *Infect. Immun.* **15**:760.

Rest, R. H., Cooney, M. H., and Spitznagel, J. K., 1977, Susceptibility of lipopolysaccharide mutants to the bactericidal action of human neutrophil lysosomal fractions, *Infect. Immun.* **16**:145.

Rest, R. H., Cooney, M. H., and Spitznagel, J. K., 1978, Bactericidal activity of specific and azurophil granules from human neutrophils: Studies with outer membrane mutants of *Salmonella typhimurium* LT-2, *Infect. Immun.* **19**:131.

Richardson, W. P., and Sadoff, J. C., 1977, Production of a capsule by *Neisseria gonorrhoeae, Infect. Immun.* **15**:663.

Robbins, J. B., McCracken, G. H., Gotschlich, E. C., Ørksov, F., Ørksov, I., and Hanson, L. A., 1974, *Escherichia coli* K capsular polysaccharide associated with neonatal meningitis, *N. Engl. J. Med.* **290**:1216.

Roberts, R. B., 1967, The interaction *in vitro* between group B meningococci and rabbit polymorphonuclear leukocytes. Demonstration of type specific opsonins and bacteriocidins, *J. Exp. Med.* **126**:795.

Roberts, R. B., 1970, The relationship between group A and group C meningococcal polysaccharide and serum opsonins in man, *J. Exp. Med.* **131**:499.

Rogers, D. E., and Tompsett, R., 1952, The survival of staphylococci within human leukocytes, *J. Exp. Med.* **94**:209.

Rottini, G., Dri, P., Sorango, M. R., and Patriarca, P., 1975, Correlation between phagocytic activity and metabolic response of polymorphonuclear leukocytes toward different strains of *Escherichia coli, Infect. Immun.* **11**:417.

Roos, D., Goldstein, I. M., Kaplan, H. B., and Weissmann, G., 1976, Dissociation of phagocytosis, metabolic stimulation, and lysosomal enzyme release in human leukocytes, *Agents Actions* **6**:256.

Sawyer, W. D., 1969, Interaction of influenza virus with leukocytes and its effect on phagocytosis, *J. Infect Dis.* **119**:541.

Scharmann, W., Jacob., F., and Porstendorfer, J., 1976, The cytotoxic action of leucocidin from *Pseudomonas aeruginosa* on human polymorphonuclear leukocytes, *J. Gen. Microbiol.* **93**:303.

Schultz, D. R., and Miller, K. D., 1974, Elastase of *Pseudomonas aeruginosa*: Inactivation of complement components and complement derived chemotactic and phagocytic factors, *Infect. Immun.* **10**:128.

Senff, L. M., and Sawyer, W. D., 1977, Release of enzymes from human leukocytes during incubation with *Neisseria gonorrhoeae, Br. J. Vener. Dis.* **53**:360.

Senff, L. M., Sawyer, W. D., and Haak, R. A., 1977, Effect of gonococci on nitroxide spin-labeled membranes of human blood cells, in: *Abstracts of the Annual Meeting of the American Society of Microbiology*, p. 77, D48, American Society of Microbiology Publications Office, Washington, D.C.

Silverstein, S. C., Astell, C., Levin, D. H., Schonberg, M., and Acs, G., 1972, The mechanism of reovirus uncoating and gene activation *in vivo, Virology* **47**:797.

Simberkoff, M. S., and Elsbach, P., 1971, The interaction *in vitro* between polymorphonuclear leukocytes and mycoplasma, *J. Exp. Med.* **134**:1417.

Smith, H., 1977, Microbial surfaces in relation to pathogenicity, *Bacteriol. Rev.* **41**:475.

Smith, M. R., and Wood, W. B., Jr., 1947, Studies on the mechanism of recovery in pneumonia due to Friedlanders bacillus. III. The role of "surface phagocytosis" in the destruction of the microorganisms in the lung, *J. Exp. Med.* **86**:257.

Spitznagel, J. K., Dalldorf, F. G., Leffell, M. S., Folds, J. D., Welsh, J. R. H., Cooney, M. H., and Martin, L. A., 1974, Character of azurophil and specific granules purified from human polymorphonuclear leukocytes, *Lab. Invest.* **30**:774.

Stossell, T. P., 1974, Phagocytosis, *N. Engl. J. Med.* **290**:717, 773, 833.

Sullivan, G. W., Sullivan, J. A., and Mandell, G. L., 1978, Leukotoxic streptococci and neutrophil degranulation, *Clin. Res.* **26**:407A.

Suter, E., 1956. Interaction between phagocytes and pathogenic microorganisms, *Bacterial Rev.* **20**:94.

Swanson, J., Sparks, E., Zeligs, B., Siam, M. A., and Parrott, C., 1974, Studies on gonococcus infection. V. Observations on *in vitro* interactions of gonococci and human neutrophils, *Infect. Immun.* **10**:633.

Swanson, J., Sparks, E., Young, D., and King, G., 1975, Studies on gonococcus infection. X. Pili and leukocyte association factor as mediators of interactions between gonococci and leukocytic cells *in vitro*, *Infect. Immun.* **11**:1352.

Tenner-Racz, K., Racz, P., Myrvik, Q. N., and Faunter, L., 1977, Interaction of normal and activated alveolar macrophages of the rabbit with *Mycobacterium bovis* (BCG) and *M. smegmatis in vivo* and *in vitro*, *J. Reticuloendothel. Soc.* **22**:41A.

Thongthai, C., and Sawyer, W. D., 1973, Studies on the virulence of *Neisseria gonorrhoeae*. I. Relation of colony morphology and resistance to phagocytosis by polymorphonuclear leukocytes, *Infect. Immun.* **7**:373.

van Oss, C. J., and Gillman, C. F., 1972, Phagocytosis as a surface phenomenon: II. Contact angles and phagocytosis of encapsulated bacteria before and after opsonization by specific antiserum and complement, *J. Reticuloendothel. Soc.* **12**:497.

Ward, P. A., and Hill, J. H., 1970, C5 chemotactic fragments produced by an enzyme in lysosomal granules of neutrophils, *J. Immunol.* **104**:535.

Watt, P. J., 1970, The fate of gonococci in polymorphonuclear leukocytes, *J. Med. Microbiol.* **3**:501.

Weissmann, G., Goldstein, I., Hoffstein, S., and Tsung, P., 1975, Reciprocal effects of cAMP and cGMP on microtubule-dependent release of lysosomal enzymes, *Ann. N.Y. Acad. Sci.* **253**:750.

Weksler, B. B., and Hill, M. J., 1968, Inhibition of leukocyte migration by a staphylococcal factor, *J. Bacteriol.* **98**:1030.

Williams, R. C., Jr., Gewurz, H., and Quie, P. G., 1972, Effects of fraction I from *Yersinia pestis* on phagocytosis *in vitro*, *J. Infect Dis.* **126**:235.

Wilson, A. J., 1957, The leukotoxic action of streptococci, *J. Exp. Med.* **105**:463.

Wood, W. B., Jr., Smith, M. R., and Watson, B., 1946, Studies on the mechanism of recovery in pneumococcal pneumonia. IV. The mechanism of phagocytosis in the absence of antibody, *J. Exp. Med.* **84**:387.

Woodin, A. M., 1968, The basis of leucocidin action, in: *The Biological Basis of Medicine* (E. E. Bitter, and N. Bitter, eds.), Vol. 2, pp. 373–396, Academic Press, London.

Wright, D. G., and Gallin, J. I., 1975, Modulation of the inflammatory response by-products released from human polymorphonuclear leukocytes during phagocytosis, *Inflammation* **1**:23.

Wright, D. G., and Gallin, J. I., 1977, A functional differentiation of human neutrophil granules: Generation of C5a by a specific (secondary) granule product and inactivation of C5a by azurophil (primary) granule products, *J. Immunol.* **119**:1068.

Wright, D. G., Bralove, D. A., and Gallin, J. I., 1977, The differential mobilization of human neutrophil granules, *Am. J. Pathol.* **87**:273.

Yoshida, K., Nakamura, A., Ohtomo, T., and Jwani, S., 1974, Detection of capsular antigen production in unencapsulated strains of *Staphylococcus aureus*, *Infect. Immun.* **9**:620.

Zucker-Franklin, D., Davidson, M., and Thomas, L., 1966, The interaction of mycoplasmas with mammalian cells. I. HeLa cells, neutrophils and eosinophils, *J. Exp. Med.* **124**:521.

Zurier, R. B., Weissmann, G., Hoffstein, S., Kammerman, S., and Tai, H. H., 1974, Mechanisms of lysosomal enzyme release from human leukocytes. II. Effects of cAMP and cGMP, autonomic agonists, and agents which affect microtubule function, *J. Clin. Invest.* **53**:297.

16

Biochemical Defects of Polymorphonuclear and Mononuclear Phagocytes Associated with Disease

RICHARD B. JOHNSTON, JR.

1. INTRODUCTION

As its title indicates, the intent of this chapter is to review the known biochemical defects of human neutrophils, monocytes, and macrophages that predispose to disease, in particular, to recurrent infections. The review will be organized along the lines of the principal functions of phagocytic cells in host defense—chemotaxis, ingestion, and microbicidal activity—in an effort to place the biochemical abnormalities in the perspective of function and, thereby, to emphasize molecular–functional interactions wherever possible. The better substantiated disorders will be listed in tables, and those for which there exists some understanding of a molecular basis will be described in the text. When defects in more than one function coexist in the same disorder, somewhat arbitrary categorization will be made on the basis of apparent relative severity of the defects and the clinical pattern of infection. Since description of the clinical syndromes and functional abnormalities has greatly exceeded elucidation of their underlying molecular basis, this attempt to logically organize disorders of phagocyte function can be considered only tentative. Some description of the sites of infection and the infecting organisms will be included, in the hopes that integration of biochemical, functional, and clinical information might lend some insight into basic principles of leukocyte physiology and host defense against infection.

RICHARD B. JOHNSTON, JR. • Department of Pediatrics, National Jewish Hospital and Research Center and University of Colorado School of Medicine, Denver, Colorado 80206. Supported by USPHS grant AI 14148.

TABLE 1. DEFECTS OF CHEMOTAXIS

A. Primary cellular abnormalities
    1. Chediak–Higashi syndrome[a]
    2. Increased microtubule assembly and elevated cyclic GMP[a]
    3. Actin dysfunction[a]
    4. Lazy leukocyte syndrome
    5. Familial defects
B. Secondary cellular abnormalities
    1. Decreased chemotaxis with elevated IgE[a]
    2. Diabetes mellitus[a]
    3. Acrodermatitis enteropathica[a]
    4. Hypophosphatemia[a]
    5. Ethanol intoxication[a]
    6. Rheumatoid arthritis
    7. Mannosidosis
    8. Down syndrome
    9. Cancer
    10. Severe infections
    11. Burns
    12. Malnutrition
    13. Corticosteroid or immunosuppressive therapy
    14. Bone marrow transplantation
    15. The neonatal state
C. Humoral deficiencies
    1. Abnormalities of chemotactic factor production (e.g., C deficiencies)
    2. Absence of antagonist of chemotaxis inhibitor
D. Humoral inhibitors
    1. Chemotactic factor inhibitors: cancer, cirrhosis, sarcoidosis, leprosy, chronic granulomatous disease
    2. Elevated IgA
    3. Circulating chemotactic factors: Wiskott–Aldrich syndrome, nephritis

---

[a]Discussed in text.

2. DEFECTS OF CHEMOTAXIS

The wider availability of assays of leukocyte random motility and chemotaxis has resulted in the detection of an increasing number of abnormalities in phagocyte movement. In most cases, the underlying molecular aberration is unknown or incompletely understood. Table 1 represents an attempt to structure the clinical disorders of chemotaxis on the basis of studies to date. This group of abnormalities has been the subject of thorough and recent review (Ward, 1974; Gallin, 1975; Miller, 1975; Snyderman and Pike, 1977; Quie and Cates, 1977; Johnston and McPhail, 1980).

2.1. PRIMARY CELLULAR ABNORMALITIES

2.1.1. Chediak–Higashi Syndrome

2.1.1a. General Description. Chediak–Higashi syndrome (CHS) is a genetically determined, probably autosomal recessive disease characterized by recurrent infections; partial loss of pigment in the skin, iris, and retina; photophobia; nystagmus; giant granules in most granule-containing cells; an "accelerated phase" in which a lymphoma-like illness develops; and early death, usually in childhood. An analogous syndrome with giant granules occurs in mink, mice, cattle, cats, and killer whales (references in Bell *et al.*, 1976). Studies to date suggest that the underlying molecular defect is the same in mice and man and, thus, that the affected mice ("beige mice") are a valid model for the human disease (Bennett *et al.*, 1969; Gallin *et al.*, 1974).

Abnormally large granules are distributed in many cell types. In neutrophils they contain enzymes and proteins normally found only in azurophilic lysosomal granules (e.g., peroxidase) or in specific granules (lactoferrin) (Blume and Wolff, 1972; Rausch *et al.*, 1978). They appear to form by the progressive aggregation and fusion of azurophilic and specific granules during myelopoiesis; a similar phenomenon occurs in monocytes (Rausch *et al.*, 1978). Normal specific but not normal azurophilic granules also exist in CHS neutrophils. Giant granules are present in all neutrophils, eosinophils, and basophils, and in some lymphocytes, but not in platelets or erythrocytes (Blume and Wolff, 1972). The "partial albinism" of the syndrome is due to the aggregation of melanin into large granules (Blume and Wolff, 1972).

2.1.1b. Infections. There is little detailed information on infections. In one careful study of four patients, 29 pyogenic infections were documented (Blume and Wolff, 1972). *Staphylococcus aureus* was isolated 22 times and group A β-hemolytic streptococcus and *Haemophilus influenzae* four times each. The sites of infection were almost equally divided between the lungs, the skin and subcutaneous tissues, and the upper respiratory tract (including otitis media and sinusitis).

2.1.1c. Functional Phagocyte Abnormalities. The patient with CHS suffers several granulocyte defects which could predispose him to pyogenic infections. The first is neutropenia, which can be profound (Blume and Wolff, 1972). Granulocytes from patients with CHS also have impaired chemotactic ability, both *in vivo*, as shown with Rebuck inflammatory skin window techniques, and *in vitro*, utilizing the Boyden chamber (Clark and Kimball, 1971; Boxer *et al.*, 1976b). Thus, the movement of these patients' granulocytes to a site of bacterial invasion is sluggish.

Once the CHS cell reaches the site of infection, its ability to ingest the invading organism appears normal, or even slightly elevated, as demonstrated *in vitro* (Stossel *et al.*, 1972; Root *et al.*, 1972), and *in vivo* by removal of radiolabeled aggregated albumin from the bloodstream (Blume and Wolff, 1972). Degranulation of lysosomes into the phagocytic vacuole is abnormal, however. This has been demonstrated by the morphologic absence of peroxidase activity in phagosomes after phagocytosis (Root *et al.*, 1972) and by biochemical analyses showing that isolated phagosomes from CHS leukocytes are deficient in activities of certain hydrolytic enzymes normally delivered by the fusion of lysosomes with phagocytic vacuoles (Stossel *et al.*, 1972). Activity levels or presence by immunofluorescent staining of several lysosomal enzymes, including elastase, have been reported to be normal in neutrophils from these patients (Stossel *et al.*, 1972; Rausch *et al.*, 1978). However, other groups have found decreased

activity of three enzymes from azurophilic granules—myeloperoxidase and
β-glucuronidase (Kimball *et al.*, 1975) and elastase (Vassalli *et al.*, 1978).

Phagocytic oxidative metabolism in CHS is normal or elevated. Root *et al.*
(1972) demonstrated normal nitroblue tetrazolium (NBT) reduction and hydrogen peroxide (H_2O_2) production by CHS leukocytes and above normal oxygen
consumption, hexose monophosphate (HMP) shunt activation, and intracellular
iodination. In spite of this ability to respond well metabolically to phagocytosis,
granulocytes from CHS patients have impaired bactericidal ability (Root *et al.*,
1972). The defect is accentuated at early time periods and approaches normal
after extended incubation. This defective *in vitro* bactericidal ability differs from
that seen in leukocytes from patients with chronic granulomatous disease (CGD)
in that the impairment is not pronounced and it includes both catalase-positive
and catalase-negative organisms (Root *et al.*, 1972).

2.1.1d. Functional Abnormalities Predisposing to Infections. The
neutropenia apparent in most CHS patients is postulated to be caused by
intramedullary granulocyte destruction. This conclusion is based on several findings, including abnormal marrow reserves, elevated serum muramidase levels in
the absence of hypersplenism, and morphologic abnormalities seen in electron
and light microscopic studies of bone marrow specimens suggesting autophagic
destruction of myeloid precursors and mature neutrophils (Blume *et al.*, 1968;
Oberling *et al.*, 1976).

As previously noted, the defect in bactericidal activity is most pronounced
at early time periods, which parallels the delay in phagolysosomal fusion. This
results in an early deficiency in the phagocytic vacuole of certain lysosomal
enzymes, including myeloperoxidase (Stossel *et al.*, 1972). In fact, the bactericidal defect in CHS resembles in extent and kinetics that seen in patients with
myeloperoxidase deficiency (Lehrer *et al.*, 1969), which suggests that the defect
in bactericidal activity of CHS leukocytes could be due to early deficiency of
myeloperoxidase activity in the phagocytic vacuole. Absence of elastase there is
another possibility (Vassalli *et al.*, 1978). Patients with myeloperoxidase deficiency have not been as readily susceptible to bacterial infection as have those
with CHS, which favors the concept that the bactericidal defect in CHS does not
play a primary role in the predisposition to infection in this disease. It seems
likely that the neutropenia and depressed chemotactic activity are more at fault,
particularly in view of the pattern of superficial infections due primarily to
staphylococci which characterizes CHS, neutropenia (Pincus *et al.*, 1976;
Howard *et al.*, 1977), and chemotaxis defects (Johnston and McPhail, 1980).

2.1.1e. Molecular Basis for Functional Abnormalities. Recent evidence
suggests that the impaired chemotaxis and degranulation, and perhaps the giant
granules, of CHS leukocytes are due to a defect in microtubular function (Oliver,
1978; Oliver and Zurier, 1976; Hinds and Danes, 1976; Boxer *et al.*, 1976b; Boxer,
G. J., *et al.*, 1977). In normal neutrophils both degranulation and chemotaxis are
partially controlled by the activity of microtubules (Hoffstein *et al.*, 1977; Oliver,
1978). A convenient assay for microtubular function has been analysis of polarization, or capping, of fluorescein-labeled concanavalin A (Con A). Intact microtubules prevent cap formation; colchicine, which inhibits microtubule assem-

bly, permits cap formation and inhibits degranulation and chemotaxis in normal neutrophils and monocytes. Capping occurs without the addition of colchicine in CHS phagocytes (Oliver, 1978). Adherence of neutrophils to nylon fibers appears to require the integrity of microtubules, and this function is also abnormal in CHS (Boxer *et al.*, 1978).

Microtubular assembly has been reported to be depressed by increased intracellular concentrations of cyclic AMP and accentuated by increased cyclic GMP (Weissmann *et al.*, 1975). Cyclic GMP and cholinergic agents that elevate intracellular cyclic GMP (carbachol and bethanechol) have been found to inhibit Con A-induced cap formation in leukocytes from beige mice (Oliver *et al.*, 1975) and to prevent the development of characteristic giant granules in beige mouse fibroblasts cultured for 10–14 days (Oliver *et al.*, 1976). Beige mice given cholinergic agents subcutaneously or by mouth for at least 3 weeks showed normal granule morphology and a normal lack of Con A cap formation in their peripheral blood leukocytes (Oliver and Zurier, 1976). These studies were extended to man with the demonstration that leukocytes from CHS patients also show spontaneous Con A cap formation, and this can be reduced by cyclic GMP and cholinergic agents. Monocytes from patients developed abnormal granules when incubated in tissue culture, and these were diminished when cholinergic agents were added to the culture (Oliver and Zurier, 1976).

Cyclic AMP is believed to suppress and cyclic GMP to augment neutrophil degranulation and motility (Estensen *et al.*, 1973; Weissmann *et al.*, 1975; Ignarro and Cech, 1975). It is presently not clear to what extent these effects are mediated by alteration of microtubular function (Hoffstein *et al.*, 1977; Oliver, 1978). Whatever its mechanism, the addition of cyclic GMP or cholinergic agents to neutrophils from one CHS patient normalized their functional ability in assays of chemotaxis, degranulation, and bactericidal activity (Boxer, L. A., *et al.*, 1977). Intraleukocytic concentrations of cyclic AMP were markedly elevated in this same infant (Boxer *et al.*, 1976). In an attempt to offset the effect of the elevated cyclic AMP levels, ascorbate was fed to the infant since this agent has been shown to elevate intracellular cyclic GMP in monocytes (Sandler *et al.*, 1975). After 4 weeks of ascorbate therapy, leukocyte cyclic GMP levels remained within the normal range, but cyclic AMP levels were reduced to near normal values. Ascorbate administration resulted in the normalization of *in vitro* chemotaxis, degranulation, and bactericidal activity, as well as adherence to nylon fibers and Con A-induced assembly of microtubules (Boxer *et al.*, 1979a). Ascorbate also increased *in vitro* chemotaxis of normal cells. Treatment with ascorbate did not significantly improve the neutropenia in this patient nor did it eliminate the characteristic giant granules (see also Rausch *et al.*, 1978). This and a second patient have been free of abnormal infections during a year of ascorbate therapy (Boxer *et al.*, 1979a).

It now seems likely that the elevated cyclic AMP concentrations in this patient are a *consequence* of microtubule dysfunction rather than its cause, based on the findings of Malawista *et al.* (1978) showing that dissolution of microtubules results in marked elevation of neutrophil cyclic AMP after an appropriate stimulus. Thus, the reduction in cyclic AMP concentrations achieved by ascorbate therapy could have resulted from improved microtubule function, and demonstration of a direct effect of ascorbate on tubulin polymerization (Boxer *et al.*, 1979b) supports this interpretation.

In normal individuals virtually all circulating serotonin resides in platelets. Platelets from patients and animals with CHS do not aggregate normally and contain markedly decreased concentrations of serotonin (Bell *et al.*, 1976). Kaplan *et al.* (1978) reported that addition of as few as two normal platelets per leukocyte or of 1–100 μM serotonin corrected the abnormal bactericidal activity of leukocytes from beige mice. Since serotonin is known to enhance cyclic GMP formation in normal leukocytes (Sandler *et al.*, 1975), the findings of Kaplan *et al.* (1978) raise the possibility that the microtubule dysfunction of CHS results from deficiency of a circulating factor (serotonin) which normally permits cyclic GMP-mediated microtubule induction in neutrophils and monocytes.

One other possible location of the primary molecular abnormality in CHS has been suggested by investigations to date, namely, cell membranes (plasma or intracellular or both). Kanfer *et al.* (1968) showed accelerated turnover of membrane sphingolipids in CHS leukocytes, and, more recently, Berlin and colleagues have found altered physicochemical properties associated with elevated sphingomyelin levels in membranes from these cells (cited in Oliver, 1978). Abnormal transmembrane transfer of information from the binding of ligands, e.g., chemotactic factors and Con A, could result in failure to elicit microtubule assembly, either through a lack of communication with the microtubule itself or through failure to stimulate cyclic GMP. It is also possible that the membrane irregularities are a consequence of the decreased cyclic GMP or the microtubule defect.

In summary, the precise molecular cause of the abnormal phagocyte function in CHS is not known. On the basis of current knowledge, several reasonable possibilities exist, and these are schematized in Figure 1. It is clear that CHS phagocytes have defective microtubule function. This would permit the described increase in cyclic AMP levels and the decreased degranulation and chemotaxis that characterize the disease. The correction of these defects by cyclic GMP or agents that elevate cyclic GMP raises the possibility that the primary defect is abnormal function of the "cyclic GMP system" (cyclic GMP-mediated enzyme reactions), perhaps because of the unavailability of serotonin from platelets or of irregular membrane function. The relationship of lysosomal elastase deficiency to these abnormalities is obscure.

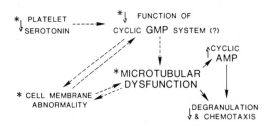

FIGURE 1. Molecular and functional interactions in Chediak–Higashi syndrome—a scheme based on current understanding. Asterisk (*) denotes possible site of primary molecular defect. Dashed arrows indicate relatively less well-defined interrelationships than those connected by solid arrows.

2.1.2. Increased Microtubule Assembly and Elevated Cyclic GMP

A second syndrome of recurrent infections and abnormalities of microtubules and cyclic GMP has been described in a single patient (Gallin *et al.*, 1978). The 7-year-old girl had digital and oral hemangiomata, difficulty in healing wounds, anemia, lymphopenia, and thrombocytopenia. Her infections included sinusitis, otitis media, lobar pneumonia, and dermatitis. Pneumococci, *H. influenzae*, staphylococci, and streptococci were repeatedly isolated. Percentages of T and B lymphocytes were subnormal, but immunoglobulin levels were normal or elevated. T-lymphocyte function was abnormal, as indicated by absent delayed cutaneous hypersensitivity and depressed transformation of lymphocytes stimulated by antigens or mitogens. The patient's cells did not stimulate cells of other individuals in a mixed leukocyte reaction, and HLA-A and HLA-B antigens could not be detected on the patient's leukocytes on two occasions. She died at 8½ years old of disseminated chicken pox, presumably as a result of her depressed cell-mediated immunity.

The patient's neutrophils and monocytes were markedly abnormal in assays of random migration and chemotaxis, showed decreased adherence to nylon wool, and had decreased spreading and pseudopod formation on glass. Neutrophil bactericidal activity was abnormal at early timepoints but normal by 90 min. Resting and stimulated reduction of nitroblue tetrazolium (NBT) was normal, but stimulated lysozyme release was 35–50% of normal. Ingestion was not fully studied.

Electron microscopic analyses of neutrophils under conditions of directed migration were particularly revealing: There was a lack of normal directional orientation in the patient's cells, associated with a threefold increase in the number of centriole-associated microtubules. Cyclic GMP levels in unstimulated mononuclear cells were four times normal. Thus, this patient exhibited a second disorder of leukocyte movement in which abnormalities of microtubules and cyclic nucleotides coexist, emphasizing the importance of these interrelationships.

2.2. SECONDARY CELLULAR ABNORMALITIES

Another group of chemotactic disorders consists of conditions in which patients' cells when removed for study are abnormal, patients' sera have no detrimental effect on the chemotaxis of normal cells, and an underlying disease state exists which apparently causes the cellular abnormality.

2.2.1. Decreased Chemotaxis with Elevated Immunoglobulin (Ig) E

A number of patients have been reported who have had decreased neutrophil chemotaxis and hyperimmunoglobulinemia E in association with a variety of clinical syndromes and increased susceptibility to infection. A common example of these syndromes has been chronic mucocutaneous candidiasis, often with negative *Candida* skin tests and diminished lymphocyte responses to *Candida* antigen (Clark *et al.*, 1973; Van Scoy *et al.*, 1975). Another such syndrome has included chronic eczema and recurrent staphylococcal abscesses. In its severe form the disorder was originally reported as "Job syndrome" (Hill and Quie, 1974; Hill *et al.*, 1974a; Fontan *et al.*, 1976; Dahl *et al.*, 1976). A syndrome with ichthyosis rather than eczema has been described (Pincus *et al.*, 1975). Other infections in these patients have included furunculosis, cellulitis, pneumonia, empyema, frequent otitis media, and, occasionally, a deep abscess or septicemia. *S. aureus* has been the most common pathogen, but group A β-hemolytic streptococcus has been a frequent isolate from abscesses, and *Candida* species have been isolated from the skin, nails, or mouth. The syndrome of Buckley *et al.* (1972; Buckley and Fiscus, 1975) with elevated IgE, pyoderma, subcutaneous abscesses, and depressed cell-mediated immunity appears to be related since chemotaxis of neutrophils or monocytes is abnormal in some of these patients (Church *et al.*, 1976; Snyderman and Buckley, 1975). The chemotactic defect may not be persistent in all patients with the eczema–hyperimmunoglobulinemia E syndrome (Fontan *et al.*, 1976; Snyderman and Buckley, 1975). Further, it is impossible to be certain in some of these syndromes that the *in vitro* chemotactic defect is a cause rather than a result of the infections, or that the chemotactic defect actually increases the patient's risk of infection.

The cell defect in these syndromes is intrinsic to the polymorphonuclear leukocyte (PMN) when it is removed for study, in that thorough washing does not improve its chemotactic response. Generally, incubating normal cells in patients' sera does not adversely affect their function. It seems unlikely that elevated IgE concentrations are directly responsible for the neutrophil abnormality here: Not all patients with atopic dermatitis and hyperimmunoglobulinemia E have abnormal chemotaxis (Dahl *et al.*, 1976; Snyderman *et al.*, 1977), and in one case, the chemotaxis defect and clinical symptoms were eliminated by an allergen-free diet while the serum IgE concentration remained high (Fontan *et al.*, 1976). On the other hand, histamine released from IgE-coated basophils and mast cells could be responsible. Histamine inhibits chemotaxis of normal neutrophils *in vitro* (Hill and Quie, 1974; Anderson *et al.*, 1977). Inhibition presumably occurs through the capacity of histamine to elevate intracellular cyclic AMP, since cyclic AMP itself has been shown to inhibit the chemotactic response (Tse *et al.*, 1972; Rivkin *et al.*, 1975). In agreement with this possibility, treatment of three patients' neutrophils with burimamide, a competitive inhibitor of the binding of histamine to cells, produced a significant increase in chemotaxis (Hill *et al.*, 1976). In addition, the anthelminthic levamisole added *in vitro* to leukocytes from normals or from patients with the hyperimmunoglobulinemia E syndrome has induced a significant improvement in neutrophil and monocyte chemotaxis and elevation of intracellular concentrations of cyclic GMP (Wright *et al.*, 1977; Hogan and Hill, 1978). Levamisole also has been reported to reduce neutrophil cyclic AMP (Anderson *et al.*, 1977; Hogan and Hill, 1978), and the presumed mechanism for its effect on chemotaxis has been modulation of intracellular cyclic nucleotide balance.

2.2.2. Diabetes Mellitus

Patients with diabetes mellitus have been reported to have reduced leukocyte function. Impaired chemotaxis has been demonstrated *in vitro* with the Boyden chamber (Miller and Baker, 1972; Hill *et al.*, 1974b; Mowat and Baum, 1971) and *in vivo* with the inflammatory skin window technique (Perillie *et al.*, 1962; Kontras and Bodenbender, 1968; Brayton *et al.*, 1970). The chemotaxis defect can be corrected *in vitro* by the addition of insulin (Mowat and Baum, 1971; Miller and Baker, 1972; Hill *et al.*, 1974b). Insulin is known to cause the movement of glucose and potassium into the leukocyte (Hadden *et al.*, 1972), as well as to raise intracellular cyclic GMP levels (Illiano *et al.*, 1973). Depletion of intracellular potassium has been shown to impair leukotaxis (Ward and Becker, 1970), while an increase in intracellular potassium (Showell and Becker, 1976) or in cyclic GMP (Estensen *et al.*, 1973) has been found to enhance the chemotactic response. In addition, glycolysis is a major source of energy for cell motility (reviewed by Wilkinson, 1974). Thus insulin deficiency theoretically could result in defective neutrophil function through potassium or glucose deficiency or a relative decrease in cyclic GMP.

Ingestion by diabetic leukocytes was shown to be impaired by several investigators (reviewed by Johnston and McPhail, 1980). Chemotaxis and ingestion involve related mechanisms, and it seems plausible that both processes are impaired in cells from some diabetics. However, a cause-and-effect relationship has not been shown between phagocyte dysfunction and infections in diabetic humans or animals.

2.2.3. Acrodermatitis Enteropathica

Two individuals with acrodermatitis enteropathica and zinc deficiency had abnormal monocyte chemotaxis that was corrected by dietary zinc supplementation (Weston *et al.*, 1977).

2.2.4. Hypophosphatemia

Severe hypophosphatemia induced by hyperalimentation was shown to result in decreased chemotaxis and ingestion and depressed levels of intracellular ATP (Craddock *et al.*, 1974). The chemotaxis defect could be reversed by preincubation of cells with adenosine and phosphate *in vitro* or by phosphate supplementation of parenteral fluids.

2.2.5. Ethanol Intoxication

Brayton *et al.* (1970) and Gluckman and MacGregor (1978) have reported markedly depressed influx of neutrophils into areas of skin abrasion in adult volunteers given ethanol. *In vitro* experiments have shown that cells removed from such individuals migrate poorly (Phelps and Stanislaw, 1969) and adhere to nylon wool abnormally (Gluckman and MacGregor, 1978). Administration of the beta adrenergic blocker propranolol to rabbits largely reversed the effects of ethanol intoxication on neutrophil adherence and death from pneumococcal peritonitis, presumably by blocking ethanol-induced elevation of cyclic AMP (reviewed in Buckley *et al.*, 1978).

3. DEFECTS OF PHAGOCYTOSIS

Defects in the process of ingestion can occur as a consequence of deficiency in the principal serum factors that promote phagocytosis (antibody and C3), deficiency of the phagocytic cells themselves (neutropenia or hyposplenia), or an intrinsic cellular defect (primary or secondary), as outlined in Table 2. These defects have been reviewed in detail (Singer, 1973; Pincus *et al.*, 1976; Howard *et al.*, 1977; Stossel, 1977; Johnston and McPhail, 1980).

3.1. ACTIN DYSFUNCTION

Abnormal chemotaxis is almost invariably present in reported cases of defective ingestion due to a primary cellular defect, supporting evidence from studies of normal cells suggesting that the functions of chemotaxis and ingestion are linked in the neutrophil. The work of Boxer *et al.* (1974) on the role of actin polymerization in these two functions supports this concept and provides a molecular basis for these phenomena. These investigators described an infant with recurrent bacterial infections that lacked pus. *S. aureus* was repeatedly

TABLE 2. DEFECTS OF PHAGOCYTOSIS

A. Deficient opsonization
1. Antibody deficiency
2. C3 deficiency
B. Neutropenia
C. Hyposplenia
D. Primary cellular abnormalities
Actin dysfunction[a]
E. Secondary cellular abnormalities
1. Hypophosphatemia[a]
2. Diabetes mellitus[a]
3. Corticosteroid therapy
4. Galactosemia
5. Viral infection
6. Circulating immune complexes
7. Macrophage "blockade" (e.g., with damaged erythrocytes)

[a]Discussed in the text.

cultured from vesicular skin lesions. Vesicles and an abdominal abscess contained macrophages and lymphocytes but no neutrophils. Chemotaxis was abnormal, and his neutrophils ingested particles at only 15% of the normal rate. Release of granule contents into phagosomes was greater than normal. Electron micrographs of his neutrophils showed decreased numbers of microfilaments, which are believed to consist of polymers of the protein actin. Although neutrophils from the patient and normals contained equal quantities of actin, the patient's protein did not polymerize normally, suggesting that microfilaments are required for normal locomotion and ingestion. The increase in degranulation noted in this patient's neutrophils and in cells treated with the microfilament-disrupting agent cytochalasin B (Henson and Oades, 1973) indicate that microfilaments exert a controlling influence on degranulation.

4.   DEFECTS OF MICROBICIDAL ACTIVITY

Defects of phagocytic microbicidal activity (Table 3) have been reviewed by Babior (1978) and Johnston and McPhail (1980).

4.1.   CHRONIC GRANULOMATOUS DISEASE

4.1.1.   Definition

The term chronic granulomatous disease (CGD) has been applied to a syndrome of recurrent purulent infections of the skin, lymph nodes, liver, and lungs associated with an inability of patients' phagocytes to kill fungi and bacteria that do not produce H_2O_2. Patients' phagocytes can ingest microorganisms normally but cannot convert oxygen into metabolites that contribute to bacterial

TABLE 3. DEFECTS OF MICROBICIDAL ACTIVITY

A.	Chronic granulomatous disease[a]
B.	G-6-PD deficiency[a]
C.	Myeloperoxidase deficiency[a]
D.	Alkaline phosphatase deficiency[a]
E.	Pyruvate kinase deficiency[a]
F.	Malakoplakia[a]
G.	Granule defects
	1. Chediak–Higashi syndrome[a]
	2. Absence of specific granules
	3. Myelogenous or histiocytic leukemia
H.	Felty syndrome
I.	Acquired defects of undetermined cause
J.	Leukemia
K.	Viral infections

[a]Discussed in the text.

death. Recent evidence suggests that the syndrome can be caused by different molecular defects.

4.1.2. Clinical Picture

Table 4 describes the common clinical findings of CGD. Material for the table was obtained from a recent review (Johnston and McPhail, 1980) and additional cases reported recently and in adequate detail (Kobayashi *et al.*, 1978). Some of the articles reviewed apparently reported only the most striking findings. Thus, the table serves to list the most common signs and symptoms of CGD but illustrates only in general terms their relative frequency. The most characteristic abnormalities reflect involvement of the RES. Lymphadenopathy has been described in almost all patients who lived beyond infancy. In most instances the nodes have become purulent and have drained pus. Hepatomegaly, splenomegaly, and hepatic or perihepatic abscesses have been common. All of these findings reflect the survival of bacteria or fungi in phagocytic cells that cannot kill them, with the compensatory accumulation of more phagocytes, so that hyperplasia, if not abscess formation, ensues.

The second major group of signs reflects the inability of circulating phagocytes to kill invading bacteria at sites of penetration beneath the skin and mucous membranes. This group includes pneumonitis, subcutaneous abscesses, and furunculosis, all of which occur sooner or later in most patients with CGD, as well as osteomyelitis and perianal abscesses. Although the serious infections in CGD are typically localized, septicemia or meningitis has occurred in 32 cases (see also Lazarus and Neu, 1975).

The eczematoid or seborrheic dermatitis seen in CGD, especially on the eyelids or around the nares or mouth, is like that seen in Wiskott–Aldrich syndrome and severe combined immunodeficiency disease, and may represent response to lowgrade infection or to microbial products. The same might be said for the persistent diarrhea and rhinitis and the polyarthritis seen in CGD. The onset of CGD has been heralded by dermatitis or lymphadenitis in most of the

TABLE 4. MAJOR SIGNS AND SYMPTOMS IN 189 PATIENTS
WITH CHRONIC GRANULOMATOUS DISEASE

Finding	Number of patients involved	Finding	Number of patients involved
Marked lymphadenopathy	156	Onset with lymphadenitis	41
Pneumonitis	151	Facial periorificial dermatitis	39
Dermatitis	140	Persistent diarrhea	38
Onset by 1 year old	127	Perianal abscess	35
Hepatomegaly	126	Persistent rhinitis	33
Suppuration of nodes	119	Septicemia or meningitis	32
Splenomegaly	105	Ulcerative stomatitis	30
Hepatic-perihepatic abscess	80	Conjunctivitis	29
Osteomyelitis	59	Death from pneumonitis	27
Onset with dermatitis	45		

120 cases in which first symptoms have been described. Other, less frequent findings have been reviewed (Johnston and McPhail, 1980).

Of the 160 patients whose age at onset of disease was reported, 127 had developed their first symptom by 1 year old and 145 by 2 years old. At least 15 children had their first definite symptom of CGD in their first week of life. There have been 60 reported deaths, 45 before the age of 7 and 51 before the age of 12. However, what may be milder forms of the disease have been seen in four brothers who were alive when reported at ages 28, 30, 32, and 40 years (Dilworth and Mandell, 1977) and two female patients who lived into their thirties (Johnston and Baehner, 1971; Rodey *et al.*, 1970). Pulmonary disease was the primary cause of death in 31 of the 53 cases in which a cause was stated; 14 children died with septicemia or meningitis.

4.1.3. Infecting Organisms

Information about infecting microorganisms has been reported in reasonable detail for 137 patients (Johnston and McPhail, 1980). *Staphylococcus aureus* has been the predominant pathogen (94 patients), but *Klebsiella-Aerobacter* (29 patients), *Escherichia coli* (26 patients), and other enteric bacteria have been commonly isolated. The fungi *Aspergillus* and *Candida* have been cultured from 14 patients each. Five patients have been infected with mycobacteria (three of these with the BCG strain). Proven infections by *H. influenzae* and pneumococci have not been reported, and streptococci have been isolated infrequently, which correlates with the ability of CGD phagocytes to kill these catalase-negative, peroxide-producing organisms *in vitro*.

4.1.4. Molecular Basis for Defective Phagocyte Function

Evidence for X-linked and autosomal recessive modes of inheritance of CGD has been reviewed (Johnston and Newman, 1977). Although one cannot exclude the possibility that female patients are heterozygous carriers of the X-linked defect, the absence thus far of detectable heterozygotes in the families of female patients makes this possibility unlikely. The existence of different "varieties" of CGD in addition to those represented by different modes of inheritance seems probable, but little proof for such exists. Three sisters with the neutrophil biochemical and bactericidal defects of CGD, but a much milder course and absence of granuloma formation (Rodey *et al.*, 1970), and two brothers with mild disease and a partial defect in phagocyte function (Ochs and Igo, 1973) could well represent variant disorders. However, variability in the phagocytic bactericidal capacity (Repine and Clawson, 1977), phagocytic oxidative response (Briggs *et al.*, 1977), and clinical picture (Dilworth and Mandell, 1977; Clark and Klebanoff, 1978) of the disease has recently been emphasized. Separation of any atypical cases as a distinct entity will require identification of a precise molecular defect that can explain the biochemical and functional abnormalities. The search for this basic defect has been an integral part of the effort to understand oxidative metabolism in the normal phagocyte.

Cells from patients with CGD do not undergo the phagocytosis-associated increase in oxygen consumption, superoxide anion (O_2^-) and H_2O_2 generation, HMP shunt activation, NBT reduction, or chemiluminescence that characterizes normal cells. Presumably, the basic molecular defect of CGD is deficient activity of an enzyme responsible for conversion of oxygen to bactericidal species (O_2^-, H_2O_2, hydroxyl radical and, perhaps, singlet oxygen). NADH oxidase, NADPH oxidase, glutathione peroxidase, and cytochrome b have each been proposed as the deficient enzyme in CGD, but none has been clearly proven as such. Evidence that NADPH oxidase is this enzyme (and the enzyme responsible for the phagocytosis-associated burst of oxidative metabolism in normal cells) seems reasonable: NADPH oxidase activity in a pellet obtained by centrifugation of granulocyte homogenates at 27,000g is increased significantly if the cells have phagocytized particles before being homogenized (DeChatelet *et al.*, 1975; Hohn and Lehrer, 1975; Patriarca *et al.*, 1971). (Electron micrographs of this pellet show lysosomal granules and fragments of membrane.) Such an increase after phagocytosis was not obtained in similar fractions of cells from patients with X-linked or apparently non-X-linked forms of CGD (DeChatelet *et al.*, 1975; Hohn and Lehrer, 1975; McPhail *et al.*, 1977). Other experimental and theoretical considerations that favor NADPH oxidase as the enzyme missing in CGD have been presented by Roos and Weening (1980).

A second leading candidate for the missing enzyme is NADH oxidase, which has been reported abnormally low in homogenates and membrane fractions of CGD cells that have not been activated by phagocytosis (Baehner and Karnovsky, 1968; Segal and Peters, 1976) and in phagocytizing CGD neutrophils studied by cytochemical techniques (Briggs *et al.*, 1977). A third possibility, glutathione peroxidase, was reported to be about one-third the normal value when tested in cells from two girls and two boys with CGD (including one brother–sister pair) (Holmes *et al.*, 1970; Malawista and Gifford, 1975; Matsuda *et al.*, 1976), but normal activity of this enzyme has been described in seven other patients with both inheritance patterns (DeChatelet *et al.*, 1976; Windhorst and Katz, 1972). The enzyme appears to be an important means of removing excess H_2O_2 (Baehner *et al.*, 1975), but on the basis of current knowledge, it is difficult to assign it a primary role in initiating the phagocytic "respiratory burst." It has been proposed that glutathione peroxidase levels might be influenced by H_2O_2 production and, therefore, secondarily deficient in CGD (DeChatelet *et al.*, 1976).

A recent candidate for the deficient oxidase is cytochrome b. Segal *et al.* (1978) have reported the absence or abnormality in CGD cells of a cytochrome b said to become incorporated into the phagocytic vacuole of human neutrophils. They have proposed analogy of this cytochrome to "cytochrome o" of bacteria, an oxidase coupled to a proton pump. Their suggestion that the decreased microbicidal activity of CGD phagocytes might be due, at least in part, to abnormal acidification of the phagocytic vacuole appears unlikely in view of the work of Jacques and Bainton (1978) showing that phagocytic vacuoles of CGD neutrophils acidify more rapidly and to a greater extent than do those of normal cells.

Weening *et al.* (1976) have described what appears to be the first true molecular variant of classical CGD. Neutrophils from a sister and brother with reasonably typical clinical CGD could ingest staphylococci but not kill them (van der Meer *et al.*, 1975). Neutrophil oxygen consumption, O_2^- production, HMP shunt activation, and iodination of ingested particles were all severely deficient on ingestion of serum-opsonized zymosan or latex. However, phagocytosis of latex particles heavily coated with IgG or of IgG aggregates initiated normal activity of the respiratory burst. Latex and IgG–latex were ingested equally well, and binding of IgG- and C3b-coated erythrocytes by patients' neutrophils was normal. Thus, it appeared that the children's phagocytes possessed normal oxidative enzymatic machinery but an abnormal "trigger" mechanism for its activation. How common the defect is among patients with "CGD" is not clear, but the authors did not find it in six other patients studied (three boys and three girls).

Klebanoff and White (1969) and, later, Giblett *et al.* (1971) drew attention to the fact that erythrocytes from some boys with CGD carry the very rare null Kell blood group phenotype K_0, in which all antigenic products of the Kell locus are absent. More recently, Marsh and associates reported that neutrophils from all normals tested possessed an antigen, K_X, that appears to represent precursor material in the biosynthetic pathway for Kell group antigens (Marsh *et al.*, 1975, 1976, 1977). All nine boys with CGD that were tested lacked this antigen (Marsh *et al.*, 1976). [Girls with CGD had K_X (Marsh *et al.*, 1976), as did one boy with non-X-linked disease (Clark and Klebanoff, 1978).] This highly significant association raises the possibility that the absence of a membrane structure on X-linked CGD neutrophils is involved in their abnormal function. Perhaps like the abnormality in the patients of Weening *et al.* (1976), this could consist of failure of activation, rather than absence, of the enzyme responsible for the conversion of oxygen to microbicidal products.

Although the molecular defects that result in CGD remain to be fully elucidated, the disease has taught us a great deal about the physiology of normal cells. It has shown that phagocytic bactericidal activity cannot be replaced in host defense by antibody, complement, or other factors; that a commensal relationship with many bacteria of low virulence is maintained only if phagocyte function is normal; that phagocytic oxygen metabolism is essential for the killing of many, if not most, bacteria and fungi; and that lysosomal degranulation is not enough by itself to accomplish killing. The occurrence of granulomas in CGD has enforced the concept that these structures result from the prolonged intraphagocytic residence of materials not easily digested, whether the nondigestion is due to the nature of the material itself or to a faulty cell. Awareness of this relationship allows the clinician or pathologist to narrow possible causative agents of granuloma formation to particles and parasites that could be resistant to intracellular digestion.

4.2. GLUCOSE-6-PHOSPHATE DEHYDROGENASE (G-6-PD) DEFICIENCY

Deficiency of G-6-PD involving leukocytes as well as erythrocytes could be considered a variant of CGD. At least, it represents the first enzyme defect shown clearly to cause the CGD syndrome as defined above. When neutrophil

G-6-PD levels are less than 1% of normal, the patient suffers a clinical syndrome that mimics CGD in type and location of infections but is much milder (Cooper *et al.*, 1972; Gray *et al.*, 1973). These patients' cells do not reduce NBT or generate H_2O_2, which has been attributed to a lack of substrate (NADH and NADPH) for the enzymatic conversion of oxygen (Baehner *et al.*, 1972). Increased lability of G-6-PD (Bellanti *et al.*, 1970) and altered levels of G-6-PD stablizing factors (Erickson *et al.*, 1972) have been described in CGD but do not appear to represent the primary underlying defect.

4.3. MYELOPEROXIDASE DEFICIENCY

Primary myeloperoxidase (MPO) deficiency has been described in at least eight patients (Grignaschi *et al.*, 1963; Lehrer and Cline, 1969; Lehrer *et al.*, 1969; Patriarca *et al.*, 1975; Rosen and Klebanoff, 1976; Undritz, 1966). Peroxidase activity is absent from neutrophils and monocytes but is present in eosinophils, which correlates with other observations that eosinophil peroxidase is a different enzyme (Archer *et al.*, 1965). The deficiency has been shown histochemically, biochemically, and immunologically (Salmon *et al.*, 1970).

Patients are generally not abnormally susceptible to disease. However, disseminated candidiasis has been reported in one patient (Lehrer and Cline, 1969) and acne vulgaris in two patients (Rosen and Klebanoff, 1976; Patriarca *et al.*, 1975), one of whom also had a history of frequent upper respiratory infections in childhood.

MPO deficiency of variable degree has also been reported associated with isolated cases of certain leukemias (Davis *et al.*, 1971; Repine *et al.*, 1976a; El-Maalem and Fletcher, 1976), anemias (Arakawa *et al.*, 1965; Higashi *et al.*, 1965; Lehrer *et al.*, 1972), generalized pustular psoriasis (Stendahl and Lindgren, 1976), and neuronal storage disease (Armstrong *et al.*, 1974a,b). The lack of MPO is probably a secondary phenomenon in most, if not all, of these cases.

The MPO-deficient leukocyte has impaired microbicidal activity *in vitro* which is most apparent at early time periods (Lehrer *et al.*, 1969; Lehrer and Cline, 1969; Klebanoff, 1970; Stendahl and Lindgren, 1976) but is not as severe as that in CGD at any time during 120–180 min of incubation. MPO-deficient phagocytes kill *C. albicans* poorly, which correlates with the occurrence of disseminated candidiasis due to this strain in one MPO-deficient patient (Lehrer and Cline, 1969). However, effective nonoxidative mechanisms for killing other fungal strains exist in MPO-deficient neutrophils (Lehrer, 1972, 1975; Lehrer *et al.*, 1975).

Oxidative metabolism of MPO-deficient granulocytes has been reported as significantly elevated (Klebanoff and Pincus, 1971; Rosen and Klebanoff, 1976; Stendahl and Lindgren, 1976). It has been suggested that this represents a compensatory means of preserving oxygen-dependent but MPO-independent bactericidal mechanisms (Klebanoff, 1970; Klebanoff and Pincus, 1971; Rosen and Klebanoff, 1976; Stendahl and Lindgren, 1976). Recent evidence suggests that at least part of the augmentation may be due to increased survival of the O_2^- generating system, which appears to be inactivated by $MPO-H_2O_2$ attack (Jandl *et al.*, 1978).

4.4. ALKALINE PHOSPHATASE DEFICIENCY

Several patients with leukocyte alkaline phosphatase deficiency and repeated infections have had an associated bactericidal defect (Strauss *et al.*, 1974; Repine *et al.*, 1976a,b). However, the impaired microbial killing in these individuals was approximately that seen in carriers of X-linked CGD, who are normally not unduly susceptible to infection, so that the relationship of this defect to the propensity to infections is unclear.

4.5. PYRUVATE KINASE DEFICIENCY

A woman was described with the clinical syndrome of severe CGD and abnormally rapid decay *in vitro* of the neutrophil glycolytic enzyme pyruvate kinase (Burge *et al.*, 1976). Ingestion of fungi, reduction of NBT, and activation of the HMP shunt were normal, but killing of *S. aureus,* her infecting organism, was abnormal. Ingestion of bacteria and levels of granule enzymes were not studied. It is not clear what relationship the pyruvate kinase abnormality bears to the bactericidal defect.

4.6. MALAKOPLAKIA

Malakoplakia is a histopathologic entity characterized by inflammatory granulomatous lesions containing macrophages with large intracytoplasmic inclusions. Electron micrographs of these macrophages have shown undigested or partially digested bacteria in phagocytic vacuoles, suggesting the existence of a bactericidal defect. Patients with this entity have had chronic bacterial infections, especially of the genitourinary tract. One 28-year-old man with malakoplakia and panhypogammaglobulinemia had a history of retroperitoneal and splenic abscesses, a chronic subcutaneous abscess, recurrent pneumonia, giardiasis, and a urinary tract infection with retrovesical fistula. His blood mononuclear cells (approximately 70% lymphocytes and 25% monocytes) had 20% of the normal concentration of cyclic GMP, and release of lysosomal β-glucuronidase from blood leukocytes was decreased (Abdou *et al.*, 1977). Chemotaxis of neutrophils and monocytes and microtubular assembly of neutrophils were normal. The basis for the predisposition to infections in this patient was not evident from the *in vitro* studies reported, and the possibility was not excluded that the cellular defects were secondary to chronic infection. However, daily administration of bethanecol by mouth resulted in normalization of the mononuclear cell cyclic GMP level, a marked decrease in the number of large inclusions in monocytes, and a clearing of the infections, none of which was achieved by administration of γ-globulin alone.

5. PHAGOCYTE BIOCHEMICAL DEFECTS NOT ASSOCIATED WITH RECURRENT INFECTIONS

5.1. GLUTATHIONE REDUCTASE DEFICIENCY AND GLUTATHIONE SYNTHETASE DEFICIENCY

Two recently elucidated enzyme defects deserve at least brief mention because they have significantly expanded knowledge of the mechanisms of oxidant removal in neutrophils, though they have not yet been associated with a predisposition to infection. Both involve the H_2O_2-stimulated cycle (Figure 2) by which glutathione peroxidase catalyzes the peroxidation of reduced glutathione (GSH) to oxidized glutathione (GSSG), which, in turn, is returned to GSH by glutathione reductase, generating NADP$^+$ to drive the HMP shunt (Reed, 1969). This mechanism was hypothesized to protect the cell by removing excess H_2O_2 (and also, thereby, H_2O_2-derived radicals) generated during phagocytosis (references in Johnston, 1978). As might have been predicted by this hypothesis, deficiency in glutathione reductase (Loos *et al.*, 1976) or deficiency in glutathione synthetase (Spielberg *et al.*, 1977), have resulted in hemolytic anemia; and one patient with the latter disorder has had profound neutropenia in association with infections (Spielberg *et al.*, 1979). The presumption that the accelerated hemolysis is H_2O_2-derived has been substantiated by the improvement achieved in the patient's hemolytic anemia by oral administration of vitamin E, an H_2O_2 scavenger (Oliver, personal communication).

Functional studies of glutathione reductase-deficient neutrophils indicate that the phagocytosis-associated respiratory burst proceeds at a normal rate for 5–10 minutes, then rapidly declines to zero (Roos and Weening, 1979). Studies of glutathione synthetase-deficient neutrophils gave somewhat different results

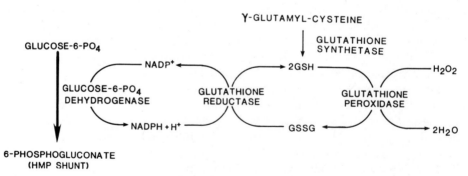

FIGURE 2. Schematic representation of the glutathione oxidation–reduction cycle. GSH, reduced glutathione; GSSG, oxidized glutathione.

(Spielberg *et al.*, 1979), but neither cell type killed staphylococci normally, especially at a high bacteria-to-phagocyte ratio. Electron micrographs of post-phagocytic neutrophils from the patient with glutathione synthetase deficiency revealed abnormal damage to microtubules and membranous structures (Spielberg *et al.*, 1979). It seems highly likely that the neutrophil dysfunction seen in patients with both of these disorders results from oxidative damage. Studies of microtubule function in glutathione-synthetase-deficient neutrophils emphasize an important role for glutathione in protecting tubulin from oxidant damage (Oliver *et al.*, 1978).

REFERENCES

Abdou, N. I., NaPombejara, C., Sagawa, A., Ragland, C., Stechschulte, D. J., Nilsson, U., Gourley, W., Watanabe, I., Lindsey, N. J., and Allen, M. S., 1977, Malakoplakia: Evidence for monocyte lysosomal abnormality correctable by cholinergic agonist *in vitro* and *in vivo*, *N. Engl. J. Med.* **297**:1413.

Anderson, R., Glover, A., and Rabson, A. R., 1977, The *in vitro* effects of histamine and metiamide on neutrophil motility and their relationship to intracellular cyclic nucleotide levels, *J. Immunol.* **118**:1690.

Arakawa, T., Wada, Y., Hayashi, T., Kakizaki, R., Chida, N., Chiba, R., Konno, T., and Shioura, H., 1965, Uracil-uric refractory anemia with peroxidase negative neutrophils, *Tohoku J. Exp. Med.* **87**:52.

Archer, G. T., Air, G., Jackas, M., and Morell, D. M., 1965, Studies on rat eosinophil peroxidase, *Biochim. Biophys. Acta* **99**:96.

Armstrong, D., Dimmitt, S., Boehme, D. H., Leonberg, S. C., Jr., and Vogel, W., 1974a, Leukocyte peroxidase deficiency in a family with a dominant form of Kuf's disease, *Science* **186**:155.

Armstrong, D., Dimmitt, S., and van Wormer, D. E., 1974b, Studies in Batten disease. I. Peroxidase deficiency in granulocytes, *Arch. Neurol.* **30**:144.

Babior, B. M., 1978, Oxygen-dependent microbial killing by phagocytes, *N. Engl. J. Med.* **298**:721.

Baehner, R. L., and Karnovsky, M. L., 1968, Deficiency of reduced nicotinamide-adenine dinucleotide oxidase in chronic granulomatous disease, *Science* **162**:1277.

Baehner, R. L., Johnston, R. B., Jr., and Nathan, D. G., 1972, Comparative study of the metabolic and bactericidal characteristics of severely glucose-6-phosphate dehydrogenase-deficient polymorphonuclear leukocytes and leukocytes from children with chronic granulomatous disease, *J. Reticuloendothel. Soc.* **12**:150.

Baehner, R. L., Murrmann, S. K., Davis, J., and Johnston, R. B., Jr., 1975, The role of superoxide anion and hydrogen peroxide in phagocytosis-associated oxidative metabolic reactions, *J. Clin. Invest.* **56**:571.

Bell, T. G., Meyers, K. M., Prieur, D. J., Fauci, A. S., Wolff, S. M., and Padgett, G. A., 1976, Decreased nucleotide and serotonin storage associated with defective function in Chediak–Higashi syndrome cattle and human platelets, *Blood* **48**:175.

Bellanti, J. A., Cantz, B. E., and Schlegel, R. J., 1970, Accelerated decay of glucose-6-phosphate dehydrogenase activity in chronic granulomatous disease, *Pediatr. Res.* **4**:405.

Bennett, J. M., Blume, R. S., and Wolff, S. M., 1969, Characterization and significance of abnormal leukocyte granules in the beige mouse: A possible homologue for Chediak-Higashi Aleutian trait, *J. Lab. Clin. Med.* **73**:235.

Blume, R. S., and Wolff, S. M., 1972, The Chediak–Higashi syndrome: Studies in four patients and a review of the literature, *Medicine* **51**:247.

Blume, R. S., Bennett, J. M., Yankee, R. A., and Wolff, S. M., 1968, Defective granulocyte regulation in the Chediak–Higashi syndrome, *N. Engl. J. Med.* **279**:1009.

Boxer, G. J., Holmsen, H., Robkin, L., Bang, N. U., Boxer, L. A., and Baehner, R. L., 1977, Abnormal platelet function in Chediak–Higashi syndrome, *Br. J. Haematol.* **35**:521.

Boxer, L. A., Hedley-Whyte, E. T., and Stossel, T. P., 1974, Neutrophil actin dysfunction and abnormal neutrophil behavior, *N. Engl. J. Med.* **291**:1093.

Boxer, L. A., Watanabe, A. M., Rister, M., Besch, H. R., Allen, J., and Baehner, R. L., 1976, Correction of leukocyte function in Chediak–Higashi syndrome by ascorbate, *N. Engl. J. Med.* **295**:1041.

Boxer, L. A., Rister, M., Allen, J. M., and Baehner, R. L., 1977, Improvement of Chediak–Higashi leukocyte function by cyclic guanosine monophosphate, *Blood* **49**:9.

Boxer, L. A., Allen, J. M., Watanabe, A. M., Besch, H. R., Jr., and Baehner, R. L., 1978, Role of microtubules in granulocyte adherence, *Blood* **51**:1045.

Boxer, L. A., Albertini, D. F., Baehner, R. L., and Oliver, J. M., 1979a, Impaired microtubule assembly and polymorphonuclear leukocyte function in the Chediak–Higashi syndrome correctable by ascorbic acid, *Br. J. Haematol.* **43**:207.

Boxer, L. A., Vanderbilt, B., Bonsib, S., Jersild, R., Yang, H.-H., and Baehner, R. L., 1979b, Enhancement of chemotactic response and microtubule assembly in human leukocytes by ascorbic acid, *J. Cell. Physiol.* **100**:119.

Brayton, R. G., Stokes, P. E., Schwartz, M. S., and Louria, D. B., 1970, Effect of alcohol and various diseases on leukocyte mobilization, phagocytosis and intracellular bacterial killing, *N. Engl. J. Med.* **282**:123.

Briggs, R. T., Karnovsky, M. L., and Karnovsky, M. J., 1977, Hydrogen peroxide production in chronic granulomatous disease: A cytochemical study of reduced pyridine nucleotide oxidases, *J. Clin. Invest.* **59**:1088.

Buckley, R. H., and Fiscus, S. A., 1975, Serum IgD and IgE concentrations in immunodeficiency diseases, *J. Clin. Invest.* **55**:157.

Buckley, R. H., Wray, B. B., and Belmaker, E. Z., 1972, Extreme hyperimmunoglobulinemia E and undue susceptibility to infection, *Pediatrics* **49**:59.

Buckley, R. M., Ventura, E. S., and MacGregor, R. R., 1978, Propranolol antagonizes the anti-inflammatory effect of alcohol and improves survival of infected intoxicated rabbits, *J. Clin. Invest.* **62**:554.

Burge, P. S., Johnson, W. S., and Hayward, A. R., 1976, Neutrophil pyruvate kinase deficiency with recurrent staphylococcal infections: First reported case, *Br. Med. J.* **1**:742.

Church, J. A., Frenkel, L. D., Wright, D. G., and Bellanti, J. A., 1976, T lymphocyte dysfunction, hyperimmunoglobulinemia E, recurrent bacterial infections, and defective neutrophil chemotaxis in a Negro child, *J. Pediatr.* **88**:982.

Clark, R. A., and Kimball, H. R., 1971, Defective granulocyte chemotaxis in the Chediak-Higashi syndrome, *J. Clin. Invest.* **50**:2645.

Clark, R. A., and Klebanoff, S. J., 1978, Chronic granulomatous disease: Studies of a family with impaired neutrophil chemotactic, metabolic and bactericidal function, *Am. J. Med.* **65**:941.

Clark, R. A., Root, R. K., Kimball, H. R., and Kirkpatrick, C. H., 1973, Defective neutrophil chemotaxis and cellular immunity in a child with recurrent infections, *Ann. Int. Med.* **78**:515.

Cooper, M. R., DeChatelet, L. R., McCall, C. E., LaVia, M. F., Spurr, C. L., and Baehner, R. L., 1972, Complete deficiency of leukocyte glucose-6-phosphate dehydrogenase with defective bactericidal activity, *J. Clin. Invest.* **51**:769.

Craddock, P. R., Yawata, Y., Van Santen, L., Gilberstadt, S., Silvis, S., and Jacob, H. S., 1974, Acquired phagocyte dysfunction: A complication of the hypophosphatemia of parenteral hyperalimentation, *N. Engl. J. Med.* **290**:1403.

Dahl, M. V., Greene, W. H., Jr., and Quie, P. G., 1976, Infection, dermatitis, increased IgE, and impaired neutrophil chemotaxis, *Arch. Dermatol.* **112**:1387.

Davis, A. T., Brunning, R. D., and Quie, P. G., 1971, Polymorphonuclear leukocyte myeloperoxidase deficiency in a patient with myelomonocytic leukemia, *N. Engl. J. Med.* **285**:789.

DeChatelet, L. R., McPahil, L. C., Mullikin, D., and McCall, C. E., 1975, An isotopic assay for NADPH oxidase activity and some characteristics of the enzyme from human polymorphonuclear leukocytes, *J. Clin. Invest.* **55**:714.

DeChatelet, L. R., Shirley, P. S., and McPhail, L. C., 1976, Normal leukocyte glutathione peroxidase activity in patients with chronic granulomatous disease, *J. Pediatr.* **89**:598.

Dilworth, J. A., and Mandell, G. L., 1977, Adults with chronic granulomatous disease of "childhood," *Am. J. Med.* **63**:233.

El-Maalem, H., and Fletcher, J., 1976, Impaired neutrophil function and myeloperoxidase deficiency in myeloid metaplasia, *Br. J. Haematol.* **33**:144.

Erickson, R. P., Stites, D. P., Fudenberg, H. H., and Epstein, C. J., 1972, Altered levels of glucose-6-phosphate dehydrogenase stabilizing factors in X-linked chronic granulomatous disease, *J. Lab. Clin. Med.* **80**:644.

Estensen, R. D., Hill, H. R., Quie, P. G., Hogan, N., and Goldberg, N. D., 1973, Cyclic GMP and cell movement, *Nature* **245**:458.

Fontan, G., Lorente, F., Garcia Rodriguez, M. C., and Ojeda, J. A., 1976, Defective neutrophil chemotaxis and hyperimmunoglobulinemia E—a reversible defect?, *Acta Paediatr. Scand.* **65**:509.

Gallin, J. I., 1975, Abnormal chemotaxis: Cellular and humoral components, in: *The Phagocytic Cell in Host Resistance* (J. A. Bellanti and D. H. Dayton, eds.), pp. 227–243, Raven Press, New York.

Gallin, J. I., Bujak, J. S., Patten, E., and Wolff, S. M., 1974, Granulocyte function in the Chediak-Higashi syndrome of mice, *Blood* **43**:201.

Gallin, J. I., Malech, H. L., Wright, D. G., Whisnant, J. K., and Kirkpatrick, C. H., 1978, Recurrent severe infections in a child with abnormal leukocyte function: Possible relationship to increased microtubule assembly, *Blood* **51**:919.

Giblett, E. R., Klebanoff, S. J., Pincus, S. H., Swanson, J., Park, B. H., McCullough, J., 1971, Kell phenotypes in chronic granulomatous disease: A potential transfusion hazard, *Lancet* **1**:1235.

Gluckman, S. J., and MacGregor, R. R., 1978, Effect of acute alcohol intoxication on granulocyte mobilization and kinetics, *Blood* **52**:551.

Gray, G. R., Stamatoyannopoulos, G., Naiman, S. C., Kliman, M. R., Klebanoff, S. J., Austin, T., Yoshida, A., and Robinson, G. C. F., 1973, Neutrophil dysfunction, chronic granulomatous disease, and non-spherocytic haemolytic anaemia caused by complete deficiency of glucose-6-phosphate dehydrogenase, *Lancet* **2**:530.

Grignaschi, V. J., Sperperato, A. M., Etcheverry, M. J., and Macario, A. J. L., 1963, Un nuevo cuadro citoquimico negatividad espontanea de las reacciones de peroxidasas, oxidasas y lipido en la progenie neutrofila y en los monocitos de dos hermanos, *Rev. Asoc. Med. Argent.* **77**:218.

Hadden, J. W., Hadden, E. M., Wilson, E. E., Good, R. A., and Coffey, R. G., 1972, Direct action of insulin on plasma membrane ATPase activity in human lymphocytes, *Nature New Biol.* **235**:174.

Henson, P. M., and Oades, Z. G., 1973, Enhancement of immunologically induced granule exocytosis from neutrophils by cytochalasin B, *J. Cell Biol.* **110**:290.

Higashi, O., Katsuyama, N., and Satodate, R., 1965, A case with hematological abnormality characterized by the absence of peroxidase activity in blood polymorphonuclear leukocytes, *Tohoku J. Exp. Med.* **87**:77.

Hill, H. R., and Quie, P. G., 1974, Raised serum IgE levels and defective neutrophil chemotaxis in three children with eczema and recurrent bacterial infections, *Lancet* **1**:183.

Hill, H. R., Ochs, H. D., Quie, P. G., Clark, R. A., Pabst, H. F., Klebanoff, S. J., and Wedgwood, R. J., 1974a, Defect in neutrophil granulocyte chemotaxis in Job's syndrome of recurrent "cold" staphylococcal abscesses, *Lancet* **2**:617.

Hill, H. R., Sauls, H. S., Dettloff, J. L., and Quie, P. G., 1974b, Impaired leukotactic responsiveness in patients with juvenile diabetes mellitus, *Clin. Immunol. Immunopathol.* **2**:395.

Hill, H. R., Estensen, R. D., Hogan, N. A., and Quie, P. G., 1976, Severe staphylococcal disease associated with allergic manifestations, hyperimmunoglobulinemia E, and defective neutrophil chemotaxis, *J. Lab. Clin. Med.* **88**:796.

Hinds, K., and Danes, B. S., 1976, Microtubular defect in Chediak-Higashi syndrome, *Lancet* **2**:146.

Hoffstein, S., Goldstein, I. M., and Weissmann, G., 1977, Role of microtubular assembly in lysosomal enzyme secretion from human polymorphonuclear leukocytes: A reappraisal, *J. Cell Biol.* **73**:242.

Hogan, N. A., and Hill, H. R., 1978, Enhancement of neutrophil chemotaxis and alteration of levels of cellular cyclic nucleotides by levamisole, *J. Infect. Dis.* **138**:437.

Hohn, D. C., and Lehrer, R. I., 1975, NADPH oxidase deficiency in X-linked chronic granulomatous disease, *J. Clin. Invest.* **55**:707.

Holmes, B., Park, B. H., Malawista, S. E., Quie, P. G., Nelson, D. L., and Good, R. A., 1970,

Chronic granulomatous disease in females: A deficiency of leukocyte glutathione peroxidase, *N. Engl. J. Med.* **283**:217.

Howard, M. W., Strauss, R. G., and Johnston, R. B., Jr., 1977, Infections in patients with neutropenia, *Am. J. Dis. Child.* **131**:788.

Ignarro, L. J., and Cech, S. Y., 1975, Bidirectional regulation of lysosomal enzyme secretion and phagocytosis in human neutrophils by guanosine 3′,5′-monophosphate and adenosine 3′,5′-monophosphate, *Proc. Soc. Exp. Biol. Med.* **151**:448.

Illiano, G., Tell, G. P. E., Siegel, M. I., and Cuatrecasas, P., 1973, Guanosine 3′:5′-cyclic monophosphate and the action of insulin, *Proc. Natl. Acad. Sci. USA* **70**:2443.

Jacques, Y. V., and Bainton, D. F., 1978, Changes in pH within the phagocytic vacuoles of human neutrophils and monocytes, *Lab. Invest.* **39**:179.

Jandl, R. C., André-Schwartz, J., Borges-DuBois, L., Kipnes, R. S., McMurrich, B. J., and Babior, B. M., 1978, Termination of the respiratory burst in human neutrophils, *J. Clin. Invest.* **61**:1176.

Johnston, R. B., Jr., 1978, Oxygen metabolism and the microbicidal activity of macrophages, *Fed. Proc.* **37**:2759.

Johnston, R. B., Jr., and Baehner, R. L., 1971, Chronic granulomatous disease: Correlation between pathogenesis and clinical findings, *Pediatrics* **48**:730.

Johnston, R. B., Jr., and McPhail, L. C., 1980, The patient with impaired phagocyte function, in: *Comprehensive Immunology*, Vol. 8, *Immunology of Human Infection* (A. J. Nahmias and R. J. O'Reilly, eds.), Plenum Press, New York (in press).

Johnston, R. B., Jr., and Newman, S. L., 1977, Chronic granulomatous disease, *Pediatr. Clin. North Am.* **24**:365.

Kanfer, J. N., Blume, R. S., Yankee, R. A., and Wolff, S. M., 1968, Alteration of sphingolipid metabolism in leukocytes from patients with the Chediak-Higashi syndrome, *N. Engl. J. Med.* **279**:410.

Kaplan, S. S., Boggs, S. S., Nardi, M. A., Basford, R. E., and Holland, J. M., 1978, Leukocyte–platelet interactions in a murine model of Chediak–Higashi syndrome, *Blood* **52**:719.

Kimball, H. R., Ford, G. H., and Wolff, S. M., 1975, Lysosomal enzymes in normal and Chediak–Higashi blood leukocytes, *J. Lab. Clin. Med.* **86**:616.

Klebanoff, S. J., 1970, Myeloperoxidase: Contribution to the microbicidal activity of intact leukocytes, *Science* **169**:1095.

Klebanoff, S. J., and Pincus, S. H., 1971, Hydrogen peroxide utilization in myeloperoxidase-deficient leukocytes: A possible microbicidal control mechanism, *J. Clin. Invest.* **50**:2226.

Klebanoff, S. J., and White, L. R., 1969, Iodination defect in the leukocytes of a patient with chronic granulomatous disease of childhood, *N. Engl. J. Med.* **280**;460.

Kobayashi, Y., Amano, D., Ueda, K., Kagosaki, Y., and Usui, T., 1978, Treatment of seven cases of chronic granulomatous disease with sulfamethoxazole-trimethoprim (SMX-TMP), *Eur, J. Pediatr.* **127**:247.

Kontras, S. B., and Bodenbender, J. G., 1968, Studies of the inflammatory cycle in juvenile diabetes, *Am. J. Dis. Child.* **116**:130.

Lazarus, G. M., and Neu, H. C., 1975, Agents responsible for infection in chronic granulomatous disease of childhood, *J. Pediatr.* **86**:415.

Lehrer, R. I., 1972, Functional aspects of a second mechanism of candidacidal activity by human neutrophils, *J. Clin. Invest.* **51**:2566.

Lehrer, R. I., 1975, The fungicidal mechanisms of human monocytes. I. Evidence for myeloperoxidase-linked and myeloperoxidase-independent candidacidal mechanisms, *J. Clin. Invest.* **55**:338.

Lehrer, R. I., and Cline, M. J., 1969, Leukocyte myeloperoxidase deficiency and disseminated candidiasis: The role of myeloperoxidase in resistance to *Candida* infection, *J. Clin. Invest.* **48**:1478.

Lehrer, R. I., Hanifin, J., and Cline, M. J., 1969, Defective bactericidal activity in myeloperoxidase-deficient human neutrophils, *Nature* **223**:78.

Lehrer, R. I., Goldberg, L. S., Apple, M. A., and Rosenthal, N. P., 1972, Refractory megaloblastic anemia with myeloperoxidase-deficient neutrophils, *Ann. Int. Med.* **76**:447.

Lehrer, R. I., Ladra, K. M., and Hake, R. B., 1975, Nonoxidative fungicidal mechanisms of mammalian granulocytes: Demonstration of components with candidacidal activity in human, rabbit, and guinea pig leukocytes, *Infect. Immun.* **11**:1226.

Loos, J. A., Roos, D., Weening, R. S., and Houwerzijl, J., 1976, Familial deficiency of glutathione reductase in human blood cells, *Blood* **48**:53.

Malawista, S. E., and Gifford, R. H., 1975, Chronic granulomatous disease of childhood (CGD) with leukocyte glutathione peroxidase (LGP) deficiency in a brother and sister: A likely autosomal recessive inheritance, *Clin. Res.* **23**:416.

Malawista, S. E., Oliver, J. M., and Rudolph, S. A., 1978, Microtubules and cyclic AMP in human leukocytes: On the order of things, *J. Cell Biol.* **77**:881.

Marsh, W. L., and Kimball, L. F., 1977, The Kell blood group, Kx antigen, and chronic granulomatous disease, *Mayo Clin. Proc.* **52**:150.

Marsh, W. L., Uretsky, S. C., and Douglas, S. D., 1975, Antigens of the Kell blood group system on neutrophils and monocytes: Their relation to chronic granulomatous disease, *J. Pediatr.* **87**:1117.

Marsh, W. L., Øyen, R., and Nichols, M. E., 1976, Kx antigen, the McCleod phenotype, and chronic granulomatous disease: Further studies, *Vox Sang.* **31**:356.

Matsuda, I., Oka, Y., Taniguchi, N., Furuyama, M., Kodama, S., Arashima, S., and Mitsuyama, T., 1976, Leukocyte glutathione peroxidase deficiency in a male patient with chronic granulomatous disease, *J. Pediatr.* **88**:581.

McPhail, L. C., DeChatelet, L. R., Shirley, P. S., Wilfert, C., Johnston, R. B., Jr., and McCall, C. E., 1977, Deficiency of NADPH oxidase activity in chronic granulomatous disease, *J. Pediatr.* **90**:213.

Miller, M. E., 1975, Pathology of chemotaxis and random mobility, *Semin. Hematol.* **12**:59.

Miller, M. E., and Baker, L., 1972, Leukocyte functions in juvenile diabetes mellitus: Humoral and cellular aspects, *J. Pediatr.* **81**:979.

Mowat, A. G., and Baum, J., 1971, Chemotaxis of polymorphonuclear leukocytes from patients with diabetes mellitus, *N. Engl. J. Med.* **284**:621.

Oberling, F., Lang, J. M., Juif, J. G., and Luckel, J. C., 1976, Autophagia in myeloid precursors: An explanation for neutropenia in Chediak–Higashi syndrome?, *Scand. J. Haematol.* **17**:105.

Ochs, H. D., and Igo, R. P., 1973, The NBT slide test: A simple screening method for detecting chronic granulomatous disease and female carriers, *J. Pediatr.* **83**:77.

Oliver, J. M., 1978, Cell biology of leukocyte abnormalities: Membrane and cytoskeletal function in normal and defective cells, *Am. J. Pathol.* **93**:221.

Oliver, J. M., and Zurier, R. B., 1976, Correction of characteristic abnormalities of microtubule function and granule morphology in Chediak–Higashi syndrome with cholinergic agonists: Studies *in vitro* in man and *in vivo* in the beige mouse, *J. Clin. Invest.* **57**:1239.

Oliver, J. M., Zurier, R. B., and Berlin, R. D., 1975, Concanavalin A cap formation on polymorphonuclear leukocytes of normal and beige (Chediak–Higashi) mice, *Nature* **253**:471.

Oliver, J. M., Krawiec, J. A., and Berlin, R. D., 1976, Carbamylcholine prevents giant granule formation in cultured fibroblasts from beige (Chediak–Higashi) mice, *J. Cell Biol.* **69**:205.

Oliver, J. M., Spielberg, S. P., Pearson, C. B., and Schulman, J. D., 1978, Microtubule assembly and function in normal and glutathione synthetase-deficient polymorphonuclear leukocytes, *J. Immunol.* **120**:1181.

Patriarca, P., Cramer, R., Moncalvo, S., Rossi, F., and Romeo, D., 1971, Enzymatic basis of metabolic stimulation in leucocytes during phagocytosis: The role of activated NADPH oxidase, *Arch. Biochem. Biophys.* **145**:255.

Patriarca, P., Cramer, R., Tedesco, F., and Kakinuma, K., 1975, Studies on the mechanism of metabolic stimulation in polymorphonuclear leukocytes during phagocytosis. II. Presence of the NADPH$_2$ oxidizing activity in a myeloperoxidase-deficient subject, *Biochim. Biophys. Acta* **385**:387.

Perillie, P. E., Nolan, J. P., and Finch, S. C., 1962, Studies of the resistance to infection in diabetes mellitus: Local exudative cellular response, *J. Lab. Clin. Med.* **59**:1008.

Phelps, P., and Stanislaw, D., 1969, Polymorphonuclear leukocyte motility *in vitro*. I. Effect of pH, temperature, ethyl alcohol, and caffeine, using a modified Boyden chamber technic, *Arth. Rheum.* **12**:181.

Pincus, S. H., Thomas, I. T., Clark, R. A., and Ochs, H. D., 1975, Defective neutrophil chemotaxis with variant ichthyosis, hyperimmunoglobulinemia E, and recurrent infections, *J. Pediatr.* **87**:908.

Pincus, S. H., Boxer, L. A., and Stossel, T. P., 1976, Chronic neutropenia in childhood: Analysis of 16 cases and a review of the literature, *Am. J. Med.* **61**:849.

Quie, P. G., and Cates, K. L., 1977, Clinical conditions associated with defective polymorphonuclear leukocyte chemotaxis, *Am. J. Pathol.* **88**:711.

Rausch, P. G., Pryzwansky, K. B., and Spitznagel, J. K., 1978, Immunocytochemical identification of azurophilic and specific granule markers in the giant granules of Chediak–Higashi neutrophils, *N. Engl. J. Med.* **298**:693.

Reed, P. W., 1969, Glutathione and the hexose monophosphate shunt in phagocytizing and hydrogen peroxide-treated rat leukocytes, *J. Biol. Chem.* **244**:2459.

Repine, J. E., and Clawson, C. C., 1977, Quantitative measurement of the bactericidal capability of neutrophils from patients and carriers of chronic granulomatous disease, *J. Lab. Clin. Med.* **90**:522.

Repine, J. E., Clawson, C. C., and Brunning, R. D., 1976a, Abnormal pattern of bactericidal activity of neutrophils deficient in granules, myeloperoxidase, and alkaline phosphatase, *J. Lab. Clin. Med.* **88**:788.

Repine, J. E., Clawson, C. C., and Brunning, R. D., 1976b, Primary leukocyte alkaline phosphatase deficiency in an adult with repeated infections, *Br. J. Haematol.* **34**:87.

Rivkin, I., Rosenblatt, J., and Becker, E. L., 1975, The role of cyclic AMP in the chemotactic responsiveness and spontaneous motility of rabbit peritoneal neutrophils, *J. Immunol.* **115**:1126.

Rodey, G. E., Park, B. H., Ford, D. K., Gray, B. H., Good, R. A., 1970, Defective bactericidal activity of peripheral blood leukocytes in lipochrome histiocytosis, *Am. J. Med.* **49**:322.

Roos, D., and Weening, R. S., 1980, Defects in the oxidative killing of microorganisms by phagocytic leukocytes, in: *Ciba Symposium: Oxygen Free Radicals and Tissue Damage* (in press).

Root, R. K., Rosenthal, A. S., and Balestra, D. J., 1972, Abnormal bactericidal, metabolic, and lysosomal functions of Chediak–Higashi syndrome leukocytes, *J. Clin. Invest.* **51**:649.

Rosen, H., and Klebanoff, S. J., 1976, Chemiluminescence and superoxide production by myeloperoxidase-deficient leukocytes, *J. Clin. Invest.* **58**:50.

Salmon, S. E., Cline, M. J., Schultz, J., and Lehrer, R. I., 1970, Myeloperoxidase deficiency: Immunologic study of a genetic leukocyte defect, *N. Engl. J. Med.* **282**:250.

Sandler, J. A., Gallin, J. I., and Vaughan, M., 1975, Effects of serotonin, carbamylcholine, and ascorbic acid on leukocyte cyclic GMP and chemotaxis, *J. Cell Biol.* **67**:480.

Segal, A. W., and Peters, T. J., 1976, Characterisation of the enzyme defect in chronic granulomatous disease, *Lancet* **1**:1363.

Segal, A. W., Jones, O. T. G., Webster, D., and Allison, A. C., 1978, Absence of a newly described cytochrome b from neutrophils of patients with chronic granulomatous disease, *Lancet* **2**:446.

Showell, H. J., and Becker, E. L., 1976, The effects of external K^+ and Na^+ on the chemotaxis of rabbit peritoneal neutrophils, *J. Immunol.* **116**:99.

Singer, D. B., 1973, Postsplenectomy sepsis, *Perspectives Pediatr. Pathol.* **1**:285.

Snyderman, R., and Buckley, R. H., 1975, Defects of monocyte chemotaxis in patients with hyperimmunoglobulinemia E and undue susceptibility to infection, *J. Allergy Clin. Immunol.* **55**:102.

Snyderman, R., and Pike, M., 1977, Disorders of leukocyte chemotaxis, *Pediatr. Clin. North Am.* **24**:377.

Snyderman, R., Rogers, E., and Buckley, R. H., 1977, Abnormalities of leukotaxis in atopic dermatitis, *J. Allergy Clin. Immunol.* **80**:121.

Spielberg, S. P., Kramer, L. I., Goodman, S. I., Butler, J., Tietze, F., Quinn, P., and Schulman, J. D., 1977, 5-Oxoprolinuria: Biochemical observations and case report, *J. Pediatr.* **91**:237.

Spielberg, S. P., Boxer, L. A., Oliver, J. M., Allen, J. M., and Schulman, J. D., 1979, Oxidative damage to neutrophils in glutathione synthetase deficiency, *Br. J. Haematol.* **42**:215.

Stendahl, O., and Lindgren, S., 1976, Function of granulocytes with deficient myeloperoxidase-mediated iodination in a patient with generalized pustular psoriasis, *Scand. J. Haematol.* **16**:144.

Stossel, T. P., 1977, Phagocytosis, *Am. J. Pathol.* **88**:741.

Stossel, T. P., Root, R. K., and Vaughan, M., 1972, Phagocytosis in chronic granulomatous disease and the Chediak–Higashi syndrome, *N. Engl. J. Med.* **286**:120.

Strauss, R. G., Bove, K. E., Jones, J. F., Mauer, A. M., and Fulginiti, V. A., 1974, An anomaly of neutrophil morphology with impaired function, *N. Engl. J. Med.* **290**:478.

Tse, R. L., Phelps, P., and Urban, D., 1972, Polymorphonuclear leukocyte motility *in vitro*. VI. Effect of purine and pyrimidine analogues: Possible role of cyclic AMP, *J. Lab. Clin. Med.* **80**:264.

Undritz, V. E., 1966, Die Alius-Grignaschi anomalie: Der erblich-konstitutionelle Peroxydasedefekt der Neutrophilen und Monozyten, *Blut* **14**:129.

van der Meer, J. W. M., van Zwet, T. L., van Furth, R., and Weemaes, C. M. R., 1975, New familial defect in microbicidal function of polymorphonuclear leucocytes, *Lancet* **2**:630.

Van Scoy, R. E., Hill, H. R., Ritts, R. E., Jr., and Quie, P. G., 1975, Familial neutrophil chemotaxis defect, recurrent bacterial infections, mucocutaneous candidiasis, and hyperimmuno-globulinemia E, *Ann. Int. Med.* **82**:766.

Vassalli, J.-D., Granelli-Piperno, A., Griscelli, C., and Reich, E., 1978, Specific protease deficiency in polymorphonuclear leukocytes of Chediak–Higashi syndrome and beige mice, *J. Exp. Med.* **147**:1285.

Ward, P. A., 1974, Leukotaxis and leukotactic disorders: A review, *Am. J. Pathol.* **77**:520.

Ward, P. A., and Becker, E. L., 1970, Potassium reversible inhibition of leukotaxis by ouabain, *Life Sci.* **9**:355.

Weening, R. S., Roos, D., Weemaes, C. M. R., Homan-Müller, J. W. T., and van Schaik, M. L. J., 1976, Defective initiation of the metabolic stimulation in phagocytizing granulocytes: A new congenital defect, *J. Lab. Clin. Med.* **88**:757.

Weissmann, G., Goldstein, I., Hoffstein, S., and Tsung, P.-K., 1975, Reciprocal effects of cAMP and cGMP on microtubule-dependent release of lysosomal enzymes, *Ann. N.Y. Acad. Sci.* **253**:750.

Weston, W. L., Huff, J. C., Humbert, J. R., Hambidge, K. M., Neldner, K. H., and Walravens, P. A., 1977, Zinc correction of defective chemotaxis in acrodermatitis enteropathica, *Arch. Dermatol.* **113**:422.

Wilkinson, P. C., 1974, *Chemotaxis and Inflammation*, p. 62, Churchill Livingstone, Edinburgh.

Windhorst, D. B., and Katz, E. D., 1972, Normal enzyme activities in chronic granulomatous disease leukocytes, *J. Reticuloendothel. Soc.* **11**:400.

Wright, D. G., Kirkpatrick, C. H., and Gallin, J. I., 1977, Effects of levamisole on normal and abnormal leukocyte locomotion, *J. Clin. Invest.* **59**:941.

Index